Novel Food Preservation and Microbial Assessment Techniques

Novel Food Preservation and Microbial Assessment Techniques

Editor

Ioannis S. Boziaris

School of Agricultural Sciences
University of Thessaly
Volos, Greece

CRC Press
Taylor & Francis Group
Boca Raton London New York

CRC Press is an imprint of the
Taylor & Francis Group, an **informa** business
A SCIENCE PUBLISHERS BOOK

CRC Press
Taylor & Francis Group
6000 Broken Sound Parkway NW, Suite 300
Boca Raton, FL 33487-2742

Version Date: 20140318

International Standard Book Number-13: 978-1-4665-8076-3 (eBook - PDF)

Visit the Taylor & Francis Web site at
http://www.taylorandfrancis.com

and the CRC Press Web site at
http://www.crcpress.com

Preface

The introduction of microbial ecology and 'hurdle theory' concepts in food microbiology during late 1980s contributed greatly to its evolution and the understanding of food preservation mechanisms. The improvement of microbiological methods have made possible useful tools for monitoring microbiological quality and safety of foods. The demand for minimal processing has promoted the development of innovative non-thermal processing methods and application of mild preservation strategies for microbial control in foods. Furthermore, the application of molecular genetic methods, have provided a powerful tool for better exploration and understanding of food microbial diversity.

The aim of this book is to provide a thorough presentation of microbiology of novel food preservation and microbial assessment techniques. The book is divided in two parts. The first part discusses the microbiology of novel food preservation methods, such as the effect of these treatments on survival and inactivation of microorganisms in various foods, mode of action, modelling aspects etc. Some of those methods have already numerous applications, such as high pressure processing (HPP), while others appear to be very promising for future. The methodologies presented, apart from HPP, include high-voltage pulsed electric fields (PEF), UV and pulsed UV-light technologies, ultrasound treatment, decontamination treatments with ozone and non-thermal plasma (NTP), photosensitization, the encapsulation and delivery of antimicrobial compounds and the application of antimicrobial edible films and coatings in foods.

The second part is dedicated on the newly developed applications of molecular genetics and instrumental analysis for detection and monitoring of microorganisms and/or their activity in food systems. Various culture-depended and culture-independent molecular methodologies with special emphasis on PCR-based techniques for microbial community analysis, quantification of spoilage microorganisms and detection of food-borne pathogens are presented. Regarding the newly emerged instrumental analysis methods, MALDI-TOF Mass Spectrometry for rapid identification of bacterial species, vibrational spectroscopy (FT-IR & Raman) and

Hyperspectral Imaging for monitoring food quality related to microbial spoilage activity in foods are also described.

Many of these aspects have not been covered in one book so far. This volume will be a useful reference book for university students, academics, researchers, food processors and food microbiologists and technologists professionals.

Ioannis S. Boziaris

Contents

Preface v

Part A: Modern Preservation Methods

1. **High Pressure Microbial Inactivation: Modelling and its Features** 3
Shigenobu Koseki and *Kazutaka Yamamoto*

2. **Implementation of Microbially Safe Foods with Pulsed Electric Fields** 25
Mansel W. Griffiths and *Markus Walkling-Ribeiro*

3. **Applications and Modeling Aspects of UV and Pulsed UV-Light for Food Decontamination** 67
Nene M. Keklik and *Ali Demirci*

4. **Ultrasound Processing for Food Safety and Preservation** 102
Daniela A. Abela, David Millan Sango, Anna McElhatton and *Vasilis P. Valdramidis*

5. **Ozone: A Novel Microbial Inactivation Process** 126
S. Patil, P.J. Cullen and *P. Bourke*

6. **Nonthermal Plasma Technology for Decontamination of Foods** 155
N.N. Misra, Lu Han, B.K. Tiwari, Paula Bourke and *P.J. Cullen*

7. **Novel Approach to Control Pathogenic and Harmful Microorganisms in Nonthermal Way: Photosensitization** 184
Zivile Luksiene

8. **Encapsulation and Delivery of Antimicrobial Compounds** 218
Atul Kumar Singh, Preetam Sarkar, Srinivas Janaswamy, Yuan Yao and *Arun Kumar Bhunia*

9. Application of Edible Films and Coatings on Food 237
Anastasia E. Kapetanakou, Stavros G. Manios and *Panagiotis N. Skandamis*

Part B: Modern Techniques in Food Microbiology

10. Old Targets, New Weapons: Food Microbial Communities Revealed With Molecular Tools 277
Aspasia Nisiotou, Foteini F. Parlapani, Konstantinos A. Kormas and *Ioannis S. Boziaris*

11. The Use of qPCR-based Methods to Identify and Quantify Food Spoilage Microorganisms 313
Daniel M. Linares, Beatriz del Rio, Begoña Redruello, María Fernández, María Cruz Martín, Victor Ladero and *Miguel A. Alvarez*

12. Advances in the Detection of Food-borne Pathogens Using Molecular Methods 335
Luca Cocolin and *Kalliopi Rantsiou*

13. MALDI-TOF Mass Spectrometry, a Rapid and Reliable Method for the Identification of Bacterial Species in Food-Microbiology Laboratories 353
M. Quintela-Baluja, K. Böhme, I.C. Fernández-No, M.E. Alnakip, S. Caamaño, J. Barros-Velázquez and *Pilar Calo-Mata*

14. Monitoring Microbial Spoilage of Foods by Vibrational Spectroscopy (FT-IR & Raman) 386
Anthoula A. Argyri, Efstathios Z. Panagou and *George–John Nychas*

15. The Potential of Hyperspectral Imaging for Monitoring Microbial Activity in Foods 435
Aoife A. Gowen, Edurne Gaston and *Vasilis Valdramidis*

Color Plate Section 453

Part A
Modern Preservation Methods

CHAPTER 1

High Pressure Microbial Inactivation: Modelling and its Features

Shigenobu Koseki[1,*] and *Kazutaka Yamamoto*[2]

Introduction

The effect of high hydrostatic pressure (HHP) on bio-materials has been studied extensively and ever since the report on the effect of pressure on the preservation of milk was published in 1899 (Hite 1899), Bridgman (1914) reported on the coagulation of egg albumen by HHP-treatment and Hite (1914) reported on the effect of pressure on microorganisms in the preservation of fruits and vegetables. Since these early reports, however, it was nearly after a century that HHP-treatment was first introduced to the food industry (Hayashi et al. 1987). Thereafter, research and development on HHP food processing accelerated in Japan, and also worldwide. In 1990, the first HHP-treated foods (fruit jams) were commercialized in Japan, and other HHP-treated foods followed in the rest of the world. Recently, various HHP-treated foods have been put in the market: jams, hams and sausages, instant rice packs, shucked oysters and clams, smoothies (fruit beverages), etc. (San Martín et al. 2002).

[1] Hokkaido University, Kita 9, Nishi 9, Kita-ku, Sapporo 060-8589, Japan.
Email: koseki@bpe.agr.hokudai.ac.jp
[2] National Food Research Institute, Kannondai, Tsukuba, Ibaraki 305-8642, Japan.
Email: kazutaka@affrc.go.jp
* Correspondent author

HHP-treatment has been investigated as an attractive non-thermal processing technique for producing minimally processed high quality food (Hoover 1993, Knorr 1993). HHP-treatment allows inactivation of pathogenic and spoilage bacteria in food with minimal changes in color, flavor, and texture as compared to conventional thermal processing (Knorr 1993, Cheftel 1995). The effect of HHP-treatment on foodborne pathogens has been extensively studied (Shigehisa et al. 1991, Kalchayanand et al. 1998, Linton et al. 1999, Zook et al. 1999, Alpas et al. 2000). The level of pressure, pressurization time, temperature, cell growth phase, type of microorganisms, and suspending solution influence the effectiveness of HHP-treatment on the inactivation of microbes (Patterson et al. 1995, McClements et al. 2001).

Factors influencing the effect of HHP-treatment on microbial inactivation are more complex and require greater research than thermal inactivation, since temperature needs to be considered as a process parameter in addition to pressure and time. Moreover, it has been reported that food constituents such as sodium chloride and sugars (sucrose, fructose and glucose) in the media and also the pH, influence the effect of HHP-induced inactivation (Horie et al. 1991, Palou et al. 1997, Van Opstal et al. 2003, Molina-Hoppner et al. 2004). To apply HHP-treatment to a real food production appropriately, organized knowledge regarding the characteristics of microbial inactivation using HHP-treatment is needed.

In this chapter, knowledge about the characteristics of microbial inactivation using HHP-treatment is described based on reports published so far. In addition, to optimize the HHP-treatment, predictive modeling for HHP induced microbial inactivation will be a useful tool. The predictive modeling of HHP induced microbial inactivation is also discussed. Furthermore, the phenomena of injured-recovery of microbial cells after HHP-treatment are discussed for realising the limitations of HHP-treatment.

Characteristics of Microbial Inactivation using High Pressure Processing

Factors influencing HHP microbial inactivation

Consumers the world over are today demanding fresh and/or minimally processed foods. Because conventional thermally processed foods loose the organoleptic and nutritional properties of foods, there has been a great need for non-thermal microbial inactivation methods. Among the non-thermal processing technologies, HHP-treatment is one of the most popular since the late 1980s, as mentioned before. There are now several products using HHP-treatment such as fruit jams, orange juice, sliced ham, avocado puree, etc. However, since there are many factors influencing HHP-induced microbial inactivation such as pressure, temperature, treatment time, pH, water activity, etc., systematic understanding of these factors' effect on microbial inactivation is essential. For appropriate process control, here we shall discuss major environmental factors on the effect of HHP-induced microbial inactivation.

Effect of temperature

Numerous studies have demonstrated the dependence on temperature of the antimicrobial effects of high pressure (Ludwig et al. 1992, Sonoike et al. 1992, Moerman et al. 2001). However, the effect of temperature on HHP inactivation of bacterial cells is complex. While it is documented that elevated temperature (30–50°C) promotes HHP inactivation of microorganisms, the effect of low temperature (< 20°C) on inactivation is less clear. The combined effects of high pressure and low or subzero temperatures on microbial inactivation have been studied by some authors. A synergistic effect between these parameters has generally been reported in the inactivation of microorganisms in the vegetative state (Sonoike et al. 1992, Takahashi 1992, Hashizume et al. 1995, Hayakawa et al. 1998, Perrier-Cornet et al. 2005). In some cases, the initial microbial populations were completely inactivated with a combined treatment of high pressure and low or subzero temperature, whereas only a slight microbial inactivation was achieved under the same pressure conditions at room temperature (Perrier-Cornet et al. 2005). The magnitude of this synergistic effect is strongly dependent on the type of microorganism (Takahashi 1992).

Effect of pH

The influences of the pH of the media used during HHP-treatment on the effect of HHP-induced inactivation have been extensively reported. Early studies exhibited two absolutely different results. One group of reports showed that the pH of media ranging from 3 to 8 did not affect HHP-induced inactivation of fungi such as *Saccharomyces bayanus, S. cerevisiae, Mucor plumbeus, Zygosaccharomyces bailii,* and *Rhodotorula rubra* (Ogawa et al. 1990, Horie et al. 1991, Oxen and Knorr 1993, Pandya et al. 1995, Reyns et al. 2000). In contrast, other results indicated that the pH of media ranging from 4 to 6 greatly affected the efficacy of HHP-induced inactivation of bacteria such as *Listeria monocytogenes, Staphylococcus aureus, Escherichia coli* O157:H7, and *Salmonella* spp. (Mackey et al. 1995, Stewart et al. 1997, Alpas et al. 2000). These differences of susceptibility to pH suggested that the effect of HHP-induced inactivation of microbes is dependent on the kind of microorganisms subjected to HHP-treatment.

Several researchers reported the influence of pH of the media on HHP-induced inactivation of bacteria. Mackey et al. (1995) reported that HHP-treatment at 304 MPa resulted in an additional 1.8 log reduction in *L. monocytogenes* after when the media pH was reduced from 7.1 to 5.3. Stewart et al. (1997) demonstrated an additional 3 log cycle reduction in *L. monocytogenes* CA when HHP-treated at a pH of 4.0 in comparison to treatment at a pH of 6.0 at 353 MPa and 45°C for 10 min. Alpas et al. (2000) showed that HHP-treatment at 345 MPa and a temperature of 25 or 35°C for 5 min in the presence of either citric or lactic acid increased the bactericidal effect on *S. aureus*, *E. coli* O157:H7, *L. monocytogenes*, and *Salmonella* spp. by an additional 1.2–3.9 log reduction at a pH of 4.5 when compared with a pH of 6.5. While these earlier studies investigated the effect of pH on HHP-treatment in a buffer system that did not possess a significant baroprotective effect, Koseki and Yamamoto (2006a) investigated the effect of pH in a baroprotective medium in order to determine the dominant parameter. Their results revealed that reducing pH (4 or below) has a predominant influence on the effect of HHP-induced inactivation of *L. monocytogenes*. At a relatively low pressure of 300 MPa, an additional 8 log cycle reduction of *L. monocytogenes* in a low concentration of buffered peptone water (BPW) (0.1 to 5% w/v) was observed when HHP-treatment was applied at a pH of 4 in comparison to a pH of 7. Furthermore, the study determined the effect of reducing pH on HHP-induced inactivation in baroprotective media as represented by a high concentration of 10% w/v BPW. Even though HHP-treatment at 600 MPa reduced the number of *L. monocytogenes* by only 2 log cycles in 10% w/v BPW when trails conducted at the pH ranging from 5 to 7, complete inactivation (more than an 8 log reduction) was observed in 10% w/v BPW at a pH of 4. This result suggests that reducing the pH (4 or less) of the medium predominantly influences the effect of HHP-induced inactivation of *L. monocytogenes*. Since there are many other factors influencing the baroprotection such as fat, minerals, and the kind of acid, further investigation will be needed in future. However, knowledge of the pH of the media or food will help greatly with the formulation of a protocol to obtain sufficient bacterial inactivation with a minimal level of HHP-treatment.

Effect of water activity (a_w)

It has been reported that food constituents such as sodium chloride and sugars (sucrose, fructose, and glucose) in the media inhibit the effect of HHP-induced inactivation (Horie et al. 1991, Palou et al. 1997, Van Opstal et al. 2003, Molina-Hoppner et al. 2004). In order to generalize the effect of solute concentration, some researchers applied water activity (a_w) as a parameter of HHP-induced inactivation. In contrast, Molina-Guiterrez et al. (2002) and Molina-Hoppner et al. (2004) reported that the influence of sodium chloride or sugars on a_w of foods or suspension media does not explain the baroprotective effects of these solutes.

Some researchers reported that sucrose concentrations above 30% (w/v), which is almost 0.9 M, influenced the effect of HHP-induced inactivation of *Rhodotrula rubra, Zygosaccharomyces bailii,* and *Escherichia coli.* Molina-Guiterrez et al. (2002) and Molina-Hoppner et al. (2004) demonstrated that HHP-induced inactivation of *Lactococcus lactis* was inhibited in 4 M sodium chloride solution. Koseki and Yamamoto (2007a) illustrated that the baroprotective results on *L. monocytogenes* in sodium chloride and sucrose solutions were almost consistent with those of earlier studies. They also reported that the baroprotective effect was also shown in phosphate buffer with higher concentrations. There was a significant difference in HHP-induced inactivation of *L. monocytogenes* at 400 MPa and 25°C for 10 min in phosphate buffer at concentrations between 0.01 M and 0.1 M. Although phosphate buffer (0.01 M or 0.1 M) has been widely used as a medium of microorganism suspension in HHP-induced inactivation studies, the effect of buffer concentration on HHP-induced inactivation has not yet been investigated. It might be necessary to take the phosphate buffer concentration into consideration when the effect of HHP-induced inactivation is discussed.

Baroprotective effect differed depending on the kind of solute even at the same a_w. This result showed no consistent correlation between a_w and solute concentration in terms of the baroprotective effect. Attempts have been made to utilize a_w for generalization of results concerning the effect of solute concentration on HHP-induced inactivation. Oxen and Knorr (1993) demonstrated the baroprotective effect of sucrose, glucose, fructose, and sodium chloride on HHP-induced inactivation of *R. rubra*. The data in their study indicated that there was a difference in the baroprotective effect among the four solutions when $a_w = 0.94$. While the effect of HHP-induced inactivation of *R. rubra* in sucrose at $a_w = 0.94$ showed a 1 log reduction, the reduction in sodium chloride solution at $a_w = 0.94$ was 7 logs. This result closely agreed with the results of Koseki and Yamamoto (2007a) on the different effects on HHP-induced inactivation of *L. monocytogenes* of sucrose and sodium chloride at $a_w = 0.93$. However, Oxen and Knorr (1993) concluded that the baroprotective effect started at $a_w < 0.92$, regardless of the solute type without discussing differences in the baroprotective effect between sucrose and sodium chloride at $a_w = 0.94$.

Palou et al. (1997) demonstrated a linear relationship between a reduced a_w adjusted by sucrose and the inhibition effect of HHP-induced inactivation of *Z. bailii*. They suggested that the a_w of food or media would be a parameter of inhibition of HHP-inactivation. However, the relationship between a_w and HHP-induced inactivation was obtained for sucrose only. The effect of solute type with adjusted a_w was not considered. Recently, Molina-Hoppner et al. (2004) reported on the influence of sugars or salts on the a_w of media, but could not explain the protective effect of these solutes in terms of HHP-induced inactivation. Koseki and Yamamoto (2007a) illustrated that there are significant differences on the effect of HHP-induced inactivation of *L. monocytogenes* even at the same a_w in solutions of sodium chloride, sucrose, and phosphate buffer. Accordingly, the a_w of food or media will be limited as an index of inhibition of HHP-induced inactivation.

As an alternative approach, the saturation of suspension was used for evaluating the effect of HHP-induced inactivation of *L. monocytogenes* (Koseki and Yamamoto 2007a). As the saturation of suspension increased, the effect of HHP-induced inactivation of *L. monocytogenes* decreased regardless of the kind of solute. Furthermore, they found that a higher concentration of each solution of 5.0 M sodium chloride, 2.0 M sucrose, and 1.0 M phosphate buffer, which are over-saturated, completely inhibited the inactivation of *L. monocytogenes* by HHP-treatment even at 600 MPa. When the solution is over-saturated, i.e., solute molecules exist excessively, pressure transmission to the water might be inhibited. As a result, bacterial cells in higher saturation solution would not be pressurized sufficiently and could survive after HHP-treatment.

Predictive modeling

Kinetic approach

The number of studies examining the microbial inactivation kinetics of HHP-treatment has increased (Buzrul and Alpas 2004, Chen et al. 2004, Buzrul et al. 2005, Chen et al. 2005, Doona et al. 2005, Kilimann et al. 2005, Ross et al. 2005, Yamamoto et al. 2005). The patterns of the inactivation kinetics of microorganisms by HHP-treatment do not follow first-order kinetics such as thermal inactivation, but are generally non-linear curves with tailing. In order to describe the non-linear kinetics of HHP-induced microbial inactivation, appropriate functions such as log-logistic, modified Gompertz, or Weibull have been applied to the inactivation curve (Buzrul and Alpas 2004, Buzrul et al. 2005, Yamamoto et al. 2005). Among these functions, the Weibull function has fitted experimental data better than the other functions. These fitting procedures are effective for the evaluation of HHP-induced microbial inactivation kinetics. However, these procedures have merely represented fitting, not prediction. The inactivation kinetics for conditions other than those examined have not been predicted by curve-fitting procedures.

Koseki and Yamamoto (2007b) illustrated an alternative modeling procedure. They assumed that the inactivation kinetics during HHP-treatment on a semi-log plot consist of two parts. One represents an exponential death phase (linear portion) followed by the tailing (Fig. 1). The curve is symmetrical in form or a 'mirror image' of a microbial growth curve with no lag phase. This curve will be described by the following differential equation which is a modification of the Baranyi and Roberts' model (Baranyi and Roberts 1994):

$$\frac{dN}{dt} = -k_{max}\left(1-\left(\frac{N_{min}}{N}\right)^{m}\right)N \quad (1)$$

where $N(t)$ denotes the cell number (arithmetic number) at time t. k_{max} is the maximum inactivation rate, N_{min} represents the minimum cell number at the end point of HHP-treatment, m is the parameter that characterizes the curve shape. When N nears N_{min} during HHP-treatment, the value of the term $\left(1-\left(\frac{N_{min}}{N}\right)^{m}\right)$ becomes very small, thus making the rate of inactivation, which is represented by $\frac{dN}{dt}$, very low. From a theoretical point of view, the loss of inactivation potency that represents a tailing in a curve could be due to the presence of an inherently resistant sub-population initially present among the population that were exposed to HHP-treatment. The inactivation curve can be interpreted as indicative of the existence of a mixed population whose sub-populations all have first-order mortality kinetics with different rate constants. Equation (1) would be described well using the above interpretation.

Parameter k_{max} was set to be a measured rate constant of inactivation. The effect of pressure on the absolute value of the maximum inactivation

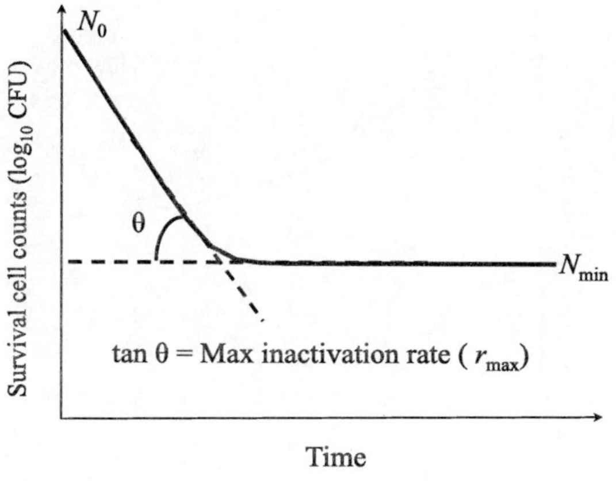

Figure 1. Concept of modeling for high hydrostatic pressure-induced microbial inactivation [adapted from Koseki and Yamamoto (2007b)].

rate (k_{max}) of a microbial population can be described using the following simple square root type model (Ratkowsky et al. 1982):

$$\sqrt{k_{max}} = b(P - P_{min}) \tag{2}$$

where b is a constant, P is the pressure and P_{min} is the conceptual minimum pressure for microbial inactivation. The model for k_{max} was substituted into the above differential Eq. (1), and the pressure designated as time-dependent. The system was solved numerically as a means of obtaining predictions of bacterial concentration during time-dependent pressure changes.

The resulting model was compared with the *E. coli* inactivation kinetics at various constant pressures using the fitted Weibull (Peleg and Cole 1998), modified Gompertz (Linton et al. 1995), and linear functions (Fig. 2). The proposed model [Eqs. (1) and (2)] showed comparable goodness-of-fit to the observed experimental data. In addition, the inactivation kinetics were predicted under dynamic pressure conditions (Fig. 3). The newly developed model successfully simulated the changes in bacterial numbers during dynamic pressure conditions including the come-up time and pressure-release time. These predictions under dynamic pressure conditions cannot be achieved by the conventional curve-fitting procedure. These results indicate that the new model has significant flexibility for corresponding to various conditions of HHP.

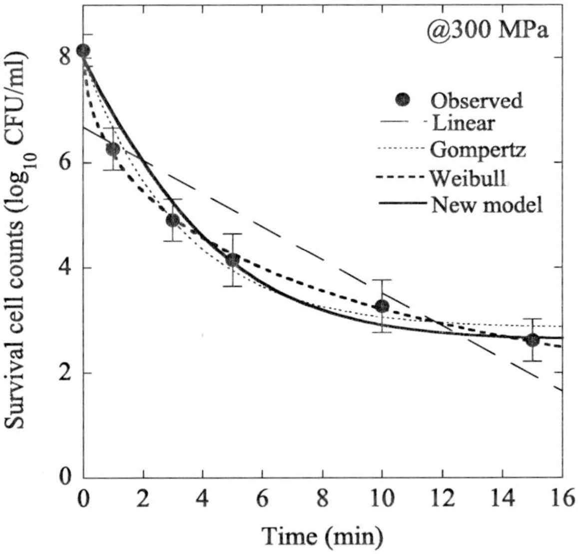

Figure 2. Comparison of prediction accuracy of high pressure-induced *E. coli* (ATCC 25922) inactivation by the newly developed model with those obtained by fitting using Weibull, modified Gompertz, and linear functions under various constant pressure conditions [adapted from Koseki and Yamamoto (2007b)].

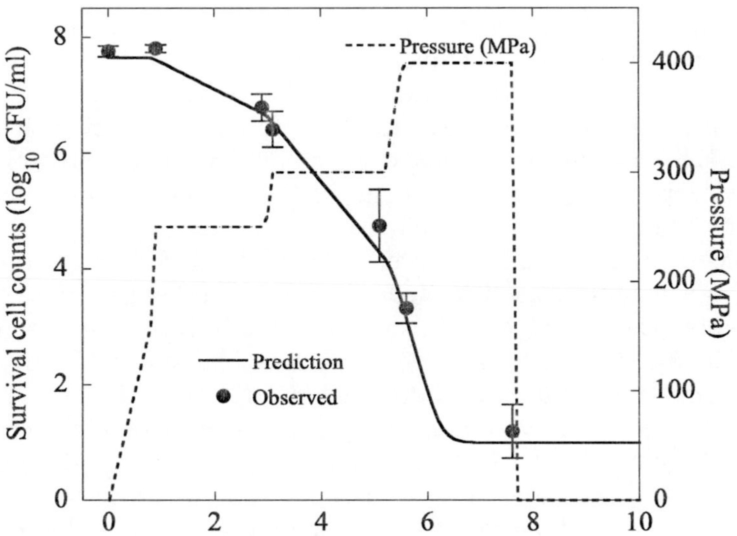

Figure 3. Inactivation kinetics of *E. coli* (ATCC 25922) under dynamic pressure. The curves were calculated using the newly developed model and related to the pressure changes. Observed data are mean values of triplicate trials. Error bars indicate standard deviation [adapted from Koseki and Yamamoto (2007b)].

Probabilistic approach

The main concern for food processors in ensuring microbiological safety is to set processing criteria for achieving a required log reduction of the microbial population. This point is also the focus of concepts such as the food safety objective (FSO), performance objective (PO), and performance criterion suggested by the International Commission on Microbiological Specifications for Foods and Codex Alimentarius Commission (ICMSF 2001, CAC 2005). The performance criterion concept signifies the change required to reach a hazard level at each step of the food chain in order to meet a PO or FSO. The determination of the *D*-value, the time required at a specific temperature to obtain a 1-log reduction, and the *z*-value, the temperature increase required to decrease the *D*-value by 90%, has been widely applied to thermal inactivation processes to assess the inactivation effect and set a processing condition for achieving a required log reduction. These concept values are calculated for inactivated microorganisms that follow log-linear kinetics. However, these values are not applicable to HPP-inactivated microorganisms that display nonlinear kinetics. The calculation of a *D*-value from nonlinear inactivation kinetics results in an underestimation or overestimation of the log reduction, depending on the calculation method used. This is because the *D*-value for nonlinear kinetics is not constant and the total time to achieve a required log reduction is not proportional to the level of reduction (Peleg and Cole 1998, Tassou et al. 2008). Although some models based on the Weibull model were introduced to determine a required log reduction from nonlinear inactivation kinetics, one needs to calculate a more elaborate model and choose an appropriate equation depending on the curve shapes (Peleg 1999). It is not easy for a non-expert to master the calculation procedure.

Recently, a survival/death interface model, which is a new predictive modeling procedure used to determine bacterial behavior after HHP inactivation as a probability of survival or death, was developed (Koseki and Yamamoto 2007c). In this procedure, microbial cells in a medium at an arbitrary initial inoculum level are treated by HHP and then viable cells are either detected (survival) or not detected (at < 1 CFU/ml; death). The probability of death after HHP is then modeled using logistic regression. The modeling procedure is used to predict a minimal processing condition to achieve a required log reduction, which represents a net log reduction that takes into account the inoculum level, independent of the underlying inactivation kinetics. In addition, the certainty of the predicted inactivation effect under the predicted processing condition can be estimated simultaneously, because of the probability-based approach.

As described in the previous report (Koseki et al. 2009), infant formula (IF) inoculated with *Cronobacter sakazakii* at different inoculum levels (3, 5, and 7 log CFU/ml) was treated by HHP ranging from 400 to 600 MPa at 50 MPa increments at different temperatures (25 and 40°C). For each replicate response of *C. sakazakii*, survival and death were scored with values 0 and 1 respectively. The data were fitted to the following simple logistic regression model using R statistical software (R Development Core Team 2012).

$$Logit(P) = a_0 + a_1 Press + a_2 \ln(Time) + a_3 Temp + a_4 IC \tag{3}$$

where Logit(P) represents ln [P/(1-P)], ln is the natural logarithm, P is the probability of survival (range: 0–1), a_i are coefficients to be estimated, *Press* is the applied pressure (MPa), *Time* is the amount of time that the pressure is held (min), *Temp* is the temperature at which the treatment was conducted (°C), and *IC* is the inoculum level ($\log_{10}$ CFU/ml) of *C. sakazakii* in the reconstituted IF. Because the developed model expresses the natural logarithm of the odds (P/(1-P)) of the survival of *C. sakazakii*, the probability of survival can be described as following transformation of Eq. (3):

$$P = \frac{1}{1 + \exp\left(-\left(a_0 + a_1 Press + a_2 \ln(Time) + a_3 Temp + a_4 IC\right)\right)} \tag{4}$$

The probability of survival of *C. sakazakii* (5 $\log_{10}$ CFU/ml) is calculated with respect to the pressure-holding time under different pressure conditions, as shown in Fig. 4. The results show that as the pressure-holding time is increased, the probability of inactivation increases. Furthermore, if Eq. (4) was differentiated with respect to time, the following equation is obtained:

$$\frac{dP}{dt}=\frac{d}{dt}\left(\frac{1}{1+\exp\left(-\left(a_0+a_1Press+a_2\ln(Time)+a_3Temp+a_4IC\right)\right)}\right) \quad (5)$$

According to Eq. (5), the probability density distribution could be obtained as shown in Fig. 5. The probability distribution of the required time for inactivation became narrower as the applied pressure increased. This result might indicate that the variability and/or uncertainty in the inactivation effect becomes small as the processing magnitude increases. In addition, from the probability density distribution, we could calculate probabilities by integrating the targeted area.

The shaded region represents the 95% prediction interval that may result in a 5-log reduction (Fig. 6). From the kinetics study, it was estimated that the time required for a 5-log reduction in reconstituted IF at 600 MPa was approximately 4.9 min. Thus, the observed pressure-holding time for a 5-log reduction fell within the 95% prediction interval. The survival/death interface modelling procedure yields results that are consistent with the experimental kinetic data, indicating the equivalence of the evaluation of the kinetic and probabilistic modelling procedures.

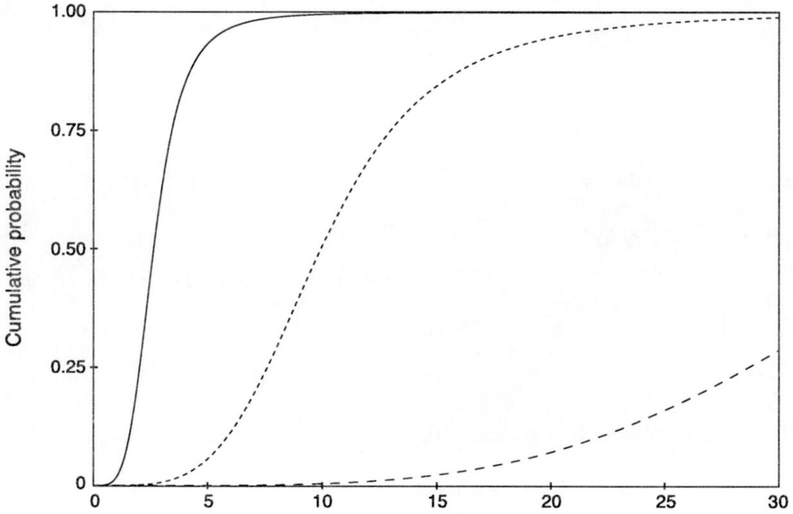

Figure 4. Changes in the cumulative probability for the inactivation of *C. sakazakii* corresponding to a 5-log reduction in reconstituted infant formula under different pressures at 25°C. Solid line: 600 MPa, dotted line: 500 MPa, dashed line: 400 MPa [adapted from Koseki (2013)].

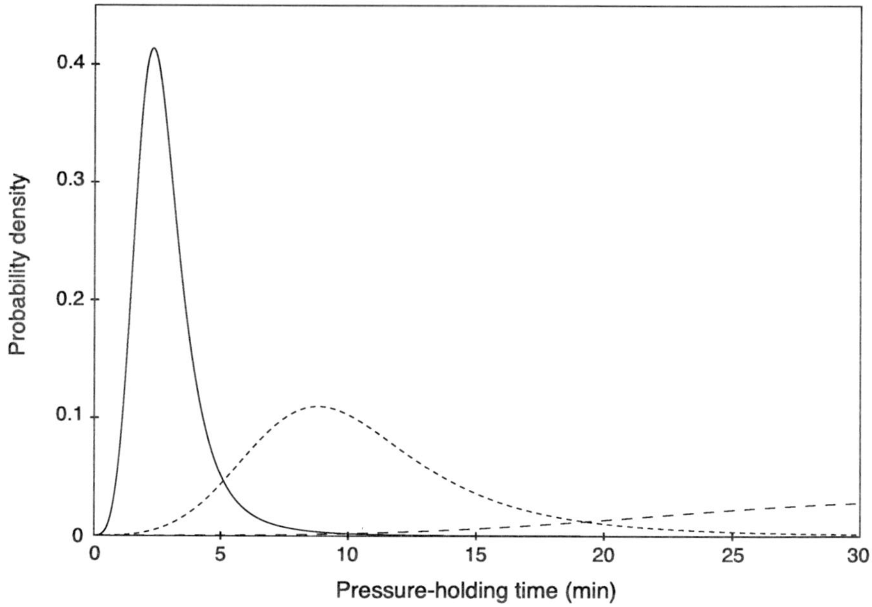

Figure 5. Probability density distribution for the inactivation of *C. sakazakii* corresponding to a 5-log reduction in reconstituted infant formula under different pressures at 25°C. Solid line: 600 MPa, dotted line: 500 MPa, dashed line: 400 MPa [adapted from Koseki (2013)].

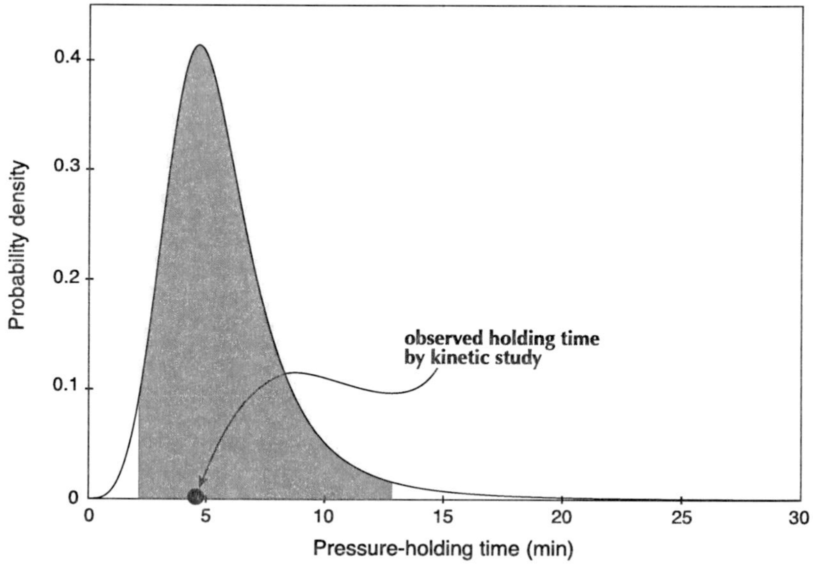

Figure 6. Probability density distribution of the time required for the inactivation of *C. sakazakii* corresponding to a 5-log reduction in reconstituted infant formula at 600 MPa and 25°C. Comparison of the inactivation results between a kinetic study (•, 4.9 min) and the present probabilistic study (shaded region represents the 95% prediction interval) [adapted from Koseki (2013)].

The application of the new model described herein would be useful for food processing in terms of microbiological food safety requirements based on the concepts of FSOs and POs because the microbial inactivation conditions that result in the required log reduction can be determined, along with the probability that those conditions will yield the desired results. The modelling procedure, which uses a new and different type of microbial inactivation model, will contribute to progress in the field of predictive microbiology.

Consideration of Pressure-injured Bacterial Cell Recovery

Recovery phenomena of bacterial cells after HHP-treatment

Damage to the cell membrane is believed to be one of the critical factors leading to the death of HHP-treated bacteria (Ritz et al. 2001, Russell 2002). However, some cells can survive even after HHP-treatment at lethal levels (Patterson et al. 1995, Simpson and Gilmour 1997, McClements et al. 2001, Ritz et al. 2001, Chen and Hoover 2003). Sub-lethal damage caused by HHP-treatment may be repairable, and the cells could potentially proliferate after repairing the site of injury during storage (Jordan et al. 2001, Russell 2002). Recently, some researchers have reported recovery of pathogenic and spoilage bacteria sub-lethally damaged by HHP-treatment during storage at several temperatures in broth, milk, and ground pork within 6 h to 4 weeks (Ellenberg and Hoover 1999, Chilton et al. 2001, Bozoglu et al. 2004, Bull et al. 2005). Bozoglu et al. (2004) have shown that even though injured cells were not detected immediately after HHP-treatment, these were recovered after repair within 1 to 15 days and could potentially recover in food.

Most publications on the recovery of bacteria after HHP treatment are based on qualitative assessment by streaking the sample containing HHP-treated bacteria onto solid media (Bozoglu et al. 2004, Bull et al. 2005) and do not involve a count of colony number. Details of recovery kinetics have not been examined from a quantitative point of view. Furthermore, studies on the recovery of injured bacteria have been investigated in broth, milk, or ground pork, which are nutrient-rich media. Thus, bacterial proliferation could not be evaluated separately to distinguish between repair of injured cells and bacterial growth after HHP-treatment. The recovery kinetics of HHP-treated cells in a nutrient-free medium would provide information to determine the shelf life of HHP-processed foods. Furthermore, an assessment of the microbial risks of HHP-processed foods will be supported by knowledge of the recovery kinetics of HHP-treated bacteria.

Koseki and Yamamoto (2006b) attempted to examine the recovery of bacteria in a nutrient-free medium (phosphate buffered saline, PBS) in order to distinguish between repair and growth. As shown in Fig. 7, recovery of

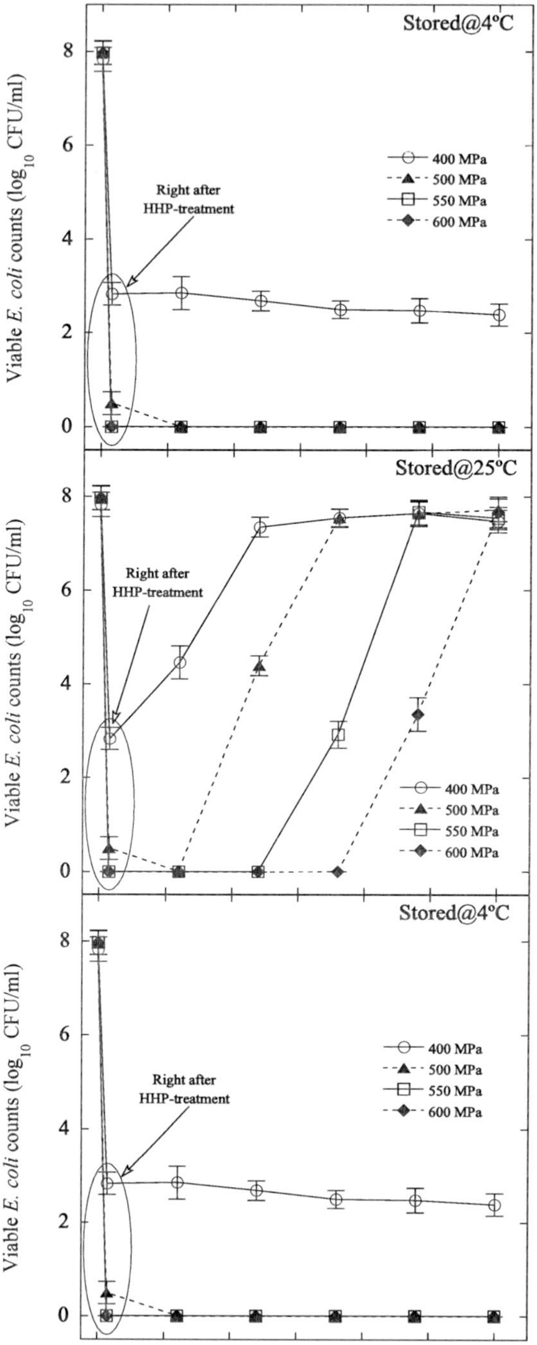

Figure 7. Changes in the number of *E. coli* in phosphate buffered saline (PBS) solution at 4, 25, and 37°C for 120 h followed by high hydrostatic pressure (HHP) treatment. Results are mean values of triplicate trials ± standard error [adapted from Koseki and Yamamoto (2006b)].

E. coli in PBS was observed during storage at 25°C after HHP treatment. During storage at 25°C, *E. coli* recovered from below the detectable level to the level of initial cell counts within 48 h regardless of the magnitude of the treatment pressure. In contrast, no recovery was observed during storage at 4 and 37°C. The number of *E. coli* cells (approximate 10^3 CFU/ml) without HHP-treatment did not increase at all in PBS between 4 and 37°C for a 120 h period. These results indicated that *E. coli* did not grow during storage, but that it repaired the injury and/or recovered its cell proliferating-ability. Furthermore, results suggested that the recovery of bacteria depend greatly on the storage temperature.

Koseki and Yamamoto (2006b) determined the recovery of *E. coli* in nutrient-rich medium (TSB) during incubation after HHP-treatment. Significant recovery was observed in TSB during storage at 25°C, as well as in PBS. The recovery in TSB was faster than that observed in PBS, as bacterial growth in TSB assisted faster recovery. During storage at 37°C, which is an optimal growth temperature for *E. coli* (Salter et al. 1998), no recovery was observed in TSB after HHP-treatment with 550 or 600 MPa, indicating that the optimal recovery temperature is lower than the optimal growth temperature. Bull et al. (2005) reported that the optimal recovery temperature of HHP-treated *L. monocytogenes* was 15°C, which was lower than the optimal growth temperature observed at around 30°C. Speculation concerning the mechanism underlying this phenomenon asserts that the membrane repair of pressure-induced pores would be harder to accomplish due to weakened intermolecular forces among membrane molecules as a result of warming (Russell 2002). However, the findings are not clear at present, and further clarification of the phenomenon is necessary using various bacterial strains.

Since the results suggested that recovery of bacteria was greatly affected by storage temperatures, Koseki and Yamamoto (2006b) investigated the effect of temperature change during storage on recovery (Fig. 8). No recovery was observed in PBS during storage at 4 and 37°C for 120 h; when the storage-temperatures were shifted to 25°C, the number of *E. coli* initially stored at 4°C increased approximately to the initial cell number of 10^8 CFU/ ml within 48 h regardless of the pressure level. In contrast, *E. coli* initially stored at 37°C showed no recovery. These results indicated that the possibility of bacterial recovery remained during storage below ambient temperature after HHP-treatment for at least 10 days.

Recovery in real food after HHP-treatment

HPP has been considered as a promising post-processing technique for use after the slicing and packaging of cooked ham (Cheftel and Culioli 1997). Some previous reports have examined the effect of HPP on bacterial

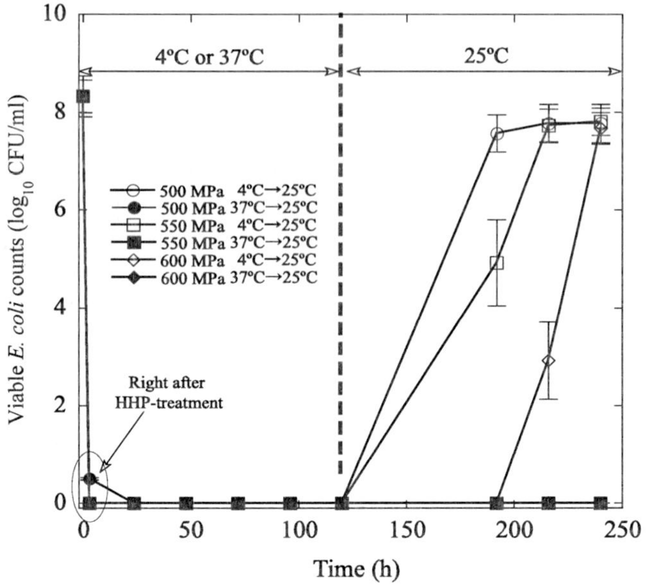

Figure 8. Changes in the number of *E. coli* in phosphate buffered saline (PBS) solution treated with high hydrostatic pressure (HHP) during storage of 4 or 37°C for 120 h and then shifted 25°C for 120 h. Results are mean values of triplicate trials ± standard error [adapted from Koseki and Yamamoto (2006b)].

inactivation on sliced cooked ham (Lopez-Caballero et al. 1999, Hayman et al. 2004, Aymerich et al. 2005, Morales et al. 2006). *L. monocytogenes* cells inoculated onto sliced cooked ham (10^4 CFU/g) were inactivated by HPP at 600 MPa and 20°C for 3 min (Hayman et al. 2004), and at 400 MPa and 17°C for 10 min (Aymerich et al. 2005). However, while the number of sub-lethally injured *L. monocytogenes* cells did not recover during 42 days of storage at 1°C or 6°C, the bacterial count subsequently increased to 10^6 CFU/g at 84 days of storage regardless of the temperature (Aymerich et al. 2005). Furthermore, although lactic acid bacteria on sliced cooked ham were inactivated by HPP at 400 MPa and 7°C for 20 min, the sub-lethally injured survivors started to recover after 21 days of storage at 2°C (Lopez-Caballero et al. 1999). The recovery of bacteria after HPP has also been reported on other kinds of food (Bozoglu et al. 2004, Bull et al. 2005, Chilton et al. 2001, Ellenberg and Hoover 1999, Koseki and Yamamoto 2006b). Accurate estimation of bacterial recovery is therefore important for ensuring the safety of high-pressure-processed foods during their distribution. Predictive modeling of the recovery of bacteria after HPP will thus be a useful tool for estimating the risks associated with these products.

Koseki et al. (2007d) reported the development of a bacterial recovery model after HPP. The predictive model is not just a simple tool for describing the time course of changes in the number of bacteria; rather, they have used the assessment of recovery/no recovery at arbitrary periods during storage as an indicator of the bacterial recovery after HPP. The target bacterium, *L. monocytogenes*, is a foodborne pathogen that is considered to pose a very low risk of listeriosis at the level of <10^2 CFU/g in foods at the time of consumption (Farber et al. 1996, Buchanan et al. 1997, The Commission of the European Communities 2005). Therefore, the time taken for the cell count to reach 10^2 CFU/g is important for a predictive model aimed at ensuring food safety with respect to *L. monocytogenes*. After reaching 10^2 CFU/g, the further increase in numbers of *L. monocytogenes* could be predicted with reference to the known growth kinetics of this organism in ham without HPP.

HPP above 550 MPa at 25°C for 5 min reduced *L. monocytogenes* cells by 7 $\log_{10}$ CFU/ml in milk (Koseki et al. 2008); however, during storage at 25°C after HPP, the injured survivors recovered and the population increased beyond the initial level within 7 days. HPP (550 MPa for 5 min) followed by mild-heat treatments (37–50°C) completely inhibited bacterial recovery during storage at 25°C for up to 70 days. Thereby, a predictive model for estimating the mild-heat treatment conditions required to prevent the recovery of *L. monocytogenes* in milk after HPP (550MPa for 5min) was developed using logistic regression procedure.

Bacterial recovery during storage after HPP was promoted at a temperature lower than the optimal temperature of bacterial growth as mentioned above. Accordingly, the storage temperature after HPP would play an important role in bacterial recovery. However, to date, the conditions for inhibition of bacterial recovery after HPP have not been investigated in detail. Koseki et al. (2008) examined the appropriate mild-heat treatment conditions after HPP (550MPa for 5min) from the point of view of controlling the recovery of *L. monocytogenes* in milk during storage. Furthermore, they developed a predictive model that can estimate the mild-heat treatment conditions necessary to inhibit the recovery of *L. monocytogenes* in milk during storage after HPP (550MPa for 5min). The model could identify various combinations of temperature and temperature-holding time that inhibit bacterial recovery. Speculation concerning the mechanism underlying the recovery phenomenon asserts that the membrane repair of pressure-induced destruction would be more difficult to accomplish due to weakened intermolecular forces among membrane molecules as a result of warming (Russell 2002). However, the findings are not clear at present, and further clarification of the mechanism of the phenomenon is needed from a molecular-level study. This would lead to clarification of the mechanisms of HPP-induced microbial inactivation, and promote certainty about the microbiological control technique using HPP.

In conclusion, the studies so far revealed that the increase in bacterial number during storage after HHP treatment resulted from the repair of injured cells and not bacterial growth. Furthermore, it was shown that the storage temperature during storage after HHP treatment played a critical role in the recovery of bacteria. Although the recovery potential of HHP-treated bacteria is problematic, the results of the studies indicate that proper temperature management may optimize the recovery of pathogenic and spoilage bacteria and thus prevent overestimation of the lethal effect of HHP treatment.

Conclusion

The characteristics of HHP treatment have been extensively investigated so far. In particular, HHP treatment has been investigated as a non-thermal microbial inactivation technique to produce high quality food. To meet consumers' demands for high quality, natural, healthy food, HHP treatment will be a promising technique in the future. However, there is still room for investigation into the effect of HHP treatment under sub-zero temperature on various aspects including microbial inactivation and also retention of organoleptic quality. This point of HHP treatment might lead to one more advanced stage of the HHP application on food processing. Extensive studies and also appropriate extension of the HHP treatment will play a key role in the future food industry.

Acknowledgement

This work was supported by a grant from the Japan Society for the Promotion of Science (JSPS) for Young Scientists. Furthermore, this study was conducted as part of the project "Development of evaluation and manage-ment methods for supply of safe, reliable and functional food and farm produce" funded by the National Food Research Institute.

Keywords: High pressure processing, non-thermal inactivation, injured recovery, survival model, probabilistic model, logistic regression.

References

Alpas, H., N. Kalchayanand, F. Bozoglu and B. Ray. 2000. Interactions of high hydrostatic pressure, pressurization temperature and pH on death and injury of pressure-resistant and pressure-sensitive strains of foodborne pathogens. Int. J. Food Microbiol. 60: 33–42.

Aymerich, T., A. Jofre, M. Garriga and M. Hugas. 2005. Inhibition of *Listeria monocytogenes* and *Salmonella* by natural antimicrobials and high hydrostatic pressure in sliced cooked ham. J. Food Prot. 68: 173–177.

Baranyi, J. and T.A. Roberts. 1994. A dynamic approach to predicting bacterial-growth in food. Int. J. Food Microbiol. 23: 277–294.

Bozoglu, F., H. Alpas and G. Kaletunc. 2004. Injury recovery of foodborne pathogens in high hydrostatic pressure treated milk during storage. FEMS Immunol. Med. Microbiol. 40: 243–247.

Bridgman, P.W. 1914. The coagulation of albumen by pressure. J. Biol. Chem. 19: 511–512.

Buchanan, R.L., W.G. Damert, R.C. Whiting and M. van Schothorst. 1997. Use of epidemiologic and food survey data to estimate a purposefully conservative dose-response relationship for *Listeria monocytogenes* levels and incidence of listeriosis. J. Food Prot. 60: 918–922.

Bull, M.K., M.M. Hayman, C.M. Stewart, E.A. Szabo and S.J. Knabel. 2005. Effect of prior growth temperature, type of enrichment medium, and temperature and time of storage on recovery of *Listeria monocytogenes* following high pressure processing of milk. Int. J. Food Microbiol. 101: 53–61.

Buzrul, S. and H. Alpas. 2004. Modeling the synergistic effect of high pressure and heat on inactivation kinetics of *Listeria innocua*: a preliminary study. FEMS Microbiol. Lett. 238: 29–36.

Buzrul, S., H. Alpas and F. Bozoghu. 2005. Use of Weibull frequency distribution model to describe the inactivation of *Alicyclobacillus acidoterrestris* by high pressure at different temperatures. Food Res. Int. 38: 151–157.

Cheftel, J.C. 1995. Review: High-pressure, microbial inactivation and food preservation. Food Sci. Technol. Int. 1: 75–90.

Cheftel, J.C. and J. Culioli. 1997. Effects of high pressure on meat: A review. Meat Sci. 46: 211–236.

Chen, H.Q. and D.G. Hoover. 2003. Pressure inactivation kinetics of *Yersinia enterocolitica* ATCC 35669. Int. J. Food Microbiol. 87: 161–171.

Chen, H.Q., D.G. Hoover and D.H. Kingsley. 2005. Temperature and treatment time influence high hydrostatic pressure inactivation of *feline calicivirus*, a *norovirus* surrogate. J. Food Prot. 68: 2389–2394.

Chen, H.Q., R.D. Joerger, D.H. Kingsley and D.G. Hoover. 2004. Pressure inactivation kinetics of phage lambda cI 857. J. Food Prot. 67: 505–511.

Chilton, P., N.S. Isaacs, P. Manas and B.M. Mackey. 2001. Biosynthetic requirements for the repair of membrane damage in pressure-treated *Escherichia coli*. Int. J. Food Microbiol. 71: 101–104.

Codex Alimentarius Commission. 2005. Report of the 22nd Session of the Codex Committee on General Principles (No. ALINORM 05/28/33A). Paris.

Doona, C.J., F.E. Feeherry and E.W. Ross. 2005. A quasi-chemical model for the growth and death of microorganisms in foods by non-thermal and high-pressure processing. Int. J. Food Microbiol. 100: 21–32.

Ellenberg, L. and D.G. Hoover. 1999. Injury and survival of *Aeromonas hydrophila* 7965 and *Yersinia enterocolitica* 9610 from high hydrostatic pressure. J. Food Safety 19: 263–276.

Farber, J.M., W.H. Ross and J. Harwig. 1996. Health risk assessment of *Listeria monocytogenes* in Canada. Int. J. Food Microbiol. 30: 145–156.

Hashizume, C., K. Kimura and R. Hayashi. 1995. Kinetic analysis of yeast inactivation by high pressure treatment at low temperatures. Biosci. Biotechnol. Biochem. 59: 1455–1458.

Hayakawa, K., Y. Ueno, S. Kawamura, T. Kato and R. Hayashi. 1998. Microorganism inactivation using high-pressure generation in sealed vessels under sub-zero temperature. Appl. Microbiol. Biotechnol. 50: 415–418.

Hayashi, R., Y. Kawamura and S. Kunugi. 1987. Introduction of high pressure to food processing: preferential proteolysis of β-lactoglobulin in milk whey. J. Food Sci. 52: 1107–1108.

Hayman, M.M., I. Baxter, P.J. O'Riordan and C.M. Stewart. 2004. Effects of high-pressure processing on the safety, quality, and shelf life of ready-to-eat meats. J. Food Prot. 67: 1709–1718.

Hite, B.H. 1899. The effect of pressure in the preservation of milk: A preliminary report. Bulletin, West Virginia University. Agricultural Experiment Station 58: 15–36.

Hite, B.H. 1914. The effect of pressure on certain micro-organisms encountered in the preservation of fruits and vegetables. Bulletin, West Virginia University. Agricultural Experiment Station 146: 3–67.

Hoover, D.G. 1993. Pressure effects on biological-systems. Food Technol. 47: 150–155.

Horie, Y., K. Kimura, M. Ida, Y. Yosida and K. Ohki. 1991. Jam preparation by pressurization. Nippon Nogeikagaku Kaishi-J. Jpn. Soc. Biosci. Biotechol. Agrochem. 65: 975–980.

Interantional Commission on Microbiological Specifications for Foods (ICMSF). 2001. Microorganisms in Foods; 7. Microbiological testing in food safety management. Kluwer Academic/Plenum Publishers, New York, pp. 23–44.

Jordan, S.L., C. Pascual, E. Bracey and B.M. Mackey. 2001. Inactivation and injury of pressure-resistant strains of *Escherichia coli* O157 and *Listeria monocytogenes* in fruit juices. J. Appl. Microbiol. 91: 463–469.

Kalchayanand, N., A. Sikes, C.P. Dunne and B. Ray. 1998. Interaction of hydrostatic pressure, time and temperature of pressurization and pediocin AcH on inactivation of foodborne bacteria. J. Food Prot. 61: 425–431.

Kilimann, K.V., C. Hartmann, A. Delgado, R.F. Vogel and M.G. Gänzle. 2005. A fuzzy logic-based model for the multistage high-pressure inactivation of *Lactococcus lactis* ssp. cremoris MG 1363. Int. J. Food Microbiol. 98: 89–105.

Knorr, D. 1993. Effects of high-hydrostatic-pressure processes on food safety and quality. Food Technol. 47: 156–161.

Koseki, S., M. Matsubara and K. Yamamoto. 2009. Prediction of a required log reduction with probability for *Enterobacter sakazakii* during high-pressure processing, using a survival/death interface model. Appl. Environ. Microbiol. 75: 1885–1891.

Koseki, S., Y. Mizuno and K. Yamamoto. 2008. Use of mild-heat treatment following high-pressure processing to prevent recovery of pressure-injured *Listeria monocytogenes* in milk. Food Microbiol. 25: 288–293.

Koseki, S. and K. Yamamoto. 2006a. pH and solute concentration of suspension media affect the outcome of high hydrostatic pressure treatment of *Listeria monocytogenes*. Int. J. Food Microbiol. 111: 175–179.

Koseki, S. and K. Yamamoto. 2006b. Recovery of *Escherichia coli* ATCC 25922 in phosphate buffered saline after treatment with high hydrostatic pressure. Int. J. Food Microbiol. 110: 108–111.

Koseki, S. and K. Yamamoto. 2007a. Water activity of bacterial suspension media unable to account for the baroprotective effect of solute concentration on the inactivation of *Listeria monocytogenes* by high hydrostatic pressure. Int. J. Food Microbiol. 115: 43–47.

Koseki, S. and K. Yamamoto. 2007b. A novel approach to predicting microbial inactivation kinetics during high pressure processing. Int. J. Food Microbiol. 116: 275–282.

Koseki, S. and K. Yamamoto. 2007c. Modelling the bacterial survival/death interface induced by high pressure processing. Int. J. Food Microbiol. 116: 136–143.

Koseki, S., Y. Mizuno and K. Yamamoto. 2007d. Predictive modelling of the recovery of *Listeria monocytogenes* on sliced cooked ham after high pressure processing. Int. J. Food Microbiol. 119: 300–307.

Koseki, S. 2013. Alternative approaches to predicting microbial behaviour: A probabilistic modelling approach for microbial inactivation and a revised web-tool, the Microbial Responses Viewer. Food Control. 29: 416–421.

Linton, M., J.M.J. McClements and M.F. Patterson. 1999. Inactivation of *Escherichia coli* O157:H7 in orange juice using a combination of high pressure and mild heat. J. Food Prot. 62: 277–279.

Linton, R.H., W.H. Carter, M.D. Pierson and C.R. Hackney. 1995. Use of A modified Gompertz equation to model nonlinear survival curves for *Listeria monocytogenes* Scott-A. J. Food Prot. 58: 946–954.

Lopez-Caballero, M.E., J. Carballo and F. Jimenez-Colmenero. 1999. Microbiological changes in pressurized, prepackaged sliced cooked ham. J. Food Prot. 62: 1411–1415.

Ludwig, H., C. Bieler, K. Hallbauer and W. Scigalla. 1992. Inactivation of microorganisms by hydrostatic pressure. Col-loque INSERM. John Libbey and Co., Ltd, London, pp. 25–32.

Mackey, B.M., K. Forestière and N. Isaacs. 1995. Factors affecting the resistance of *Listeria monocytogenes* to high hydrostatic pressure. Food Biotechnol. 9: 1–11.

McClements, J.M.J., M.F. Patterson and M. Linton. 2001. The effect of growth stage and growth temperature on high hydrostatic pressure inactivation of some psychrotrophic bacteria in milk. J. Food Prot. 64: 514–522.

Moerman, F., B. Mertens, L. Demey and A. Huyghebaert. 2001. Reduction of *Bacillus subtilis, Bacillus stearothermophilus* and *Streptococcus faecalis* in meat batters by temperature-high hydrostatic pressure pasteurization. Meat Sci. 59: 115–125.

Molina-Guiterrez, A., B. Rademacher, M.G. Gänzle and R.F. Vogel. 2002. Effect of sucrose and sodium chloride on the survival and metabolic activity of *Lactococcus lactis* under high-pressure conditions. Elsevier, Amsterdam, pp. 295–302.

Molina-Hoppner, A., W. Doster, R.F. Vogel and M.G. Gänzle. 2004. Protective effect of sucrose and sodium chloride for *Lactococcus lactis* during sublethal and lethal high-pressure treatments. Appl. Environ. Microbiol. 70: 2013–2020.

Morales, P., J. Calzada and M. Nunez. 2006. Effect of high-pressure treatment on the survival of *Listeria monocytogenes* Scott A in sliced vacuum-packaged Iberian and Serrano cured hams. J. Food Prot. 69: 2539–2543.

Ogawa, H., K. Fukuhisa, Y. Kubo and H. Fukumoto. 1990. Pressure inactivation of yeasts, molds, and pectinesterase in satsuma mandarin juice: effects of juice concentration, pH, and organic acids, and comparison with heat sanitation. Agricultural and biological chemistry 54: 1219–1225.

Oxen, P. and D. Knorr. 1993. ScienceDirect.com—LWT—Food Science and Technology—Baroprotective Effects of High Solute Concentrations Against Inactivation of *Rhodotorula rubra*. LWT - Food Science and Technology 26: 220–223.

Palou, E., A. LopezMalo, G.V. BarbosaCanovas, J. WeltiChanes and B.G. Swanson. 1997. Effect of water activity on high hydrostatic pressure inhibition of *Zygosaccharomyces bailii*. Lett. Appl. Microbiol. 24: 417–420.

Pandya, Y., F.F. Jewett and D.G. Hoover. 1995. Concurrent effects of high hydrostatic pressure, acidity and heat on the destruction and injury of yeasts. J. Food Prot. 58: 301–304.

Patterson, M.F., M. Quinn, R. Simpson and A. Gilmour. 1995. Sensitivity of vegetative pathogens to high hydrostatic-pressure treatment in phosphate-buffered saline and foods. J. Food Prot. 58: 524–529.

Peleg, M. 1999. On calculating sterility in thermal and non-thermal preservation methods. Food Res. Int. 32: 271–278.

Peleg, M. and M.B. Cole. 1998. Reinterpretation of microbial survival curves. Crit. Rev. Food Sci. Nutr. 38: 353–380.

Perrier-Cornet, J.-M., S. Tapin, S. Gaeta and P. Gervais. 2005. High-pressure inactivation of *Saccharomyces cerevisiae* and *Lactobacillus plantarum* at subzero temperatures. J. Biotechnol. 115: 405–412.

R Development Core Team. 2012. In R: A language and environment for statistical computing (2.14.1 ed.). Vienna, Austria: R Foundation for Statistical Computing.

Ratkowsky, D.A., J. Olley, T.A. McMeekin and A. Ball. 1982. Relationship between temperature and growth rate of bacterial cultures. J. Bacteriol. 149: 1–5.

Reyns, K.M.F.A., C.C.F. Soontjens, K. Cornelis, C.A. Weemaes, M.E. Hendrickx and C.W. Michiels. 2000. Kinetic analysis and modelling of combined high-pressure–temperature inactivation of the yeast *Zygosaccharomyces bailii*. Int. J. Food Microbiol. 56: 199–210.

Ritz, M., J.L. Tholozan, M. Federighi and M.F. Pilet. 2001. Morphological and physiological characterization of *Listeria monocytogenes* subjected to high hydrostatic pressure. Appl. Environ. Microbiol. 67: 2240–2247.

Ross, E.W., I.A. Taub, C.J. Doona, F.E. Feeherry and K. Kustin. 2005. The mathematical properties of the quasi-chemical model for microorganism growth-death kinetics in foods. Int. J. Food Microbiol. 99: 157–171.

Russell, N.J. 2002. Bacterial membranes: the effects of chill storage and food processing. An overview. Int. J. Food Microbiol. 79: 27–34.

Salter, M.A., T. Ross and T.A. McMeekin. 1998. Applicability of a model for non-pathogenic *Escherichia coli* for predicting the growth of pathogenic *Escherichia coli*. J. Appl. Microbiol. 85: 357–364.

San Martín, M.F., G.V. Barbosa-Cánovas and B.G. Swanson. 2002. Food processing by high hydrostatic pressure. Crit. Rev. Food Sci. Nutr. 42: 627–645.

Shigehisa, T., T. Ohmori, A. Saito, S. Taji and R. Hayashi. 1991. Effects of high hydrostatic-pressure on characteristics of pork slurries and inactivation of microorganisms associated with meat and meat-products. Int. J. Food Microbiol. 12: 207–216.

Simpson, R.K. and A. Gilmour. 1997. The resistance of *Listeria monocytogenes* to high hydrostatic pressure in foods. Food Microbiol. 14: 567–573.

Sonoike, K., T. Setoyama, Y. Kuma and S. Kobayashi. 1992. Effect of pressure and temperature on the death rates of *Lactobacillus casei* and *Escherichia coli*. Col- loque INSERM. John Libbey and Co., Ltd, London, pp. 297–301.

Stewart, C.M., F.F. Jewett, C.P. Dunne and D.G. Hoover. 1997. Effect of concurrent high hydrostatic pressure, acidity and heat on the injury and destruction of *Listeria monocytogenes*. J. Food Safety 17: 23–36.

Takahashi, K. 1992. Sterilization of microorganisms by hydrostatic pressure at low temperature. Col- loque INSERM. John Libbey and Co., Ltd, London, pp. 303–307.

Tassou, C.C., E.Z. Panagou, F.J. Samaras, P. Galiatsatou and C.G. Mallidis. 2008. Temperature-assisted high hydrostatic pressure inactivation of *Staphylococcus aureus* in a ham model system: evaluation in selective and nonselective medium. J. Appl. Microbiol. 104: 1764–1773.

The Commission of the European Communities. 2005. Commission Regulation (EC) No 2073/2005. Official Journal of the European Union.

Van Opstal, I., S.C.M. Vanmuysen and C.W. Michiels. 2003. High sucrose concentration protects *E. coli* against high pressure inactivation but not against high pressure sensitization to the lactoperoxidase system. Int. J. Food Microbiol. 88: 1–9.

Yamamoto, K., M. Matsubara, S. Kawasaki, M.L. Bari and S. Kawamoto. 2005. Modeling the pressure inactivation dynamics of *Escherichia coli*. Brazilian J. Med. Biol. Res. 38: 1253–1257.

Zook, C.D., M.E. Parish, R.J. Braddock and M.O. Balaban. 1999. High pressure inactivation kinetics of *Saccharomyces cerevisiae* ascospores in orange and apple juices. J. Food Sci. 64: 533–535.

CHAPTER 2

Implementation of Microbially Safe Foods with Pulsed Electric Fields

Mansel W. Griffiths[1] and *Markus Walkling-Ribeiro*[2,*]

Introduction

In the last two decades high-voltage pulsed electric fields (PEF) technology has gradually evolved from a novel, research-based system to one of the three major 'non-thermal candidates' for food processing, only surpassed by high hydrostatic pressure and in part irradiation treatments. Similar to other non-thermal techniques PEF enables treatment of food and biomaterials at sub-pasteurization conditions, as reflected by shorter treatment times, lower processing temperatures or both, thereby, allowing production of microbiologically safe and shelf-stable products of higher nutritional and sensory quality than conventional heat pasteurization. At present the fields of application for PEF in food production and research mainly focus on the preservation for pumpable liquids (Pina-Perez et al. 2009a, Altuntas et al. 2011, Chen et al. 2013) as well as for dehydration (Gachovska et al. 2009, Donsi et al. 2010, Wiktor et al. 2013) and extraction of solids (Grimi

[1] Canadian Research Institute for Food Safety, Department of Food Science, Ontario Agricultural College, University of Guelph, 43 McGilvray Street, Guelph, Ontario, N1G 2W1 Canada.
Email: mgriffit@uoguelph.ca
[2] Department of Food Science, College of Agriculture and Life Sciences, Cornell University, 212 Stocking Hall, Ithaca, New York, 14853 USA.
Email: mw757@cornell.edu
* Corresponding author

et al. 2011, Toepfl and Heinz 2011, Puértolas et al. 2013), although there is a growing interest in the use of this emerging technology for food structure modification to generate new functional food ingredients (Yu et al. 2009, Han et al. 2012, Zhao and Yang 2013). Other bioelectric, non-food applications of PEF comprise biotechnological processes such as gene transfer (Marjanovic et al. 2010) and gel electrophoresis in proteomics (Li et al. 2010), biomedical treatments such as drug delivery (Yan et al. 2010), tissue engineering (Griffiths et al. 2012) and nano-scale material production (Virwani et al. 2010), biofouling (Pérez-Roa et al. 2006), biofuel extraction from plant cells (DeBoer et al. 2012), and further breakdown of activated waste water sludge for enhanced anaerobic digestion and production of biogas (Salerno et al. 2009). With the overall progress in semi-conductor technology and the addition of new areas for the implementation of PEF, its affordability and demand have increased, respectively. Besides its exclusive mode of action for the inactivation of microorganisms (electropermeabilization) other examples whereby PEF offers distinct advantages over other non-thermal technologies for liquid food preservation are its applicability for continuous-flow processing, efficacy being independent of fluid transparency or penetration depth, environmental friendliness as a residue-free, low-energy technology, combined with the fact that it is a relatively well-researched technology. Research activity has increased considerably over time, as indicated in Table 1, peaking between 2010 and 2011 so far. It is also noteworthy, that not only has the number of published work increased in total, but also that the research contributions are spread more evenly over time and geographic location, from a few groups in North America and Europe studying PEF before the turn of the millennium (e.g., Ohio State University, Washington State University, Berlin University of Technology, University of Guelph, University of Montpellier) to a steadily growing global community of food research institutions (e.g., University of Lleida, McGill University, University of Zaragoza, Commonwealth Scientific and Industrial Research Organisation, University College Dublin, Jiangnan University, Kharagpur Indian Institute of Technology), equipment suppliers (e.g., German Institute of Food Technologies, Diversified Technologies, PurePulse), and food manufacturers (e.g., Becker's Bester, Hoogesteger, Wernsing Feinkost GmbH). However, the latter indicates that the technology is still in its infancy for industrial applications and that there is still much room for process developments and optimization as well as marketing efforts in order to fully commercialize PEF technology in both food and bioprocess industries worldwide.

Table 1. Overview of published research classified by time period, amount of entries, and top 10 contributing authors, and listed in the database Food Science and Technology Abstracts based on the results for the search keywords "pulsed electric fields" and "food".

Publication Period	Publication Entries	Top 10 Listed Authors
1969–1995	46	Barbosa-Cánovas, G.V. (16); Swanson, B.G. (16); Zhang, Q.H. (11); Knorr, D. (5); Pothakamury, U.R. (5); Qin, B.L. (5); Monsalve-González, A. (3); Sastry, S.K. (3); Vega-Mercado, H. (3); Yousef, A.E. (3)
1996–1997	44	Barbosa-Cánovas, G.V. (12); Swanson, B.G. (12); Zhang, Q.H. (10); Chang, F.J. (8); Vega-Mercado, H. (8); Martín-Belloso, O. (7); Qin, B.L. (7); Pothakamury, U.R. (4); Sastry, S.K. (4); Yousef, A.E. (3)
1998–1999	42	Barbosa-Cánovas, G.V. (8); Swanson, B.G. (7); Calderón-Miranda, M.L. (3); Góngora-Nieto, M.M. (3); Zhang, Q.H. (3); Angersbach, A. (2); Barsotti, L. (2); Cheftel, J.C. (2); Knorr, D. (2); Pothakamury, U.R. (2)
2000–2001	64	Álvarez, I. (4); Barbosa-Cánovas, G.V. (4); Mastwijk, H.C. (4); Swanson, B.G. (4); Angersbach, A. (3); Knorr, D. (3); Mittal, G.S. (3); Pol, I.E. (3); Raso, J. (3); Zhang, Q.H. (3)
2002–2003	81	Barbosa-Cánovas, G.V. (10); Raso, J. (7); Condón, S. (6); Álvarez, I. (5); Mittal, G.S. (5); Zhang, Q.H. (5); Góngora-Nieto, M.M. (4); Griffiths, M.W. (4); Knorr, D. (4); Pagán, R. (4)
2004–2005	59	Zhang, Q.H. (7); Barbosa-Cánovas, G.V. (4); Evrendilek, G.A. (4); Garcia, D. (4); Pagán, R. (4); Chen, F. (3); Condón, S. (3); Elez-Martínez, P. (3); Gómez, N. (3); Hu, X. (3); Liao, X. (3); Lebovka, N.I. (3); Mañas, P. (3); Martín-Belloso, O. (3); Vorobiev. E. (3); Zhong, K. (3)
2006–2007	76	Martín-Belloso, O. (6); Mañas, P. (5); Rodrigo, D. (5); Rodrigo, M. (5); Martínez-López, A. (4); Sampedro, F. (4); Amiali, M. (3); Barbosa-Cánovas, G.V. (3); Condón, S. (3); Elez-Martínez, P. (3); Garcia, D. (3); Ngadi, M.O. (3); Pagán, R. (4); Raso, J. (3); Rivas, A. (3)
2008–2009	66	Martín-Belloso, O. (12); Knorr, D. (6); Condón, S. (5); Cronin, D.A. (5); Lyng, J.G. (5); Morgan, D.J. (5); Noci, F. (5); Pagán, R. (5); Walkling-Ribeiro, M. (5); Garcia, D. (4); Mañas, P. (4); Sobrino-López, A. (4)
2010–2011	115	Condón, S. (11); Álvarez, I. (10); Raso, J. (10); Knorr, D. (9); Martín-Belloso, O. (8); Pagán, R. (5); Meneses, N. (6); Saldaña, G. (6); Zhang, Q.H. (6); Jaeger, H. (5); Monfort, S. (5); Puértolas, E. (5)
2012–2013	64	Zhao, W. (6); Raso, J. (5); Yang, R. (5); Álvarez, I. (4); Martín-Belloso, O. (4); Condón, S. (3); Cronin, D.A. (3); Lyng, J.G. (3); Morgan, D.J. (3); Pagán, R. (3); Whyte, P. (3)

Data was collected prior to 2 April 2013 and is based on the search algorithm of the Food Science and Technology Abstracts database only, not representing the entirety of publications on pulsed electric fields produced during the investigated time period and relying on the database operator updates and listings' selection criteria. The numbers in parentheses following the names indicate the entries associated with the respective authors during a given period of publication.

In the first part of this chapter the basic principles of PEF processing, which cover engineering, physical, chemical and microbiological aspects, will be presented and, thereafter, the effectiveness of PEF for the inactivation of specific foodborne microorganisms will be discussed. A different approach is adopted in the subsequent section, illuminating the effect of different food matrices on the reduction of microbial loads by PEF. Moreover, observation-based modeling of PEF inactivation is briefly reviewed and a summary of current trends and future perspectives of PEF technology for food preservation is given at the end of this chapter.

PEF Treatment Parameters and Mode of Action

A network, consisting of a power supply providing high voltage direct current, a set of capacitors for storage of the electrical energy and a pulse forming network or a switch for the release of high voltage pulses between a pair of electrodes located in a chamber for high voltage product treatment, is used to generate PEF. Electricity is converted by rectifier transformers from alternating to direct current to enable high voltages to be achieved. Transformer power rating determines the charge/discharge operations of capacitors during a specific period of time. Power requirements are based on electrical resistance of the charging resistor and the capacitive load with power supplies ranging from 1.5 kW (laboratory-scale units) to 3.2 MW (industrial-scale units) (Barbosa-Cánovas and Sepúlveda 2005). Capacitor number and size affect power requirements and charging time and the capacitance determines the storage level of electrical energy. With a discharge switch (e.g., vacuum or spark gaps, electron tubes, high power transistors, semiconductors) the release of electrical energy is controlled, specifying maximum pulse repetition rate as well as the current and voltage amounts that can be handled and impact on the process.

Process Factors

Pulse properties and treatment time

Two main waveforms are used in PEF processing: exponential decay pulses, which are still used in older research equipment and square or rectangular wave pulses, which are generated in state-of-the-art PEF systems for research and industry. Both allow for generation of monopolar and bipolar wave types, without and with interchange between charging and grounding electrodes during the process, respectively, with the latter being more effective (Qin et al. 1994). The reason for it is repeated polarity change which causes an alteration in the direction of charges on the inside and the outside of the cell membrane and initiates reorientation of charged groups on the membrane (Chang 1989), thus, creating a mechanical membrane oscillation. By this additional alternating stress, the cell is more prone to membrane breakdown. Decreased occurrence of electrolytic reactions due to less build up of deposits on the electrode surfaces is another ancillary effect resulting from polarity changes. To date, it has not been determined whether or not so-called "instant-charge-reversal" pulses (bipolar pulses without relaxation times between positive and negative phases) enhance the efficacy of the PEF process. Generation of exponential decay pulses is relatively inexpensive but also less efficient as high level voltage is not maintained at peak level for the pulse duration but tails exponentially, allowing for no pulse interruption between the capacitor and the pulse generating network until depletion of the pulse energy. In spite of maintaining higher currents only low pulse frequencies can be implemented and life expectancy of such switches is comparatively short. By contrast, square wave pulses formed with solid state switches allows for interruptions, so that electrical energy is sustained at the peak voltage throughout most of the pulse duration (Barsotti et al. 1999), thereby, offering better operability, higher pulse frequencies, absence of mechanical parts and greater efficiency in switching and conducting; resulting in longer durability. However, due to the capability of managing only low current and voltage configurations, solid state switch units in series and in parallel are required, accounting for considerably higher PEF system costs. It is also noteworthy that more excess heat produced by exponential decay pulses is dissipated into the food without electroporation, achieving less microbial inactivation, providing lower energy efficiency, and demanding more cooling for the processing of heat sensitive products (Zhang et al. 1995a).

Pulse width of square wave shaped pulses is determined by the real exposure time of the sample at maximum voltage. In contrast, pulse width of exponential decay pulses is measured as the time required for exponential decay of the electric field to 37% of its initial peak strength (Zhang et al. 1995a), which allows no exact determination due to fluctuations in the electric field strength. Based on a defined pulse width, the treatment time, t_t [μ_s], can be calculated as the product of pulse width t_p [μ_s] and pulse number n_p as shown in the following equation:

$$t_t = t_p \cdot n_p \tag{1}$$

It is noteworthy that treatment time is also affected by flow velocity, with the latter being dependent on treatment chamber design. Slower and quicker product residence times and, consequently, exposure to PEF result from different hydrodynamic flow profiles, which are more turbulent close to the chamber wall and increasingly laminar towards the flow centre, respectively. Nanosecond range charging of the capacitors is required to reach the critical cell membrane potential (10 ns to 500 ns) in contrast to the pulse width, which is formed in the microsecond range and mostly accounts for the treatment effect of PEF. In the case of a higher pulse frequency, a larger pulse number crossing the interelectrode gap characterized by minimal relaxation times and increased overall high voltage pulse width and exposure time, the electropermeabilization is more likely to cause irreversible membrane pore formation that is followed by death of the microbial cell. It should also be borne in mind that cell components are subjected to different treatment intensities dependent on the pulse width, as the outer membrane is more strained by longer pulses than the substructures of the microbial cell, which in turn are more sensitive to shorter pulses (Schoenbach et al. 1997). Thus, a compromise between pulse width and pulse number has to be made depending on the specific field of application.

Electric field strength and energy density

The build-up of an external electrical field affects membrane permeabilization of microorganisms treated with PEF. When the field strength of the electric field E exceeds the critical field strength (E_c), the transmembrane potential (TMP; approximately 1V) is reached causing electropermeabilization (Neumann et al. 1989). Due to their ionic concentrations the majority of liquid foods are conductive and, hence, a high electric field strength is generated when a strong current flow is applied. The charging voltage of the high voltage pulse generating unit U [kV] as well as the gap between the parallel electrodes d [cm] are the parameters that determine the electrical field strength E [kV/cm], and the latter can be calculated with the subsequent Eq. (2) for a parallel electrode arrangement:

$$E = \frac{U}{d} \tag{2}$$

Microorganisms within the food product are not inactivated if the critical electric field strength E_c is not reached because membrane pore formation is not triggered in this case. It is the reason for a lower treatment efficiency of exponential decay pulses, where E_c is exceeded initially for a short peak fraction of the pulse width, in comparison to square wave pulses during which E_c is constantly surpassed over the duration of a pulse, thereby, sustaining a high voltage for much longer.

Effective and uniform food treatment with PEF is characterized by a homogeneous distribution of the electrical field that needs to be established in order to prevent local peaks in field strength. Generation of higher electric field strengths also relies on the PEF system and its power rating, because the maximum electric field strength is determined by the capacities of the pulse-generating equipment. With higher field strengths at the same level of specific energy input the effectiveness of the electroporation is clearly enhanced (Heinz et al. 1999), but this is also limited due to the dielectric properties of the product under treatment (Ryynaenen 1995). By determination of the specific energy input W_{PEF} [kJ/L] the effectiveness of different PEF processes and systems can be compared, analyzing the energy requirements needed per mass of food product using the following equation (Zhang et al. 1995a):

$$W_{PEF} = \frac{U^2 \cdot t_t}{R_C \cdot V_C} \tag{3}$$

where R_C [Ω] and V_C [cm³] denote the resistance and volume of the PEF treatment chamber, respectively. High intensity electric field strength in combination with a high energy input increases the chance that larger pores above critical diameters are formed in the cell membrane that induce irreversible damage and, ultimately, cell death.

Processing temperature

Higher pulse frequencies and larger pulse widths cause a rise in the PEF processing temperature and induce changes in the electrical conductivity of the food, thus affecting the maximum current flow in the treatment chamber during pulsing. The specific temperature increase ΔT [°C/l] that occurs during food processing with PEF is described by the following equation:

$$\Delta T = \frac{n \cdot W_{PEF}}{m \cdot c_p} \tag{4}$$

where the product volume and the specific heat capacity of the sample at constant pressure are indicated by *m* [l] and *cp* [kJ/l °C], respectively. Energy dissipation during PEF treatment, which is particularly interesting with regard to foods sensitive to thermal stress, can be assessed by calculating the temperature difference, thereby, revealing the suitability of PEF as an alternative preservation method. Inactivation of microorganisms by PEF is critically dependent on the product temperature prior to processing as the phospholipid layer conformation within microbial cells is altered from gel structure at low temperatures to crystalline structures at elevated temperatures. As a consequence of these conformational modifications cell functions and stability are changed. Electropermeabilization is strongly affected by the processing temperature and is likely to occur at low field strengths and relatively high temperatures due to the thinning of the phospholipid membrane layer during gel to crystalline phase transition (Coster and Zimmermann 1975). Such transitions, initiated by electrocompressive forces, are generated when the critical temperature for phase transformation has been reached. In several studies increased effectiveness of PEF at treatment temperatures above ambient level has been reported (Evrendilek and Zhang 2003, Li et al. 2005, Floury et al. 2006a, Fox et al. 2008). On the other hand, a higher treatment temperature before and during the PEF processing can amplify the thermal treatment effect, also referred to as *Joule heating*, which may surpass and negate the actual nonthermal effect of PEF, thus, thwarting the advantage over conventional heat preservation.

Treatment chamber design

Initially, PEF technology was applied in batch operation mode with static treatment chambers (Sale and Hamilton 1967, Dunn and Pearlman 1987, Zhang et al. 1995b). Once continuous type PEF systems, that feature higher throughputs, more convenient processing conditions, improved quality retention due to better heat management and more product-optimized setup options, were introduced successfully, batch type systems have generally been replaced for preservation-type applications. Based on the electrode configurations three different types of chamber layouts have been established over the years.

Treatment homogeneity is determined by the electric field formed between the electrodes. The electrode arrangement in the chamber is critical for uniform treatment of the product volume flow. An electrode configuration that simplifies the design of treatment chambers has electrodes located parallel to each other, separated by a spacer of insulating material. Homogeneous electric field distribution achieved by the parallel electrode arrangement is an important advantage over a co-axial or co-linear electrode configurations (Jeyamkondan et al. 1999), enabling simple and reliable determination of the electric field strength by calculating the quotient of the applied voltage and the inter-electrode gap. When a co-axial electrode arrangement is applied for the design of a treatment chamber the food product is passed through a radial cross section consisting of an inner and an outer cylindrical electrode. Optimizing turbulent flow zones that occur at the inlet and the outlet of a co-axial chamber is important and the radial geometry makes the implementation of homogeneous field distribution difficult (Zhang et al. 1995a), which has to be taken into account before treatment chamber construction. In addition to attaining homogeneous field distribution another engineering challenge is the electrode positioning as a narrow gap formed between radially arranged electrodes is more likely to cause blockages in a co-axial chamber when highly viscous or particulate foods are processed. At present most treatment chambers commonly feature a co-linear electrode arrangement, characterized by two ring-shaped electrodes mounted in the tubular chamber wall which also functions as an insulating spacer (Barsotti et al. 1999). Comparatively high product flow rates can be achieved with this chamber construction design. A drawback of co-linear design chambers is dead spaces, where the field is weakened and the product is inadequately treated based on the non-homogeneous electric field distribution. This has led to a strategy of using multiple chambers in series to minimize the odds that product portions are untreated or under-treated, so that recontamination of the effectively treated food, due to intermixing with insufficiently treated fractions from the dead spaces that limit the antimicrobial treatment effect, can be avoided. However, at the same time, the difference between minimally treated product portions and over-treated product portions is increased with such an approach, thus, augmenting the occurrence of adverse effects on product quality. Moreover, the process effectiveness is reduced, resulting in higher current losses due to power transmission to more than one chamber.

With regard to electrodes a suitable food-grade material such as stainless steel or platinized titanium (Ho and Mittal 2000) should be selected to avoid electrochemical reactions from the electrode surface (i.e., migration of metallic traces into the treated food as a result of electrolysis or electrode fouling). Research on novel PEF electrode materials, for example, silver(-alloy), and germanium(-alloy) (Eaton et al. 2013) or polymer composites (Roodenburg et al. 2010) is still in progress. In order to permit uniform field distribution the interface between electrode and insulating material should be manufactured at an angle of 90°. Moreover, rounded electrode edges are ideally fabricated to create higher field strengths for a wider and more efficacious PEF processing range and to reduce the chance of dielectric breakdown. The latter can occur at regional peak values within the electrical field distribution, causing a local temperature maximum due to intensified energy input that initiates ohmic heating. In these regions of maximized temperature, gaseous bubbles are formed, that can also occur if a PEF system is poorly sealed, and, if these bubbles are struck by electric field pulses, an arcing is produced that causes thermal product degradation and ozone release due to oxygen atom splitting.

Product Factors

The suitability as well as the efficacy of PEF processing is determined by both food properties and composition. Physico-chemical parameters of significance for the application of PEF are electrical conductivity, pH, water activity, and viscosity of the product, as well as the activity of enzymes that can affect product shelf-life.

Electrical conductivity

In liquid foods high ionic strength is intrinsic due to their complex composition, thus, electrical conductivity is dependent on the processing temperature applied. With a rise in temperature a reduction of the electric resistance of the food is commonly obtained. In addition, higher electrical conductivity of the food product makes the generation of an electric field challenging because of the cumulative resistance of the treatment chamber. However, the effectiveness of PEF is limited at increasing product conductivity because the electric field strength is reduced, causing a lower energy input per pulse (Vega-Mercado et al. 1996a). Modifications of the treatment chamber geometry, an increase of U or the addition of processing aids, like salts (increasing conductivity) or de-ionized water (decreasing conductivity) can be used to compensate and enhance PEF efficacy. The addition of processing auxiliaries will alter food composition though and are therefore not deemed suitable from a benefit and cost perspective.

Water activity

Activity of water can be defined as the ratio of the water vapour pressure above the food and the water vapour pressure of pure water at the same temperature. With a decrease in availability of free water due to it being bound by food constituents such as salt, sugar or proteins overall less water is available for microbial metabolism, thus, inhibiting cell growth and propagation. It is well established that a low water activity is associated with low microbial activity but at the same time it allows for formation of toxins. Moreover, environmental factors such as pH (i.e., decrease of pH causes rise of water activity), temperature (i.e., increase in temperature causes reduction of water activity) and partial oxygen pressure can also influence toxin production, e.g., under aerobic conditions and low water activity toxin production is promoted more than under anaerobic conditions (Kraemer 2007). Nevertheless, a lower effectiveness of PEF has been reported at lower water activity in foods (Aronsson and Roenner 2001, Álvarez et al. 2002, Min et al. 2002, Somolinos et al. 2008).

pH

Earlier studies (Sale and Hamilton 1967, Huelsheger et al. 1981) found that the pH level is questionable as contributing process parameter to the overall PEF microbial inactivation. Recently, it was suggested that higher (Jeantet et al. 1999) or lower (Aronsson and Roenner 2001) pH can have a considerable effect on the microbial reduction achieved with PEF, although some microorganisms have been reported to remain unaffected by a pH change (Heinz and Knorr 2000). The extent of the membrane damage caused by the electroporation is likely to provide an additional impact of pH during PEF processing. Hence, the treatment conditions, the product composition and the residing microbial species are decisive factors of influence.

Viscosity

A change in viscosity affects both consumer product perception and the handling and processing of the product. Most of the liquid foods processed with PEF exhibit a Newtonian-like fluid behavior. The viscosity of Newtonian fluids is determined by their chemical composition, temperature and pressure, in contrast to the physical stress imposed on them. Liquid foods with Newtonian-like fluid properties have been reported to exhibit no change in viscosity following treatment with electrical pulses (Cserhalmi et al. 2006). Albeit, some researchers reported that PEF appears to impact on rheological properties of more structured liquid foods, e.g., producing higher product viscosity in comparison to thermal pasteurization (Walkling-Ribeiro et al. 2010) or no treatment (Aguilo-Aguayo et al. 2008, Xiang et al. 2011) and, by contrast, reduces the viscosity of liquid foods (Floury et al. 2006b), while other researchers observed no viscosity changes (Min et al. 2003, Li and Zhang 2004). For the majority of liquid foods processed with PEF, no limitations in terms of viscosity have to be taken into account, but a product-specific approach seems necessary for more structured liquid foods in order to enable process optimization.

Enzymes

Enzymatic activity in liquid foods is usually undesirable, due to adverse effects on food quality, unless they have been added intentionally as a processing aid (e.g., enhancement of juice yield and clarification, milk coagulation, beverage fermentation). It is worth noting that after PEF treatment of fruit juices (Espachs-Barroso et al. 2006) or milk (Van Loey et al. 2001), enzymes such as alkaline phosphatase and peroxidase remain active, which suggests that some enzymes that could potentially impair product quality are unaffected by PEF. By contrast, promising results indicating partial or complete inactivation of enzymes usually present in juices or milk using PEF, particularly in combination with heat, have also been obtained, for example, with peroxidase and polyphenoloxidase (Zhong et al. 2007, Schilling et al. 2008), pectin methyl esterase (Rodrigo et al. 2003), lipoxygenase (Li et al. 2008), lipase (Bendicho et al. 2002) and protease (Vega-Mercado et al. 2001).

Microbiological Factors

Several microbiological factors, which relate to the species-dependent differences in morphology and physiology of microorganisms, determine the antimicrobial efficacy of PEF.

Organism type and growth stage

Typically, food is spoiled or contaminated by microorganisms introduced, for instance, in harvested raw materials, stored food products or during the food manufacturing processes. Occurrence and activity of these detrimental microorganisms depend both on the raw material and food properties as well as the methods and conditions applied for their reduction or complete inactivation. For example, the cell state of bacteria is essential for their resistance to food preservation treatments. Vegetative bacterial forms are far more prone to preservation treatments than bacterial endospores, which are generally regarded as comparatively resistant to PEF, as shown in the literature (Knorr et al. 1994, Grahl and Maerkl 1996, Pagán et al. 1998, Cserhalmi et al. 2002), although a few studies also suggested the opposite (Marquez et al. 1997, Dantzer et al. 1999) when treated in saline solutions. The growth stage of microorganisms is also a factor that affects the inactivation of microorganisms by PEF. Four different growth phases exist: the lag phase, the log phase or, alternatively known as, exponential phase, the stationary phase and the death or decline phase. Microbial cells in the lag phase have usually been introduced into a food medium and need to adapt, thus, showing very slow or even no growth. After the adaptation process is completed the exponential growth phase begins during which the fastest microbial growth can be observed. There are limitations in terms of logarithmic growth though because of a shortage of some substrates (i.e., oxygen demand for aerobic growth or metabolic by-products which inhibit, in particular, anaerobic cultures), environmental factors blocking further growth (e.g., external stress conditions like lethal radiation or constant presence of mechanical force) or the strain-specific maximum cell concentration. With the end of the exponential growth phase, usually a short decline, during which the specific growth rate steadily decreases to zero, the cell culture enters the stationary phase. During the stationary phase, the cell population consists of growing and lysing cells establishing an equilibrium, with possible growth depending on the rate of cell death. The final stage in the microbial growth curve is the death phase when the number of dying cells surpasses the number of viable cells. The cell population gradually declines until microbial cells cease to exist in the medium. Many studies have reported that vegetative bacterial cells are generally more sensitive during the exponential growth phase than cells harvested in the stationary phase (Huelsheger et al. 1983, Pothakamury et al. 1996, Qin et al. 1996). In addition, it was suggested that the inoculum size affects the impact of the bioelectrical effect of PEF (Huelsheger and Niemann 1980).

Organism size and bearing

Microbial cell size is critical with regard to the impact of PEF. Several decades ago it was discovered that yeast cells are more sensitive to electric pulses than smaller bacterial cells (Sale and Hamilton 1967). Due to the lower transmembrane potential of larger yeast cells compared to that of smaller bacterial cells, PEF at decreased intensity is sufficient for achieving yeast cell collapse than for bacteria. In addition, the level of the TMP differs between spherical and elliptical (Zimmermann et al. 1974) cells depending on the specific cell dimensions, the cell radius and the shorter half-axis of the cell. Moreover, the cell orientation within the electric field has an impact on E_c. The latter is increased at a low level of specific cell geometry, which is reflected by the fact that electric field strengths five times higher are needed to inactivate rod-shaped cells than spherical cells. Within a population of microorganisms, that contains many different cell sizes and cell shapes, making a prediction on the efficacy of PEF is complicated and, thus, the magnitude of E_c can only be estimated. Besides, the nonhomogeneous distribution and orientation of cells is responsible for a portion of the microbial population remaining unaffected and active following PEF treatment (Heinz et al. 2001). It is also noteworthy, that gram-negative bacterial cells have been found to be more susceptible to PEF than gram-positive ones (Barsotti and Cheftel 1999) which can also be attributed to the differences in cell structure.

Electropermeabilization

The method of action, that forms the basis for liquid food preservation by high voltage PEF, is called electropermeabilization. The latter, also known as electroporation or electropercolation, comes about when the TMP is reached because of charge accumulation along the membrane of a microbial cell. The most widely accepted electropermeabilization theory is based on electromechanical compression of the viscoelastic cell membrane that brings about pore formation during which membrane permeabilization takes place (Zimmermann et al. 1974), thereby, allowing an unregulated transfer of substances between the cell and the liquid surrounding it. Intensity of the electric field and the duration of the high voltage PEF determine whether or not the electropermeabilization is reversible or permanent (Schoenbach et al. 2002). After pore formation, a pore inversion could take place in the cell membrane from a reversible, lipophilic into an irreversible, hydrophilic pore state. Based on thermal transitions induced in the lipid bilayer from rigid gel into liquid crystalline structures membrane breakdown may occur that renders cells nonviable (Jeyamkondan et al. 1999). In order to inactivate microorganisms it is important that both electric pulse intensity and treatment duration are sufficient as a single supercritical pulse does not ensure irreversible pore formation (Knorr et al. 2001) due to potential cell membrane recovery within a few seconds after pulse expiration.

Efficacy of PEF against Important Microbial Groups

Foodborne microorganisms that have a detrimental effect on foods can be divided into spoilage organisms, that degrade the quality of foods, rendering them unacceptable for consumption, and pathogens, whose intake with foods pose a serious health risk for consumers. A wide variety of food microbiological studies are available in the literature, indicating the capability of PEF for successful inactivation of both spoilage organisms and pathogens. Based on the overview of original peer-reviewed research papers that looked into PEF efficacy against specific microorganisms in foods, presented in Table 2, the three bacteria that have attracted the greatest interest and have often been associated with foodborne outbreaks were *Escherichia coli*, *Listeria monocytogenes*, and *Salmonella* species, which have also been identified as being among the top zoonotic organisms responsible for foodborne illness in humans (European Food Safety Authority 2007). To a certain extent PEF inactivation of nonpathogenic surrogate organisms, for instance, generic *E. coli* (Gurtler et al. 2011) and *L. innocua* (Waite-Cusic et al. 2011), has also been investigated as these nonhazardous strains can be considered as an alternative in research or industrial facilities. Although these model organisms cannot replace research on actual pathogens and their application remains debatable, surrogate strains have been used to simulate and estimate the treatment resistance or other features of their similarly structured "pathogenic cousins" of the same genus. Lactobacilli were found to be the food spoilage bacteria deemed as most relevant by PEF researchers; however, to date, research on the spoilage yeast *Saccharomyces cerevisiae* seems to have more appeal than for *Lactobacillus plantarum*. A different approach was used in a study on *L. acidophilus* strains with regard to their beneficial, probiotic character following exposure to PEF (Cueva and Aryana 2012), reporting no effect on the bile tolerance of the bacterium at low and a decrease at moderate electric field strengths (5 and 15–25 kV/cm, respectively). For the effect of PEF against spore-forming bacteria *Bacillus* species were mostly used, while *Botryitis cinerea* has been the most studied mold with regard to PEF inactivation. Based on the data collected it can be concluded that PEF research on pathogenic microorganisms had a higher

Table 2. Overview of the 10 most published microorganism genera including their species or serovars, amount of entries, and most involved authors listed in the database Web of Knowledge based on the results for the search keywords "pulsed electric fields" and "microorganism inactivation".

Microorganism Genera	Microorganism Species/Serovars	Publication Entries	Most Published Authors
Escherichia	*E. coli* (88)	88	Zhang, Q.H. (13)
Listeria	*L. monocytogenes* (29) *L. innocua* (21)	49	Barbosa-Cánovas, G.V. (9), Raso, J. (9)
Salmonella	*S.* Typhimurium (18) *S.* Enteritidis (15) *S.* Senftenberg (7) *S.* Choleraesuis (2) *S.* Dublin (1) *S.* Virchow (1) *S.* Typhi (1)	36	Raso, J. (13)
Lactobacillus	*L. plantarum* (21) *L. brevis* (3) *L. acidophilus* (2) *L. hilgardii* (2) *L. leichmannii* (2) *L. bulgaricus* (1) *L. delbrueckii* (1) *L. diolivorans* (1) *L. rhamnosus* (1)	29	Raso, J. (4), Wouters, P.C. (4), Yousef, A.E. (4)
Staphylococcus	*S. aureus* (28) *S. gallinarum* (1) *S. xylosus* (1)	29	Raso, J. (5), Zhang, Q.H. (5)
Saccharomyces	*S. cerevisiae* (26) *S. fermentati* (1) *S. uvarum* (1)	27	Barbosa-Cánovas, G.V. (6), Swanson, B.G. (6)
Bacillus	*B. subtilis* (12) *B. cereus* (11) *B. stearothermophilus* (1)	24	Barbosa-Cánovas, G.V. (5)
Pseudomonas	*P. fluorescens* (8) *P. aeruginosa* (6) *P. syringae* (3) *P. putida* (1)	18	Zhang, Q.H. (4)
Cronobacter/ Enterobacter	*C. sakazakii* (4) *E. aerogenes* (1) *E. cloacae* (1)	6	Arroyo, C. (2), Cebrián, G. (2), Condón, S. (2), Martínez López, A. (2), Pagán, R. (2), Pina-Pérez, M.C. (2), Rodrigo, D. (2)
Botrytis	*B. cinerea* (4)	4	Evrendilek, G. (4)

Data was collected until 16 May 2013 and is based on the search algorithm of the Web of Knowledge database only, excluding review papers and, therefore, not representing the entirety of publications on pulsed electric fields and relying on operator updates and listings' selection criteria. The numbers in parentheses following the microorganism species/serovars and author names indicate the publication entries associated with the respective microorganisms and authors.

priority than that of spoilage organisms, clearly indicating that food safety is paramount and precedes concerns for product quality, shelf stability, and economic losses associated with food spoilage.

Bacteria

Bacilli. The resistance of *Bacillus* spp. to PEF was studied in various food model systems as bacilli are potentially pathogenic and among the most recalcitrant microorganisms due to their endospore-forming capability. Peptone water contaminated with *B. cereus* and treated with PEF at 30 kV/cm for 6000 (Rowan et al. 2000), 7000 to 8000 (Yaqub et al. 2004), and for 15000 (MacGregor et al. 2000) pulses using a frequency of 5 Hz resulted in respective reductions of up to 6.1, 5.1, and 7.5 $\log_{10}$ at processing temperatures up to 30°C. Other researchers assessed the efficacy of PEF in combination with other antimicrobial hurdles against the pathogen in model media, achieving reductions of *B. cereus* spores of up to 0.5 $\log_{10}$ after 20 electric pulses at 25 kV/cm as well as more than 3.0 $\log_{10}$ by PEF under same processing conditions with subsequent exposure to supercritical CO_2 at 200 bar and 40°C for 24 h in glycerol solution (Spilimbergo et al. 2003). A 1.2 $\log_{10}$ *B. cereus* reduction was achieved when PEF was applied at 17 kV/cm for 100 μs, which was further enhanced to 3.8 $\log_{10}$ with the addition of 0.06 μg of the bacteriocin nisin per mL buffer (Pol et al. 2000). However, Pol et al. (2001) suggested that mitigation of vegetative *B. cereus* cells by a hurdle technology comprising nisin (0.06 to 0.08 μg per mL liquid) and PEF (17 to 20 kV/cm, 60 to 100 μs) in diluted milk is more challenging than in a buffer system, due to reduced effectiveness of nisin in milk and not because of the protein content. Spores of *B. cereus* in UHT skim milk (Grahl and Maerkl 1996) were investigated, and no significant effect on the viability of the endospores after 30 pulses at 22 kV/cm with stand-alone PEF in skim milk was observed. Combining PEF with heat or nisin appeared to be the only effective strategy for reducing *B. cereus* spores in pasteurized skim and whole milk (Bermúdez-Aguirre et al. 2012) using PEF at 40 KV/cm in conjunction with heat treatments up to 65°C for 60 min for spore reductions in excess of 6.0 $\log_{10}$ or high nisin concentrations of 50 IU/mL for reductions above 3.0 $\log_{10}$. Similar effects were observed with regard to other *Bacillus* species such as a 2.1 $\log_{10}$ reduction of PEF-treated (53 kV/cm) vegetative *B. subtilis* cells in a mixture of buffer and Ringer's solution (Heinz et al. 1999), an inactivation of up 6.7 $\log_{10}$ spores of *B. subtilis* spores in an aqueous solution by PEF (83 kV/cm, 60°C) in a 0.5 MPa pressurized flow system (Choi et al. 2008), and decrease of 3.0 $\log_{10}$ in *B. stearothermophilus* cells in UHT whole milk (Shin et al. 2007). However, it is very likely that the processing conditions and requirements of PEF approaches enhanced by heat or antimicrobials to target *Bacilli* and other spore-forming bacteria will be at the expense of product quality or economic sustainability, but research on and development of sterilization processes that involve PEF are still ongoing.

Campylobacters. Although *Camplyobacter* is a pathogen that has frequently been identified as a cause for foodborne outbreaks in recent years, very little PEF research has been published on this bacterium which is most likely due to the fact that its occurrence in liquid foods is comparatively low. *C. jejuni* was studied, in a buffer that was subjected to PEF (15 to 35 kV/cm, for up to 300 µs, not exceeding 35°C) and heat (52 to 60°C), by Sagarzazu et al. (2010), who reported better inactivation by PEF in alkaline buffer (pH 7) of close to 6.0 $\log_{10}$, in contrast to the heat treatment which showed greater effectiveness in acidic (pH 4.5) buffer. Haughton et al. (2012) analysed the inactivation of *C. jejeuni* and *C. coli* isolates in liquid media after PEF processing for up to 30 s at a field intensity of up to 65 kV/cm and achieved a reduction in count up to 7.2 $\log_{10}$ of the pathogenic bacteria, thereby, indicating that *Campylobacter* are more susceptible to PEF than *E. coli* and *S. Enteritidis* investigated under the same processing conditions. A different research group (Shin et al. 2011) reported up to 5.8 $\log_{10}$ reduction in *Campylobacter* spp. counts in contaminated poultry chiller (4°C) and scalder (55°C) water after exposure to PEF for 200 s.

Cronobacters/Enterobacters. Cronobacter sakazakii—in the former nomenclature known as *Enterobacter sakazakii*—is a pathogen that has been mostly associated with outbreaks from contaminated powdered infant formula. Arroyo et al. (2012) examined the impact of environmental stresses on cells of *C. sakazakii* that were subsequently exposed to PEF (25 kV/cm, 400 µs, 35°C) in pH 7 buffer, which achieved a reduction in count of about 4.5 $\log_{10}$ without pre-stress and rendering the treatment less effective if cells were acid or heat stressed prior to PEF exposure. Another study (Arroyo et al. 2010) that used pH 4 buffer and the same PEF treatment conditions, sublethal injury was investigated by means of selective recovery media, reported *C. sakazakii* inactivation of less than 1.0 $\log_{10}$ after PEF treatment and recovery on non-selective medium due to lower pH of the food model system, but the treatment effect was enhanced synergistically to 3.0 $\log_{10}$, when PEF was combined with concurrent addition of 200 µl/L citral. When reconstituted infant formula inoculated with *C. sakazakii* was subjected to PEF Pina-Pérez et al. (2013) it was observed that combining PEF (15 kV/cm, 3000 µs) with another processing hurdle, cocoa powder, had a more pronounced bactericidal effect depending on the product contact time and concentration of the powder, increasing the reduction in bacteria cells from 1.3 up to 4.4 $\log_{10}$. *E. aerogenes*, and *E. cloacae* are pathogenic species mainly of clinical importance, that rarely cause foodborne infections but do affect urinary and respiratory tracts, whereas, the *E. cloacae* B29 strain has been linked to human obesity by an endotoxin-induced, inflammation-mediated mechanism (Fei and Zhao 2013). A maximum reduction of 1.1 $\log_{10}$ in PEF-treated (30 kV/cm, 190 µs, 30°C) tryptone soya broth that contained *E. aerogenes* was reported by Valero et al. (2003) and indicated limited suitability when PEF alone was used for the inactivation of the pathogen. Similarly, a high resistance of *E. cloacae* to PEF was also observed in chocolate liquor Min et al. (2002) processed at 25 kV/cm for 320 µs which resulted in a maximum decrease in count of 1.3 $\log_{10}$ bacteria cells.

Clostridia. While bacteria of the genus *Clostridium* can be regarded as the most important anaerobic spore-formers, to date only Grahl and Maerkl (1996) have investigated the resistance of a *Clostridium* strain from *C. tyrobutyricum* to PEF (30 pulses, 22 kV/cm) in UHT skim milk and observed no bactericidal effect. The unavailability of PEF research on *Clostridia* could be due to the fact that relatively few foodborne outbreaks are associated with *Clostridium* and because PEF is not considered to be an alternative to heat sterilization.

Escherichia. The microorganism that has by far been researched most extensively with regard to PEF application is *E. coli*. In early studies Qin et al. investigated inactivation of *E. coli* first in a static (1994) and then in a continuous flow treatment chamber (1998) achieving up to 2.5 $\log_{10}$ (40 kV/cm, 180 µs, ≤30°C) and more than 6.0 $\log_{10}$ (60 kV/cm, 50 pulses, ≤40°C) reduction in count in simulated milk ultrafiltrate (SMUF), respectively. Grahl and Maerkl (1996) obtained a 4.0 $\log_{10}$ reduction of *E. coli* cells by PEF (22.4 kV/cm, 20 pulses) in UHT skim milk. With increasing emergence and significance of *E. coli* O157:H7 strains more studies looked into their inactivation by PEF. The studies showed reductions in counts up to 2.6 and 2.0 $\log_{10}$ in apple juice (31 kV/cm, 202 µs) and skim milk (24 kV/cm, 141 µs), respectively (Evrendilek and Zhang 2005), reducing the pathogen up to 2.0 and 3.2 $\log_{10}$ (30 kV/cm, 25 pulses, 1 Hz, <35°C) in pH4 and pH7 buffers, respectively (Ait-Ouazzou et al. 2012), and reducing the enterobacteria (15kV/cm, 500 pulses, 1 Hz) in dialyzed egg white, egg yolk, and whole egg by 1.0, 3.0, and 3.5 $\log_{10}$, respectively (Amiali et al. 2004). The sub-lethal effect of PEF on O157 and non-O157 *E. coli* has also been studied and indicates that no sublethal injury was observed following PEF treatment (Dutreux et al. 2000, Iu et al. 2001, Aronsson et al. 2004, Walkling-Ribeiro et al. 2008a). However, other authors have suggested that sublethal damage at cellular level does occur following exposure to high voltage electric pulses (García et al. 2006, Saldaña et al. 2009, Somolinos et al. 2010a, Zhao et al. 2011). Thus, processing conditions and factors that allow for repair and recovery of injured cells and subsequent survival in various food systems have yet to be identified.

Lactic Acid Bacteria. Bacterial food spoilage is commonly observed after contamination with lactic acid bacteria. The majority of PEF research on lactic acid bacteria has focussed on the genus *Lactobacillus*, with *L. plantarum* being clearly the most researched member of the genus. Inactivation of the latter by PEF was reported by Roodenburg et al. (2010), who showed reductions in count of 2.1 $\log_{10}$ (28 kV/cm, 30 μs) in phosphate buffer and by Seratlić et al. (2013) who achieved a 2.65 $\log_{10}$ (32 kV/cm, 500 μs) reduction in count in DeMan, Rogosa and Sharpe broth and observed an increasing resistance to PEF of surviving subpopulations of the bacterium. For *L. plantarum*, *L. leichmannii*, *L. bulgaricus*, and *L. acidophilus* strains that were PEF-treated under the same conditions (25 kV/cm, 144 μs) reductions of 0.7, 4.6, 3.9 and 5.3 $\log_{10}$, respectively, in saline solution were obtained by Waite-Cusic et al. (2011), identifying *L. plantarum* as the least sensitive *Lactobacillus* species to PEF. Other studies on *Lactobacillus* species including *L. delbrueckii* (Pothakamury et al. 1995), *L. rhamnosus* (Jaeger et al. 2009a), *L. diolivorans* (Martínez Viedma et al. 2009), and *L. brevis* (Elez-Martínez et al. 2005) processed with PEF led to inactivation of at least 4.0 $\log_{10}$ (16 kV/cm, 40 pulses) in SMUF, 5.5 $\log_{10}$ in Ringer's solution and 2.0 $\log_{10}$ in raw milk (30 kV/cm, 115 kJ/kg). When PEF was performed in combination with a bacteriocin reductions in count of 4.9 $\log_{10}$ (35 kV/cm, 1000 μs, in combination 2 μL/mL with enterocin AS 48) were achieved in apple juice, and 5.8 $\log_{10}$ (35 kV/cm, 1000 μs, ≤32°C) in orange juice, respectively. Nonetheless, bacteria of other genera present in food such as *Lactococcus lactis* in milk (Michalac 2003), *Leuconostoc mesenteroides* in saline solution (Aronsson et al. 2001) and orange juice (McDonald et al. 2000), *Pediococcus acidilactici* in saline solution (Waite-Cusic et al. 2011), *P. parvalus* in apple juice (Martínez Viedma et al. 2010), and *P. damnosus* in beer (Evrendilek et al. 2004) have been also been investigated for their recalcitrance to PEF, exhibiting respective reductions of up to 0.3 (35 kV/cm, 188 μs), 7.0 (35 kV/cm, 160 μs), 5.0 (30 kV/cm, 15 μs, 55°C), 1.3 (25 kV/cm, 144 μs), 6.6 (35 kV/cm, 1000 μs, combined with 0.613 AU/mL enterocin AS 48), and 5.8 (41 kV/cm, 175 μs) $\log_{10}$ CFU/mL.

Listeria. As an important emerging pathogen in contaminated foods *L. monocytogenes* and other non-pathogenic *Listeria* species, mostly *L. innocua*, that serve as surrogates, have attracted the interest of many research groups (Wouters et al. 1999, Aronsson and Roenner 2001, Lado and Yousef 2003, Garcia et al. 2005, Sepulveda et al. 2006, El-Hag et al. 2011, Munoz et al. 2012), who investigated their sensitivity to PEF and subsequent survivability. In a paper by Waite-Cusic et al. (2011) a screening for adequate *L. monocytogenes* surrogates, based on the resistance of strains from different bacteria to PEF treatment, was carried out, indicating *P. acidilactici* as a suitable candidate due to its small cell size, but also noting that the relatively large *L. plantarum* was most resistant, thus, concluding cellular structure and membrane composition as pivotal factors in efficacy of PEF rather than cell size. Inactivation of 1.5, 3.9 and 5.1 $\log_{10}$ of *L. monocytogenes* cells in PEF-treated (25kV/cm, 800 µs) buffers of pH 7, 5.4 and 3.8, respectively, was reported by Álvarez et al. (2002). Combining PEF (30 kV/cm, 25 pulses, 1 Hz, <35°C) with antimicrobials (24 h exposure with 0.2 µl/mL thymol, carvacrol, linalool, α-terpinyl-acetate, p-cymene, and α-pinene) achieved a 5.0 log10 inactivation of *L. monocytogenes* in pH 4 buffer when counts were performed using nonselective recovery media (Ait-Ouazzou et al. 2011), while PEF combined with carvacrol was most effective against the pathogen in pH 7 buffer under the same treatment conditions, achieving reductions up to 2.0 $\log_{10}$. *Listeria* have also been investigated for sublethal cell injury after PEF treatment, with studies both reporting the presence of sublethally damaged cells (Unal et al. 2002, García et al. 2005, Saldaña et al. 2009, Somolinos et al. 2010b, Zhao et al. 2013) and those that could not observe their incidence (Simpson et al. 1999, Dutreux et al. 2000, Russell et al. 2000, Aronsson et al. 2001, Yaqub et al. 2004), suggesting an 'all-or-nothing effect' of PEF on microorganisms. There is also disagreement about the methodology for the detection of sublethal cell damage as some researchers deem traditional microbiological methods, using comparisons between non-selective and selective recovery media, are ineffective or inconclusive, while others seem to base their findings on processing conditions that are highly specific with regard to the processing parameters and 'tailor-made' for the achievement of the desired results and not realistic from a commercial scale PEF processing perspective.

Pseudomonas. Albeit *Pseudomonas* is particularly adaptive to environmental stress conditions and due to its resistance to antibiotics is a key pathogen in clinical environments, once in a while foodborne outbreaks associated with this reported pathogen transmitted via contaminated water sources have been reported. However, pseudomonads are the predominant spoilage organisms in refrigerated foods stored under aerobic conditions. As indicated in Table 2, the efficacy of PEF against this psychrotroph was mostly tested with *P. fluorescens* and *P. aeruginosa* strains. In a research note by Fernández-Molina et al. (2005) *P. fluorescens* in PEF-processed (38 kV/cm, 60 µs) skim milk was reduced by up to 1.9 $\log_{10}$, also suggesting that application of a single PEF treatment was more efficient than a combination with acetic or propionic acid that was also analysed in the study. By contrast, a higher PEF reduction exceeding 6.0 $\log_{10}$ of *P. fluorescens* in aqueous solutions was reported by Ho et al. (1995) using 10 kV/cm for 20 µs. MacGregor et al. (2000) achieved a 3.0 $\log_{10}$ reduction of *P. aeruginosa* load inoculated in peptone water using PEF at 30 kV/cm and 1500 µs while others (Rieder et al. 2008) reported a PEF inactivation up to 5.5 $\log_{10}$ in clinical waste water at an energy input of 162 J/mL, thereby, indicating a great potential of PEF for decontamination of pseudomonads in aqueous solutions.

Salmonella. Over the years *Salmonella* species have consistently posed a challenge from a food safety perspective and for this reason the assessment of PEF as a non-thermal processing alternative for foods has been thoroughly studied. Álvarez et al. (2000) investigated the impact of the growth phase, cell concentration, pH and conductivity on the effectiveness of PEF against *S.* Senftenberg, suggesting that cells harvested in the exponential growth phase were more prone to PEF than those from the stationary growth phase, that initial cell concentration did not affect the overall reduction by PEF, that the higher pH of the PEF-treated product allowed for a greater inactivation of the bacteria by PEF, and that conductivity had no effect on PEF treatment effectiveness. In a follow-up study on *Salmonella enterica* serovars in buffer treated with PEF (Álvarez et al. 2003), a reduction of more than 6.5 $\log_{10}$ was achieved for both *S.* Typhimurium and *S.* Enteritidis at 28 kV/cm over 1400 and 800 μs, respectively, showing higher resistance of the *S.* Typhimurium strain against PEF that was also highlighted after modeling the data. When the time for a 5.0 $\log_{10}$ reduction under these PEF conditions was analysed, inactivation of *S.* Typhimurium and *S.* Senftenberg required a treatment time twice as long as that required to inactivate the *S.* Enteritidis strain. However, in a representative food system such as liquid egg, a reduced effectiveness of PEF against these pathogens can be expected as, for example, indicated by Amiali et al. (2006), Jin et al. (2009), and Monfort et al. (2010a), who achieved respective *Salmonella* reductions of up to 3.7 (30 kV/cm, 120 μs), 3.0 (25 kV/cm, 250 μs, 55°C) and 4.0 (45 kV/cm, 30 μs) $\log_{10}$. A more powerful approach was also tested by the latter research group (Monfort et al. 2012), who compared inactivation of different *Salmonella* serovars in liquid egg processed with PEF (25 kV/cm, 100 kJ/kg), subsequent heating (up to 60°C for 1 min) and antimicrobial agents (10 mM EDTA or 2% triethyl citrate). These researchers also reported a 9.0 log inactivation of *S.* Enteritidis as well as a reduction of more than 5.0 log10 in *S.* Dublin, *S.* Typhimurium, *S.* Typhi, *S.* Senftenberg, and *S.* Virchow serovars. The PEF treatments resulted in a similar or lower decrease in soluble protein content compared to thermal treatments at 60 and 64°C with product heat exposure times of 3.5 and 2.5 min, respectively. Similar to *E. coli* and *Listeria*, sublethal cell damage following PEF treatment has been studied for *Salmonella*, and also resulted in conflicting findings that suggest the presence (Monfort et al. 2010a, Saldaña et al. 2010) and absence (Russell et al. 2000, Wuytack et al. 2003) of sublethally injured bacteria. While a combination of treatment processes in a hurdle technology is a more complex approach, it might be an overall more effective strategy because it is likely to eliminate the potential problem associated with microorganisms capable of recovery from sublethal injury and able to subsequently survive in food, although these hurdles alone and in combination will have to be researched as to whether or not microbial cells are rendered non-viable by them.

Staphylococci. Following inadequate storage of foods contaminated with *Staphylococci*, food poisoning can occur from formation of staphylococcal toxins. Decontamination of circa 2.0 $\log_{10}$ of *S. aureus* in peptone water using low intensity PEF conditions (20 kV/cm, 60 pulses) was reported by Hizal et al. (2002), in contrast to a more effective approach that combines PEF (30 kV/cm, 99µs, 50°C) and nisin (200 µg/µL) for the reduction of *S. aureus* cells in pH 3.5 buffer by up to 4.5 $\log_{10}$ (Saldaña et al. 2011). Furthermore, reductions in count of 3.0. 3.6 and 4.2 $\log_{10}$ in *S. xylosus*, *S. gallinarium* and *S. aureus* cells suspended in NaCl solution and treated with PEF (up to 30 kV/cm, 200–300 pulses) were observed by Węsierska and Trziszka (2007). *Staphylococci* were also PEF-processed in food systems, resulting in an inactivation of 5.5 $\log_{10}$ cells by PEF (40 kV/cm, 150 µs) as well as 6.8 $\log_{10}$ by means of PEF in conjunction with thermosonication (30 kHz, 10 min, 55°C) in orange juice (Walkling-Ribeiro et al. 2009a). A reduction in count of 1.0 $\log_{10}$ for *S. aureus* cells in UHT skim milk after exposure to PEF (35 kV/cm, 2400 µs) as well as up to 6.0 $\log_{10}$ of the pathogen following PEF (35 kV/cm, 2400 µs) and nisin (200 ppm) treatments were also reported (Sobrino-Lopez and Martin-Belloso 2006), thus indicating the greater efficacy of PEF-based hurdle strategies over stand-alone PEF processing.

Yeasts

Due to its ubiquity as a natural spoilage organism in food environments and its industrial importance *Saccharomyces cerevisiae* has by far been the most researched yeast with regard to PEF. For example, MacGregor et al. (2000) reported a 3.5 $\log_{10}$ reduction in *S. cerevisiae* cells suspended in peptone water and subsequently exposed to PEF (30 kV/cm, 3000 pulses), which contrasts with the findings of a study conducted by Aronsson et al. (2005), who found a decrease of 6.8 $\log_{10}$ in *S. cerevisiae* contaminated pH 7 buffer after PEF treatment (30 kV/cm, 85 µs). These researchers used propidium iodide uptake to quantify electropermeabilization and showed that uptake of the dye was not necessary to inflict lethal damage to the yeast but was necessary, for *E. coli* and *L. innocua*. When investigating the treatment effectiveness of PEF against *S. cerevisiae* cells (Pataro et al. 2011), it was found that increased flow rate or a recirculation of the product will result in a better distribution of the yeast cells by inducing reduced cluster formation of the microorganisms in the PEF-treated product; however, it should also be taken into account that the PEF processing homogeneity is likely to be reduced by higher flow rates and greater turbulences as by recirculating the product over- or undertreated portions of the product may occur because the flow profile will not be perfectly laminar. In apple juice inoculated with *S. cerevisiae* a 6.0 $\log_{10}$ reduction in count following PEF treatment was obtained by Qin et al. (1995a), while *S. cerevisiae* ascospores were reported to exhibit a much higher resistance to PEF (50 kV/cm, 18µs, 50 °C) that limited inactivation to a maximum of 2.5 $\log_{10}$ of the spores in orange juice (McDonald et al. 2000). Other yeasts that have been studied with regard to PEF inactivation include *Dekkera* species (Puértolas et al. 2009), *Kloeckera apiculata* (Marsellés-Fontanet et al. 2011), *Pichia fermentans* (McNamee et al. 2010), *Rhodotorula rubra* (Evrendilek et al. 2004), and *Zygosaccharomyces bailii* (Beveridge et al. 2004).

Moulds

Fungal contamination of food not only results in spoilage but also may represent a health risk to consumers as toxic or carcinogenic by-products (mycotoxins) can be formed. Most of the studies on moulds have been conducted with *Botrytis cinerea* in fruit juices and nectars, looking, for instance, into the tube elongation, spore germination and structural changes in conidia and hyphae induced by PEF (Evrendilek et al. 2008a). Similarly, *Penicillium expansum* has been investigated in the same commodities and by the same researchers (Evrendilek et al. 2008b), also determining morphological changes and degradation of the fungal structure that indicate PEF as an efficacious method for mould inactivation. In addition, the effect of PEF on other plant pathogens such as *Aspergillus niger* (MacGregor et al. 2000), *Neosartoya fischeri* and *Byssochlamys fulva* (Raso et al. 1998) as well as *B. nivea* (Grahl and Maerkl 1996) has been studied.

PEF Effectiveness in Different Food Matrixes

Liquid foods of plant origin

Fruit Juices. The most popular and commercially important fruit juice types are apple and orange juices. Early PEF studies on apple juice by Qin et al. (1995a) and Zhang et al. (1994a) suggested inactivation of *S. cerevisiae* by up to 6.0 $\log_{10}$ and, similarly, in later publications (Iu et al. 2001, Sen Gupta et al. 2003) respective reductions in count of *E. coli* of up to 6.0 and 5.4 $\log_{10}$ were reported, thereby, fulfilling the 5.0 $\log_{10}$ inactivation required by the US Food and Drug Administration (FDA 2001) for commercial fruit juice production that was introduced as a consequence of outbreaks linked to contaminated apple cider in the United States. An even greater processing efficacy and food safety assurance that led to comparable juice quality was obtained using hurdle technologies that, for example, combined PEF with heating (Heinz et al. 2003) and ultraviolet light (Walkling-Ribeiro et al. 2008b) and led to more effective mitigation of both pathogenic and spoilage organisms. In addition to spiking aseptic apple juice with a certain concentration of microorganisms some researchers have also observed that PEF alone and in combination with hurdles was an effective intervention method against native microbiota that developed during juice storage (Liang et al. 2003, Noci et al. 2008). Comparable findings were obtained for PEF-treated orange juice by Grahl and Maerkl (1996) and Molinari et al. (2004), who showed that their treatments were capable of inactivating more than 4.0 and 5.5 $\log_{10}$ of *S. cerevisiae*, respectively, and reduced the counts of *S.* Typhimurium by 3.5 $\log_{10}$ and *E. coli* O157:H7 by nearly 4.0 $\log_{10}$ (Gurtler et al. 2010), and *S. aureus* by nearly 5.5 $\log_{10}$ (Walkling-Ribeiro et al. 2009a). With regard to shelf life of orange juice, microbial growth in orange juice was reported to be held at 1.0 $\log_{10}$ or below by heat-assisted PEF (35 kV/cm, 59 µs, 60°C) over 112 d storage at 4, 22, or 37°C based on total aerobic plate and total yeast and mould counts (Yeom et al. 2000), and total microbial growth did not exceed 0.4 $\log_{10}$ during product storage at 4°C for 28 d when addition of nisin (100 U nisin/mL) was combined with PEF (80 kV/cm, 20 pulses, 44°C) by Hodgins et al. (2002). Microbial counts were also maintained below 2.6 $\log_{10}$ by thermosonication (30 kHz, 55°C, 10 min) and subsequent PEF (40 kV/cm, 100 µs) when orange juice was stored for up to 168 d at 25°C (Walkling-Ribeiro et al. 2009b). The storage life of juices and nectars of other fruit processed with PEF, such as apricot (Evrendilek et al. 2013), cranberry (Jin and Zhang 1999), ginkgo (Zhang et al. 2007), grape (Garde-Cerdán et al. 2007), longan (Zhang et al. 2010), mango (Ait-Ouazzou et al. 2013), melon and watermelon (Mosqueda-Melgar et al. 2007), pear (Zhao et al. 2008a), pineapple (Raso et al. 1998), sour cherry and peach (Evrendilek et al. 2008b), and strawberry (Gurtler et al. 2011) could also be extended, thus, highlighting the potential of this non-thermal technology to this sector of the food industry.

Vegetables Juices. The efficacy of PEF for the treatment of tomato juice, although technically a fruit juice, and carrot juice has been studied by Mosqueda-Melgar et al. (2008), indicating that levels of *S.* Enteritidis in tomato juice could be reduced by 4.2 $\log_{10}$ with stand-alone PEF and in combination with antimicrobials (citric acid and cinnamon bark oil), a reduction in count of 5.1 and 6.0 $\log_{10}$, respectively, was achieved. Similarly, Nguyen and Mittal (2007) reported inactivation of naturally occurring microorganisms in tomato juice using PEF (80 kV/cm, 20 pulses, 50°C) in conjunction with nisin (100 U/mL) by up to 4.4 $\log_{10}$ and achieved a shelf-life of up to 28 d under refrigeration. In a different study (Raso et al. 1998) the reduction of *B. fulva* conidiospores in different juices by PEF treatment was compared, and the organism exhibited a higher microbial resistance when present in tomato juice than in fruit juices. However, contrasting results were obtained by Ait-Ouazzou et al. (2013), who determined that a 7 min PEF treatment period was necessary in order to decrease *E. coli* O157:H7 cells by 5.0 $\log_{10}$ in tomato juice, whereas, for the same level of pathogen inactivation, a PEF processing duration of 8 min was required with orange juice. In PEF-processed carrot juice reduction of *E. coli* counts was limited to approximately 3.5 $\log_{10}$ (Zhong et al. 2005), following intense PEF treatment conditions (25 kV/cm for 2174 μs).

Beer and Wine. Research on alcoholic beverages treated with PEF was first started about a decade ago by Ulmer et al. (2002), who investigated reversible and irreversible cell damage in *L. plantarum* cells in a model beer. They did not detect sublethally injured cells, and achieved up to 1.5 $\log_{10}$ higher inactivation of the *Lactobacillus* when hop extract (100 ppm) was added prior to high intensity PEF (35 kV/cm, 63 kJ/kg) for a total reduction of 5.5 $\log_{10}$ CFU/mL in lactic acid bacteria. In a different study on PEF-treated beer (Evrendilek et al. 2004), several typical microorganisms associated with the product were analyzed, and extensive inactivation of *P. damnosus, L. plantarum*, vegetative cells of *B. subtilis, R. rubra, S. uvarum* cells (5.8, 4.8, 4.7, 4.3, and 4.1 $\log_{10}$, respectively) was observed, while native microflora were found to be significantly more resistant (0.5 $\log_{10}$ reduction in count) to PEF. Walkling-Ribeiro et al. (2011a) used response surface methodology to compare the microbial inactivation of *S. cerevisiae, L. plantarum, S.* Choleraesuis, and *B. subtilis* in PEF-processed lager beer at ambient and pre-cooled processing temperatures. They achieved microbial reductions similar to conventional pasteurization for the ambient temperature PEF treatments and a minimal inactivation of 4.0 $\log_{10}$ was obtained using the pre-cooled PEF processing conditions, which complied with the requirements for high-quality, minimal processing of beer. Moreover, based on the response surface modeling the same researchers (Walkling-Ribeiro et al. 2011a) suggested, that *B. subtilis* and *S.* Choleraesuis had the highest resistance to PEF. In addition, it was reported that *L. plantarum* was more susceptible to PEF treatment in beer with higher alcohol content, and that *S.* Choleraesuis was more sensitive to PEF in carbonated than in degassed beer. The affect of PEF treatment on *L. brevis* and *Z. bailii* in both Guinness beer and Sauvignon blanc wine was studied by Beveridge et al. (2004), who achieved complete inactivation of *L. brevis* in stout beer and white wine after respective 600 and 270 μs PEF treatments at 26 kV/cm, compared to *Z. bailii*, which was reduced to undetectable levels following 270 and 90 μs processing with PEF at the same field strength for the respective alcoholic beverages. This confirmed the previous work that indicated higher PEF efficacy against the yeast. Must and wine of red grape were also compared regarding the microbial inactivation achievable by means of PEF (Puértolas et al. 2009), and this work determined that greater reductions in count of *Dekkera*, *Lactobacillus* spp. and *S. bayanus* were obtained in wine than in must when PEF was applied with 31 kV/cm and energy densities above 300 kJ/L. Optimum PEF treatment conditions at 29 kV/cm with 186 kJ/kg were calculated based on a Weibull distribution modeling.

Other Plant-Based Liquids. In 2008 a study by Zhao et al. (2008b) examined PEF inactivation of bacteria in green tea extracts, and reported a 5.6 (38 kV/cm, 160μs) and 4.9 (38 kV/cm, 200 μs) $\log_{10}$ reduction in *E. coli* and *S. aureus* counts as well as a refrigerated shelf life of PEF-processed tea of up to 6 months. Reduction of *E. coli* in soymilk with PEF has been investigated (Li and Zhang 2004) and indicated a 5.7 $\log_{10}$ reduction in populations of coliforms after exposure to PEF (41 kV/cm, 54 μs), leading to a shelf life extension at 4°C. PEF treatment of another plant-based liquid product, horchata, which is a low acid, traditional Spanish beverage extracted from tigernut tubers, resulted in minimal reduction in counts (<1.1 $\log_{10}$ CFU/ml) of both *E. aerogenes* (Valero et al. 2003) and *L. monocytogenes* (Selma et al. 2005), indicating that stand-alone PEF did not represent an efficacious technique for reducing the bacteria load in this beverage. The treatment was more effective when the horchata was stored at lower temperatures and initial microbial count was low. PEF treatment of pea soup has also been conducted (Vega-Mercado et al. 1996b), and a reduction of 6.5 $\log_{10}$ CFU/mL of *E. coli* and 5.3 $\log_{10}$ CFU/mL of *B. subtilis* cells was achieved when 30 kV/cm with 60 μs pulses was applied.

Liquid foods of animal origin

Milk-Based Products. Together with fruit juices, milk-based products make up the most common commodity group for the application of PEF, which is exemplified by the research literature available. In a study on plain, low-fat yogurt mixed with fruit jelly and corn syrup that was pre-heated (60°C, 30 s) before PEF treatment (30 kV/cm, 32 μs) (Yeom et al. 2004), the microbial shelf stability, following storage at 4°C was extended considerably as the total aerobic and total yeast and mould counts did not increase beyond 1.5 and 1.0 $\log_{10}$ CFU/mL, respectively, when the product was stored for 91 d. During that time total aerobic and total yeast and mould counts exceeded 7.0 and 4.0 $\log_{10}$, respectively, in the yoghurt that was only pre-heated.

To date, most PEF research has been conducted on fluid milk, beginning in 1994 when Zhang et al. (1994b) reported a circa 2.5 $\log_{10}$ reduction in count of *E. coli* in skim milk after processing with PEF (40kV/cm, 195 μs). A few years later other studies (Martin et al. 1997, Calderón-Miranda et al. 1999a) confirmed that milk was a more demanding food matrix than juice for PEF treatments with regard to the achievement of sufficient levels of microbial inactivation to assure adequate food safety. Grahl and Maerkl (1996) suggested that microorganisms were better protected against PEF in milk with an increasing fat content but the opposite was reported as Picart et al. (2002) did not determine a better protection of microorganisms in dairy cream against PEF. This led to the suggestion (Walkling-Ribeiro et al. 2009c) that there might be an 'optimum' fat level below that of cream providing maximum microbial protection from PEF. Walkling-Ribeiro et al. (2009c) used moderately pre-heated (50°C) PEF (40 kV/cm, 60 µs) and obtained a reduction of the native milk biota by 6.0 $\log_{10}$ in both low-fat UHT milk and raw whole bovine milk, in agreement with other research groups (Jaeger et al. 2009b, Guerrero-Beltran et al. 2010) who confirmed higher antimicrobial efficacy of PEF treatments when combined with thermal effects or treatments. The shelf life of whole milk treated with PEF and subsequently refrigerated was at least 7 d longer than that of thermally pasteurized (72°C, 26 s) whole milk stored under the same conditions. Qin et al. (1995b), Fernandez-Molina et al. (2005), and Sepulveda et al. (2009) obtained similar results, indicating the suitability of heat-assisted PEF treatment for milk processing. Combining PEF (35 kV/cm, 2400 µs) with the lantibiotic nisin (20 ppm) was suggested as another promising strategy for processing of milk, showing a synergistic effect and reducing populations of *S. aureus* cells by up to 6.0 $\log_{10}$ (Sobrino-Lopez and Martin-Belloso 2006). An even more effective approach was reported by Smith et al. (2002) who combined PEF (80 kV/cm, 160 µs), moderate heat (52°C), and two antimicrobials (38 IU/mL nisin and 1638 IU/mL lysozyme) to achieve inactivation of the native skim milk microflora by 7.0 $\log_{10}$. In recent years novel hurdle technologies have been studied as potential alternatives to thermal pasteurization, for example, a combination of thermosonication (24 kHz, 117 µm, 85W/cm^2, 80 s) and PEF (50 kV/cm, 38 µs) (Noci et al. 2009) reduced levels of *L. innocua* by up to 6.9 $\log_{10}$ CFU/ml in low-fat UHT milk or cross-flow microfiltration (1.4 µm) (Rodríguez-Gonzalez et al. 2011) in conjunction with PEF (48 kV/cm, 459 µs), which achieved a reduction in count of at least 4.7 $\log_{10}$ for mesophilic bacteria and coliforms and at least 3.7 $\log_{10}$ for psychrotrophs, which was comparable to the microbial inactivation achieved with high-temperature, short-time pasteurization. This hurdle strategy is particularly interesting for the food industry as Bermúdez-Aguirre et al. (2012) identified stand-alone PEF processing or heat-assisted PEF as insufficient to inactivate *B. cereus* spores in milk without compromising milk quality and to be effective, higher processing temperatures, that would negate the non-thermal processing advantage over conventionally pasteurized milk, are required. In addition to removing the majority of spore-forming bacteria to levels below the detection limit, Walkling-Ribeiro et al. (2011b) reported that microfiltration (1.4 µm, 35°C) followed by high-intensity PEF (42 kV/cm, 612 µs, 49°C) was less effective than when they were applied in reversed sequence. The latter processing combination achieved a maximum reduction of 7.1 $\log_{10}$ CFU/ml for total mesophilic aerobic counts and a skim milk shelf life similar to that obtained following thermal pasteurization. The effect of PEF (43 kV/cm, 870 μs) on *E. coli* in simulated milk ultra-filtrate (SMUF), a milk model food system, was investigated by Alkhafaji and Farid (2007). These researchers reported reduction in *E. coli* counts exceeding 6.0 $\log_{10}$ CFU/ml, which indicated that the improved efficacy of PEF was due to the lower complexity of the SMUF food matrix. In a similar model system, 50% diluted acid whey, the sensitivity of *L. monocytogenes* to PEF (25 kV/cm, 48 μs), was also studied (Waite-Cusic et al. 2011), allowing for a strain-dependant inactivation between 1.8 and 3.6 $\log_{10}$ CFU/mL, which can in part also be ascribed to the effect of the composition of the food on the efficacy of PEF.

Egg-Based Products. Different research groups (Malicki et al. 2004, Bazhal et al. 2006, Jin et al. 2009, Monfort et al. 2010) have comprehensively studied PEF resistance of different *Salmonella* serovars but also that of *E. coli* in liquid egg. Hurdle technologies such as heat-assisted PEF (30 kV/cm, 210 μs, 40°C), PEF (25 kV/cm, 48 μs) combined with subsequent moderate heating (55°C, 2 min), heat-assisted PEF (50 kV/cm, 64 μs, 55°C) in conjunction with nisin (100 IU/mL), and PEF (57 kV/cm, 60 μs) paired either with high pressure (138 MPa, 8 min, 20°C) or with thermosonication (40 W, 20 kHz, 55°C, 5 min) were reported to be most effective, achieving reductions of approximately 5.0 $\log_{10}$ CFU/mL for both *S.* Enteritidis and *E. coli* O157:H7 counts in liquid egg yolk (Amiali et al. 2007), inactivating up to 8.0 $\log_{10}$ of *S.* Enteritidis cells in liquid whole egg (Monfort et al. 2011), reducing *L. innocua* populations by 5.5 $\log_{10}$ in liquid whole egg (Calderón-Miranda et al. 1999b). Combining PEF and thermosonication resulted in a reduction of *S.* Enteritidis counts of 2.3 $\log_{10}$ CFU/ml in liquid whole egg, and similar reduction in counts of this organism were achieved by a combination of high pressure and PEF (Huang et al. 2006), respectively. These findings clearly demonstrated the suitability of PEF-based hurdle processes for liquid egg processing.

Brine. Limited studies are available on brine decontamination with PEF. Rojas et al. (2007) reported an *E. coli* reduction of up to 2.0 $\log_{10}$ CFU/mL in meat injection brine using low-intensity PEF (7 kV/cm), but also suggested that no additional antimicrobial effect could be expected in the presence of sodium lactate. Similar results were reported when PEF (60 kV/cm) was used to decontaminate bacon smokehouse waste brine (Ho and Mittal 2001).

Mixed liquid foods of plant and animal origin

Over the years different blended liquid food products have been investigated such as orange juice-milk (Sampedro et al. 2009), orange-carrot juice (Rodrigo et al. 2003), smoothie type beverage (Walkling-Ribeiro et al. 2008a), apple-cranberry juice (Palgan et al. 2011), model salad dressing (Li et al. 2005), and egg-milk beverage (Pina-Pérez et al. 2009b). Freshly prepared smoothies are an example of a product group that has rapidly increased in consumer popularity in the recent years and is, in addition, characterized by a more viscous structure than other foods, which makes homogeneous processing with emerging technologies more challenging but necessary to avoid over-processing and possible loss in its high nutritional value. However, a combination of moderate heat (55°C, 60 s) and PEF (34 kV/cm, 60 μs) was applied to fresh fruit smoothie (Walkling-Ribeiro et al. 2010), resulted in an initial reduction of total aerobic bacteria and total yeast and moulds by 3.4 and 4.0 $\log_{10}$ CFU/ml, respectively, and achieved a considerable extension of refrigerated shelf life by 7 additional days on top of that achieved by mild heat pasteurization (72°C, 15 s). The quality parameters of the hurdle-treated smoothie were comparable to or better than those of the mildly pasteurized product.

Current Trends and Future Research Outlook

At the present time the efforts of researchers seem to have finally been rewarded by paving the way for a non-thermal processing technology, which advances safety and preservation of liquid foods and has achieved a commercial status. While application of PEF in the original stand-alone mode is sufficient to ensure microbiological safety for a wide range of products, the implementation of hurdle strategies to achieve minimal processing of foods is consistently drawing more attention by researchers, legislators, and decision-makers in the industry. Combining PEF with more traditional methods of decontamination such as moderate heating or antimicrobial agents results in powerful hurdle technologies, potentially achieving synergistic treatment effects. In addition, novel approaches in which PEF is combined, sequentially or simultaneously, with other emerging processing technologies, for example, heat- or pressure-assisted ultrasound, ultraviolet irradiation, pulsed high intensity light, high hydrostatic pressure, and supercritical fluids, have been studied increasingly and found to further enhance the processing effectiveness based on their specific mode of action and interactions with one another. Naturally, there is a limitation for each single process in terms of its applicability and efficacy.

Mathematical modeling of PEF, other hurdles, and their interactive effects allow for optimization of the processing parameters and identification of conditions to maximize the impact on detrimental microorganisms in a product-specific manner. Moreover, it may also shed light on how these processes can be further developed to promote energy efficiency and environmental sustainability and, thus, it is likely that research in this area will intensify in the future. With a further introduction of PEF technology in commercial scale food production, both new opportunities and challenges will arise that may increase the need for more applied research, for instance, regarding PEF processing conditions in specific food systems, for structural modification of food components or simply resulting from maintenance and scale-up issues. However, the importance of fundamental PEF research has also been emphasized in this chapter, when the occurrence of sublethal cell injury or microbial stress adaptation was discussed, as there is still a lack of understanding of many PEF-induced changes at the cellular and molecular level. In addition to addressing issues that originate from novel products and emerging pathogens there is also a need for safe and wholesome convenience foods for consumers due to steadily decreasing time resources in modern work environments, thus, necessitating more studies that look into stabilizing and modifying foods and food ingredients to enhance shelf life and render these products less prone to microbial contaminations. A further, systematic standardization of research parameters (i.e., product and product factors) and data (e.g., research with a variety of strains or different species rather than using one specific microorganism) is needed to allow highly innovative approaches to be implemented. Overall, the areas of application for PEF are well-defined and over the years the technology has become more understandable and manageable, but from a food microbiological perspective there are still a lot of questions that have remained unanswered so far and in many cases it is still a long way to full understanding and optimization of PEF for commercial food processing to be able to meet the present and upcoming demands so that the technology becomes used routinely. The interest in and amount of research being carried out as well as the manner in which such research is being organized, planned, financially supported and marketed is likely to decide the rate at which PEF development and implementation will progress.

Acknowledgements

The authors would like to acknowledge Dr. Michael O. Ngadi in the Department of Bioresource Engineering at McGill University, Montreal, QC, Canada and Dr. Roman Buckow in the Food and Nutritional Science Division of the Commonwealth Scientific and Industrial Research

Organisation, Werribee, VC, Australia for taking the time to proof-read the present book chapter and for their esteemed comments and advice.

Keywords: pulsed electric fields (PEF), nonthermal preservation, food process engineering, food safety engineering, foodborne pathogens, spoilage microorganisms, minimal processing, hurdle technologies, sublethal cell injury, microbial shelf stability

References

Aguilo-Aguayo, I., R. Soliva-Fortuny and O. Martín-Belloso. 2008. Comparative study on color, viscosity and related enzymes of tomato juice treated by high intensity pulsed electric fields or heat. Eur. Food Res. Technol. 227(2): 599–606.

Ait-Ouazzou, A., L. Cherrat, L. Espina, S. Lorán, C. Rota and R. Pagán. 2011. The antimicrobial activity of hydrophobic essential oil constituents acting alone or in combined processes of food preservation. Innovat. Food Sci. Emerg. Technol. 12(3): 320–329.

Ait-Ouazzou, A., L. Espina, L. Cherrat, M. Hassani, A. Laglaoui, P. Conchello and R. Pagán. 2012. Synergistic combination of essential oils from Morocco and physical treatments for microbial inactivation. Innovat. Food Sci. Emerg. Technol. 16: 283–290.

Ait-Ouazzou, A., L. Espina, D. García-Gonzalo and R. Pagán. 2013. Synergistic combination of physical treatments and carvacrol for *Escherichia coli* O157 : H7 inactivation in apple, mango, orange, and tomato juices. Food Control 32(1): 159–167.

Alkhafaji, S.R. and M. Farid. 2007. An investigation on pulsed electric fields technology using new treatment chamber design. Innovat. Food Sci. Emerg. Technol. 8(2): 205–212.

Altuntas, J., G.A. Evrendilek, M.K. Sangun and H.Q. Zhang. 2011. Processing of peach nectar by pulsed electric fields with respect to physical and chemical properties and microbial inactivation. J. Food Process Eng. 34(5): 1506–1522.

Álvarez, I., P. Manas, S. Condón and J. Raso. 2003. Resistance variation of *Salmonella enterica* serovars to pulsed electric fields treatments. J. Food Sci. 68(7): 2316–2320.

Álvarez, I., R. Pagán, J. Raso and S. Condón. 2002. Environmental factors influencing the inactivation of *Listeria monocytogenes* by pulsed electric fields. Lett. Appl. Microbiol. 35(6): 489–493.

Álvarez, I., J. Raso, A. Palop and F.J. Sala. 2000. Influence of different factors on the inactivation of *Salmonella* Senftenberg by pulsed electric fields. Int. J. Food Microbiol. 55(1–3): 143–146.

Amiali, M., M.O. Ngadi, V.G.S. Raghavan and J.P. Smith. 2004. Inactivation of *Eschericha coli* O157:H7 in liquid dialyzed egg using pulsed electric fields. Food Bioprod. Process. 82(C2): 151–156.

Amiali, M., M.O. Ngadi, J.P. Smith and G.S.V. Raghavan. 2007. Synergistic effect of temperature and pulsed electric field on inactivation of *Escherichia coli* O157:H7 and *Salmonella* Enteritidis in liquid egg yolk. J. Food Eng. 79(2): 689–694.

Amiali, M., M.O. Ngadi, J.P. Smith and V.G.S. Raghavan. 2006. Inactivation of *Escherichia coli* O157:H7 and *Salmonella* Enteritidis in liquid egg white using pulsed electric field. J. Food Sci. 71(3): M88–94.

Aronsson, K., E. Borch, B. Stenlof and U. Roenner. 2004. Growth of pulsed electric field exposed *Escherichia coli* in relation to inactivation and environmental factors. Int. J. Food Microbiol. 93(1): 1–10.

Aronsson, K., M. Lindgren, B.R. Johansson and U. Roenner. 2001. Inactivation of microorganisms using pulsed electric fields: the influence of process parameters on *Escherichia coli, Listeria innocua, Leuconostoc mesenteroides* and *Saccharomyces cerevisiae*. Innovat. Food Sci. Emerg. Technol. 2(1): 41–54.

Aronsson, K. and U. Roenner. 2001. Influence of pH, water activity and temperature on the inactivation of *Escherichia coli* and *Saccharomyces cerevisiae* by pulsed electric fields. Innovat. Food Sci. Emerg. Technol. 2(2): 105–112.

Aronsson, K., U. Roenner and E. Borch. 2005. Inactivation of *Escherichia coli, Listeria innocua* and *Saccharomyces cerevisiae* in relation to membrane permeabilization and subsequent leakage of intracellular compounds due to pulsed electric field processing. Int. J. Food Microbiol. 99(1): 19–32.

Arroyo, C., G. Cebrián, S. Condón and R. Pagán. 2012. Development of resistance in *Cronobacter sakazakii* ATCC 29544 to thermal and nonthermal processes after exposure to stressing environmental conditions. J. Appl. Microbiol. 112(3): 561–70.

Arroyo, C., M. Somolinos, G. Cebrian, S. Condon and R. Pagan. 2010. Pulsed electric fields cause sublethal injuries in the outer membrane of *Enterobacter sakazakii* facilitating the antimicrobial activity of citral. Lett. Appl. Microbiol. 51(5): 525–531.

Barbosa-Cánovas, G.V. and D. Sepúlveda. 2005. Present status and the future of PEF technology. *In*: G.V. Barbosa-Cánovas, M.S. Tapia and M.P. Cano (eds.). Novel Food Processing Technologies. CRC Press, Boca Raton, FL, USA, pp. 1–44.

Barsotti, L. and J.C. Cheftel. 1999. Food processing by pulsed electric fields. II. Biological aspects. Food Rev. Int. 15(2): 181–213.

Barsotti, L., P. Merle and J.C. Cheftel. 1999. Food processing by pulsed electric fields. I. Physical aspects. Food Rev. Int. 15(2): 163–180.

Bazhal, M.I., M.O. Ngadi, G.S.V. Raghavan and J.P. Smith. 2006. Inactivation of *Escherichia coli* O157:H7 in liquid whole egg using combined pulsed electric field and thermal treatments. LWT—Food Sci. Technol. 39(4): 420–426.

Bendicho, S., C. Estela, J. Giner, G.V. Barbosa-Cánovas and O. Martin. 2002. Effects of high intensity pulsed electric field and thermal treatments on a lipase from *Pseudomonas fluorescens*. J. Dairy Sci. 85(1): 19–27.

Bermúdez-Aguirre, D., C.P. Dunne and G.V. Barbosa-Cánovas. 2012. Effect of processing parameters on inactivation of *Bacillus cereus* spores in milk using pulsed electric fields. Int. Dairy J. 24(1): 13–21.

Beveridge, J.R., K. Wall, S.J. MacGregor, J.G. Anderson and N.J. Rowan. 2004. Pulsed electric field inactivation of spoilage microorganisms in alcoholic beverages. Proceedings of the IEEE 92(7): 1138–1143.

Calderón-Miranda, M.L., G.V. Barbosa-Canovas and B.G. Swanson. 1999a. Inactivation of *Listeria innocua* in skim milk by pulsed electric fields and nisin. Int. J. Food Microbiol. 51(1): 19–30.

Calderón-Miranda, M.L., G.V. Barbosa-Cánovas and B.G. Swanson. 1999b. Inactivation of Listeria innocua in liquid whole egg by pulsed electric fields and nisin. Int. J. Food Microbiol. 51(1): 7–17.

Chang, D.C. 1989. Cell poration and cell fusion using an oscillating electric field. Biophys. J. 56(4): 641–652.

Chen, Y., L.J. Yu and H.P.V. Rupasinghe. 2013. Effect of thermal and non-thermal pasteurisation on the microbial inactivation and phenolic degradation in fruit juice: a mini-review. J. Sci. Food Agr. 93(5): 981–986.

Choi, J., D. Wang, T. Namihira, S. Katsuki, H. Akiyama, X. Lin, H. Sato, H. Seta, H. Matsubara and T. Saeki. 2008. Inactivation of spores using pulsed electric field in a pressurized flow system. J. Appl. Phys. 104(9): 094701.

Coster, H.G.L. and U. Zimmermann. 1975. The mechanism of electric breakdown in the membranes of Valonia utricularis. J. Membr. Biol. 22(1): 73–90.

Cserhalmi, Z., A. Sass-Kiss, M. Tóth-Markus and N. Lechner. 2006. Study of pulsed electric field treated citrus juices. Innovat. Food Sci. Emerg. Technol. 7(1–2): 49–54.

Cserhalmi, Z., I. Vidacs, J. Beczner and B. Czukor. 2002. Inactivation of *Saccharomyces cerevisiae* and *Bacillus cereus* by pulsed electric fields technology. Innovat. Food Sci. Emerg. Technol. 3(1): 41–45.

Cueva, O. and K.J. Aryana. 2012. Bile tolerance of *Lactobacillus acidophilus* LA-K as influenced by various pulsed electric field conditions. Adv. Biosci. Biotechnol. 3(5): 620–625.

Dantzer, W.D., N. Hermawan and Q.H. Zhang. 1999. Standardized procedure for inactivation of spores by pulsed electric fields. *In*: IFT Annual Meeting Book of Abstracts. Session 83 A-9.

De Boer, K., N.R. Moheimani, M.A. Borowitzka and P.A. Bahri. 2012. Extraction and conversion pathways for microalgae to biodiesel: a review focused on energy consumption. J. Appl. Phycol. 24 (6): 1681–1698.

Donsi, F., G. Ferrari and G. Pataro. 2010. Applications of pulsed electric field treatments for the enhancement of mass transfer from vegetable tissue. Food Eng. Rev. 2(2): 109–130.

Dunn, J. and J. Pearlman. 1987. Methods and apparatus for extending the shelf life of fluid food products. United States Patent No. US 4695472. Maxwell Laboratories Inc., San Diego, CA, USA.

Dutreux, N., S. Notermans, T. Wijtzes, M.M. Gongora-Nieto, G.V. Barbosa-Canovas and B.G. Swanson. 2000. Pulsed electric fields inactivation of attached and free-living *Escherichia coli* and *Listeria innocua* under several conditions. Int. J. Food Microbiol. 54(1–2): 91–98.

Eaton, J.A., Y. El-Shoubary, H. Moise and C.M. Stewart. 2013. Silver and germanium electrodes in ohmic and PEF heating. United States Patent No. US 2013/0084371A1. PepsiCo Inc., Purchase, NY, USA.

Elez-Martínez, P., J. Escolà-Hernández, R.C. Soliva-Fortuny and O. Martín-Belloso. 2005. Inactivation of *Lactobacillus brevis* in orange juice by high-intensity pulsed electric fields. Food Microbiol. 22(4): 311–319.

El-Hag, A.H., S.H. Jayaram, O. Rodríguez-González and M.W. Griffiths. 2011. The influence of size and shape of microorganism on pulsed electric field inactivation. IEEE Transact. Nanobiosci. 10(3): 133–138.

Espachs-Barroso, A., A. Van Loey, M. Hendrickx and O. Martin-Belloso. 2006. Inactivation of plant pectin methylesterase by thermal or high intensity pulsed electric field treatments. Innovat. Food Sci. Emerg. Technol. 7(1–2): 40–48.

European Food Safety Authority. 2007. The community summary report on trends and sources of zoonoses, zoonotic agents, antimicrobial resistance and foodborne outbreaks in the European Union in 2006. EFSA Journal 130: 1–352.

Evrendilek, G.A., J. Altuntas, M.K. Sangun and H.Q. Zhang. 2013. Apricot nectar processing by pulsed electric fields. Int. J. Food Prop. 16(1): 216–227.

Evrendilek, G.A., S. Li, W.R. Dantzer and Q.H. Zhang. 2004. Pulsed electric field processing of beer: Microbial, sensory, and quality analyses. J. Food Sci. 69(8): M228–232.

Evrendilek, G.A., F.M. Tok, E.M. Soylu and S. Soylu. 2008a. Effect of pulsed electric fields on germination tube elongation and spore germination of *Botrytis cinerea* inoculated into sour cherry juice, apricot and peach nectars. Ital. J. Food Sci. 21(2): 171–182.

Evrendilek, G.A., F.M. Tok, E.M. Soylu and S. Soylu. 2008b. Inactivation of *Penicillum expansum* in sour cherry juice, peach and apricot nectars by pulsed electric fields. Food Microbiol. 25(5): 662–667.

Evrendilek, G.A. and Q.H. Zhang. 2003. Effects of pH, temperature, and pre-pulsed electric field treatment on pulsed electric field and heat inactivation of *Escherichia coli* O157:H7. J. Food Prot. 66(5): 755–759.

Evrendilek, G.A. and Q.H. Zhang. 2005. Effects of pulse polarity and pulse delaying time on pulsed electric fields-induced pasteurization of *E. coli* O157:H7. J. Food Eng. 68(2): 271–276.

FDA. 2001. *CFR—Code of federal regulations title 21 CFR120*. Federal Register Final Rule.

Fei, N. and L. Zhao. 2013. An opportunistic pathogen isolated from the gut of an obese human causes obesity in germfree mice. The ISME Journal 7: 880–884.

Fernández-Molina, J.J., B. Altunakar, D. Bermúdez-Aguirre, B.G. Swanson and G.V. Barbosa-Cánovas. 2005. Inactivation of *Pseudomonas fluorescens* in skim milk by combinations of pulsed electric fields and organic acids. J. Food Prot. 68(6): 1232–1235.

Fernandez-Molina, J.J., G.V. Barbosa-Canovas and B.G. Swanson. 2005. Skim milk processing by combining pulsed electric fields and thermal treatments. J. Food Process. Pres. 29(5–6): 291–306.

Floury, J., N. Grosset, E. Lesne and R. Jeantet. 2006a. Continuous processing of skim milk by a combination of pulsed electric fields and conventional heat treatments: does a synergetic effect on microbial inactivation exist? Lait 86(3): 203–211.

Floury, J., N. Grosset, N. Leconte, M. Pasco, M.N. Madec and R. Jeantet. 2006b. Continuous raw skim milk processing by pulsed electric field at non-lethal temperature: effect on microbial inactivation and functional properties. Lait 86(1): 43–57.

Fox, M.B., D.C. Esveld, H. Mastwijk and R.M. Boom. 2008. Inactivation of *L. plantarum* in a PEF microreactor. The effect of pulse width and temperature on the inactivation. Innovat. Food Sci. Emerg. Technol. 9(1): 101–108.

Gachovska, T.K., M.V. Simpson, M.O. Ngadi and G.S.V. Raghavan. 2009. Pulsed electric field treatment of carrots before drying and rehydration. J. Sci. Food Agr. 89(14): 2372–2376.

García, D., N. Gómez, P. Mañas, S. Condón, J. Raso and R. Pagán. 2005. Occurrence of sublethal injury after pulsed electric fields depending on the micro-organism, the treatment medium ph and the intensity of the treatment investigated. J. Appl. Microbiol. 99(1): 94–104.

García, D., P. Mañas, N. Gómez, J. Raso and R. Pagán. 2006. Biosynthetic requirements for the repair of sublethal membrane damage in Escherichia coli cells after pulsed electric fields. J. Appl. Microbiol. 100(3): 428–35.

Garde-Cerdán, T., M. Arias-Gil, A.R. Marsellés-Fontanet, C. Ancín-Azpilicueta and O. Martín-Belloso. 2007. Effects of thermal and non-thermal processing treatments on fatty acids and free amino acids of grape juice. Food Control 18(5): 473–479.

Grahl, T. and H. Maerkl. 1996. Killing of microorganisms by pulsed electric fields. Appl. Microbiol. Biotechnol. 45(1–2): 148–157.

Griffiths, S., M. MacLean, J.G. Anderson, S.J. MacGregor and M.H. Grant. 2012. Inactivation of microorganisms within collagen gel biomatrices using pulsed electric field treatment. J. Mater. Sci.–Mater. Med. 23(2): 507–515.

Grimi, N., F. Mamouni, N. Lebovka, E. Vorobiev and J. Vaxelaire. 2011. Impact of apple processing modes on extracted juice quality: Pressing assisted by pulsed electric fields. J. Food Eng. 103(1): 52–61.

Guerrero-Beltran, J.A., D.R. Sepulveda, M.M. Gongora-Nieto, B. Swanson and G.V. Barbosa-Canovas. 2010. Milk thermization by pulsed electric fields (PEF) and electrically induced heat. J. Food Eng. 100(1): 56–60.

Sen Gupta, B., F. Masterson and T.R.A. Magee. 2003. Inactivation of *E. coli* K12 in apple juice by high voltage pulsed electric field. Eur. Food Res. Technol. 217(5): 434–437.

Gurtler, J.B., R.B. Bailey, D.J. Geveke and H.Q. Zhang. 2011. Pulsed electric field inactivation of *E. coli* O157:H7 and non-pathogenic surrogate *E. coli* in strawberry juice as influenced by sodium benzoate, potassium sorbate, and citric acid. Food Control 22(10): 1689–1694.

Gurtler, J.B., R.B. Rivera, H.Q. Zhang and D.J. Geveke. 2010. Selection of surrogate bacteria in place of *E. coli* O157:H7 and *Salmonella* Typhimurium for pulsed electric field treatment of orange juice. Int. J. Food Microbiol. 139(1–2): 1–8.

Han, Z., X.A. Zeng, N. Fu, S.J. Yu, X.D. Chen and J.F. Kennedy. 2012. Effects of pulsed electric field treatments on some properties of tapioca starch. Carbohydr. Polymer. 89(4): 1012–1017.

Haughton, P.N., J.G. Lyng, D.A. Cronin, D.J. Morgan, S. Fanning and P. Whyte. 2012. Efficacy of pulsed electric fields for the inactivation of indicator microorganisms and foodborne pathogens in liquids and raw chicken. Food Contr. 25(1): 131–135.

Heinz, V., I. Álvarez, A. Angersbach and D. Knorr 2001. Preservation of liquid foods by high intensity pulsed electric fields—basic concepts for process design. Trends Food Sci. Technol. 12(3–4): 103–111.

Heinz, V. and D. Knorr. 2000. Effect of pH, ethanol addition and high hydrostatic pressure on the inactivation of *Bacillus subtilis* by pulsed electric fields. Innovat. Food Sci. Emerg. Technol. 1(2): 151–159.

Heinz, V., S.T. Phillips, M. Zenker and D. Knorr. 1999. Inactivation of *Bacillus subtilis* by high intensity pulsed electric fields under close to isothermal conditions. Food Biotechnol. 13(2): 155–168.

Heinz, V., S. Toepfl and D. Knorr. 2003. Impact of temperature on lethality and energy efficiency of apple juice pasteurization by pulsed electric fields treatment. Innovat. Food Sci. Emerg. Technol. 4(2): 167–175.

Hizal, M., A. Bayindirli, S. Damar and F. Bozog. 2002. Inactivation and injury of *Escherichia coli* O157 : H7 and *Staphylococcus aureus* by pulsed electric fields. World J. Microb. Biotechnol. 18: 1–6.

Ho, S.Y. and G.S. Mittal. 2000. High voltage pulsed electrical field for liquid food pasteurisation. Food Rev. Int. 16(4): 395–434.

Ho, S.Y. and G.S. Mittal. 2001. Non-thermal microbial inactivation in waste brine using high-voltage low-energy electrical pulses. Innovat. Food Sci. Emerg. Technol. 2(4): 251–259.

Ho, S.Y., G.S. Mittal, J.D. Cross and M.W. Griffiths. 1995. Inactivation of *Pseudomonas fluorescens* by high voltage electric pulses. J. Food Sci. 60(6): 1337–1343.

Hodgins, A.M., G.S. Mittal and M.W. Griffiths. 2002. Pasteurization of fresh orange juice using low-energy pulsed electrical field. J. Food Sci. 67(6): 2294–2299.

Huang, E., G.S. Mittal and M.W. Griffiths. 2006. Inactivation of *Salmonella* Enteritidis in liquid whole egg using combination treatments of pulsed electric field, high pressure and ultrasound. Biosystems Eng. 94(3): 403–413.

Huelsheger, H. and E.G. Niemann. 1980. Lethal effects of high-voltage pulses on *E. coli* K12. Radiat. Environ. Biophys. 18(4): 281–288.

Huelsheger, H., J. Potel and E.G. Niemann. 1981. Killing of bacteria with electric pulses of high field strength. Radiat. Environ. Biophys. 20(1): 53–66.

Huelsheger, H., J. Potel and E.G. Niemann. 1983. Electric field effects on bacteria and yeast cells. Radiat. Environ. Biophys. 22(2): 149–162.

Iu, J., G.S. Mittal and M.W. Griffiths. 2001. Reduction in levels of *Escherichia coli* O157 : H7 in apple cider by pulsed electric fields. J. Food Prot. 64(7): 964–969.

Jaeger, H., A. Schulz, N. Karapetkov and D. Knorr. 2009a. Protective effect of milk constituents and sublethal injuries limiting process effectiveness during PEF inactivation of Lb. rhamnosus. Int. J. Food Microbiol. 134(1–2): 154–61.

Jaeger, H., N. Meneses and D. Knorr. 2009b. Impact of PEF treatment inhomogeneity such as electric field distribution, flow characteristics and temperature effects on the inactivation of *E. coli* and milk alkaline phosphatase. Innovat. Food Sci. Emerg. Technol. 10(4): 470–480.

Jeantet, R., F. Baron, F. Nau, M. Roignant and G. Brulé. 1999. High intensity pulsed electric fields applied to egg white: effect on *Salmonella* Enteritidis inactivation and protein denaturation. J. Food Prot. 62(12): 1381–1386.

Jeyamkondan, S., D.S. Jayas and R.A. Holley. 1999. Pulsed electric field processing of foods: a review. J. Food Prot. 62(9): 1088–1096.

Jin, T., H. Zhang, N. Hermawan and W. Dantzer. 2009. Effects of pH and temperature on inactivation of *Salmonella* Typhimurium DT104 in liquid whole egg by pulsed electric fields. Int. J. Food Sci. Technol. 44(2): 367–372.

Jin, Z.T. and Q.H. Zhang. 1999. Pulsed electric field inactivation of microorganisms and preservation of quality of cranberry juice. J. Food Process. Pres. 23(6): 481–497.

Knorr, D., A. Angersbach, M.N. Eshtiaghi, V. Heinz and D.U. Lee. 2001. Processing concepts based on high intensity electric field pulses. Trends Food Sci. Technol. 12(3–4): 129–135.

Knorr, D., M. Geulen, T. Grahl and W. Sitzmann. 1994. Food application of high electric field pulses. Trends Food Sci. Technol. 5(3): 71–75.

Kraemer, J. 2007. Lebensmittel-Mikrobiologie. Stuttgart, Germany: UTB GmbH.

Lado, B.H. and A.E. Yousef. 2003. Selection and identification of a *Listeria monocytogenes* target strain for pulsed electric field process optimization. Appl. Environ. Microbiol. 69(4): 2223–2229.

Li, S.Q. and Q.H. Zhang. 2004. Inactivation of *E. coli* 8739 in enriched soymilk using pulsed electric fields. J. Food Sci. 69(7): M169–M174.
Li, S.Q., Q.H. Zhang, T.Z. Jin, E.J. Turek and M.H. Lau. 2005. Elimination of *Lactobacillus plantarum* and achievement of shelf stable model salad dressing by pilot scale pulsed electric fields combined with mild heat. Innovat. Food Sci. Emerg. Technol. 6(2): 125–133.
Li, Y.Q., Q. Chen, X.H. Liu and Z.X. Chen. 2008. Inactivation of soybean lipoxygenase in soymilk by pulsed electric fields. Food Chem. 109(2): 408–414.
Li, Z.Q., X.M. Dou, Y. Ni, K. Sumitomo and Y. Yamaguchi. 2010. Acetic acid denaturing pulsed field capillary electrophoresis for RNA separation. Electrophoresis 31(21): 3531–3536.
Liang, Z., Z. Cheng and G.S. Mittal. 2003. Inactivation of microorganisms in apple cider using spice powders, extracts and oils as antimicrobials with and without low-energy pulsed electric field. J. Food Agr. Environ. 1(2): 28–33.
Van Loey, A., B. Verachtert and M. Hendrickx. 2001. Effects of high electric field pulses on enzymes. Trends Food Sci. Technol. 12(3–4): 94–102.
MacGregor, S.J., O. Farish, S. Member, R. Fouracre, N.J. Rowan and J.G. Anderson. 2000. Inactivation of pathogenic and spoilage microorganisms in a test liquid using pulsed electric fields. IEEE Transact. Plasma Sci. 28(1): 144–149.
Malicki, A., M. Oziemblowski, J. Molenda, T. Trziszka and S. Brużewicz. 2004. Effect of pulsed electric field (PEF) on *Escherichia coli* within the liquid whole egg. Bull. Vet. Inst. Pulawy 48(4): 371–373.
Marjanovic, I., S. Haberl, D. Miklavcic, M. Kanduser and M. Pavlin. 2010. Analysis and comparison of electrical pulse parameters for gene electrotransfer of two different cell lines. J. Membr. Biol. 236(1): 97–105.
Marquez, V.O., G.S. Mittal and M.W. Griffiths. 1997. Destruction and inhibition of bacterial spores by high voltage pulsed electric field. J. Food Sci. 62(2): 399–401, 409.
Marsellés-Fontanet, Á.R., A. Puig-Pujol, P. Olmos, S. Mínguez-Sanz and O. Martín-Belloso. 2011. A comparison of the effects of pulsed electric field and thermal treatments on grape juice. Food Bioprocess Technol. 6(4): 978–987.
Martin, O., B.L. Qin, F.J. Chang, G.V. Barbosa-Canovas and B.G. Swanson. 1997. Inactivation of *Escherichia coli* in skim milk by high intensity pulsed electric fields. J. Food Process Eng. 20(4): 317–336.
Martínez Viedma, P., H. Abriouel, A. Sobrino López, N. Ben Omar, R. Lucas López, E. Valdivia, O. Martín Belloso and A. Gálvez, 2009. Effect of enterocin AS-48 in combination with high-intensity pulsed-electric field treatment against the spoilage bacterium *Lactobacillus diolivorans* in apple juice. Food Microbiol. 26(5): 491–496.
Martínez Viedma, P., A. Sobrino López, N. Ben Omar, H. Abriouel, R. Lucas López, O. Martín Belloso and A. Gálvez. 2010. Increased inactivation of exopolysaccharide-producing *Pediococcus parvulus* in apple juice by combined treatment with enterocin AS-48 and high-intensity pulsed electric field. J. Food Prot. 73(1): 39–43.
McDonald, C.J., S.W. Lloyd, M.A. Vitale, K. Petersson and F. Innings. 2000. Effects of pulsed electric fields on microorganisms in orange juice using electric field strengths of 30 and 50 kV/cm. J. Food Sci. 65(6): 984–989.
McNamee, C., F. Noci, D.A. Cronin, J.G. Lyng, D.J. Morgan and A.G.M. Scannell. 2010. PEF based hurdle strategy to control *Pichia fermentans*, *Listeria innocua* and *Escherichia coli* K12 in orange juice. Int. J. Food Microbiol. 138(1–2): 13–8.
Michalac, S. 2003. Inactivation of selected microorganisms and properties of pulsed electric field processed milk. J. Food Process. Pres. 27(2): 137–151.
Min, S., L. Reina and Q.H. Zhang. 2002. Water activity and the inactivation of *Enterobacter cloacae* inoculated in chocolate liquor and a model system by pulsed electric field treatment. J. Food Process. Pres. 26(5): 323–337.
Min, S., Z.T. Jin and Q.H. Zhang. 2003. Commercial scale pulsed electric field processing of tomato juice. J. Agr. Food Chem. 51(11): 3338–3344.

Molinari, P., A.M.R. Pilosof and R.J. Jagus. 2004. Effect of growth phase and inoculum size on the inactivation of *Saccharomyces cerevisiae* in fruit juices, by pulsed electric fields. Food Res. Int. 37(8): 793–798.

Monfort, S., E. Gayán, S. Condón, J. Raso and I. Alvarez. 2011. Design of a combined process for the inactivation of *Salmonella* Enteritidis in liquid whole egg at 55°C. Int. J. Food Microbiol. 145(2–3): 476–482.

Monfort, S., E. Gayán, G. Saldaña, E. Puértolas, S. Condón, J. Raso and I. Álvarez. 2010a. Inactivation of *Salmonella* Typhimurium and *Staphylococcus aureus* by pulsed electric fields in liquid whole egg. Innovat. Food Sci. Emerg. Technol. 11(2): 306–313.

Monfort, S., E. Gayán, J. Raso, S. Condón and I. Alvarez. 2010b. Evaluation of pulsed electric fields technology for liquid whole egg pasteurization. Food Microbiol. 27(7): 845–852.

Monfort, S., G. Saldaña, S. Condón, J. Raso and I. Álvarez. 2012. Inactivation of *Salmonella* spp. in liquid whole egg using pulsed electric fields, heat, and additives. Food Microbiol. 30(2): 393–399.

Mosqueda-Melgar, J., R.M. Raybaudi-Massilia and O. Martín-Belloso. 2007. Influence of treatment time and pulse frequency on *Salmonella* Enteritidis, *Escherichia coli and Listeria monocytogenes* populations inoculated in melon and watermelon juices treated by pulsed electric fields. Int. J. Food Microbiol. 117(2): 192–200.

Mosqueda-Melgar, J., R.M. Raybaudi-Massilia and O. Martín-Belloso. 2008. Inactivation of *Salmonella* enterica Ser. Enteritidis in tomato juice by combining of high-intensity pulsed electric fields with natural antimicrobials. J. Food Sci. 73(2): M47–53.

Munoz, A., I. Palgan, F. Noci, D.A. Cronin, D.J. Morgan, P. Whyte and J.G. Lyng. 2012. Combinations of selected non-thermal technologies and antimicrobials for microbial inactivation in a buffer system. Food Res. Int. 47: 100–105.

Neumann, E., A.E. Sowers and C.A. Jordan. 1989. Electroporation and electrofusion in cell biology. New York, NY, USA: Plenum Press.

Nguyen, P. and G.S. Mittal. 2007. Inactivation of naturally occurring microorganisms in tomato juice using pulsed electric field (PEF) with and without antimicrobials. Chem. Eng. Process. : Process Intens. 46(4): 360–365.

Noci, F., J. Riener, M. Walkling-Ribeiro, D.A. Cronin, D.J. Morgan and J.G. Lyng. 2008. Ultraviolet irradiation and pulsed electric fields (PEF) in a hurdle strategy for the preservation of fresh apple juice. J. Food Eng. 85(1): 141–146.

Noci, F., M. Walkling-Ribeiro, D.A. Cronin, D.J. Morgan and J.G. Lyng. 2009. Effect of thermosonication, pulsed electric field and their combination on inactivation of *Listeria innocua* in milk. Int. Dairy J. 19(1): 30–35.

Pagán, R., S. Esplugas, M.M. Gongora-Nieto, G.V. Barbosa-Cánovas and B.G. Swanson. 1998. Inactivation of *Bacillus subtilis* spores using high intensity pulsed electric fields in combination with other food conservation technologies. Food Sci. Technol. Int. 4(1): 33–44.

Palgan, I., I.M. Caminiti, A. Muñoz, F. Noci, P. Whyte, D.J. Morgan, D.A. Cronin and J.G. Lyng. 2011. Combined effect of selected non-thermal technologies on *Escherichia coli* and *Pichia fermentans* inactivation in an apple and cranberry juice blend and on product shelf life. Int. J. Food Microbiol. 151(1): 1–6.

Pataro, G., B. Senatore, G. Donsì and G. Ferrari. 2011. Effect of electric and flow parameters on PEF treatment efficiency. J. Food Eng. 105(1): 79–88.

Pérez-Roa, R.E., D.T. Tompkins, M. Paulose, C.A. Grimes, M.A. Anderson and D.R. Noguera. 2006. Effects of localised, low-voltage pulsed electric fields on the development and inhibition of *Pseudomonas aeruginosa* biofilms. Biofouling 22(6): 383–390.

Picart, L., E. Dumay and J.C. Cheftel. 2002. Inactivation of *Listeria innocua* in dairy fluids by pulsed electric fields: influence of electric parameters and food composition. Innovat. Food Sci. Emerg. Technol. 3(4): 357–369.

Pina-Pérez, M.C., A. Martínez-López and D. Rodrigo. 2013. Cocoa powder as a natural ingredient revealing an enhancing effect to inactivate *Cronobacter sakazakii* cells treated by pulsed electric fields in infant milk formula. Food Control 32(1): 87–92.

Pina-Perez, M.C., D. Rodrigo and A. Martinez-Lopez. 2009a. Sub-lethal damage in *Cronobacter sakazakii* subsp. *sakazakii* cells after different pulsed electric field treatments in infant formula milk. Food Control 20(12): 1145–1150.

Pina-Pérez, M.C., A.B. Silva-Angulo, D. Rodrigo and A. Martínez-López. 2009b. Synergistic effect of pulsed electric fields and CocoanOX 12% on the inactivation kinetics of *Bacillus cereus* in a mixed beverage of liquid whole egg and skim milk. Int. J. Food Microbiol. 130(3): 196–204.

Pol, I.E., H.C. Mastwijk, P.V. Bartels and E.J. Smid. 2000. Pulsed-electric field treatment enhances the bactericidal action of nisin against *Bacillus cereus*. Appl. Environ. Microbiol. 66(1): 428–430.

Pol, I.E., H.C. Mastwijk, R.A. Slump, M.E. Popa and E.J. Smid. 2001. Influence of food matrix on inactivation of *Bacillus cereus* by combinations of nisin, pulsed electric field treatment, and carvacrol. J. Food Prot. 64(7): 1012–1018. http://apps.webofknowledge.com.proxy.library.cornell.edu/full_record.do?product=UA&search_mode=GeneralSearch&qid=6&SID=3BaSGfGWW1UguFHF1x7&page=1&doc=3.

Pothakamury, U.R., A. Monsalve Gonzalez, G.V. Barbosa-Canovas and B.G. Swanson. 1995. High-voltage pulsed electric field inactivation of *Bacillus subtilis* and *Lactobacillus delbrueckii*. Revista Espanola De Ciencia Y Tecnologia De Alimentos 35(1): 101–107.

Pothakamury, U.R., H. Vega, Q.H. Zhang and G.V. Barbosa-Cánovas and B.G. Swanson. 1996. Effect of growth stage and processing temperature on the inactivation of *E. coli* by pulsed electric fields. J. Food Prot. 59(11): 1167–1171.

Puértolas, E., O. Cregenzan, E. Luengo, I. Álvarez and J. Raso. 2013. Pulsed-electric-field-assisted extraction of anthocyanins from purple-fleshed potato. Food Chem. 136(3–4): 1330–1336.

Puértolas, E., N. López, S. Condón, J. Raso and I. Alvarez. 2009. Pulsed electric fields inactivation of wine spoilage yeast and bacteria. Int. J. Food Microbiol. 130(1): 49–55.

Qin, B.L., Q.H. Zhang, G. Barbosa-Cánovas, B.G. Swanson and P.D. Pedrow. 1994. Inactivation of microorganisms by pulsed electric fields of different voltage waveforms. IEEE Transact. Dielectr. Electr. Insul. 1(6): 1047–1057.

Qin, B.L., F.J. Chang, G.V. Barbosa-Cánovas and B.G. Swanson. 1995a. Nonthermal inactivation of *Saccharomyces cerevisiae* in apple juice using pulsed electric fields. LWT - Food Sci. Technol. 28(6): 564–568.

Qin, B.L., U.R. Pothakamury, H. Vega, O. Martin, G.V. Barbosa-Canovas and B.G. Swanson. 1995b. Food pasteurization using high-intensity pulsed electric fields. Food Technol. 49(12): 55–60.

Qin, B.L., U.R. Pothakamury, G.V. Barbosa-Cánovas, B.G. Swanson and M. Peleg. 1996. Nonthermal pasteurization of liquid foods using high-intensity pulsed electric fields. Crit. Rev. Food Sci. Nutr. 36(6): 603–627.

Qin, B.L., G.V. Barbosa-Canovas, B.G. Swanson, P.D. Pedrow and R.G. Olsen. 1998. Inactivating microorganisms using a pulsed electric field continuous treatment system. IEEE Transact. Industry Appl. 34(1): 43–50.

Raso, J., G. Barbosa-Canovas, G. Barry, L. Calderon and M. Gongora. 1998. Inactivation of mold ascospores and conidiospores suspended in fruit juices by pulsed electric fields. LWT - Food Sci. Technol. 31(7–8): 668–672.

Rieder, A., T. Schwartz, K. Schön-Hölz, S.M. Marten, J. Süss, C. Gusbeth, W. Kohnen, W. Swoboda, U. Obst and W. Frey. 2008. Molecular monitoring of inactivation efficiencies of bacteria during pulsed electric field treatment of clinical wastewater. J. Appl. Microbiol. 105(6): 2035–2045.

Rodrigo, D., G.V. Barbosa-Canovas, A. Martinez and M. Rodrigo. 2003. Pectin methyl esterase and natural microflora of fresh mixed orange and carrot juice treated with pulsed electric fields. J. Food Prot. 66(12): 2336–2342.

Rodríguez-Gonzalez, O., M. Walkling-Ribeiro, S. Jayaram and M.W. Griffiths. 2011. Factors affecting the inactivation of the natural microbiota of milk processed by pulsed electric fields and cross-flow microfiltration. J. Dairy Res. 78(3): 270–278.

Rojas, M.C., S.E. Martin, R.A. Wicklund, D.D. Paulson, F.A. Desantos and M.S. Brewer. 2007. Effect of high-intensity pulsed electric fields on survival of *Escherichia Coli* K-12 suspended in meat injection solutions. J. Food Safety 27(4): 411–425.

Roodenburg, B., S.W.H. De Haan, L.B.J. Van Boxtel, V. Hatt, P.C. Wouters, P. Coronel and J.A. Ferreira. 2010. Conductive plastic film electrodes for pulsed electric field (PEF) treatment —A proof of principle. Innovat. Food Sci. Emerg. Technol. 11(2): 274–282.

Rowan, N.J., S.J. MacGregor, J.G. Anderson, R.A. Fouracre and O. Farish. 2000. Pulsed electric field inactivation of diarrhoeagenic *Bacillus cereus* through irreversible electroporation. Lett. Appl. Microbiol. 31(2): 110–4.

Russell, N.J., M. Colley, R.K. Simpson, A.J. Trivett and R.I. Evans. 2000. Mechanism of action of pulsed high electric field (PHEF) on the membranes of food-poisoning bacteria is an "all-or-nothing" effect. Int. J. Food Microbiol. 55(1–3): 133–136.

Ryynaenen, S. 1995. The electromagnetic properties of food materials: a review basic principles. J. Food Eng. 26(4): 409–429.

Sagarzazu, N., G. Cebrián, R. Pagán, S. Condón and P. Mañas. 2010. Resistance of *Campylobacter jejuni* to heat and to pulsed electric fields. Innovat. Food Sci. Emerg. Technol. 11(2): 283–289.

Saldaña, G., H. Minor-Perez, J. Raso and I. Alvarez. 2011. Combined effect of temperature, pH, and presence of nisin on inactivation of *Staphylococcus aureus* and *Listeria monocytogenes* by pulsed electric fields. Foodborne Pathog. Dis. 8(7): 797–802.

Saldaña, G., E. Puertolas, S. Condon, I. Alvarez and J. Raso. 2010. Modeling inactivation kinetics and occurrence of sublethal injury of a pulsed electric field-resistant strain of *Escherichia coli* and *Salmonella* Typhimurium in media of different pH. Innovat. Food Sci. Emerg. Technol. 11(2): 290–298.

Saldaña, G., E. Puértolas, N. López, D. García, I. Álvarez and J. Raso. 2009. Comparing the PEF resistance and occurrence of sublethal injury on different strains of *Escherichia coli, Salmonella* Typhimurium, *Listeria monocytogenes* and *Staphylococcus aureus* in media of pH 4 and 7. Innovat. Food Sci. Emerg. Technol. 10(2): 160–165.

Sale, A.J.H. and W.A. Hamilton. 1967. Effects of high electric fields on microorganisms: Killing of bacteria and yeasts. Biochim. Biophys. Acta 148(3): 781–788.

Salerno, M.B., H.S. Lee, P. Parameswaran and B.E. Rittmann. 2009. Using a pulsed electric field as a pretreatment for improved biosolids digestion and methanogenesis. Water Environ. Res. 81(8): 831–839.

Sampedro, F., D.J. Geveke, X. Fan, D. Rodrigo and Q.H. Zhang. 2009. Shelf-life study of an orange juice-milk based beverage after PEF and thermal processing. J. Food Sci. 74(2): S107–112.

Schilling, S., S. Schmid, H. Jaeger, M. Ludwig, H. Dietrich, S. Toepfl, D. Knorr, S. Neidhart, A. Schieber and R. Carlet. 2008. Comparative study of pulsed electric field and thermal processing of apple juice with particular consideration of juice quality and enzyme deactivation. J. Agr. Food Chem. 56(12): 4545–4554.

Schoenbach, K.H., S. Katsuki, R.H. Stark, E.S. Buescher and S.J. Beebe. 2002. Bioelectrics—new applications for pulsed power technology. IEEE Transact. Plasma Sci. 30(1): 293–300.

Schoenbach, K.H., F.E. Peterkin, R.W. Alden and S.J. Beebe. 1997. The effect of pulsed electric fields on biological cells: Experiments and applications. IEEE Transact. Plasma Sci. 25(2): 284–292.

Selma, M.V., M.C. Salmeron, M. Valero and P.S. Fernandez. 2005. Efficacy of pulsed electric fields for *Listeria monocytogenes* inactivation and control in horchata. J. Food Saf. 26(2): 137–149.

Sepulveda, D.R., M.M. Góngora-Nieto, J.A. Guerrero and G.V. Barbosa-Cánovas. 2009. Shelf life of whole milk processed by pulsed electric fields in combination with PEF-generated heat. LWT - Food Sci. Technol. 42(3): 735–739.

Sepulveda, D.R., J.A. Guerrero and G.V. Barbosa-Cánovas. 2006. Influence of electric current density on the bactericidal effectiveness of pulsed electric field treatments. J. Food Eng. 76(4): 656–663.

Seratlić, S., B. Bugarski, V. Nedović, Z. Radulović, L. Wadsö, P. Dejmek and F.G. Galindo. 2013. Behavior of the surviving population of *Lactobacillus plantarum* 564 upon the application of pulsed electric fields. Innovat. Food Sci. Emerg. Technol. 17: 93–98.

Shin, D., B.C. Martin and M.X. Sánchez-Plata. 2011. Pulsed electric field effects to reduce the level of *Campylobacter* spp. in scalder and chiller water during broiler chicken processing. Asian-Australasian J. Anim. Sci. 24(9): 1314–1317.

Shin, J.K., K.J. Jung, Y.R. Pyun and M.S. Chung. 2007. Application of pulsed electric fields with square wave pulse to milk inoculated with *E. coli*, *P. fluorescens*, and *B. stearothermophilus*. Food Sci. Biotechnol. 16(6): 1082–1084.

Simpson, R.K., R. Whittington, R.G. Earnshaw and N.J. Russell. 1999. Pulsed high electric field causes "all or nothing" membrane damage in *Listeria monocytogenes* and *Salmonella* Typhimurium, but membrane H+-ATPase is not a primary target. Int. J. Food Microbiol. 48(1): 1–10.

Smith, K., G.S. Mittal and M.W. Griffiths. 2002. Pasteurization of milk using pulsed electrical field and antimicrobials. J. Food Sci. 67(6): 2304–2308.

Sobrino-Lopez, A. and O. Martin-Belloso. 2006. Enhancing inactivation of *Staphylococcus aureus* in skim milk by combining high-intensity pulsed electric fields and nisin. J. Food Prot. 69(2): 345–353.

Somolinos, M., D. Garcia, P. Manas, S. Condon and R. Pagan. 2008. Effect of environmental factors and cell physiological state on pulsed electric fields resistance and repair capacity of various strains of Escherichia coli. Int. J. Food Microbiol. 124: 260–267.

Somolinos, M., D. García, P. Mañas, S. Condón and R. Pagán. 2010a. Organic acids make *Escherichia coli* more resistant to pulsed electric fields at acid pH. Int. J. Food Microbiol. 136(3): 381–384.

Somolinos, M., L. Espina, R. Pagán and D. Garcia. 2010b. sigB absence decreased *Listeria monocytogenes* EGD-e heat resistance but not its pulsed electric fields resistance. Int. J. Food Microbiol. 141(1–2): 32–38.

Spilimbergo, S., F. Dehghani, A. Bertucco and N.R. Foster. 2003. Inactivation of bacteria and spores by pulse electric field and high pressure CO2 at low temperature. Biotechnol. Bioeng. 82(1): 118–25.

Toepfl, S. and V. Heinz. 2011. Pulsed electric field assisted extraction—a case study. *In*: H.Q. Zhang, G.V. Barbosa-Canovas, V.M. Balasubramaniam, C.P. Dunne, D.F. Farkas and J.T.C. Yuan (eds.). Nonthermal Processing Technologies for Food. Wiley-Blackwell, Ames, IA, USA, pp. 190–200.

Ulmer, H.M., V. Heinz, M.G. Gänzle, D. Knorr and R.F. Vogel. 2002. Effects of pulsed electric fields on inactivation and metabolic activity of *Lactobacillus plantarum* in model beer. J. Appl. Microbiol. 93(2): 326–335.

Unal, R., A.E. Yousef and C.P. Dunne. 2002. Spectrofluorimetric assessment of bacterial cell membrane damage by pulsed electric field. Innov. Food Sci. Emerg. Technol. 3(3): 247–254.

Valero, M., M.C. Salmer, M.V. Selma and P.S. Fern. 2003. Control of *Enterobacter aerogenes* by high-intensity, pulsed electric fields in horchata, a Spanish low-acid vegetable beverage. Food Microbiol. 20(1): 105–110.

Vega-Mercado, H., U.R. Pothakamury, F.J. Chang, G.V. Barbosa-Cánovas and B.G. Swanson. 1996a. Inactivation of *Escherichia coli* by combining pH, ionic strength and pulsed electric fields hurdles. Food Res. Int. 29(2): 117–121.

Vega-Mercado, H., O. Martin-Belloso, F.J. Chang, G.V. Barbosa-Canovas and B.G. Swanson. 1996b. Inactivation of *Escherichia coli* and *Bacillus subtilis* suspended in pea soup using pulsed electric fields. J. Food Process. Pres. 20(6): 501–510.

Vega-Mercado, H., J.R. Powers, O. Martín-Belloso, G.V. Barbosa-Cánovas, L. Luedecke and B.G. Swanson. 2001. Change in susceptibility of proteins to proteolysis and the inactivation of an extracellular protease from *Pseudomonas fluorescens* M3/6 when exposed to pulsed electric fields. *In*: G.V. Barbosa-Cánovas and Q.H. Zhang (eds.). Pulsed electric fields in food processing: fundamental aspects and applications. Technomic Publishing Company, Inc., Lancaster, PA, USA, pp. 105–120.

Virwani, K.R., A.P. Malshe and K.P. Rajurkar. 2010. Pulse preakdown in sub-20 nm organic dielectrics for nanoscale-electromachining (nano-EM). J. Manuf. Sci. Eng. 132(3): 030915–(1–8).

Waite-Cusic, J.G., B.H.S. Diono and A.E. Yousef. 2011. Screening for *Listeria monocytogenes* surrogate strains applicable to food processing by ultrahigh pressure and pulsed electric field. J. Food Prot. 74 (10): 1655–61.

Walkling-Ribeiro, M., F. Noci, D.A. Cronin, J.G. Lyng and D.J. Morgan. 2008a. Inactivation of *Escherichia coli* in a tropical fruit smoothie by a combination of heat and pulsed electric fields. J. Food Sci. 73(8): M395–399.

Walkling-Ribeiro, M., F. Noci, D.A. Cronin, J. Riener, J.G. Lyng and D.J. Morgan. 2008b. Reduction of *Staphylococcus aureus* and quality changes in apple juice processed by ultraviolet irradiation, pre-heating and pulsed electric fields. J. Food Eng. 89(3): 267–273.

Walkling-Ribeiro, M., F. Noci, D.A. Cronin, J.G. Lyng and D.J. Morgan. 2009a. Shelf life and sensory evaluation of orange juice after exposure to thermosonication and pulsed electric fields. Food Bioprod. Process. 87(2): 102–107.

Walkling-Ribeiro, M., F. Noci, J. Riener, D.A. Cronin, J.G. Lyng and D.J. Morgan. 2009b. The impact of thermosonication and pulsed electric fields on *Staphylococcus aureus* inactivation and selected quality parameters in orange juice. Food Bioprocess Technol. 2(4): 422–430.

Walkling-Ribeiro, M., F. Noci, D.A. Cronin, J.G. Lyng and D.J. Morgan. 2009c. Antimicrobial effect and shelf-life extension by combined thermal and pulsed electric field treatment of milk. J. Appl. Microbiol. 106(1): 241–248.

Walkling-Ribeiro, M., F. Noci, D.A. Cronin, J.G. Lyng and D.J. Morgan. 2010. Shelf life and sensory attributes of a fruit smoothie-type beverage processed with moderate heat and pulsed electric fields. LWT—Food Sci. Technol. 43(7): 1067–1073.

Walkling-Ribeiro, M., O. Rodríguez-Gonzalez, S.H. Jayaram and M.W. Griffiths. 2011a. Processing temperature, alcohol and carbonation levels and their impact on pulsed electric fields (PEF) mitigation of selected characteristic microorganisms in beer. Food Res. Int. 44(8): 2524–2533.

Walkling-Ribeiro, M., O. Rodríguez-González, S. Jayaram and M.W. Griffiths. 2011b. Microbial inactivation and shelf life comparison of "cold" hurdle processing with pulsed electric fields and microfiltration, and conventional thermal pasteurisation in skim milk. Int. J. Food Microbiol. 144(3): 379–386.

Węsierska, E. and T. Trziszka. 2007. Evaluation of the use of pulsed electrical field as a factor with antimicrobial activity. J. Food Eng. 78(4): 1320–1325.

Wiktor, A., M. Iwaniuk, M. Sledz, M. Nowacka, T. Chudoba and D. Witrowa-Rajchert. 2013. Drying kinetics of apple tissue treated by pulsed electric field. Drying Technol. 31(1): 112–119.

Wouters, P.C., N. Dutreux, J.P.P.M. Smelt and H.L.M. Lelieveld. 1999. Effects of pulsed electric fields on inactivation kinetics of Listeria innocua. Appl. Environ. Microbiol. 65(12): 5364–5371.

Wuytack, E.Y., L.D.T. Phuong, A. Aertsen, K.M.F. Reyns, D. Marquenie, B. De Ketelaere, B. Masschalck, I. Van Opstal, A.M.J. Diels and C.W. Michiels. 2003. Comparison of sublethal injury induced in *Salmonella enterica* serovar Typhimurium by heat and by different nonthermal treatments. J. Food Prot. 66(1): 31–37.

Xiang, B.Y., M.V. Simpson, M.O. Ngadi and B.K. Simpson. 2011. Flow behaviour and viscosity of reconstituted skimmed milk treated with pulsed electric field. Biosystems Eng. 109(3): 228–234.

Yan, K., H. Todo and K. Sugibayashi. 2010. Transdermal drug delivery by in-skin electroporation using a microneedle array. Int. J. Pharm. 397(1–2): 77–83.

Yaqub, S., J.G. Anderson, S.J. MacGregor and N.J. Rowan. 2004. Use of a fluorescent viability stain to assess lethal and sublethal injury in food-borne bacteria exposed to high-intensity pulsed electric fields. Lett. Appl. Microbiol. 39(3): 246–51.

Yeom, H.W., G.A. Evrendelik, Z.T. Jin and Q.H. Zhang. 2004. Processing of yogurt-based products with pulsed electric fields: microbial, sensory and physical evaluations. J. Food Process. Pres. 28(3): 161–178.

Yeom, H.W., C.B. Streaker, Q.H. Zhang and D.B. Min. 2000. Effects of pulsed electric fields on the quality of orange juice and comparison with heat pasteurization. J. Agr. Food Chem. 48(10): 4597–605.

Yu, L.J., M. Ngadi and G.S.V. Raghavan. 2009. Effect of temperature and pulsed electric field treatment on rennet coagulation properties of milk. J. Food Eng. 95(1): 115–118.

Zhang, H., Z. Wang, R.-J. Yang and S.-Y. Xu. 2007. Inactivation of microorganisms in cloudy ginkgo (*Ginkgo biloba Linn.*) juice by pulsed electric fields. Food Sci. Technol. Int. 13(2): 83–90.

Zhang, Q., A. Monsalve-Gonzalez, B.L. Qin, G.V. Barbosa-Canovas and B.G. Swanson. 1994a. Inactivation of *Saccharomyces cerevisiae* in apple juice by square-wave and exponential-decay pulse electric fields. J. Food Process Eng. 17(4): 469–478.

Zhang, Q., F.J. Chang, G.V. Barbosa-Canovas and B.G. Swanson. 1994b. Inactivation of microorganisms in a semisolid model food using high voltage pulsed electric fields. LWT - Food Sci. Technol. 27(6): 538–543.

Zhang, Q.H., G.V. Barbosa-Cánovas and B.G. Swanson. 1995a. Engineering aspects of pulsed electric field pasteurization. J. Food Eng. 25(2): 261–281.

Zhang, Q.H., B.L. Qin, G.V. Barbosa-Canovas and B.G. Swanson. 1995b. Inactivation of *E. coli* for food pasteurization by high-strength pulsed electric fields. J. Food Process. Pres. 19(2): 103–118.

Zhang, Y.I., B.E.I. Gao, M. Zhang, J. Shi and Y. Xu. 2010. Pulsed electric field processing effects on physicochemical properties, flavor compounds and microorganisms of longan juice. J. Food Process. Pres. 34(6): 1121–1138.

Zhao, J., R. Yang, W. Zhao, S. Ma and Y.F. Wang. 2008a. Effect of pulsed electric fields on inactivation of microbe and quality of freshly-squeezed pear juice. Transact. Chin. Soc. Agr. Eng. 24(6): 239–244.

Zhao, W., R. Yang, R. Lu, M. Wang, P. Qian and W. Yang. 2008b. Effect of PEF on microbial inactivation and physical–chemical properties of green tea extracts. LWT - Food Sci. Technol. 41(3): 425–431.

Zhao, W., R. Yang, X. Shen, S. Zhang and X. Chen. 2013. Lethal and sublethal injury and kinetics of *Escherichia coli*, *Listeria monocytogenes* and *Staphylococcus aureus* in milk by pulsed electric fields. Food Control 32(1): 6–12.

Zhao, W., R. Yang, H.Q. Zhang, W. Zhang, X. Hua and Y. Tang. 2011. Quantitative and real time detection of pulsed electric field induced damage on *Escherichia coli* cells and sublethally injured microbial cells using flow cytometry in combination with fluorescent techniques. Food Control 22(3–4): 566–573.

Zhao, W. and R.J. Yang. 2013. Pulsed electric field induced aggregation of food proteins: Ovalbumin and bovine serum albumin. Food Bioprocess Technol. 5(5): 1706–1714.

Zhong, K., F. Chen, Z. Wang, J. Wu, X. Liao and X. Hu. 2005. Inactivation and kinetic model for the *Escherichia coli* treated by a co-axial pulsed electric field. Eur. Food Res. Technol. 221(6): 752–758.

Zhong, K., J.H. Wu, Z.F. Wang, F. Chen, X.J. Liao, X.S. Hu and Z.H. Zhang. 2007. Inactivation kinetics and secondary structural change of PEF-treated POD and PPO. Food Chem. 100(1): 115–123.

Zimmermann, U., G. Pilwat and F. Riemann. 1974. Dielectric breakdown in cell membranes. Biophys. J. 14(11): 881–899.

CHAPTER 3

Applications and Modeling Aspects of UV and Pulsed UV-Light for Food Decontamination

Nene M. Keklik[1,*] and *Ali Demirci*[2]

Introduction

Food safety has always been one of the major issues of our societies. Despite the technological advances, industrial developments, and broad education, millions of people throughout the world continue to suffer from foodborne illnesses associated with the consumption of contaminated food and water (WHO 2007). Besides the health issues, which sometimes lead to complications or even deaths, the foodborne disease outbreaks also cause huge economic losses for the medical and food sectors. Therefore, the food industry and researchers in the food safety area strive to find methods to assure the safety of food products.

Thermal processing is the most common decontamination method used in the food industry to pasteurize or sterilize food products. The basic principle of this technique is to heat the food until its central point reaches a certain temperature, and hold the food at that temperature for sufficient

[1] Cumhuriyet University, Faculty of Engineering, Department of Food Engineering, 58140 Sivas, Turkey.
Email: meltemkeklik@gmail.com
[2] Department of Agricultural and Biological Engineering, The Pennsylvania State University, University Park, PA 16802 USA.
Email: demirci@psu.edu
* Corresponding author

amount of time to eliminate the spoilage and pathogenic microorganisms. Thermal processing has successfully been adapted to the food industry thanks to the development of various types of efficient heat exchangers and jacketed tanks. Also, the widely known method of *D* and *z* values has proved useful particularly in the canning industry (van Boekel 2002).

Despite its success in food decontamination, thermal treatment has certain drawbacks. When subjected to elevated temperatures, the nutrients, especially the vitamins, in the food can be destructed (Lešková et al. 2006). On the other hand, some undesirable compounds such as oxidation and polymerization byproducts and products can be formed in the presence of heat. Another issue with the thermal processing is its inapplicability to the raw, fresh, and minimally processed foods. Applying heat to these kinds of food may result in cooked, withered, or discolored products.

In order to avoid these unwanted effects of thermal processing on food quality, researchers have focused on decontamination methods which do not or minimally involve heat. Among these techniques are high hydrostatic pressure (HHP) or high pressure processing (HPP), pulsed electric field (PEF), ultrasound, oscillating magnetic field, ozone, supercritical carbon dioxide, ionizing radiation, ultraviolet (UV) light, and pulsed UV-light. These techniques present certain advantages and disadvantages over one another, and the efficiency of a particular method is dependent on the characteristics of the food and the microorganisms to be eliminated.

UV-light technology is one of the most promising non-thermal methods for food decontamination. As of now, it is widely used in the treatment of drinking water and wastewater (Wright and Cairns 1998). UV light is defined as the electromagnetic radiation in the wavelength range of 100–400 nm. Microbial cells are inactivated when the UV radiation damages the microbial DNA by means of photochemical action (Miller et al. 1999). UV light can eliminate a wide range of microorganisms including bacteria, fungi, bacterial and fungal spores, viruses, protozoa, and algae. Another advantage of UV-light treatment is that it is a dry process, i.e., no water is involved, and thus the operation and cleaning of the treatment unit is quite easy. The biggest drawback of UV-light technology is its limited penetration into the food. The more transparent the food, the more UV-light is able to pass through it. Researchers have managed to improve the transmission of UV-light through liquid and particulate foods (e.g., grain) to an extent (Koutchma et al. 2006, Hidaka and Kubota 2006); however, for solid foods, UV-light is considered just as a surface decontamination method for now.

Pulsed UV-light is an advanced UV-light technology, which involves light pulses in the wavelength range of 100–1100 nm delivered by an inert gas lamp. Even though it has a broad wavelength range, it is called pulsed UV-light, because the majority of the light is in the UV region.

Also, the pulses are so intense that they kill microorganisms by a more complex inactivation mechanism than the UV-light alone. In addition to the photochemical action, pulsed UV-light has been demonstrated to have photothermal and photophysical effects on microbial cells (Fine and Gervais 2004, Wekhof 2003, Krishnamurthy et al. 2010). This complex inactivation mechanism prevents the repair of DNA by the microbial cells using their self-defense mechanisms. Pulsed UV-light is faster and thus more efficient than the conventional UV-light. The main issue with pulsed UV-light is that heat accumulates over time due to the infrared portion of the radiation (700–1100 nm), raising the temperature of the food significantly after prolonged treatment (Bialka 2006). However, pulsed UV-light is considered to be non-thermal for short treatment times. Another issue is again the limited penetration. Although pulsed UV-light can penetrate better into the food compared to the conventional UV-light (Bialka 2006, Xenon 2007a), it may still not able to reach the deeper parts of the food.

A number of researchers have studied the application of UV and pulsed UV-light for food decontamination, and demonstrated the effectiveness of these methods on various types of microorganisms on/in liquid, solid, and powder foods. In addition to these studies, researchers have also focused on the description of the microbial inactivation by UV and pulsed UV-light. To do so, various mathematical models have been used to predict the microbial survival after UV and pulsed UV treatments. The accurate prediction of microbial inactivation would let us acquire knowledge on how microbial populations respond to UV or pulsed UV-light, which in turn would ease the assessment of microbiological hazard risks during processing and storage of foods. In other words, the careful estimation of microbial survival would give us more control over the decontamination process.

In this chapter, both UV and pulsed UV-light technologies will be explained in terms of their principles, inactivation mechanisms, and applications for food decontamination. Then, the modeling aspects of these technologies will be discussed with focus on different examples of inactivation models, evaluation/validation methods, and model applications. We hope that this chapter will provide the readers with the information they seek, and that it will be a useful resource for the future studies on the application of UV and pulsed UV-light technologies for food decontamination.

Basic Principles of UV and Pulsed UV-light Technologies

Sunlight has long been known to have bactericidal effects ever since English scientists Downes and Blunt demonstrated in 1877 that the growth of bacteria in solutions was inhibited when exposed to sunlight (Reed 2010). Later in 1892, Marshall Ward found out the main reason behind the bactericidal effect of sunlight to be the UV part of the spectrum (Bischof 1994). Since then, UV light has been considered as an effective tool to eliminate undesirable microorganisms on various surfaces, in water and foods. Thanks to the developments in UV-light technology, there are now many types of lamps capable of producing monochromatic and polychromatic UV light. Unless otherwise stated, UV-light refers to the light produced by low or medium mercury arc lamps, which is also called continuous-wave (CW) or conventional UV light. Pulsed UV (PUV) light or intense light pulses (ILP), on the other hand, is produced by inert gas lamps operating in pulsing mode, whose spectrum involves portions of UV, visible, and infrared light. Below, the basic principles of both UV and pulsed UV-light are explained in detail.

UV light

UV light is defined as the electromagnetic radiation in the wavelength range of 100–400 nm, which is further divided into four sections: UV-A (315–400 nm), UV-B (280–315 nm), UV-C (200–280 nm), and vacuum-UV (100–200 nm). About 10 per cent of total solar radiation is in the UV spectrum (El-Boury et al. 2007, Melquiades et al. 2008). The germicidal properties of UV light are mostly attributed to UV-C and vacuum UV portions, which possess higher photon energies compared to UV-A and UV-B. The photon energy decreases as the wavelength increases, and ranges from 124 to 3.10 eV between 100 and ~400 nm (Krishnamurthy et al. 2008). Unlike gamma and X-rays, UV-light does not cause ionization of molecules, and thus is a type of nonionizing radiation.

Light can cause formation or dissociation of chemical bonds if sufficient amount of photon energy is absorbed by atoms or molecules (Blatchley and Peel 2001). Thus, UV-light is capable of inducing photochemical reactions in living cells, which eventually leads to the inactivation or death of the microorganisms. However, in order for UV-light to be absorbed by the microbial cells, the medium (e.g., food) carrying the cells needs to transmit as much light as possible. Unfortunately, UV-light is little or not absorbed by opaque materials, which is why it is only effective on the microorganisms mostly present on the surface.

UV-light has been utilized as a disinfectant since the 1900s (Wright and Cairns 1998), and is commonly used for the treatment of drinking water and wastewater. With the production of more durable and reliable UV lamps, the production of UV equipment started to rise after about 1970 (Solsona and Mendez 2003). Examples of commercial UV disinfection chambers can be seen in Fig. 1. UV treatment is an alternative to chemical disinfectants such as chlorine, and sometimes used in combination with it for the treatment of drinking water (Wright and Cairns 1998).

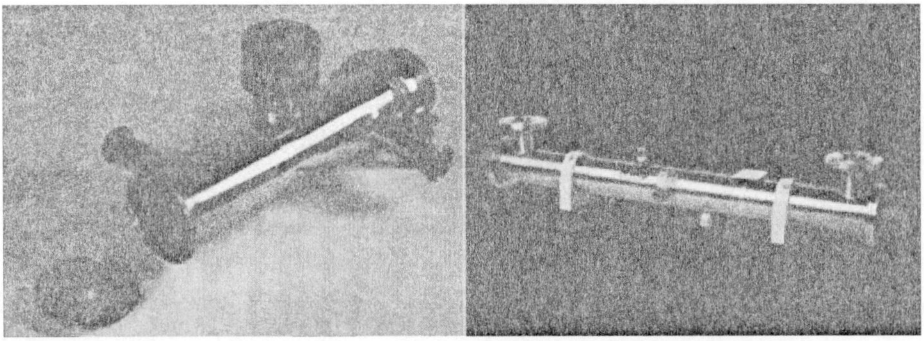

Figure 1. Examples of UV disinfection chambers. Source: atg UV Technology, England, UK.

There are several types of lamps capable of producing UV light, among which the most commonly used ones are the low- and medium-pressure mercury arc lamps. Low pressure mercury arc lamps yield monochromatic radiation with 85–90 per cent of the emission at 253.7 nm, while medium pressure mercury arc lamps emit polychromatic radiation in the wavelength range of 185–600 nm (Linden 2004, Bohrerova et al. 2008). These lamps consist of UV-transmittant silica or quartz tube containing a mixture of mercury and inert argon vapor. At both ends in the tube are the electrodes that are often made of tungsten with a mixture of alkaline earth metals, which assist in arc formation when high voltage is applied across the electrodes (Wright and Cairns 1998). With the formation of the electric arc, the mercury vapor becomes excited, and emits UV light until getting back to a lower energy level. The argon in the vapor mixture helps the start-up of the lamp, extends the electrode life, and decreases thermal losses without affecting the spectral output of the lamp (Wright and Cairns 1998). While producing a more continuum light with higher UV intensity (McDonald et al. 2000, Koutchma et al. 2009), medium pressure mercury arc lamps are germicidally less efficient compared to low pressure mercury lamps (Linden 2004).

There are some other types of lamps alternative to low and medium pressure mercury gas lamps such as microwave UV lamps, low-pressure amalgam lamps, and excimer lamps. Microwave UV lamps involve no electrodes to excite the mercury gas, and thus the lamp warm up time is shorter and the lamp life is longer as compared to low or medium pressure mercury lamps (Koutchma et al. 2009). Low-pressure amalgam lamps also have long life and high efficiency with negligible heat generation (Koutchma et al. 2009). Excimer lamps produce monochromatic light by using a mixture of rare gases, which determines the characteristic wavelength of the emission (Blatchley and Peel 2001).

In the food industry, UV light is utilized for disinfection of air, decontamination on the surfaces of the equipment and packaging materials as well as on fruits and vegetables during post-harvest storage to extend their shelf-life by inactivating spoilage fungi (Begum et al. 2009). The poor penetration ability of UV light, however, limits the adaptation of UV technology to the commercial food decontamination. The FDA approved UV radiation under certain conditions in three areas of food processing: surface decontamination of food and food products, disinfection of potable water used in food production, and reduction of microorganisms in juice products (FDA 2011a). The first condition is that the radiation sources consist of low pressure mercury lamps of which 90 per cent of the emission is at a wavelength of 253.7 nm. Other limitations are listed below (FDA 2011a):

a) Surface decontamination of food and food products: no ozone production; foods with high fat content irradiated in vacuum or in an inert atmosphere; intensity of radiation, 1 W (of 253.7 nm) per 5–10 ft^2.
b) Disinfection of potable water used in food production: no ozone production; coefficient of absorption, 0.19 per cm or less; flow rate, 100 gal/h per watt of 253.7 nm radiation; water depth, 1 cm or less; lamp-operating temperature, 36–46°C.
c) Reduction of microorganisms in juice products: Turbulent flow through tubes with a minimum Reynolds number of 2,200.

In order to inactivate microorganisms in a food product, the UV radiation needs to be at least 400 J/m^2 in all parts of the product (FDA 2000). The critical factors for UV processing of foods include the transmissivity of the product, geometric properties of the system, power, wavelength, physical arrangement of UV lamp(s), the flow profile, and the radiation path length (FDA 2000).

Pulsed UV-light

Pulsed UV-light technology utilizes UV-light, but is considered to be more advantageous in terms of efficacy and safety compared to conventional or continuous UV light. Pulsed UV light was initially used in Japan in the late 1970s for microbial inactivation (Oms-Oliu et al. 2010). However, it was not until 1999 that the FDA granted approval for the pulsed light treatment of foods (Federal Register 1999). Pulsed UV light is produced by inert gas lamps as intense light pulses in a continuum broadband spectrum in the range of 100–1100 nm, which is especially rich and efficient in the UV range (Krishnamurthy et al. 2010, Demirci and Panico 2008). Like UV light, pulsed UV light is nonionizing and nonchemical, and is considered to be nonthermal for short treatment times.

In a pulsed UV-light system, the alternate current is converted into high voltage direct current (Palmieri et al. 1999). The produced electrical energy is then stored in a high energy density electrical storage capacitor, which discharges the energy to the lamp whenever the circuit is closed by means of a switch. This energy is then used to ionize the inert gas (xenon or krypton) inside the lamp, which in turn emits short intense light pulses (Palmieri et al. 1999). The pulse duration can range from 100 ns to 2 ms (Demirci and Panico 2008), and the typical pulsing rate is about 1 to 20 flashes per second (Dunn 1996, Palmieri et al. 1999). A bench-type pulsed UV-light system is shown in Fig. 2.

There are two types of pulsed UV light producing lamps: flashlamps and surface discharge lamps (Bohrerova et al. 2008). In a flashlamp, the discharge is constrained within a small envelope, whereas in a surface discharge lamp, plasma discharge is used. Pulsed surface discharge lamps operate at higher energy per pulse given the same pulse duration, and thus has a higher UV efficiency compared to flashlamps. Also, they can operate at the stress levels above the explosion limit of flashlamps (Schaefer et al. 2007). Ozone production is generally an issue with the pulsed UV lamps, which can be controlled via specific spectral cutoff configuration yielding desired wavelength characteristics (Xenon 2007b).

Pulsed UV-light is four times more efficient than continuous UV light in terms of the microbial inactivation (Bialka 2006). This is most probably due to the high peak power delivered by pulsed UV-light in a very short time compared to UV light, which delivers constant power over a longer period of time. Involving UV, visible, and infrared spectra, pulsed UV light has a more complex inactivation mechanism than UV light, as will be explained later. Also, pulsed UV lamps are safer than mercury UV lamps, because inert gases do not pose toxicity like mercury in case of lamp breakage. On

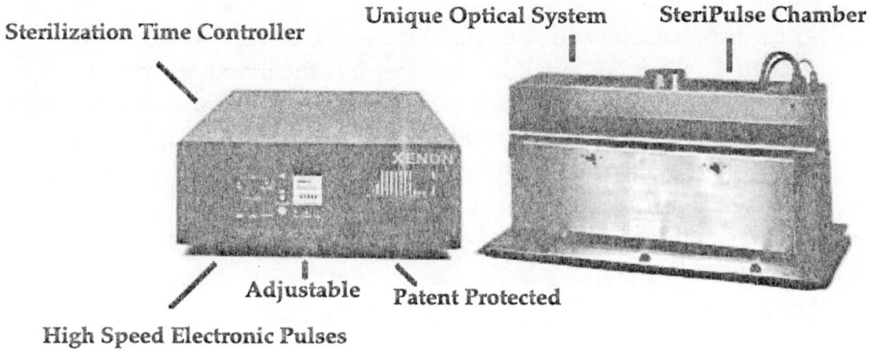

Figure 2. A bench-top pulsed UV-light system. Source: Xenon Corporation, Wilmington, MA, USA.

the other hand, pulsed UV light causes heat accumulation over time due to the infrared resulting in significant temperature increase in food product (Bialka 2006). Therefore, the choice of the optimum treatment time is critical in order to avoid thermal stress on the food. Pulsed UV-light does not penetrate much into opaque materials although its penetration ability is better compared to UV light (Dunn 1996, Xenon 2007a). However, it can be transmitted through certain transparent plastics, and thus may be used for the treatment of packaged products (Keklik et al. 2009, 2010a).

According to the FDA, pulsed light is safe to use for the treatment of foods under the following conditions (FDA 2011b):

a) Radiation source: Xenon flashlamps designed to emit broadband radiation in a wavelength range of 200–1,100 nm and with a pulse duration no longer than 2 ms;
b) Used for surface microorganism control;
c) Pulsed light treated foods shall receive the minimum treatment reasonably needed to accomplish the intended technical effect; and
d) The total cumulative dose shall not exceed 12.0 J/cm^2.

The critical process factors in pulsed light treatment include the light characteristics such as the wavelength, intensity, duration and number of pulses, packaging and food properties such as the type, transparency and color (FDA 2000). For fluid foods, the transparency and the depth of the fluid column are considered to be the critical factors (FDA 2000). Pulsed UV-light treatment takes generally shorter times than other line processes, thus the treatment is not considered to be a rate limiting factor (Dunn 1996).

Mechanisms of Microbial Inactivation by UV and Pulsed UV-Light

In order to improve the efficiency of UV and pulsed UV technologies for food decontamination, the mechanisms behind the inactivation of microbial cells by these techniques need to be understood. A number of researchers have managed to describe the actions of UV and pulsed UV light on the microbial cells. However, it is yet not known how exactly the light alters the functions of a microbial cell. Below are summarized the known (observed) mechanisms involved in the inactivation of microorganisms by UV and pulsed UV-light.

UV-light

The exposure of cells to UV light can be mutagenic and lethal (Miller et al. 1999). UV light damages the living cells by altering their nucleic acids, which include photodimerization between adjacent pyrimidine bases, photohydration of cytosine, and, less frequently, inter- and intra-strand cross-links, protein-DNA cross-links, and rare base adducts. Other than these direct effects, there are indirect effects of UV-light, which occur when reactive oxygen species are produced, which react with DNA, damage bases, break strands, and cross-link DNA and proteins (Miller et al. 1999).

The inactivation by UV light is primarily due to the formation of pyrimidine dimers between adjacent pairs of thymine or cytosine on the same DNA or RNA strand (Koutchma et al. 2009). These dimers prevent the replication of DNA, and thus leave the microbial cells inactive. The UV absorbance of nucleic acid is maximal at a wavelength of about 260 nm, at which highest damage to the microbial DNA occurs (Aguiar et al. 1996). Therefore, UV-C is the most germicidal, although other regions of UV-light also contribute to the inactivation.

Bacteria are able to repair UV-induced damage in two mechanisms classified as "light" and "dark" (Miller et al. 1999, Koutchma et al. 2009). For the "light" mechanism bacterial cells use visible light (380–430 nm) as a cofactor and the enzyme photolyase to reverse certain types of UV-induced damage. This mechanism is also know as photoreactivation. On the other hand, the damage that could not be repaired by photoreactivation is treated by "dark" mechanism, which does not require visible light as a cofactor. Although organisms are able to repair certain types of UV-induced damage, many of these repair mechansims have adverse effects on the organism itself such as retardation of the growth and mutagenicity (Miller et al. 1999).

Pulsed UV-light

Mechanisms involved in pulsed UV inactivation of microbials cells is much more complex than those in UV inactivation. This is primarily due to the broad wavelength range and the high peak power delivered by the pulsed UV-light. There are three modes of inactivation explained for pulsed UV-light so far, which are photochemical, photothermal, and photophysical effects. The photochemical effect of pulsed UV-light is primarily caused by the UV-C part, and is the same mechanism in UV inactivation involving the formation of pyrimidine dimers. Cheigh et al. (2012) reported the DNAs of *Listeria monocytogenes* and *Escherichia coli* O157:H7 cells treated with either intense pulsed light or UV-C accumulated double-strand breaks, single-strand breaks, and cyclobutane pyrimidine dimers, with more double-strand breaks detected following UV-C compared to intense pulsed light for both microorganisms.

The photothermal effect of pulsed UV light is attributed to all parts of UV light, and causes the bacterial cells to rupture (Wekhof 2000, Fine and Gervais 2004). Wekhof (2000) suggested that the bacteria are momentously overheated by absorbing all UV light from a flash lamp at high flux densities (>~0.5 J/cm^2 at the sample level). This overheating was attributed to a difference between the absorption of UV light by the bacteria and that by the surrounding medium. Accordingly, if UV light is absorbed during a short pulse, then the cooling of bacteria from surface to the surrounding medium is too slow during the same heating time that the overheating and rupture of the bacteria cannot be prevented. Photophysical effect of pulsed UV light, on the other hand, is attributed to the constant disturbance caused by the high-energy pulses, which result in damages to the cell structure (Krishnamurthy et al. 2010).

In the study of Krishnamurthy et al. (2010), *Staphylococcus aureus* cells suspended in phosphate buffer and exposed to pulsed UV-light for 5 s at a total dose of 4.95 J/cm^2 were imaged using transmission electron microscopy (TEM), in which cell wall rupture and disappearance, cytoplasm shrinkage, membrane damage, and the eventual leakage of cellular contents were clearly demonstrated. Cheigh et al. (2012) also observed by TEM that bacterial cell structures (*L. monocytogenes* and *E. coli* O157:H7) were destroyed by intense pulsed light but not by UV-C. For the cells that are ruptured due to the photothermal and photophysical effects during pulsed UV-light treatment, the repair becomes impossible unlike in photochemical mechanisms, which stands as another advantage of pulsed UV-light over UV-light.

Applications of UV and Pulsed UV-Light for Food Decontamination

There are many studies that have been conducted to investigate the effects of UV and pulsed UV-light on the microbial populations in/on foods as well as the product quality. The efficiency of UV or pulsed UV treatment of a food product depends on the characteristics of the light, food, microorganisms, and the processing conditions. Whether the food is solid or liquid, its transparency determines how well the pulsed UV light can penetrate into the food. However, if the food has flowability, such as liquid and powder foods, many particles in the food can be exposed to pulsed UV light by flowing the food under the lamp. On the other hand, for opaque solid foods, only the surface is exposed to light with little or no penetration. For this case, the exposure of the food surface to light can be increased by either rotating the food under the lamp or placing more lamps with or without reflectors around the food. In the treatment of packaged products, the packaging material adds another barrier to the penetrability of pulsed UV light. In the study of Keklik et al. (2010a), among different packaging materials, polypropylene was the most suitable packaging material for pulsed light treatment, transmitting about 75% of UV light at 260 nm.

Pulsed UV light appears to be more effective on gram-negative bacteria compared to gram-positive bacteria (Rowan et al. 1999, MacGregor et al. 1998, Anderson et al. 2000, Elmnasser et al. 2007, Keklik et al. 2012). This may be due to the thicker cell wall structure of gram-positive bacteria providing additional protection from light. Since the characteristics of a specific microorganism or food cannot be altered much, research usually focuses on how to improve the effectiveness of these methods by changing the characteristics of the light and the processing conditions. The properties of the light are determined by the type of the lamp and generator. In most of the studies, the energy dose of the light is taken as the main parameter which affects the microbial load of the food exposed to the light. The control of the energy dose also allows making comparisons between different microorganisms, foods, and processing units or conditions. Treatments for prolonged time or extremely high dosage can result in undesirable visual and chemical changes in foods, which will not be discussed here. Below are given some examples of studies on the decontamination of foods using UV and pulsed UV technologies.

UV-light

Fonseca and Rushing (2006) obtained >1 $\log_{10}$ reduction on aerobic plate counts on fresh-cut watermelons after 3 min UV treatment at 4.1 kJ/m^2. On the other hand, the treatment of watermelons in polypropylene packages resulted in <1 $\log_{10}$ reduction. Stevens et al. (1997) reported that the incidence of brown rot (*Monilinia fructicola*), green mold (*Penicillium digitatum*), and Rhizopus soft rot (*Rhizopus stolonifer*) were reduced after UV treatment of peach, tangerine, tomato, and sweet potato. In the study of Fino and Kniel (2008), UV-light treatment (40–240 mW s/cm^2) of strawberries, green onions, and lettuce yielded 1.9–2.6, 2.5–5.6, and 4.5–4.6 $\log_{10}$ reductions in the infectivity of the viruses, feline calicivirus, hepatitis A virus, and Aichi virus, respectively.

Isohanni and Lyhs (2009) investigated the effectiveness of UV light on *Campylobacter jejuni* on broiler meat. The treatments for 14–18 s at UV doses of 32.4–32.9 mW s/cm^2 resulted in the maximum $\log_{10}$ reductions of 0.7, 0.8, and 0.4 on broiler meat, skin, and carcass, respectively. In the study of Sommers et al. (2009), treatment of frankfurters containing sodium diacetate and potassium lactate with UV-light at 4.0 J/cm^2, followed by flash (steam) pasteurization (3 s steam at 121°C) yielded about 3.89 $\log_{10}$ reduction in *Listeria innocua*. *L. innocua* did not grow during storage at 8°C for 8 weeks.

UV light is able to significantly reduce bacteria and molds on shell eggs (Kuo et al. 1997). Coufal et al. (2003) constructed a UV treatment cabinet for eggs, in which the eggs were carried by a conveyor. Treatment of eggs for 4 min with UV-light at 4–14 mW/cm^2 resulted in about 1.3, 4, and 4–5 $\log_{10}$ reductions in aerobic plate counts (APC), *Salmonella typhimurium*, and *E. coli* on hatching eggs, respectively. Unluturk et al. (2010) reported that *E. coli* K12, *E. coli* O157:H7, and *L. innocua* in liquid egg were reduced by 0.896, 1.403, and 0.960 $\log_{10}$ after 20 min exposure to UV light at 1.314 mW/cm^2.

Hakguder (2009) reported more than 5 $\log_{10}$ reduction of *E. coli* K12 in white grape juice at a UV dose of 75.04 mJ/cm^2, and 1.76 $\log_{10}$ reduction of naturally grown microorganisms in fresh squeezed orange juice at a UV dose of 144.36 mJ/cm^2. In the study of Matak et al. (2005) more than 5 $\log_{10}$ reduction of *L. monocytogenes* was obtained in goat milk after 12-time passing of milk through the UV-light system with a cumulative exposure time and dose of ~18 s and 15.8 mJ/cm^2, respectively.

Pulsed UV-Light

Bialka and Demirci (2008) demonstrated that a 60 s treatment with pulsed UV-light at doses ranging from 25.7 to 72 J/cm^2 decreased *E. coli* O157:H7 and *Salmonella* spp. by 3.9 and 3.4 $\log_{10}$ cfu/g on raspberries, and by 2.1 and 2.8 $\log_{10}$ cfu/g on strawberries, respectively. Lagunas-Solar et al. (2006) obtained more than 5 $\log_{10}$ reductions in plant (fungal) pathogens on fresh fruits including apples, kiwi, lemon, nectarines, oranges, peaches, pears, raspberries, and grapes in less than 10 s. The minimal UV dose for complete fungal inactivation varied between 100 and 1900 mJ/cm^2 for each microorganism. Jun et al. (2003) reported that a 100 s treatment at a distance of 3 cm from the pulsed UV strobe could reduce *Aspergillus niger* spores in corn meal by a maximum of 4.93 $\log_{10}$ cfu/g.

Dunn (1996) reported as much as 8 $\log_{10}$ reduction of *S. enteritidis* after treatment of shell egg surfaces with 8 pulses at a total dose of 4 J/cm^2. Keklik et al. (2010b) obtained complete inactivation of *S. enteritidis* on shell-eggs at a pulsed UV dose of 23.6 J/cm^2. In the study of Hierro et al. (2009) a maximum reduction of 3.6 $\log_{10}$ CFU/egg in *S. enteritidis* was obtained on 24 to 80 per cent of the unwashed eggs.

Ozer and Demirci (2006) obtained about 1 $\log_{10}$ cfu/g reduction of *E. coli* O157:H7 and *L. monocytogenes* on salmon fillets after 60 s treatment at 8 cm from the quartz window of the pulsed UV-light system. In the study of Keklik et al. (2009), about 1.6 and 1.5 $\log_{10}$ cfu/cm^2 reductions of *L. monocytogenes* were obtained for unpackaged and vacuum-packaged chicken frankfurters after 60 s treatments at 8 cm from the quartz window corresponding to energy doses of 55.9 and 48.4 J/cm^2, respectively.

About 2 $\log_{10}$ reductions of *S. typhimurium* were obtained on both unpackaged and vacuum-packaged chicken breast after only 15 and 30 s treatments at 5 cm distance from the quartz window (Keklik et al. 2010a).

In another study of Keklik et al. (2011), the treatment of whole chicken carcasses in a pilot-scale pulsed UV-light system, in which the chickens were moved by a linear rail between two sets of pulsed UV-light systems on each side, resulted in reductions of 0.87 to 1.43 $\log_{10}$ cfu/ml rinse solution in *E. coli* K12 after 30 s and 180 s treatments, corresponding to the conveyor velocities of 78 and 13 cm/min, respectively.

Krishnamurthy et al. (2007) obtained complete inactivation of *S. aureus* using a flow-through pulsed UV-light system both after single pass at 20 ml/min flow rate at 8 cm distance from quartz window (at 0.98 W/cm^2), and after two passes at 20 ml/min at 11 cm sample distance (0.80 W/cm^2). Sauer and Moraru (2009) tested the effectiveness of pulsed UV-light on *E. coli* strains in apple juice and cider in static and turbulent flows. Low turbulence and high turbulence corresponded to velocities of 500 and 3000 rpm, respectively. At high turbulence, inactivation levels of 5.76 and 7.15 $\log_{10}$ were obtained for *E. coli* in apple cider and apple juice, below the energy dose of 12 J/cm^2.

Modeling of Microbial Inactivation by UV and Pulsed UV-Light

An important field in food microbiology area is the predictive microbiology, which aims at describing the responses of microorganisms to their environments by using mathematical models. In predictive microbiology, the responses of the microbial populations to the environmental factors are considered to be reproducible, which makes it possible to predict microbial behavior in similar environments from past observations by characterizing identifiable, dominant environmental factors (Ross and McMeekin 1994, Baranyi et al. 1999). Predictive microbiology offers many benefits such as providing tools for Hazard Analysis and Critical Control Points (HACCP) implementation, exposure assesment in microbial risk assessment, and decision support in various aspects of the management of microbiological food safety and quality (Baranyi et al. 1999, Cassin et al. 1998, Ross and McMeekin 1994, Whiting and Buchanan 1996). In this section, the modeling aspects including examples of inactivation models and their applications for UV and pulsed UV treatment of foods will be explained.

Inactivation models

Microbial inactivation by different lethal agents is predicted by mathematical models using survival curves, which are obtained by plotting the survival ratio (Eqn. 1) versus the treatment time. The fit of the predicted survival ratios to the experimentally obtained data is one of the main indicators of how good a model can estimate the inactivation. Curve fitting is performed generally using computer programs, in which regression analysis is done for a series of experimental and predicted data to find the "best fit" for a model. The goodness-of-fit is determined by using parameters such as the determination coefficient (R^2), mean square error (MSE), root mean square error (RMSE), and per cent RMSE (as the percentage of the actual mean survival ratio) (Buzrul and Alpas 2004, Chen and Hoover 2003, Keklik et al. 2012). Residual plots are also useful for the identification of any non-linearity or non-constant variance in a model (Mellefont et al. 2003). Survival ratio ($S(t)$) is defined as follows:

$$S(t) = N_t/N_0 \tag{1}$$

with N_t as the number of survivors after exposure time, t and N_0 as the initial number of microorganisms.

Lethal agents or stresses behind the inactivation can be chemical or antimicrobial agents or disinfection processes such as heat, high pressure, electric field, UV, and pulsed UV light. In the construction of primary models, the intensity of the lethal stress is kept constant to observe the survival trend over time for that specific treatment (Baranyi and Roberts 2004, Peleg 2000, Alvarez et al. 2003). For example, temperature is kept constant in modeling the inactivation by heat treatment, while for chemical or antimicrobial agents it is generally the concentration being constant. In the case of UV and pulsed UV light, the intensity refers to the irradiance, while the total energy dose or fluence delivered by the lamp is used instead of time. On the other hand, seconday models are developed to determine the effects of environmental factors on the primary model's parameters, which are usually simple, empirical, multivariate polynomials (Baranyi and Roberts 2004).

Generally, four different types of microbial survival curves can be obtained: linear, concave downward (with a shoulder), concave upward (with a tail), and sigmoidal. The non-linear types of survival curves are given in Fig. 3. The linear survival curve indicates all microbial cells in a population are assumed to have identical resistance to a lethal agent (Bialka et al. 2008a, Chen 2007). The non-linear survival curves were interpreted by Peleg (2000) as follows: Downward concavity (Fig. 3A) indicates the surviving cells are weakened by the accumulated damage increasing the destruction rate with exposure time. Upward concavity (Fig. 3B) suggests that weaker or more sensitive cells in the population are killed faster leaving more resistant survivors behind. For the sigmoidal curves, there are two scenarios. In the first case (Fig 3C), the resistant cells are weakened to a large extent by the treatment, but the lethal effect becomes noticeable only after the sensitive members of the population have been mostly destroyed. In the other case (Fig 3D), the lethal stress weakens the resistant members of the population to a much lesser extent compared to the sensitive ones.

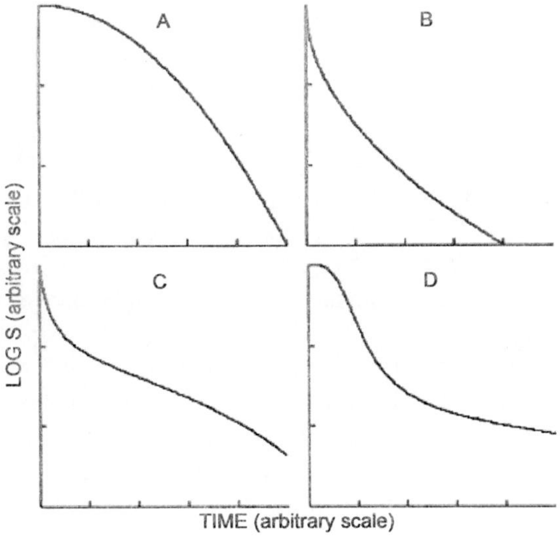

Figure 3. Typical shapes of the microbial survival curves (Reprinted from Food Research International, 32, M. Peleg, On calculating sterility in thermal and non-thermal preservation methods, 271–278, Copyright (1999), with permission from Elsevier).

Traditionally, the first order kinetics is used to decribe the microbial inactivation by heat treatment (Peleg 2003). The widely accepted method of *D* and *z* values, developed by Bigelow, Ball and Stumbo (Stumbo 1973) and commonly practiced in the canning industry, is also based on the first order kinetics (van Boekel 2002). The survival curve for this type of inactivation is linear, and the model based on the first order kinetics is called the first-order or log-linear model with the following equation:

$$logS(t) = -kt = -\frac{t}{D} \quad (2)$$

with *k* as a rate constant (*s* or min) and *D* as the decimal reduction time (1/*k*). Although the first-order kinetics is a useful method of modeling inactivation in thermal processes, many survival curves do not seem to obey the first order kinetics, which is thus invalid for clearly nonlinear survival curves (van Boekel 2002, Peleg 2003).

In order to describe nonlinear survival curves, many models have been developed. The Weibull distribution, which is one of the widely used methods in reliability engineering to describe the lifetime of electronic and mechanical systems, is also accepted as one of the most flexible and convenient functions to describe microbial inactivation (van Boekel 2002, Peleg 1999). Models based on the Weibull distribution have been reported to be successful in describing microbial survival during both thermal and non-thermal treatments (Chen 2007, Alvarez et al. 2003, Jagannath et al.

2005, Ugarte-Romero et al. 2006, San Martin et al. 2007, Buzrul et al. 2008). The Weibull model equation is given as follows:

$$logS(t) = -\frac{1}{2.303}\left(\frac{t}{\alpha}\right)^{\beta} \tag{3}$$

with α as the scale parameter (characteristic time) (s), and β as the shape parameter (unitless). β gives the direction of the concavity by indicating whether the death rate is increasing, constant, or decreasing with treatment time. The survival curve is concave upward if $\beta<1$, linear (similar to the log-linear plot) if $\beta=1$, and concave downward if $\beta>1$ (Fig. 4). At a constant value of β, a smaller α indicates a higher amount of inactivation at a specific treatment time (Keklik et al. 2012). The reliable life, t_R, is defined as the time required to achieve 1-log_{10} reduction in microbial population (analogous to the *D*-value of the log-linear model) (van Boekel 2002, Keklik et al. 2012). The typical t_R equation for the Weibull model is given as follows:

$$t_R = \alpha(2.303)^{\frac{1}{\beta}} \tag{4}$$

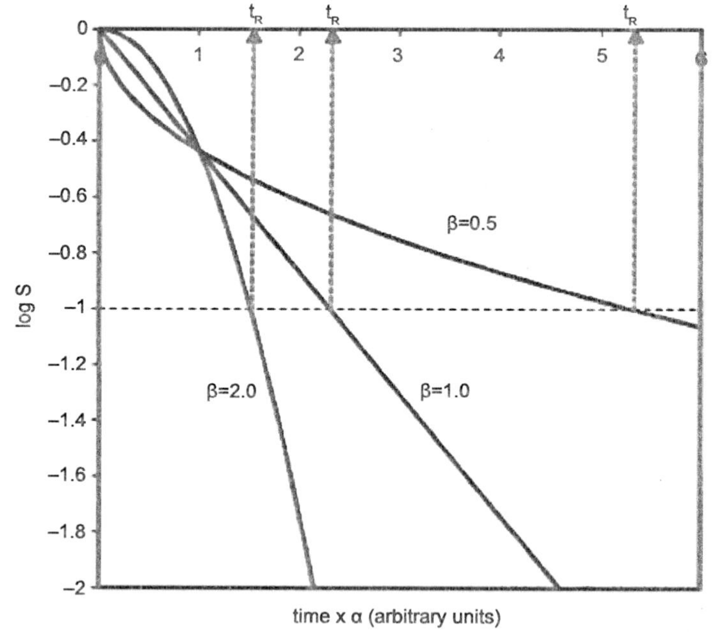

Figure 4. Schematic plot of survival curves obtained using the Weibull model with three different values of the shape parameter, β. The dashed lines indicate the time for 90 per cent reduction of bacterial counts for each β value (Reprinted from International Journal of Food Microbiology, 74, M.A.J.S van Boekel, On the use of the Weibull model to describe thermal inactivation of microbial vegetative cells, 139–159, Copyright (2002), with permission from Elsevier).

The first order model can reasonably predict the microbial inactivation in liquid foods in UV reactors when the fluence is within certain limits. However, at high or low fluence the first order model does not seem to fit the inactivation data well (Severin et al. 1984). The series-event model is a model capable of describing nonlinearity, which accounts for the lag at low fluence (Koutchma et al. 2009, Ye et al. 2007). It was proposed by Severin et al. (1983), which assumes the inactivation takes place in a stepwise fashion. At each step, the inactivation rate is the first order with respect to fluence rate, I:

$$\frac{dN_i}{dt} = k_{SE}I(N_{i-1}-N_i) \tag{5}$$

with subscript i as the event level and k_{SE} as the inactivation constant (assumed to be identical for different event levels). If k_{SE} and I are constant, the concentration of the viable or surviving microorganisms (N) can be determined by (Koutchma et al. 2009, Ye et al. 2007):

$$\frac{N}{N_0} = exp(-k_{SE}I\,t)\sum_{i=0}^{n-1}\frac{[(k)]_{SE}\,I\,t)^i}{t!} \tag{6}$$

with n as the treshold. If n is equal to 1, then Eq. 6 is reduced to the first-order model. The series-event model implies that more than one hit is needed for the inactivation of a microbial cell. Due to the low fluence, at the beginning of the UV inactivation, the probability of a microbial cell to obtain n hits is low. However, as the inactivation continues, the number of microorganisms that have received n–1 hits and require only one additional hit to be completely inactivated keeps increasing (Harm 1980), so the inactivation curve becomes steeper with higher fluence (Koutchma et al. 2009, Ye et al. 2007).

The microbial survival may show nonlinearity either due to the normal characteristics of the inactivation process or to the artifacts and limitations in the experimental procedure, e.g., variability in heating procedures, mixed culture or population, protective effect of dead cells, enumeration method, and poor statistical design (McKellar and Lu 2004). There are a number of other mathematical expressions that have been used in model development to describe the nonlinear inactivation by UV and pulsed UV light. Examples to these models will be given in the section 'Model Application'.

Evaluation/Validation methods

Fitting of a model to a set of experimental survival data is the first step in assessing that model's ability of describing the inactivation. However, the goodness-of-fit parameters are only valid for the experimental data that is used for the model development. In order to test the reliability of a model, its performance needs to be evaluated by applying the constructed model on another set of data obtained at similar environmental conditions, which can be sub-sets of the data set which the model is derived from, new data by experiments in laboratory media or in food product, other data in the literature, and trials in industry (Mellefont et al. 2003). A common method to assess the performance of a model is to plot the predicted versus observed data, and then do linear regression. The model fits perfectly if the R^2 and slope of the line are both equal to 1.

Ross (1996) proposed two indices, the bias factor and the accuracy factor to evaluate the performance of mathematical models used in predictive microbiology. The bias factor (B_f) is expressed as the antilogarithm of the average of the logarithm of the ratio between the predicted and observed values (Mellefont et al. 2003). In the sense of microbial inactivation; B_f indicates the relative average deviation of predicted $\log_{10}$ reductions from the observed values. On the other hand, the accuracy factor (A_f) indicates the spread of the results about the predicted value (Ross 1996). The equations for B_f and A_f are given below:

$$B_f = 10^{\frac{\sum log\left(\frac{predicted}{observed}\right)}{n}} \tag{7}$$

$$A_f = 10^{\frac{\sum \left|log\left(\frac{predicted}{observed}\right)\right|}{n}} \tag{8}$$

with n as the number of observations used in calculation. The value of B_f indicates how much, on average, the observed $\log_{10}$ reduction is overestimated ($B_f>1$) or underestimated ($B_f<1$) by the model, whereas A_f indicates the magnitude of the difference between the predicted and observed $\log_{10}$ reductions. The model fits perfectly if B_f and A_f are both equal to 1. Ross et al. (2000) suggested that each environmental variable in a model adds an estimated relative error of 10 per cent. Thus, an acceptible value of A_f for the model predicting the effect of pulsed UV dose or fluence on the microbial survival would be 1.1 (Keklik et al. 2012).

Mellefont et al. (2003) considered the underprediction of the generation time of *E. coli* as "fail-safe" and overprediction as "fail-dangerous". Similarly, the underprediction of the $\log_{10}$ reduction could be regarded as fail-safe as the model predicts on average lower inactivation levels than those actually observed. On the other hand, overprediction is 'fail-dangerous' because the model predicts higher inactivation levels than observed. A B_f value <1 indicates underprediction (fail-safe), while a B_f value >1 indicates overprediction (fail-dangerous). A low B_f means the prediction is fail-safe, which, however, may be due to the model being too conservative. A B_f value of 1 indicates the predicted and observed values are in perfect agreement, however, at this point the underestimated and overestimated values will tend to cancel out each other due to the opposite signs of the logarithm of the ratios. Thus, $B_f = 1$ is interpreted as indicating no systematic over or underprediction (Mellefont et al. 2003). For evaluating model performances involving pathogens, Ross (1999) proposed that B_f values in the range of 0.90–1.05 can be considered good; 0.70–0.90 or 1.06–1.15 acceptable; and <0.70 or >1.15 unacceptable.

The data used in constructing and evaluating a model are generally obtained in laboratory conditions, where laboratory culture media are frequently used instead of foods (Pinon et al. 2004). However, the inactivation model developed for a specific treatment needs to be validated for the successful application of the model for a real case of food processing. Evaluation of the models via performance tests such as those using indices such as the bias and accuracy factors, can be considered as 'a first step towards the development of an objective and useful definition of the term "validated model" in predictive food microbiology' (Ross 1996, Baranyi et al. 1999). Nevertheless, the performance evaluation of a model is often confused with the validation (Mellefont et al. 2003). The term 'validation' is more suitable for the evaluation of a model in industrial trials in the processing environment with the application of objective criteria.

Model Applications

Researchers have focused on different mathematical models to describe the microbial inactivation by UV and pulsed UV light. Since there are many factors playing a role in microbial inactivation by UV or pulsed UV-light such as the characteristics of the light, food, and microorganism as well as the processing conditions, it is hard to expect a single model to be able to fully define the UV or pulsed UV inactivation in general. Rather, a model specific to a certain treatment or processing seems to be more appropriate. In this section, various examples of model applications in UV and pulsed UV-light studies will be summarized.

UV-light

Generally, the UV dose or fluence is calculated by multiplying the intensity or the irradiance (W/cm^2) by the exposure time. However, for the processing of fluid foods, UV dose is expressed in J/ml, and calculated by determining the residence time of microorganisms inside the reactor. The residence time is calculated as the theoretical retention time (TRT) by dividing the volume of the reactor by the flow rate. However, for non-ideal reactors where a residence time distribution (RTD) exists among different fractions of the flow, TRT could lead to important errors (Gayán et al. 2011). In that case, the mean residence time of the flow needs to be calculated from experimental analyses (Gayán et al. 2011).

In the study of Ye et al. 2007, the UV inactivation kinetics of *Yersinia pseudotuberculosis* and *E. coli* K12 in annular reactors were studied. Both first order and series-event inactivation models were used to model the UV inactivation. The series-event inactivation model was found better in describing the inactivation compared to first-order model based on the R^2, standard deviation of $\log_{10}$ reductions, and the maximum absolute and relative errors.

In the study of Ngadi et al. (2003), the effects of pH, fluid depth, and UV-light dose on the inactivation of *E. coli* O157:H7 in apple juice and egg white were investigated. Instead of a first-order model, a two-phase kinetic model was used to describe the inactivation, where the inactivation rate at higher fluid depths was controlled by the second, slower inactivation phase, leading to a lower overall inactivation.

Based on this model, the UV inactivation of *E. coli* O157:H7 was expressed in terms of the inactivation ratio X_r (X/X_0; where X is the cell population surviving UV treatment and X_0 is the original cell population) as (Ngadi et al. 2003):

$$X_r = X_c\, e^{-k_1 d} + (1 - X_c)\, e^{-k_2 d} \qquad (9)$$

with X_c as the critical survival ratio k_1 and k_2 the characteristic kinetic constants for the first and second inactivation phases, respectively; d as the delivered UV dose, I as the UV intensity, t as the exposure time (min).

Although the inactivation of *E. coli* O157:H7 cells was initially rapid, tailing was observed at higher UV doses (Fig. 5). The pH difference of apple juice and egg white did not affect the inactivation. The kinetic parameters indicated that the first phase, characterized by higher reduction rates, was important for fluid depths of 1 and 3.5 mm. On the other hand, the second inactivation phase, characterized by slower inactivation rates, was dominant for fluid depths of 5 and 10 mm. Ngadi et al. (2003) attributed this kinetic pattern to the increased UV transmission with decreasing fluid depth.

In the study of Unluturk et al. (2010), the efficiency of UV-C irradiation was evaluated for liquid egg white as a non-thermal pasteurization process. The inactivation data obtained for the populations of *E. coli* K-12, *E. coli* O157:H7 and *L. innocua* were modeled using Weibull, Log-Linear, Hom, and modified Chick Watson models. Based on the values of RMSE and R^2, the

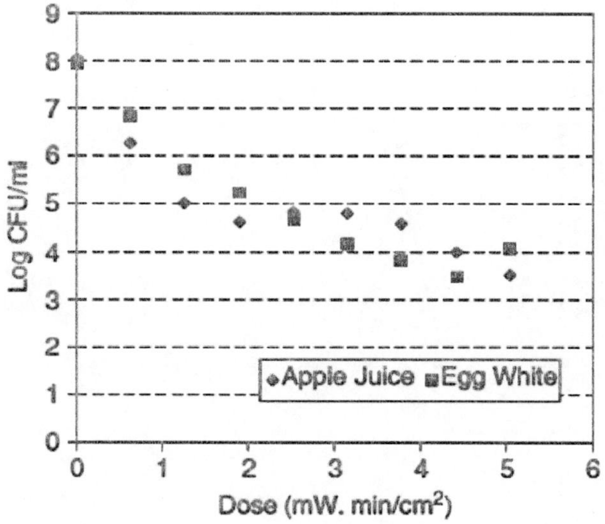

Figure 5. Typical comparison of *E. coli* O157:H7 survival in apple juice and egg white products following UV treatment (fluid depth 1 mm) (Reprinted from *Journal of the Science of Food and Agriculture*, 83, M. Ngadi, J.P. Smith, and B. Cayouette, Kinetics of ultraviolet light inactivation of *Escherichia coli* O157:H7 in liquid foods, 1551–1555, Copyright (2003), with permission from Elsevier).

inactivation of all microbial populations was best described by modified Chick Watson model, which was expressed by the following equation (Marugán et al. 2008):

$$log \frac{N}{N_0} = -k_1[1-\exp(-k_2 t)] \tag{10}$$

with k_1 as the inactivation rate constant (min^{-1}), and k_2 as the first-order UV decay constant (min^{-1}).

Quintero-Ramos et al. (2004) studied the interactions of UV-light dose and apple cider pH on the inactivation of *E. coli*, and used two kinetic models to describe the inactivation: the target model describing exponential inactivation and the logistic regression model describing a sigmoidal curve with an initial and tailing phase. The UV dose was highly significant in the inactivation, while the pH did not significantly affect the inactivation at higher UV doses. The observed survival curve was sigmoidal (Fig. 6), and thus the target theory, which is based on the first order kinetics, could not describe the inactivation. Instead, the logistic regression model adequately fitted the experimental inactivation data. The mathematical expression for this model was given by Quintero-Ramos et al. (2004) as follows:

$$Y = a + \frac{c}{1 + e^{-b(D-m)}} \tag{11}$$

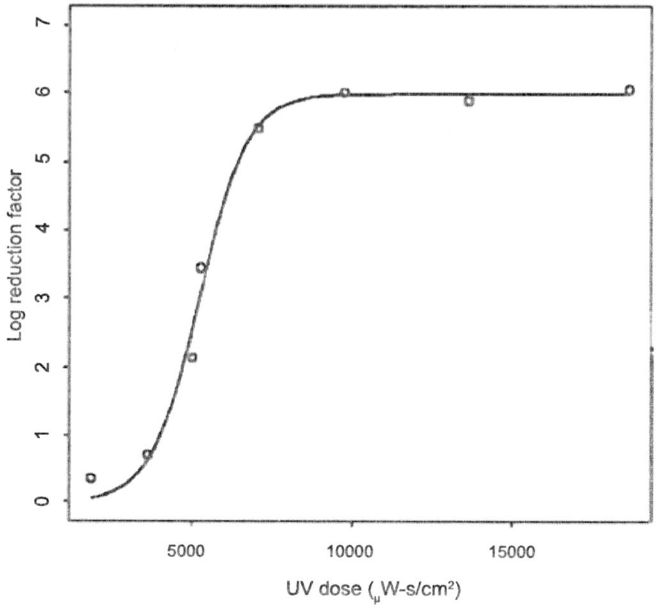

Figure 6. Fitted curve for the inactivation of *E. coli* in apple cider exposed to varying UV doses (Reprinted with permission from the Journal of Food Protection. Copyright held by the International Association for Food Protection, Des Moines, Iowa, USA. A. Quintero-Ramos, J.J. Churey, P. Hartman, J. Barnard and R.W. Worobo. 2004. Modeling of *Escherichia coli* inactivation by UV irradiation at different pH values in apple cider 67: 1153–1156).

with Y as $\log(N/N_0)$, D as the irradiation dose (mJ/cm^2), a as the initial survival fraction at zero dose, b as the slope parameter, c as the upper asymptote, and m as the UV dose given the maximum death rate. Quintero-Ramos et al. (2004) interpreted the survival curve in Fig. 6 as follows: The bacteria survive at low UV doses; however, the response curve becomes a straight line when the critical UV dose is surpassed; and the inactivation rate is decreased at higher UV doses due to the approach of the kill rate to the initial inoculation levels.

In the study of Oteiza et al. (2010), the effect of yeast concentration on ultraviolet inactivation of five strains of *E. coli* O157:H7 inoculated individually or as cocktail in orange juice was investigated. As the yeast concentration increased, higher UV doses were required to inactivate the bacterial strains due to the higher UV absorption coefficients. Different mathematical models including log-linear, logistic, modified Gompertz, and Weibull were used to describe the inactivation. While the UV inactivation of individual strains was fitted successfully by the first-order model, the inactivation of the five-strain cocktail deviated from the log-linear equation. Based on the MSE values, the logistic equation (Eqn. 12) yielded the best fit to the survival curves of the five strain cocktail (Oteiza et al. 2010).

$$\log N = \frac{\log N_0}{1 + \left(\frac{E}{E_i}\right)^b} \tag{12}$$

with E as the radiation dose (J/cm^2), E_i as the dose required to decrease the microbial population by 50 per cent, and b as the parameter called 'Hill slope' or 'Hill coefficient', which describes the steepness of the curve. The slope of the logistic equation (Eqn. 12) indicates the effect of the UV dose on the microbial population. For the strain cocktail, this slope was not constant and depended on $\log(N_0)$, b and E_i. Oteiza et al. (2010) suggested that the tailing of the survival curve might be due to the differences in the susceptibility of the multiple strains of *E. coli* O157:H7 to UV-light.

Pulsed UV-light

Sharma and Demirci (2003) investigated the effects of treatment time, distance from the strobe, and the product thickness on the reduction of *E. coli* O157:H7 on alfalfa seeds by pulsed UV light. The bacterial survival was predicted by empirical models developed using the response surface model. The regression coefficients were determined by doing response surface regression analysis using uncoded units. The general equation for the model was given as:

$$y = \alpha_0 + \alpha_1 x_1 + \alpha_2 x_2 + \alpha_3 x_1^2 + \alpha_4 x_2^2 + \alpha_5 x_1 x_2 \tag{13}$$

with y as the $\log_{10}$ reduction of *E. coli* O157:H7 on alfalfa seeds, α_0 as the constant term, $\alpha_{1\ldots5}$ as the coefficients, x_1 as the treatment time, and x_2 as the thickness of the seed layer (or distance from the UV strobe). Based on Eq. 13, two models, the "thickness model" and the "distance model", were developed. This study demonstrated that the distance, treatment time (number of pulses) and seed thickness considerably influenced the inactivation of *E. coli* O157:H7 on alfalfa seeds. The inactivation was higher when the distance from the UV strobe was shorter, treatment time was longer, and the thickness of the alfalfa seeds was smaller. Sharma and Demirci (2003) also indicated that up to 8 cfu/g $\log_{10}$ reduction of *E. coli* O157:H7 could be achieved under ideal conditions (e.g., higher initial bacterial population and no thermal effects of the pulsed UV-light) based on the predictions obtained by the model.

Krishnamurthy et al. (2007) studied the inactivation of *Staphylococcus aureus* in milk by a continuous-flow pulsed UV-light system. A surface response model with constant linear interactions and squared terms based on (Box–Behnken) method was used to predict the $\log_{10}$ reduction, and correlate it with the distance from the quartz window, number of passes, and flow rate. The equation for the model was given as follows:

$$log_{10}\, reduction = \beta_0 + \beta_1 D + \beta_2 P + \beta_3 F + \beta_{11} D^2 + \beta_{22} P^2 + \beta_{33} F^2 + \beta_{12} DP + \beta_{13} DF + \beta_{23} PF + \varepsilon \quad (14)$$

with D as the distance from the quartz window (cm), P as the number of passes, F as the flow rate (mL/min), ε as the error, and $\beta_{0...23}$ as the coefficients. The model was fit reasonably well to a second independent data set with an R^2 of 0.836.

Bialka et al. (2008b) studied the pulsed UV inactivation of *E. coli* K12 in agar and whey protein as two solid model systems for clear and opaque media, respectively. The survival curves indicated that the relationship between the energy dose and inactivation was nonlinear. Three approaches based on the Weibull model were used to describe the inactivation as a function of the energy dose, treatment time, and material depth. The first model (Weibull model 1) was based on Eq. 3 with t replaced by D, the energy dose (J/cm²):

$$log_{10}(S) = -\frac{1}{2.303}\left(\frac{D}{\alpha}\right)^{\beta} \quad (15)$$

Weibull model 1 was able to predict the reductions of *E. coli* K12 based on R^2 and RMSE values. The second model (Weibull model 2) additionally solved for the extinction coefficient value in the exponential dose model, in which the energy dose in Eq. 15 was redefined with respect to material depth and treatment time:

$$log_{10}(S) = -\left(\frac{I_0 e^{-kx} * time}{\alpha}\right)^{\beta} * \frac{1}{2.303} \quad (16)$$

with I_0 as the initial intensity at the top of the sample (J/cm²s), k as the extinction coefficient (mm^{-1}), and x as the material depth (mm). This model yielded higher R^2 values and lower RMSE values. However, it underestimated the energy dose.

Based on the issues related to the use of broadband energy dose as an indicator of microbial inactivation, a third model (Weibull model 3) was built to fit the individual depth data, in which each depth was treated as a unique scenario because of the different composition of wavelengths at each depth (Bialka et al. 2008b). Non-linear least squares regression was used to determine the Weibull distribution parameters at each depth to minimize the sum of square error. The relationship between the parameters and the material depth, which followed sigmoidal curves, was modeled by using a least squares polynomial fit. The Weibull model 3 more accurately predicted the inactivation and dose, but was overfitted.

Bialka et al. (2008b) concluded that each model has its own advantages and drawbacks, but for all three models, the broadband energy was not a suitable predictor of microbial reduction. This study also demonstrated that pulsed UV-light is able to penetrate up to 10 mm deep in opaque materials with decreasing energy levels.

Also, Bialka et al. (2008a) estimated the inactivation of *E. coli* O157:H7 and *S. enterica* on raspberries and strawberries exposed to pulsed UV-light by using log-linear and Weibull models. The models were built by using the averages of two experimental data sets, and a third data set was back-predicted by the models. The survival curves obtained exhibited concavity (Fig. 7), and thus were described more accurately by the Weibull model compared to the log-linear model based on the values of RMSE, R^2 and the slope.

Keklik et al. (2012) also used the log-linear model and the Weibull model to predict the inactivation of the pathogens on poultry products. The model equations were the same as Eqs. 2 and 3, except t (time) was replaced by D (the energy dose) in both equations. RMSE and per cent RMSE were used as goodness-of-fit parameters, while R^2, slope, bias factor (B_f), and accuracy

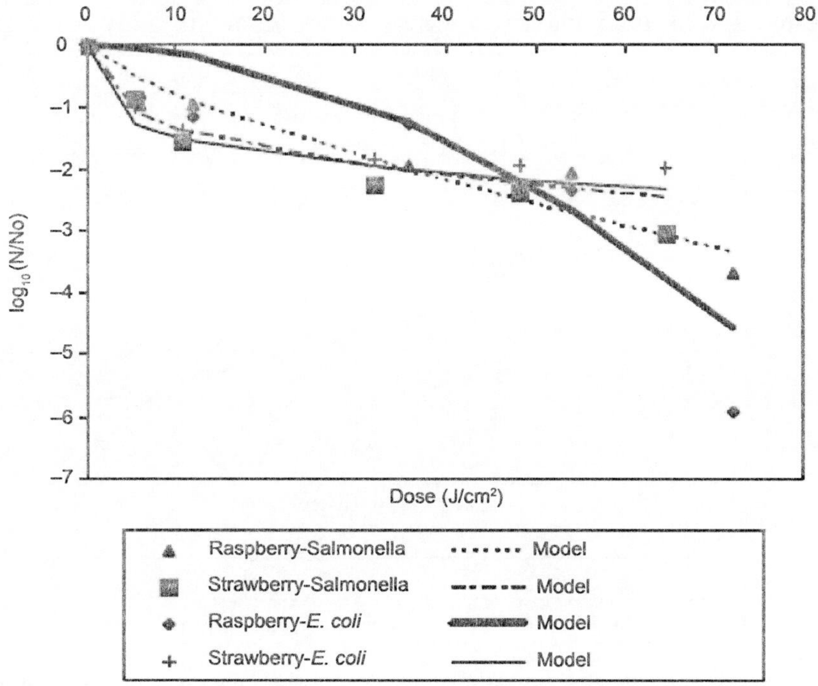

Figure 7. Weibull fit for the pulsed UV inactivation data (Reprinted from *Journal of Food Engineering*, 85, K.L. Bialka, A. Demirci, and V.M. Puri, Modeling the inactivation of *Escherichia coli* O157:H7 and *Salmonella enterica* on raspberries and strawberries resulting from exposure to ozone or pulsed UV-light, 444–449, Copyright (2008), with permission from Elsevier).

factor (A_f) were used as indices for the evaluation of model performance. The inactivation of *S. typhimurium* on unpackaged and vacuum-packaged chicken breast and *L. monocytogenes* on unpackaged and vacuum-packaged chicken frankfurters all exhibited concavity, and thus were described more accurately by the Weibull model compared to the log-linear model. On the other hand, the survival curve of *S. enteritidis* on shell-eggs followed a sigmoidal shape (Fig. 8), for which a third model was also constructed. This model named as 'the sigmoidal model' has four parameters (b, n, k_1, and k_2), and had been suggested by Peleg (2003) with the following equation:

$$\log S(t) = -\frac{[1+bt^n]t}{k_1 + k_2 t} \tag{17}$$

where t was again replaced with D. However, the model parameters in Eq. 17 do not have biological significance and are strongly correlated, which causes the parameter errors to be balanced by each other (Avsaroglu et al. 2007). Furthermore, the outcome of the regression can be extremely sensitive to a small scatter or error in the experimental data due to the involvement of four adjustable parameters, which is why the survival ratios need to be determined accurately and at close intervals (Peleg 2003). Keklik et al. (2012) suggested that the inactivation patterns of microorganisms may be affected

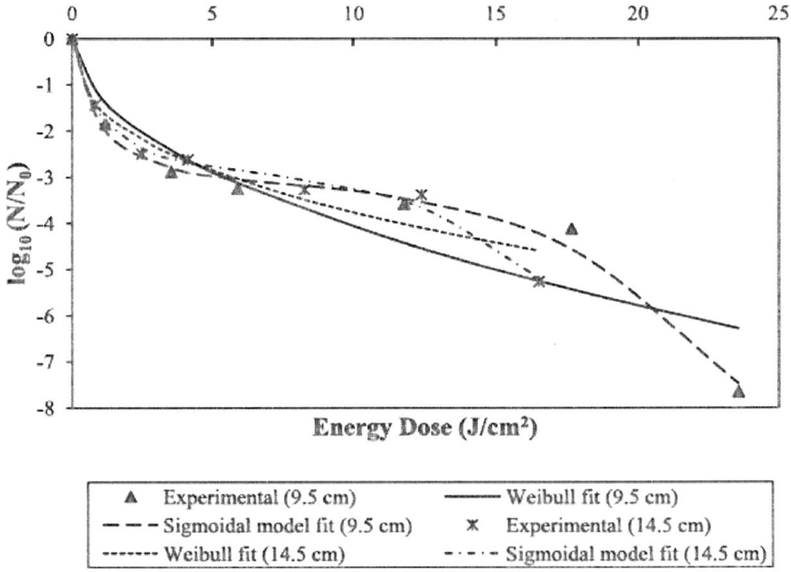

Figure 8. Fit of the Weibull and sigmoidal model to the inactivation data of *S. enteritidis* on shell-eggs exposed to pulsed UV-light (Reprinted with permission from the *Journal of Food Protection*. Copyright held by the International Association for Food Protection, Des Moines, Iowa, U.S.A. N.M. Keklik, A. Demirci, V.M. Puri and P.H. Heinemann. 2012. Modeling the inactivation of *Salmonella* Typhimurium, *Listeria monocytogenes*, and *Salmonella* Enteritidis on poultry products exposed to pulsed UV-light 75: 281–288).

by the surface characteristics of the associated food and the temperature increase during pulsed UV-light treatments.

Izquier and Gómez-López (2011) studied the pulsed light inactivation kinetics of naturally occuring microorganisms (aerobic mesophilic bacteria) in iceberg lettuce, white cabbage, and Julienne-style cut carrots. The log-linear model, Weibull model and the log-linear and tail model were used. The survival curves exhibited no shoulder but tailing. The Weibull model fitted well with the inactivation data based on the values of RMSE, R^2, and slope. For lettuce, the log-linear and tail model was also found suitable.

Hsu and Moraru (2011) developed a method for predicting levels and spatial distribution of microbial inactivation in liquid substrate by pulsed light treatment. *E. coli* or *L. innocua* inoculated into three liquids (apple juice, trypticase soy broth, and Butterfield's phosphate buffer) having different composition and optical properties were treated with pulsed light, in two different geometries (cylindrical and rectangular prismatic). The microbial survival for each organism was estimated by the Weibull model, which was coupled with previously determined fluence equations describing the spatial distribution of light in the pulsed light chamber. Both total fluence (F_{total}) and UV fluence (F_{UV}) were used as treatment dose, for which the Weibull parameters were determined. Maps of spatial distribution of microbial inactivation were generated by a simulation software, which was used to predict the average volumetric inactivation for each substrate. Hsu and Moraru (2011) found that especially in more opaque and turbid substrates (TSB and apple juice), the Weibull model based on F_{UV} predicted the microbial inactivation more accurately compared to that based on F_{total}, with a tendency towards underestimation. Therefore, the authors suggested that this approach could be useful for the processors in designing potential applications of pulsed light such as for the processing of beverages.

Future Aspects

The UV and pulsed UV-light technologies offer many benefits for food decontamination. Not only these techniques are effective on a broad range of microorganisms, but also they can be applied to a variety of food such as raw, minimally processed, cooked, or packaged products. The main challenge before the industrial application of these technologies is the limited penetrability of UV or pulsed UV light into the food. In the case of the treatment of fluid foods, the transmittivity can usually be increased by optimizing the number of lamps, residence time in the reactor, and the type of the flow (laminar or turbulent). However, for solid foods, the UV transmittivity depends mainly on the color and transparency of the food. Therefore, UV and pulsed UV-light technologies are, as of now, considered as surface disinfection techniques for solid foods, particularly with an objective of postprocessing decontamination.

The industrial applications of these technologies will only be successful if pilot-scale and full-scale trials are performed, in which the processing is fully defined with its parameters (system parameters including the wavelenth range, power, number of lamps, positions of lamps and reflectors, reactor size, treatment time); characteristics of the food such as the color, UV transmittivity, surface smoothness of solid foods; and the resistance of the target microorganims) as well as the total cost of the system, operation, and maintenance. Also, the effects of the treatment on the chemical composition and physical quality of foods need to be carefully evaluated. The mathematical modeling of the inactivation in these trials is critically important in order to understand the responses of microorganisms on/in a specific food to UV or pulsed UV-light, which would allow us to control and optimize the processing. Industrial trials need to be conducted to validate the models used to describe the microbial inactivation by UV and pulsed UV-light.

Conclusions

Due to the undesirable effects of thermal processing on the chemical and nutritive quality of foods, researchers continue to develop and evaluate nonthermal processing methods. UV-light is one of these alternative techniques that is FDA-approved for the decontamination of food or food product surfaces, disinfection of potable water used in food processing, and for the reduction of microbial activity in juice products. UV-light inactivates microbial cells by photochemical actions, in which the absorbed light damages nucleic acids, particularly the DNA. There are a number of advantages of UV-light over thermal and other nonthermal methods. First of all, UV-light does not utilize heat to inactivate microorganims, which allows it to be used for the decontamination of minimally processed and raw foods. Another advantage of UV-light is that it is effective on a wide range of microorganisms including bacteria, fungi, bacterial and fungal spores, viruses, protozoa, and algae. Since UV treatment is a dry process, i.e., no water is involved, the operation and cleaning of the processing unit becomes quite easy. The main drawback of UV-light technology is its limited permeability into the food.

Pulsed UV-light technology, on the other hand, is an advanced UV-light technology, which is faster and more efficient than conventional or continuous UV-light due to the dissipation of high peak power in very short time intervals. Moreover, the pulsed UV lamps do not contain mercury, and thus are safer in case of a lamp breakage or explosion. In a pulsed UV-light system, heat is accumulated over time due to the infrared portion of the light, but the temperature of the food product does not significantly increase if the treatment time is kept short. The penetrability of the light is still an issue with pulsed UV-light. Hence, the transparency and color of the food, and for fluid products, the flow rate and type of the flow are among the factors affecting the transmittivity of UV or pulsed UV-light. Having a broad spectrum and high peak power delivered instantanously, pulsed UV-light has a more complex inactivation mechanism involving photochemical, photothermal and photophysical effects, and prevent the self-repair of microorganims by rupturing the cells.

A number of researchers have evaluated the effects of UV and pulsed UV-light on various microorganisms in/on foods. The studies demonstrate that UV and pulsed UV-light have the potential to decontaminate foods. In order to fully integrate these technologies into the food industry, more pilot and full-scale studies need to be carried out, in which the process needs to be optimized in terms of UV transmittivity, number and position of the lamps, elimination of heat (in the case of pulsed UV-light), treatment time or conveyor speed etc.

Modeling of microbial inactivation is a powerful tool to understand the responses of microorganism to the effects of the UV or pulsed UV-light. Many studies have been conducted in which different mathematical models were applied to the microbial inactivation data obtained by UV or pulsed UV treatment of foods. Most of these studies demonstrate that the survival curves tend to be nonlinear. The biggest challenge in modeling inactivation by UV or pulsed UV-light may be that the efficacy of the processing depends on many factors, which somehow affect the microbial responses. This is especially critical for the validation of the models as the processing conditions in an industrial trial will be inevitably different from those in the laboratory.

In conclusion, UV-light and pulsed UV-light are two promising technologies for non-thermal food decontamination. Although UV-light is a more commonly known and practiced technique compared to pulsed UV-light, the latter is more efficient and safer due to the high peak power delivered, short treatment time, lamp involving inert gas instead of mercury, and inactivation mechanism partly preventing cell repair. As the efforts of researchers on the assessment and description of the microbial inactivations by these techniques continue, it will be more feasable to develop innovative processing methods based on UV or pulsed UV-light to successfully maintain the microbiological safety and quality of foods and food products.

Keywords: Pulsed UV light, food decontamination, pathogens, non-thermal processing, modeling of microbial inactivation

References

Aguiar, A., D. Bainbridge, M. Bond, C. Gupta, K. Matthews, J. Plante, E. Rantanen, C. Turner, D. Yang, W. Zhang and D. Drew. 1996. Modeling UV-damage to *E. coli* bacteria. University of Minnesota, Minneapolis, MN. Available from:http://www.ima.umn.edu/preprints/Sept96/1422g.pdf (accessed 04 June 2012).

Alvarez, I., R. Virto, J. Raso and S. Condon. 2003. Comparing predicting models for the *Escherichia coli* inactivation by pulsed electric fields. Innov. Food Sci. Emerg. Technol. 4: 195–202.

Anderson, J.G., N.J. Rowan, S.J. MacGregor, R.A. Fouracre and O. Farish. 2000. Inactivation of food-borne enteropathogenic bacteria and spoilage fungi using pulsed-light. IEEE Trans. Plasma Sci. 28: 83–88.

Avsaroglu, M.D., S. Buzrul, H. Alpas and M. Akcelik. 2007. Hypochlorite inactivation kinetics of lactococcal bacteriophages. LWT–Food Sci. Technol. 40: 1369–1375.

Baranyi, J., C. Pin and T. Ross. 1999. Validating and comparing predictive models. Int. J. Food Microbiol. 48: 159–166.

Baranyi, J. and T.A. Roberts. 2004. Predictive microbiology—quantitative microbial ecology. Culture (March). Available from: http://www.ifr.ac.uk/safety/comicro/Culture_25.pdf (accessed 04 June 2012).

Begum, M., A.D. Hocking and D. Miskelly. 2009. Inactivation of food spoilage fungi by ultra violet (UVC) irradiation, Int. J. Food Microbiol. 129: 74–77.

Bialka, K.L. 2006. Decontamination of berries with ozone and pulsed UV-light. Ph.D. Thesis, Department of Agricultural and Biological Engineering, The Pennsylvania State University, University Park, PA.

Bialka, K.L. and A. Demirci. 2008. Efficacy of pulsed UV-light for the decontamination of *Escherichia coli* O157:H7 and *Salmonella* spp. on raspberries and strawberries. J. Food Sci. 73: M201–M207.

Bialka, K.L., A. Demirci and V.M. Puri. 2008a. Modeling the inactivation of *Escherichia coli* O157:H7 and *Salmonella enterica* on raspberries and strawberries resulting from exposure to ozone or pulsed UV-light. J. Food Eng. 85: 444–449.

Bialka, K.L., P.N. Walker, V.M. Puri and A. Demirci. 2008b. Pulsed UV-light penetration of characterization and the inactivation of *Escherichia coli* K12 in solid model systems. Trans. ASABE. 51: 195–204.

Bischof, M. 1994. The history of bioelectromagnetism. *In:* M.W. Ho, F.A. Popp and U. Warnke (eds.). Bioelectrodynamics and Biocommunication. World Scientific Publishing, Singapore, pp. 1–33.

Blatchley, III E.R. and M.M. Peel. 2001. Disinfection by ultraviolet irradiation. *In:* S.S. Block [ed.]. Disinfection, Sterilization, and Preservation, 5th edn. Lippincott Williams & Wilkins, Philadelphia, PA, USA, pp. 823–851.

Bohrerova, Z., H. Shemer, R. Lantis, C.A. Impellitteri and K.G. Linden. 2008. Comparative disinfection efficiency of pulsed and continuous-wave UV irradiation technologies. Water Research 42: 2975–2982.

Buzrul, S. and H. Alpas. 2004. Modeling the synergistic effect of high pressure and heat on inactivation kinetics of *Listeria innocua*: a preliminary study. FEMS Microbiol. Lett. 238: 29–36.

Buzrul, S., H. Alpas, A. Largeteau and G. Demazeau. 2008. Modeling high pressure inactivation of *Escherichia coli* and *Listeria innocua* in whole milk. Eur. Food Res. Technol. 227: 443–448.

Cassin, M.H., A.M. Lammerding, E.C.D. Todd, W. Ross and R.S. McColl. 1998. Quantitative risk assessment for *Escherichia coli* O157-H7 in ground beef hamburgers. Int. J. Food Microbiol. 41: 21–44.

Cheigh, C.I., M.H. Park, M.S. Chung, J.K. Shin and Y.S. Park. 2012. Comparison of intense pulsed light- and ultraviolet (UVC)-induced cell damage in *Listeria monocytogenes* and *Escherichia coli* O157:H7. Food Control 25: 654–659.

Chen, H. and D.G. Hoover. 2003. Modeling the combined effect of high hydrostatic pressure and mild heat on the inactivation kinetics of *Listeria monocytogenes* Scott A in whole milk. Innov. Food Sci. Emerg. Technol. 4: 25–34.

Chen, H. 2007. Temperature-assisted pressure inactivation of *Listeria monocytogenes* in turkey breast meat. Int. J. Food Microbiol. 117: 55–60.

Coufal, C.D., C. Chavez, K.D. Knape and J.B. Carey. 2003. Evaluation of a method of ultraviolet light sanitation of broiler hatching eggs. Poultry Sci. 82: 754–759.

Demirci, A. and L. Panico. 2008. Pulsed ultraviolet light. Food Sci. Technol. Int. 14: 443–446.

Dunn, J. 1996. Pulsed light and pulsed electric field for foods and eggs. Poultry Sci. 75: 1133–1136.

El-boury, S., C. Couteau, L. Boulande, E. Paparis and L.J.M. Coiffard. 2007. Effect of the combination of organic and inorganic filters on the Sun Protection Factor (SPF) determined by *in vitro* method. Int. J. Pharm. 340: 1–5.

Elmnasser, N., S. Guillou, F. Leroi, N. Orange, A. Bakhrouf and M. Federighi. 2007. Pulsed-light system as a novel food decontamination technology: a review. Can. J. Microbiol. 53: 813–821.

FDA (Food and Drug Administration). 2000. Kinetics of microbial inactivation for alternative food processing technologies, a report of the institute of food technologists for the food and drug administration of the US department of health and human services. Available from: http://www.fda.gov/Food/ScienceResearch/ResearchAreas/SafePracticesforFoodProcesses/ucm100158.htm (accessed 04 June 2012).

FDA (Food and Drug Administration). 2011a. Ultraviolet radiation for the processing and treatment of food. *In:* Irradiation in the production, processing and handling of food. 21CFR179.39. Available from:http://www.accessdata.fda.gov/scripts/cdrh/cfdocs/cfcfr/CFRSearch.cfm?fr=179.39 (accessed 04 June 2012).

FDA (Food and Drug Administration). 2011b. Pulsed light for the treatment of food. *In:* Irradiation in the production, processing and handling of food. 21CFR179.41. Available from: http://www.accessdata.fda.gov/scripts/cdrh/cfdocs/cfcfr/CFRSearch.cfm?fr=179.39 (accessed 04 June 2012).

Federal Register. 1999. Pulsed light treatment of food. Fed. Regist. 66: 338829–338830.

Fine, F. and P. Gervais. 2004. Efficiency of pulsed UV light for microbial decontamination of food powders. J. Food Prot. 67: 787–792.

Fino, V.R. and K.E. Kniel. 2008. UV light inactivation of hepatitis A virus, Aichi virus, and feline calicivirus on strawberries, green onions, and lettuce. J. Food Prot. 71: 908–913.

Fonseca, J.M. and J.W. Rushing. 2006. Effect of ultraviolet-C light on quality and microbial population of fresh-cut watermelon. Postharv. Biol. Technol. 40: 256–261.

Gayán, E., S. Monfort, I. Álvarez and S. Condón. 2011. UV-C inactivation of *Escherichia coli* at different temperatures. Innov. Food Sci. Emerg. Technol. 12: 531–541.

Hakguder, B. 2009. UV disinfection of some of the fruit juices, decontamination of berries with ozone and pulsed UV-light. M.Sc. Thesis, Department of Food Engineering, Izmir Institute of Technology, Izmir, Turkey.

Harm, W. 1980. Biological effects of ultraviolet radiation. Cambridge University Press, Cambridge, UK.

Hidaka, Y. and K. Kubota. 2006. Study on the sterilization of grain surface using UV radiation —development and evaluation of UV irradiation equipment. Jpn. Agr. Res. Q. 40: 157–161.

Hierro, E., S. Manzano, J.A. Ordonez, L. de la hoz and M. Fernandez. 2009. Inactivation of *Salmonella enterica* serovar Enteritidis on shell eggs by pulsed light technology. Int. J. Food Microbiol. 135: 125–130.

Hsu, L. and C.I. Moraru. 2011. A numerical approach for predicting volumetric inactivation of food borne microorganisms in liquid substrates by pulsed light treatment. J. Food Eng. 105: 569–576.

Isohanni, P.M.I. and U. Lyhs. 2009. Use of ultraviolet irradiation to reduce *Campylobacter jejuni* on broiler meat. Poultry Sci. 88: 661–668.

Izquier, A. and V.M. Gómez-López. 2011. Modeling the pulsed light inactivation of microorganisms naturally occurring on vegetable substrates. Food Microbiol. 28: 1170–1174.

Jagannath, A., T. Tsuchido and J.M. Membre. 2005. Comparison of the thermal inactivation of *Bacillus subtilis* spores in foods using the modified Weibull and Bigelow equations. Food Microbiol. 22: 233–239.

Jun, S., J.M. Irudayaraj, A. Demirci and D. Geiser. 2003. Pulsed UV-light treatment of corn meal for inactivation of *Aspergillus niger* spores. Int. J. Food Sci. Technol. 38: 883–888.

Keklik, N.M., A. Demirci and V.M. Puri. 2009. Inactivation of *Listeria monocytogenes* on unpackaged and vacuum-packaged chicken frankfurters using pulsed UV-light. J. Food Sci. 74: M431–M439.

Keklik, N.M., A. Demirci and V.M. Puri. 2010a. Decontamination of unpackaged and vacuum-packaged chicken breast with pulsed UV-light. Poultry Sci. 89: 570–581.

Keklik, N.M., A. Demirci, P.H. Patterson and V.M. Puri. 2010b. Pulsed UV light inactivation of *Salmonella* Enteritidis on egg shells and its effects on egg quality. J. Food Prot. 73: 1408–1415.

Keklik, N.M., A. Demirci and R.G. Bock. 2011. Surface decontamination of whole chicken carcasses using a pilot-scale pulsed UV light system. Trans. ASABE. 54: 993–1000.

Keklik, N.M., A. Demirci, V.M. Puri and P.H. Heinemann. 2012. Modeling the inactivation of *Salmonella* Typhimurium, *Listeria monocytogenes*, and *Salmonella* Enteritidis on poultry products exposed to pulsed UV light. J. Food. Prot. 75: 281–288.

Koutchma, T., B. Parisi and S. Unluturk. 2006. Evaluation of UV dose in flow-through reactors for fresh apple juice and cider. Chem. Eng. Commun. 193: 715–728.

Koutchma, T.N., L.J. Forney and C.I. Moraru. 2009. Ultraviolet Light in Food Technology—Principles and Applications. CRC Press, Boca Raton, FL, USA.

Krishnamurthy, K., A. Demirci and J. Irudayaraj. 2007. Inactivation of *Staphylococcus aureus* in milk using flow-through pulsed UV-light treatment system. J. Food Sci. 72: M233–M239.

Krishnamurthy, K., J.C. Tewari, J. Irudayaraj and A. Demirci. 2010. Microscopic and spectroscopic evaluation of inactivation of *Staphylococcus aureus* by pulsed UV light and infrared heating. Food Bioprocess Technol. 3: 93–104.

Krishnamurthy, K., J. Irudayaraj, A. Demirci and W. Yang. 2008. UV pasteurization of food materials. *In:* S. Jun and J.M. Irudayaraj [eds.]. Food Processing Operations Modeling: Design and Analysis. CRC Press, Boca Raton, FL, USA, pp. 281–299.

Kuo, F.L., J.B. Carey and S.C. Ricke. 1997. UV irradiation of shell eggs: effect of populations of aerobes, molds, and inoculated *Salmonella* Typhimurium. J. Food Prot. 60: 639–643.

Lagunas-Solar, M.C., C. Pina, J.D. Macdonald and L. Bolkan. 2006. Development of pulsed UV light processes for surface fungal disinfection of fresh fruits. J. Food Prot. 69: 376–384.

Lešková, E., J. Kubíková, E. Kováčiková, M. Košická, J. Porubská and K. Holčíková. 2006. Vitamin losses: Retention during heat treatment and continual changes expressed by mathematical models. J. Food Compos. Anal. 19: 252–276.

Linden, K. 2004. Disinfection Efficiency and Dose Measurement of Polychromatic UV Light. IWA Publishing, London, UK, pp. 1–5.

MacGregor, S.J., N.J. Rowan, L. McIlvaney, J.G. Anderson, R.A. Fouracre and O. Farish. 1998. Light inactivation of food-related pathogenic bacteria using a pulsed power source. Lett. Appl. Microbiol. 27: 67–70.

Marugán, J., R. Grieken, C. Sordo and C. Cruz. 2008. Kinetics of the photocatalytic disinfection of *Escherichia coli* suspensions. Appl. Catal. B: Environ. 82: 27–36.

Matak, K.E., J.J. Churey, R.W. Worobo, S.S. Sumner, E. Hovingh, C.R. Hackney and M.D. Pierson. 2005. Efficacy of UV light for the reduction of *Listeria monocytogenes* in goat's milk. J. Food Prot. 68: 2212–2216.

Mcdonald, K.F., R.D. Curry, T.E. Clevenger, K. Unklesbay, A. Eisenstark, J. Golden and R.D. Morgan. 2000. A comparison of pulsed and continuous ultraviolet light sources for the decontamination of surfaces. IEEE Trans. Plasma Sci. 28: 1581–1587.

McKellar, R.C. and X. Lu. 2004. Primary models. *In:* R.C. McKellar and X. Lu [eds.]. Modeling Microbial Responses in Food. CRC Press, Boca Raton, FL, USA, pp. 21–62.

Mellefont, L.A., T.A. McMeekin and T. Ross. 2003. Performance evaluation of a model describing the effects of temperature, water activity, pH and lactic acid concentration on the growth of *Escherichia coli*. Int. J. Food Microbiol. 82: 45–58.

Melquiades, F.L., D.D. Ferreira, C.R. Appoloni, F. Lopes, A.G. Lonni, F.M. Oliveira and J.C. Duarte. 2008. Titanium dioxide determination in sunscreen by energy dispersive X-ray fluorescence methodology. Analytica Chimica Acta 613: 135–143.

Miller, R.V., J.D. Mitchell and M. Elasri. 1999. Bacterial responses to ultraviolet light. ASM News 65: 535–541.

Ngadi, M., J.P. Smith and B. Cayouette. 2003. Kinetics of ultraviolet light inactivation of *Escherichia coli* O157:H7 in liquid foods. J. Sci. Food Agric. 83: 1551–1555.

Oms-oliu, G., O. Martin-Belloso and R. Soliva-Fortuny. 2010. Pulsed light treatments for food preservation. A review. Food Bioprocess Technol. 3: 13–23.

Oteiza, J.M., L. Giannuzzi and N. Zaritzky. 2010. Ultraviolet treatment of orange juice to inactivate *E. coli* O157:H7 as affected by native microflora. Food Bioprocess Technol. 3: 603–614.

Ozer, N. and A. Demirci. 2006. Inactivation of *Escherichia coli* O157:H7 and *Listeria monocytogenes* inoculated on raw salmon fillets by pulsed UV-light treatment. Int. J. Food Sci. Technol. 41: 354–360.

Palmieri, L., D. Cacace and G. Dall'Aglio. 1999. Non-thermal methods of food preservation based on electromagnetic energy. Food Technol. Biotechnol. 37: 145–149.

Peleg, M. 1999. On calculating sterility in thermal and non-thermal preservation methods. Food Res. Int. 32: 271–278.

Peleg, M. 2000. Microbial survival curves—the reality of flat "shoulders" and absolute thermal death times. Food Res. Int. 33: 531–538.

Peleg, M. 2003. Calculation of the non-isothermal inactivation patterns of microbes having sigmoidal isothermal semi-logarithmic survival curves. Crit. Rev. Food Sci. Nutr. 43: 645–658.

Pinon, A., M. Zwietering, L. Perrier, J.M. Membre, B. Leporq, E. Mettler, D. Thuault, L. Coroller, V. Stahl and M. Vialette. 2004. Development and validation of experimental protocols for use of cardinal models for prediction of microorganism growth in food products. Appl. Environ. Microbiol. 70: 1081–1087.

Quintero-Ramos, A., J.J. Churey, P. Hartman, J. Barnard and R.W. Worobo. 2004. Modeling of *Escherichia coli* inactivation by UV irradiation at different pH values in apple cider. J. Food Prot. 67: 1153–1156.

Reed, N.G. 2010. The history of ultraviolet germicidal irradiation for air disinfection. Public Health Rep. 125: 15–27.

Ross, T. and T.A. McMeekin. 1994. Predictive microbiology—a review. Int. J. Food Microbiol. 23: 241–264.

Ross, T. 1996. Indices for performance evaluation of predictive models in food microbiology. J. Appl. Bacteriol. 81: 501–508.

Ross, T. 1999. Predictive Food Microbiology Models in the Meat Industry (MSRC.003). Meat and Livestock Australia, North Sydney, Australia.

Ross, T., J. Baranyi and T.A. McMeekin. 2000. Predictive microbiology and food safety. *In:* R. Robinson, C.A. Batt and P. Patel [eds.]. Encyclopaedia of Food Microbiology. Academic Press, London, UK, pp. 1699–1710.

Rowan, N.J., S.J. MacGregor, G. Anderson, R.A. Fouracre, L. Mcllvaney and O. Farish. 1999. Pulsed–light inactivation of food-related microorganisms. App. Environ. Microbiol. 65: 1312–1315.

San Martin, M.F., D.R. Sepulveda, B. Altunakar, M.M. Gongora-Nieto, B.G. Swanson and G.V. Barbosa-Canovas. 2007. Evaluation of selected mathematical models to predict the inactivation of *Listeria innocua* by pulsed electric fields. LWT-Food Sci. Technol. 40: 1271–1279.

Sauer, A. and C.I. Moraru. 2009. Inactivation of *Escherichia coli* ATCC 25922 and *Escherichia coli* O157:H7 in apple juice and apple cider, using pulsed light treatment. J. Food Prot. 72: 937–944.

Schaefer, R., M. Grapperhaus, I. Schaefer and K. Linden. 2007. Pulsed UV lamp performance and comparison with UV mercury lamps. J. Environ. Eng. Sci. 6: 303–310.

Sharma, R.R. and A. Demirci. 2003. Inactivation of *Escherichia coli* O157:H7 on inoculated alfalfa seeds with pulsed ultraviolet light and response surface modeling. J. Food Sci. 68: 1448–1453.

Severin, B.F., M.T. Suidan and R.S. Engelbrecht. 1983. Kinetic modeling of U.V. disinfection of water. Water Res. 17: 1669–1678.

Severin, B.F., M.T. Suidan, B.E. Rittmann and R.S. Engelbrecht. 1984. Inactivation kinetics in a flow-through UV reactor. J. WPCF. 56: 164–169.

Solsona, F. and J.P. Mendez. 2003. Ultraviolet radiation. *In:* Water Disinfection. Lima, CEPIS. Panamerican Health Organization, Washington, D.C., USA. Available from: http://www.bvsde.paho.org/bvsacg/fulltext/desinfeccioneng/chapter4.pdf (accessed 04 June 2012). pp. 71–84.

Sommers, C.H., D.J. Geveke, S. Pulsfus and B. Lemmenes. 2009. Inactivation of *Listeria innocua* on frankfurters by ultraviolet light and flash pasteurization. J. Food Sci. 74: M138–M141.

Stevens, C., V.A. Khan, J.Y. Lu, C.L. Wilson, P.L. Pusey, E.C.K. Igwegbe, K. Kabwe, Y. Mafolo, J. Liu, E. Chalutz and S. Droby. 1997. Integration of ultraviolet (UV-C) light with yeast treatment for control of postharvest storage rots of fruits and vegetables. Biol. Control. 10: 98–103.

Stumbo, C.R. 1973. Thermobacteriology in Food Processing. Academic Press, New York, USA.

Ugarte-Romero, E., H. Feng, S.E. Martin, K.R. Cadwallader and S.J. Robinson. 2006. Inactivation of *Escherichia coli* with power ultrasound in apple cider. J. Food Sci. 71: E102–E108.

Unluturk, S., M.R. Atilgan, A.H. Baysal and M.S. Unluturk. 2010. Modeling inactivation kinetics of liquid egg white exposed to UV-C irradiation. Int. J. Food Microbiol. 142: 341–347.

van Boekel, M.A.J.S. 2002. On the use of the Weibull model to describe thermal inactivation of microbial vegetative cells. Int. J. Food Microbiol. 74: 139–159.

Wekhof, A. 2000. Disinfection with flash lamps. PDA J. Pharm. Sci. Technol. 54: 264–276.

Wekhof, A. 2003. Ultra-fast sterilization by disintegration of microorganisms with intense pulsed UV light. Business Briefing: Global Surgery. Technology & Services, pp. 1–5.

Whiting, R.C. and R.L. Buchanan. 1996. Predictive modeling. *In:* N. Doyle, N. Beuchat and N. Montville [eds.]. Food Microbiology: Fundamentals and Frontiers. ASM Press, Washington, DC, USA, pp. 728–739.

WHO (World Health Organization). 2007. 'Food safety and foodborne illness'. Fact sheet No: 237. Available from: http://www.who.int/mediacentre/factsheets/fs237/en/ (accessed 24 April 2012).

Wright, H.B. and W.L. Cairns. 1998. 'Ultraviolet light', in Regional symposium on water quality: effective disinfection, Lima, CEPIS/OPS, 1–26. Available from: http://www.bvsde.paho.org/bvsacg/i/fulltext/symposium/ponen10.pdf (accessed 24 April 2012).

Ye, Z., T. Koutchma, B. Parisi, J. Larkin and L.J. Forney. 2007. Ultraviolet inactivation kinetics of *Escherichia coli* and *Yersinia pseudotuberculosis* in annular reactors. J. Food Sci. 72: E271–278.

Xenon. 2007a. Pulsed UV-light benefits. Xenon Corporation, Wilmington, MA, USA. Available from: http://www.xenoncorp.com/perf_cure.html (accessed 25 April 2012).

Xenon. 2007b. RC-800 Series. UV Curing Systems. Xenon Corporation, Wilmington, MA. Available at: http://uv-curing.xenoncorp.com/Literature/PDF/RC-800%20Series%20 Brochure.pdf (accessed 04 June 2012).

CHAPTER 4

Ultrasound Processing for Food Safety and Preservation

Daniela A. Abela,[1,2] *David Millan Sango,*[1] *Anna McElhatton*[1] and *Vasilis P. Valdramidis*[1,*]

Ultrasound in the Food Industry

Food is rich in nutrients and water causing proliferation of microorganisms present in these types of biotic environments. Food safety is an area of increasing concern as microbial contamination causes food-borne illnesses and may also lead to deterioration of food quality (Chemat and Khan 2011). Therefore, thermal treatment is commonly used to inactivate microorganisms, with the effectiveness of this treatment enhanced with increasing temperature and treatment time (Dolatowski et al. 2007). However, severe thermal treatments lead to higher loss of nutrients and of the organoleptic properties of the food products (Chandrapala et al. 2012). Nowadays consumers demand high quality food that is safe and minimally processed. Thus, there is growing interest in the development of non-thermal technologies that ensure that these demands are met (Cárcel et al. 2012, Ulusoy et al. 2007). One such technology gaining importance in the food industry is ultrasound, a process with significant advantages which include that it is relatively cheap and energy saving (Awad et al. 2012). According to Earnshaw et al.

[1] Department of Food Studies and Environmental Health, Faculty of Health Sciences, Msida MSD 2080, University of Malta, Malta.
Emails: Daniela.A.Abela@mcast.edu.mt; david.millan@um.edu.mt; anna.mcelhatton@um.edu.mt
[2] Institute of Applied Science, MCAST Main Campus, Triq Kordin, Paola PLA 9032, Malta.
* Corresponding author: vasilis.valdramidis@um.edu.mt

(1995), research about the use of ultrasound as a disinfection technology was initiated in the 1960s after the discovery that sound waves that were used in anti-submarine warfare killed fish.

Ultrasound consists of sound waves (Awad et al. 2012) of a specific frequency. This frequency refers to the number of waves passing a stationary point per second (Williams 2012). Wavelength is the distance between two points on consecutive waves while the amplitude of a wave is the height from the equilibrium position to the maximum displacement (either the top of a crest or the bottom of a trough). The sonotrode (in case of the use of ultrasonic horns) is responsible for transmitting ultrasonic vibration into the liquid.

Ultrasound treatment has different applications in the food industry depending on the operating parameters, such as the amplitude and frequency that are used. Low-power ultrasound processing is applied at a frequency of 100 kHz to 1 MHz (Soria and Villamiel 2010) while high-power ultrasound processing is applied at a frequency of 20 kHz to 100 kHz (Ulusoy et al. 2007) (Fig. 1). Low-power ultrasound is non-destructive and can be used to investigate the physicochemical properties of food products including composition and structure, and for quality control (Awad et al. 2012, Demirdöven and Baysal 2008). On the other hand, high-power ultrasound generates mechanical and shear forces, and localized high temperatures and pressures which lead to physical disruption and promotion of chemical reactions such as oxidation (McClements 1995). As a result, high-power ultrasound has several applications including disruption of cells, microbial inactivation, tenderization of meat, inactivation of enzymes and formation of emulsions (Demirdöven and Baysal 2008, McClements 1995). Factors which affect the effectiveness of ultrasound include the type of medium and microorganisms, amplitude, intensity and treatment time (Chandrapala et al. 2012). The advantages of using high-power ultrasound for microbial

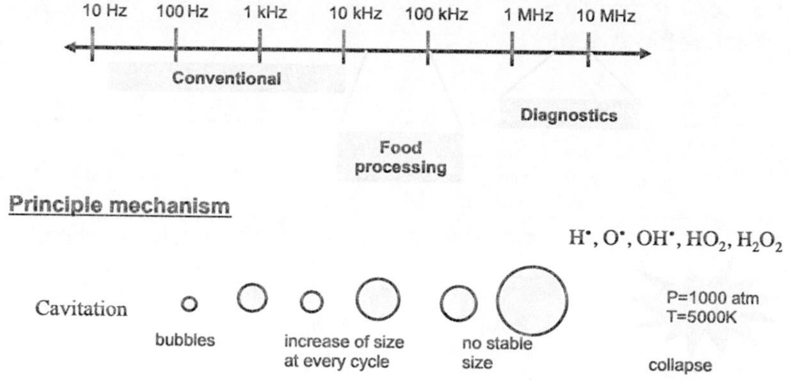

Figure 1. (Ultra-)sound frequencies and principle mechanism of power ultrasound treatments.

inactivation are that less energy is used, flavour loss in liquid foods (for example juices) is minimal, homogeneity is increased and agglomerates of bacteria are broken down (Awad et al. 2012).

Mechanism of Action of High-Power Ultrasound

The mechanism of high-power ultrasound involves a phenomenon known as cavitation which is used to inactivate microbes during food processing (Fig. 1). This phenomenon only occurs at low frequencies. During sonication, longitudinal waves are created whereby one finds areas of alternating compression and expansion. These cause the formation of cavitation bubbles, which repeatedly contract and expand until they implode. When these bubbles implode, areas of very high temperature (up to 5500°C) and pressure (up to 50 MPa) are created. The resulting pressure changes are one of the main causes of death of the bacteria. The areas of very high temperature are localized and therefore only kill bacteria present in a small area (Piyasena et al. 2003). Apart from high temperatures and pressures, high shear energy waves and turbulence are created in the cavitation zones (Soria and Villamiel 2010). Cavitation also causes the formation of free radicals from water molecules and the formation of asymmetrical micro-jets which remove contaminants from solid surfaces when created near them (Awad et al. 2012). All of these cause damage to microbial cell membranes (Patist and Bates 2008) and cause perforation of cell walls, scattering of microorganisms' clumps, and increases their sensitivity to heat (Ulusoy et al. 2007). Intracellular cavitation also occurs which disrupts the intracellular structure leading to cell lysis (Chandrapala et al. 2012). Microorganisms are killed due to the production of free radicals which damage DNA, localized heating and thinning of cell membranes (Herceg et al. 2012). Cavitation depends on the parameters of ultrasound, ambient conditions and the properties of the product (Dolatowski et al. 2007). For a summary of the mechanism, refer to Table 1.

Table 1. Summary of the antimicrobial mechanism of action of power ultrasound.

Mode of Action	Description
Mechanical forces	Generation of pressure gradients → cellular structure disruption and cell lysis (López-Malo et al. 2005)
Shear forces	Micro-streaming as a result of compression and rarefaction cycles → effect on cell membrane properties (Floros and Liang 1994)
Chemical effects	Sonolysis of water resulting in formation of free radicals ($H^\cdot$, $O^\cdot$, $OH^\cdot$, $HO_2^\cdot$) and H_2O_2 → effect on cell wall structure/ bactericide effect of H_2O_2 (Joyce and Mason 2008)

Factors affecting Ultrasound efficacy

The effectiveness of ultrasound on microbial inactivation depends on different factors including type and volume of food product, amplitude, treatment time, temperature, and type of microorganism (Dolatowski et al. 2007).

Ultrasound and type of food

The type of food used affects the efficiency of the treatment. Sams and Feria (1991) carried out investigations about the effect of ultrasound in a water bath on *Salmonella* spp. present on broiler drumstick skin. They found that a frequency of 47 kHz and a power of 200 W did not significantly reduce ($p>0.05$) the aerobic plate count after 14 days of storage, possibly due to the irregular skin surface which provides protection to the microbes from the ultrasonic waves. Another example is the case of liquid whole eggs, where the viscosity of the medium reduced the effect of cavitation (Hülsen 1999) and as a result, after 30 min treatment at 50°C, the maximum reduction was less than 1 log (Wrigley and Llorca 1992).

Ultrasound, intensity, amplitude and treatment time

The intensity level of ultrasound processing can be reported by estimating the Acoustic Energy Density (AED) and Ultrasound Intensity (UI). Both parameters can be determined calorimetrically using the following equations that were described by Mason (1990):

$$AED = \frac{P}{V} \tag{1}$$

where V [L] is the volume of the medium used.

$$UI = \frac{4P}{\pi d^2} \tag{2}$$

where d [cm] is the diameter of the sonotrode.

In both equations, P [W] represents the absolute ultrasonic power and can be defined as.

$$P = mC_p\left(\frac{dT}{dt}\right) \tag{3}$$

where, m [g] is the mass, C_p [$J \cdot g^{-1} \cdot K^{-1}$] is the specific heat capacity and dT/dt [$K \cdot s^{-1}$] is the heating rate during sonication.

The importance of these parameters in microbial disinfection has been shown in several papers. Bermúdez-Aguirre et al. (2009) reported that an increase of UI from 0.86 W/cm^2 to 1.71 W/cm^2 resulted in 3 log-reduction of *Listeria innocua* in raw whole milk at 63°C during 10 min of treatment. Adekunte et al. (2010a) found that an ultrasound system at 1500 W of a moderate temperature (39.9°C), 8 min of treatment and an AED of 0.81 W/ml, can reduce more than 5 logs the yeast *Pichia fermentans* in juice tomato. In addition, Sagong et al. (2011) found that ultrasounds caused a significant reduction of *Escherichia coli* O157:H7, *Salmonella typhimurium*, and *Listeria monocytogenes* on lettuce when the time increased from 5 min to 20 min. It is evident that the effectiveness of ultrasound in killing microorganisms increases with increasing intensity and treatment time (Scherba et al. 1991).

In addition, an increase of amplitude generates a higher efficacy of ultrasound. Raso et al. (1998) through their research on *Yersinia entercolitica*, expressed that the effectiveness of ultrasound increased when the amplitude changed from 21 μm to 150 μm at a frequency of 20 kHz, at a temperature of 30°C and pressure of 200 kPa. In a similar approach, Patil et al. (2009) found that ultrasound efficacy depends on the amplitude levels, i.e., obtaining more inactivation at higher amplitudes. According to all these results, it can be concluded that inactivation of microorganisms increased on increasing all the three parameters; temperature, treatment time and amplitude (Herceg et al. 2012).

Ultrasound in combination with temperature

The efficiency of ultrasound increases on increasing the temperature. For example, when *Salmonella typhimurium* cells were sonicated for 30 min in brain heart infusion broth they were reduced by 1 log and more than 3 log when the treatment temperature was 20°C and 40°C, respectively. In the case of skim milk, ultrasonication for 30 min decreased the number of *Salmonella typhimurium* by 2.5 log and 3.0 log at a treatment temperature of 40°C and 50°C respectively (Wrigley and Llorca 1992). It can be concluded, when the temperature is increased, the efficiency of ultrasound increases (Adekunte et al. 2010b, Arroyo et al. 2011).

When comparing ultrasound with heat as a disinfection technique, ultrasound has a number of advantages including that it is more energy efficient, and it has less effect on flavour (Piyasena et al. 2003). According to research carried out by Wordon et al. (2011), analysis of *Saccharomyces cerevisiae* cells subjected to heat showed that the number of viable cells decreased with time mainly due to membrane injury, which then led to death. On the other hand, when subjected to ultrasound using a horn type sonicator at amplitude of 124 μm and frequency of 20 kHz for a treatment of

5 min, only viable or dead cells were found, thus showing that ultrasound was rapidly microbicidal. Following the 5-min treatment, approximately 92 per cent of the cell population was killed. Sonication also caused an increase in the intracellular pressure, which forced out the cell contents (Wordon et al. 2011).

Ultrasound and type of microorganisms

The effectiveness of ultrasound also depends on the type and shape of the microorganisms. Bacterial spores and coccus-shaped bacteria are more resistant than vegetative rod-shaped bacteria (Feng et al. 2008). According to Palacios et al. (1991), ultrasound caused the spores of *Geobacillus stearothermophilus* to be less heat resistant. This may be caused by the high pressure that occurs during sonication which affected the permeability of the protoplast membrane resulting in the release of certain substances such as calcium and allowed entrance of water. Also, generally fungi are more resistant than vegetative microorganisms (Chandrapala et al. 2012).

The effectiveness of ultrasound on gram-positive and gram-negative bacteria is in some publications debated. Studies such as Scherba et al. (1991) reported that there is no significant difference between the resistance of gram-positive coccus-shaped bacteria and gram-negative rod-shaped bacteria to ultrasound. Others, such as Alliger (1975), Villamiel and de Jong (2000) and Feng et al. (2008) expressed that gram-negative and rod-shaped bacteria are more sensitive to ultrasound than gram-positive and coccus-shaped bacteria. For example, high-power ultrasound at a frequency of 20 kHz was used to inactivate *Staphylococcus aureus* and *Escherichia coli* in milk that contained 4% milk fat. During this investigation, the temperatures used were at 20°C, 40°C and 60°C, the treatment times were 6 min, 9 min and 12 min while the amplitude used were 60 µm, 90 µm and 120 µm. *S. aureus* was more resistant to the treatment compared to *E. coli*. When milk was treated with amplitude of 120 µm at 60°C, the D value was 2.78 min for *E. coli* and 4.80 min for *S. aureus* (Herceg et al. 2012).

Other researchers have reported that gram-positive bacteria are more resistant than gram-negative bacteria possibly due to the presence of a thick layer of peptidoglycan wall in gram-positive bacteria, which provides a better protection against ultrasound treatment (Drakopoulou et al. 2009). One should note that in the research carried out by Drakopoulou et al. (2009), the difference between the effect of sonication on gram-positive and gram-negative bacteria decreased on increasing treatment time. Also, gram-positive bacteria were more resistant than gram-negative bacteria when sonicated for an hour at a frequency of 24 kHz with TiO_2 particles. 99.9 per cent of gram-negative bacteria which in this case were total coliforms, faecal coliforms and *Pseudomonas* spp. were inactivated whereas in the case of gram-positive bacteria, 72.8 per cent of faecal *Streptococci* and 87.1 per cent of *Clostridium perfringens* species were inactivated.

These findings were also supported by Ananta et al. (2005), who found that the gram-negative bacteria *E. coli* was less resistant to ultrasound than the gram-positive bacteria *Lactobacillus rhamnosus*. Also, when ultrasound was used to inactivate total bacteria in milk, and *Pseudomonas fluorescens* and *Streptococcus thermophilus* in tryptic soy broth through a continuous flow process, one could observe that *P. fluorescens* which is a gram-negative bacterium was more sensitive to the treatment than *Streptococcus thermophilus* which is gram-positive (Villamiel and de Jong 2000).

Research on the Antimicrobial Use of Ultrasound

Ultrasound and surfaces disinfection

When microorganisms adhere to surfaces and colonize them, they form biofilms that can lead to food contamination (Oulahal-Lagsir et al. 2001). Ultrasound can be used to clean surfaces since the force of the fluid due to the collapse of bubbles displaces bacteria from surfaces. For example, using a probe, ultrasound at a frequency of 40 kHz for 10 s was four times more effective at removing a biofilm consisting of spores of *Bacillus stearothermophilus* variety *calidolactis* in meat from stainless steel sheets when compared to swabbing since 83 per cent and 20 per cent of the biofilm were removed respectively. This research was carried out to find a way of cleaning stainless steel surfaces in food processing equipment (Oulahal-Lagsir et al. 2001). According to the research carried out by Zips et al. (1990), the rate of detachment depends on different factors including the exposure time and the intensity. One of the advantages of using ultrasound is that it can reach crevices that are usually unreachable by conventional methods (Chemat and Khan 2011).

Ultrasound and solid food products disinfection

According to Seymour et al. (2002), high-power ultrasound can be used to decontaminate fresh produce due to the cavitation phenomena, which increases the mechanical removal of bacteria attached on the surface of fresh produce, such as lettuce. Ultrasound assisted with a bactericide can be used to disinfect the surface of hatchery eggs. Dirty eggs would otherwise have to be discarded as contamination occurs through the porous shell which kills the developing embryo. High-power ultrasound can be used for disinfection since it disrupts cell walls, however, very high intensities would be required for complete disinfection (Mason et al. 1996). Ultrasound can also be used as a disinfection technique in chocolate manufacturing. For example, when chocolate was inoculated with *Salmonella eastbourne* and subjected to ultrasound through the use of an ultrasonic probe, at a frequency of 160 kHz and power of 100 W at 5°C for 30 min, the microbial population was reduced by 74 per cent. This was partly due to the high temperature, 73°C that was produced during the treatment (Lee et al. 1989).

Ultrasound and liquid products disinfection

Ultrasound can also be used as a disinfection technology in the drinking water industry (Brown and Salveson 2006) where the localized high temperatures and pressures and the generated micro-jets cause microbial inactivation. In the case of waste water, when ultrasound generated by a horn sonotrode was used at a frequency of 20 kHz for 60 min and at an ultrasound density of 400 W/L, the amount of *Escherichia coli* was reduced by 2.9 log (Blume et al. 2002).

High-power ultrasound can potentially have applications in the wine industry to inactivate undesirable organisms during various stages of the process (Jiranek et al. 2008). For example, these organisms can be inactivated during fermentation. An advantage of this technique is that the treatment can be directly carried out in the barrel containing the wine (Yap et al. 2007).

A variety of spoilage organisms such as yeast, acetic acid bacteria and lactic acid bacteria can be found in grapes especially grapes that are damaged or over-ripe which if not removed, may negatively affect the properties of the wine (Jiranek et al. 2008). In grapes and wine, the predominant species of yeast is *Saccharomyces cerevisiae,* while in the case of bacteria one finds two genera of acetic acid bacteria (*Acetobacter* and *Gluconobacter*) and four genera of lactic acid bacteria (*Lactobacillus, Leuconostoc, Oenococcus* and *Pediococcus*) (Bartowsky 2009). Spoilage bacteria may cause a reduction in wine quality since for example, certain bacteria such as *Lactobacillus* sp. lead to the production of compounds which negatively affect the wine quality by altering its aroma and flavour (Bartowsky 2009).

When applied to date syrup, sonication decreased the amount of microorganisms present compared to thermal treatment. The traditional method used to prepare date syrup, such as Kabkab date syrup, involves mixing date fruits with water at a high temperature for one hour to extract mainly sugars. Apart from reducing the nutritional value and affecting the colour by darkening it, the conditions used do not kill the micro-organisms present in the date fruit and syrup. According to the results obtained, sonication led to a lower microbial count, a higher extraction in a shorter time and it did not affect the quality of the product compared to the thermal treatment. High-intensity ultrasound (20 kHz frequency and 125 W power) was more effective at removing micro-organisms than low-intensity ultrasound (20 kHz frequency and 50 W power). The best conditions which led to 85 per cent destruction of microorganisms after 90 min involved high-power ultrasound, at a temperature of 15°C and water: fruit ratio 3:1. On the other hand, using the traditional method, the highest percentage of microbial destruction was 52 per cent (Entezari et al. 2004).

High-power ultrasound can be used to reduce *Escherichia coli* by 5 logs in fruit juices to levels required by the FDA and also to obtain the required cloud stability (Tiwari et al. 2009). More specifically, ultrasound both alone and combined with mild temperature can reduce *Escherichia coli* (Ugarte-Romero et al. 2006) and *Listeria monocytogenes* (Baumann et al. 2005) in apple cider. Single frequency waves limit the degree of cavitation, which in turn might affect the efficiency of the treatment. On the other hand, multi-frequency ultrasound (Fig. 2) increases the degree of cavitation and thus may increase the effectiveness of microbial inactivation in fruit juices. The inactivation rates of *Escherichia coli* O157:H7, *Salmonella* spp., *Listeria monocytogenes* and spoilage yeast in cloudy apple juice were investigated using a multi-frequency ultrasound at a temperature below the lethal range to limit microbial inactivation due to temperature. The treatment time was 30 min to ensure that lethal temperature ranges were not reached. In Dynashock waves there are three frequencies at 28 kHz, 45 kHz and 100 kHz that are alternately switched on at time intervals of 1 ms. This helps to enhance the cavitation phenomenon. On the one hand, the inactivation patterns of *E. coli* O157:H7, *Salmonella* spp. and yeasts were log-linear, however in the case of yeast, the inactivation rate was slower. On the other hand, in the case of *L. monocytogenes*, the inactivation pattern was biphasic with characteristic inactivation shoulders thus showing that it takes a shorter time to inactivate them (Gabriel 2012).

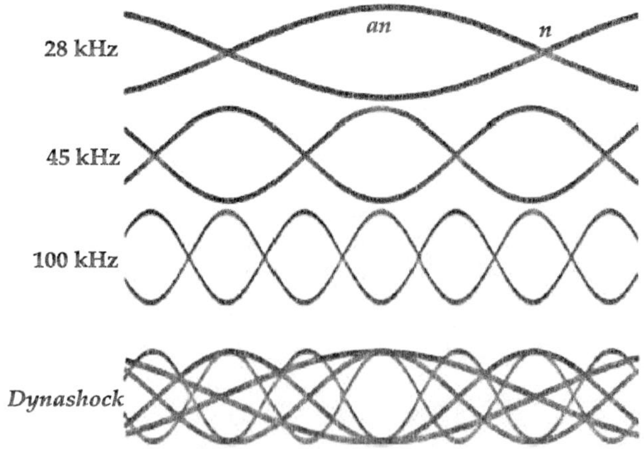

Figure 2. Difference between single frequency and Dynashock waves (Gabriel 2012).
Color image of this figure appears in the color plate section at the end of the book.

Milk pasteurization

The common method used to inactivate microorganisms in milk is pasteurization, however, the high temperatures that are used, may alter both the nutritional value and the quality of milk. Ultrasound is an alternative non-thermal treatment that can be used in pasteurization without affecting the nutritional value or quality of the milk (Herceg et al. 2012). According to research carried out by Herceg et al. (2012), shorter treatment times at 20 kHz, 600 W using a probe to generate the ultrasonic waves, are less effective in inactivating microorganisms due to decreased cavitation and microstreaming. The lowest counts of *Staphylococcus aureus* and *Escherichia coli* in this research were reported when the milk was treated for 12 min at a temperature of 59.99°C and amplitude of 117.27 µm in the case of *Staphylococcus aureus* and 110.41 µm in the case of *Escherichia coli*. The use of ultrasound in milk pasteurization was also supported by research carried out by Cameron et al. (2009). During this research, the parameters included a probe, ultrasound at a frequency of 20 kHz, power of 750 W and amplitude of 124 µm. After 10 min treatment, 100 per cent of *Escherichia coli* and 99 per cent of *Listeria monocytogenes* viable cells were inactivated. In the case of *Pseudomonas fluorescens*, 100 per cent viable cells were inactivated after 6 min treatment. Furthermore, the protein content and lactose content were not negatively affected. Through ultrasound, the legal requirements of contamination levels according to the South African laws were met, hence, it can be concluded that this technique can be used instead of thermal pasteurization.

Ultrasound Combined/Assisted Methods

Ultrasound in combination with other physical processes and antimicrobial solutions

The effectiveness of ultrasound as a disinfection technique can be increased by combining it with another treatment such as pressure, heat and antimicrobial solutions (Awad et al. 2012, Chemat and Khan 2011, Ulusoy et al. 2007, McClements 1995). Ultrasound requires high power and longer treatment times to inactivate microorganisms when compared to ultrasound combined with another technique (Chandrapala et al. 2012). Thus, assisted ultrasound is more energy efficient and since lower intensities can be used, it has less impact on the food properties (Dolatowski et al. 2007). Due to this, ultrasound combined with another technique has greater potential for industrial applications (Demirdöven and Baysal 2008). A summary of a number of combined/assisted methods is given in Table 2.

Table 2. Overview of ultrasound combined/assisted methods (AEW: acidic electrolyzed Water, POAA: peroxyacetic acid, ASC: acidified sodium chlorite).

Ultrasound assisted by	Operation Conditions		Food (model) system	Microbial reduction	References
Temperature (thermosonication)	20 kHz 150 W 118 W/cm^2 18 min	57°C	Milk	5 log CFU/mL of *L. monocytogenes*	D'Amico et al. (2006)
	20 kHz, 600 W, 120 μm, 12 min	20°C	Milk	1.49 log CFU/mL of *S. aureus* 3.07 log CFU/mL of *E. coli*	Herceg et al. (2012)
		60°C		0.94 log CFU/mL *of S. aureus* 2.49 log CFU/mL of *E. coli*	
Pressure (manosonication)	20 kHz 117 μm	200 kPa	Test tubes (TSBYE)	*D*-value up to 2.8 min of *L. monocytogenes*	Pagan et al. (1999)
		400 kPa		*D*-value up to 3.3 min of *L. monocytogenes*	
Pressure and Heat (manothermosonication)	20 kHz 61°C 4 min	100 kPa	Buffer system	Up to 5 log CFU/mL of *E. coli*	Lee et al. (2009)
		200 kPa		Up to 7 log CFU/mL of *E. coli*	
		300 kPa		Up to 7 log CFU/mL of *E. coli*	
		400 kPa		Up to 7 log CFU/mL of *E. coli*	
Osmosis (Osmosonication)	20 kHz, 50 W 48 μm 25°C, 48 hr	10.9 Pa	Orange juice	5 log CFU/mL of *Salmonella spp*	Wong et al. (2012)
Pulse Electric Fields (PEF)	20 kHz, 31 μm, 40 W/cm^2 200 kPa 52°C	34 KV/cm 32 μs	Smoothies	5.6 log CFU/mL of *Listeria monocytogenes*	Palgan et al. (2012)

	34 KV/cm² 32 µs	20 kHz, 31 µm, 40 W/cm 200 kPa 52°C		4.2 log CFU/mL of *Listeria monocytogenes*	
Pulse Electric Fields (PEF)	20 kHz 40 W 55°C, 5 min	30 pulses 5.67 KV/mm	Liquid whole eggs	2.25 log of reduction of *Salmonella enteritidis*	Huang et al. (2006a)
	30 pulses 5.67 KV/mm	20 kHz 40 W 55°C, 5 min		2.30 log of reduction of *Salmonella enteritidis*	
Ozone(Sozone)	<u>Horn</u> 22 kHz 240 W	0.5 mg/L ozone	Water	Between 95 and 9.9% of reduction of total coliforms, faecal coliforms and faecal streptococci	Jyoti and Pandit (2004)
	<u>Water Bath</u> 20.5 kHz 120 W				
UV radiation	1400 W 15 min	2 UV-C lamps of 150 W	Wastewater	Reduction of *E.coli* below limits of legislation (2 CFU/100 mL)	Naddeo et al. (2009)
	30 seconds of UV radiations (14 W of which 3W are emitted at 254 nm)	20 kHz, 5 s 50 W/L	Wastewater	3.3 log CFU/mL of fecal colifroms	Blume and Neis (2004)
		20 kHz, 5 s 310 W/L		3.7 log CFU/mL of fecal coliforms	
	20 kHz 100 W 20 min	UV at 100W at 257.7 nm	Orange juice	3.5 log CFU/mL of *E.Coli* ATCC 35218	Char et al. (2010)

Table 2. contd....

Table 2. contd.

Ultrasound assisted by		Operation Conditions		Food (model) system	Microbial reduction	References
Antimicrobial Solutions Antimicrobial solutions	Organic Acids	40 kHz 30 W/L 5 min	Malic acid at 2% (v/v)	Lettuce	2.52 log CFU/g *E. coli* O157:H7 2.73 log CFU/g of *S. typhymurium*	Sagong et al. (2011)
			Lactic acid at 2% (v/v)		2.75 log CFU/g *E. coli* O157:H7 2.71 log CFU/g of *S. typhymurium*	
			Citric acid at 2%(v/v)		2.43 log CFU/g *E. coli* O157:H7 3.18 log CFU/g of *S. typhymurium*	
	Natural antimicrobials	20 kHz, 600 W, 95.2 μm 45.2°C 8 min	Combination of 1,500 ppm of vanillin and 100 ppm of citral	Orange Juice	6 log CFU/mL of *L. monocytogenes*	Ferrante et al. (2007)
		20 kHz, 500 W, 150 μm, $40 W/cm^2$ 43°C 126 s	2.5 mg/L of nisin	Buffer	Up to 2.8 log CFU/mL of *L. monocytogenes*	Muñoz et al. (2012)
		70 kHz $20 mW/cm^2$ 2 hr	24 μml/mL of gentamicin	Biopolymeric surfaces	97% of *E. coli*	Johnson et al. (1998)
	Sanitizers	170 kHz 10 min	40 ppm of ClO_2	Apples	4 log CFU/g of *S.* Typhymurium 3.5 log CFU/g of *E. coli*	Huang et al. (2006b)
		32–40 kHz 10–15 W/L 10 min	25 ppm of Chlorine	Cut iceberg lettuce	2.7 log CFU/g of *S.* Typhymurium	Seymour et al. (2002)

		21.2 kHz 200 W/L 2 min	Chlorinated water (200 mg/L)	Spinach Leaves	3.1 log CFU/g of *E. coli* O157:H7	Zhou et al. (2009)
			AEW (80 mg/L)		3.1 log CFU/g of *E. coli* O157:H7	
			POAA (80 mg/L)		2.9 log CFU/g of *E. coli* O157:H7	
			ASC (200 mg/L)		4.0 log CFU/g of *E. coli* O157:H7	
		45 kHz 10 min	40 mg/L of peracetic acid	Cherry tomatoes	3.88 log CFU/g of *S.* Typhimurium	Brilhante Sao Jose and Dantas Vanetti (2012)
Enzymatic solutions		40 kHz 10 s	Protease type XXIII (36 U/mL)	Milk biofilms on stainless steel sheets	Up to 76% of *E. coli*	Oulahal-Lagsir et al. (2003)
			Crystalline trypsin (7,600 U/mL)		Up to 61% of *E. coli*	
			Amyloglucosidase (50 U/mL)		Up to 96% of *E. coli*	
			Papain (3 U/mL)		Up to 76% of *E. coli*	
			Lysozyme (6000 U/mL)		Up to 70% of *E. coli*	
		40 kHz 10 s EDTA (0.025 mol/L)	Crystalline trypsin (7,600U/mL) Protease type XXIII (36 U/mL)	Meat biofilms on stainless steel sheets	Up to 50% of *E. coli* Up to 79% of *S. aureus*	Oulahal et al. (2007)

Table 2. contd....

Table 2. contd.

Ultrasound assisted by		Operation Conditions		Food (model) system	Microbial reduction	References
			Crystalline trypsin (7,600 U/mL) Lysozyme (6,000 U/mL)		Up to 27% of *E. coli* Up to 100% of *S. aureus*	
			Papain (3 U/mL) Protease type XXIII (36 U/mL)		Up to 75% of *E. coli* Up to 70% of *S. aureus*	
			Protease type XXIII (36 U/mL) Lysozyme (6000 U/mL)		Up to 56% of *E. coli* Up to 74% of *S. aureus*	
			Lysozyme (6000 U/mL) Papain (3U/mL)		Up to 54% of *E. coli* Up to 30% of *S. aureus*	
			Amyloglucosidase (50 U/mL)		Up to 42% of *E. coli* Up to 100% of *S. aureus*	

Two common techniques used for microbial inactivation in food products are thermal pasteurization and sterilization. However, the effectiveness of these techniques increases on increasing treatment time and temperature, which in turn leads to deterioration of the food quality including an increase in nutrient loss. When ultrasound is combined with thermal treatment, the duration and intensity of the thermal treatment are reduced thus decreasing the effect on the food quality. Reducing thermal treatment is also more environmentally friendly and cost efficient (Piyasena et al. 2003). When ultrasound is coupled with heat, the technique is referred to as thermosonication. When ultrasound is coupled with pressure or with both heat and pressure, the techniques are referred to as manosonication and manothermosonication (MTS), respectively (Dolatowski et al. 2007, Piyasena et al. 2003). Ultrasound combined with heat, pressure or both is more effective because they cause more mechanical disruption of cells thus making them more susceptible to cavitation. Apart from this, lower temperatures and shorter treatment times can be applied (Chandrapala et al. 2012).

According to investigations carried out by D'Amico et al. (2006), ultrasound generated by a probe at 20 kHz frequency, 150 W power and 118 W/cm^2 intensity, combined with mild heat at a temperature of 57°C reduced the amount of *Listeria monocytogenes* in milk by 5 log CFU/mL when treated for 18 min. When treated with amplitude of 120 µm for 12 min at a temperature of 60°C, the maximum inactivation of *Staphylococcus aureus* in milk was achieved. Even though cavitation decreases on increasing the temperature (Raso et al. 1998), this was not observed in the research carried out by Herceg et al. (2012). In the case of milk, the solids present in the suspension may enhance cavitation. When the amplitude is increased, cavitation increases, which leads to increased microbial inactivation. In the case of *Escherichia coli*, inactivation also increased on increasing temperature and amplitude (Herceg et al. 2012). Other researches were carried out in which *Saccharomyces cerevisiae* was subjected to a combined treatment consisting of ultrasonication at 20 kHz frequency and 124 µm amplitude using a horn type sonicator for 1 min followed by addition of sterile physiological saline which was pre-heated to 55°C or 60°C. This research showed that the death rate increased since ultrasound caused intracellular injuries resulting in more heat cell susceptibility (Wordon et al. 2011).

Certain microorganisms are more resistant to ultrasound, thus, to avoid long treatment times, ultrasound can be combined with pressure and/or heat which increases the mechanical disruption of cells. To investigate this, research was carried out on *Listeria monocytogenes*. In the case of *L. monocytogenes*, ultrasound treatment using a horn-type sonicator at a frequency of 20 kHz and an amplitude of 117 µm at ambient temperature was not very effective (Pagan et al. 1999). Manosonication (the combination of pressure, and ultrasound) was more effective whereby on increasing the pressure to 200 kPa, the time taken to reduce bacterial activity by 90 per cent (D value) was reduced from 4.3 min to 1.5 min. When the pressure was increased to 400 kPa, the time taken was further reduced to 1.0 min. Temperatures above 50°C increased the inactivation rate. Furthermore, even though cultures grown at 37°C were twice as resistant to heat than those grown at 4°C, this did not affect the effectiveness of manosonication (Pagan et al. 1999). A study carried out by Lee et al. (2009) has shown the effectiveness of this technology, i.e., up to 7 logs reduction for *E. coli* in buffer systems at 500 kPa at 61°C at 20 kHz and 4 min. *Cronobacter sakazakii* is more acid-tolerant than most of the closely related microbial pathogens. The effectiveness of the ultrasound treatment on the inactivation of *Cronobacter sakazakii* in apple juice was increased when it was combined with pressure and heat. Heat makes the cells more sensitive to ultrasound. The effect of MTS depends on the type of microorganisms present depending on its degree of heat resistance, the treatment temperature and the treatment media. In the case of treatment media, lower temperatures can be used if the microorganism has a low heat resistance (Arroyo et al. 2011).

Another example of combined technology is osmosonication which involves the combination of ultrasound treatment with storage at a high osmotic pressure. According to Wong et al. (2012) this treatment increased the degree of membrane damage and viability loss of *Salmonella* spp. in orange juice when sonicated using a probe at a frequency of 20 kHz, at a power of 50 W, amplitude of 48 µm, and a treatment temperature of 25°C, and stored for 48 hours at an osmotic pressure of 10.9 MPa. This reduced the amount of bacteria by 5 log without affecting the properties of the juice.

The so-called emerging technologies can also be combined with ultrasound. Pulsed Electric Field (PEF) is a non-thermal technology able to kill microorganisms due to the electroporation of the cell membrane which is achieved by the application of short pulses of high-intensity electric fields (Jaeger et al. 2009). Application of MTS at 20 kHz of frequency, 31 µm of amplitude, 40 W/cm^2 of intensity at 200 KPa of pressure and 42°C of temperature followed by PEF at 34 KV/cm and 32 µs and PEF followed by MTS at the same conditions on smoothies, led to a reduction of 5.6 log CFU/mL and 4.2 log CFU/mL of *Listeria innocua*, respectively (Palgan et al. 2012). Huang et al. (2006a) reported that combination of US followed by PEF and PEF followed by US (US parameters were 20 kHz, 40 W at 55°C for 5 min and PEF parameters were 30 pulses of 5.67 kV/mm in both experiments) were able to reduce the level of *Salmonella enteritidis* in liquid whole eggs up to 2.25 and 2.30 log, respectively.

Synergetic effects have also been observed when ozone was combined with ultrasound (so-called sonoze). According to Jyoti and Pandit (2004) application of ultrasound either by horn at 22 kHz and 240 W or water bath at 20.5 kHz and 120 W in combination with ozone (0.5 mg/L) during 15 min were able to decreased the number of total coliforms, faecal coliforms and faecal streptococci between 95 and 99.9 per cent in water and reduced the dosage of chemical agent. This effectiveness can be related with the ability of ultrasounds to increase the ozone gas-liquid transfer and decompose the dissolved ozone (Dahl 1976).

Other relevant applications include waste water treatment with UV and ultrasound. Naddeo et al. (2009) reported that two low pressure UV-C lamps of 150 W for 15 min were ineffective to reduce *E. coli* in waste water below the limit required by the legislation (10 CFU/100 mL). However, UV radiations at the same conditions in combination with ultrasound (1400 W) were able to decrease the microbial load up to 2 CFU/100 mL. Similar results were reported by Blume and Neis (2004), where 30 seconds of UV radiations (14 W of which 3 W are emitted at 254 nm) reduced fecal coliforms up to 2.5 log in water. When the same process was followed by 5 seconds of ultrasound treatment at 50 W/L or 310 W/L microbial reductions were 3.3 and 3.7 log, respectively. In addition, a research carried by Char et al. (2010) reported that application of ultrasound at 20 kHz and 95 μm in combination with UV irradiation at 100 W at 257.7 nm during 20 minutes resulted in 3,5 log reduction of *E. coli* ATCC 35218 in orange juice. This proves a synergistic effect considering an additional 1.5 log of reduction compared with treatments performed separately or in a sequence, i.e., UV (100 W at 257.7 nm and 10 minutes) followed by US (20 kHz, 95 μm, 10 minutes). The exact mechanism of action of the synergistic effects of UV and US still remains unknown.

Ultrasound in combination with antimicrobial solutions

Ultrasound can also be combined with antimicrobial solutions for increasing its effectiveness (refer to Table 2 that summarized all applications). A wide range of antimicrobial substances has been used to inactivate microorganisms. Organic acids are well known as an additive in food preservation due to its characteristic to affect the microorganisms in two ways: (i) by decreasing the pH of the medium and (ii) by their ability to pass through the microbial membrane and dissociate inside the bacteria. Based on these phenomena, Sagong et al. (2011) reported that the effect of different organic acids (malic acid, lactic acid, citric acid) in combination with ultrasound, enhance the inactivation of *E. coli* O157:H7, *S. typhimurium* and *L. monocytogenes* on lettuce. The reduction of the pathogens was incremented to 0.8–1.0 log CFU/g reduction compared with the individual treatments.

Natural antimicrobial substances, alone or in combination with other technologies have several application in food safety. Some of them are used in the form of plant-extracted essential oils. Ferrante et al. (2007) found that the combination of several natural antimicrobials such as vanillin and citral, have a synergetic effect when combined with ultrasound and a 6 logs CFU/mL reduction of *Listeria monocytogenes* in orange juice could be achieved. Natural antimicrobial substances are able to inactivate microorganisms due to the presence of chemical compounds (e.g., poliphenols, terpenes), conferring them antimicrobial properties. Therefore, the combination of different natural antimicrobial substances with ultrasound can result in increased microbial reduction.

Other natural compounds obtained from microorganisms, such as nisin, have been widely used against gram-positive bacteria, due to the presence of the uncommon amino acid of lanthionine. A study conducted by Muñoz et al. (2012) reported that the application of nisin resulted in pronounced antimicrobial effect of the ultrasound against *Listeria innocua* in buffer systems (from 0.3 log CFU/mL without nisin to 2.8 log CFU/mL reduction with nisin). Moreover, the combination of ultrasound and nisin with high-intensity pulse light or pulsed electric fields raised the microbial inactivation to 5.6 logs and 4.8 logs, respectively. Another bacterial compound used to kill pathogens is the antibiotic of gentamicin. Its effectiveness has been reported against gram-negative bacteria. Ultrasound treatment at a frequency of 70 kHz for 2 hours on a biofilm combined with the use of gentamicin sulfate reduced the amount of *Escherichia coli* by up to 97 per cent. This occurs since ultrasound destabilizes the cell membrane thus increasing the diffusion of the antibiotic through it (Johnson et al. 1998).

Besides, ultrasound can be combined with other treatments as well. Washing with different antimicrobial sanitizers has become the most applied technique in the food industry to disinfect fresh produce. Thus, the application of ultrasound with sanitizers can enhance food disinfection. Research carried out by Huang et al. (2006b) reported that a combination of ultrasound at 170 kHz and chlorine dioxide at 10 ppm for 10 min resulted in a 4 log CFU/g reduction of *Salmonella* spp. and 3.5 log CFU/g reduction in *E.coli* O157:H7 on apples. According to Seymour et al. (2002), high-power ultrasound (32–40 kHz) increased the antimicrobial activity of chlorine since through ultrasound, cells of *Salmonella typhimurium* were removed from the surface of the fresh produce. Zhou et al. (2009) applied several sanitizer treatments such as chlorine, acidified sodium chlorite, peroxyacetic acid and acidic electrolyzed water with ultrasound to improve the reduction of *Escherichia coli* O157:H7 on spinach leaves. In the same way, a study carried out by Brilhante de São José and Dantas Vanetti (2012) reported that using 40 mg/L of peracetic acid in combination with ultrasound at 45 kHz during 10 min, is able to reduce up to 3.88 log CFU/g reduction of *Salmonella typhimurium* in whole cherry tomatoes.

Another example of combined technology involves ultrasound combined with enzymatic treatment. For example, during a research carried out by Oulahal-Lagsir et al. (2003), enzymatic treatment, sonication and a combination of these two methods were investigated for the removal of an *Escherichia coli* model biofilm that was constructed using milk on stainless steel sheets. Only 30 per cent (±7%) of the biofilm was removed when treated with ultrasound at 40 kHz for 10 s. On the other hand, when the combined treatment was used, 61 per cent to 96 per cent of the biofilm were removed according to the conditions used (Oulahal-Lagsir et al. 2003). During another research, meat biofilms consisting of *Escherichia coli* and *Staphylococcus aureus* on stainless steel sheets were investigated (Oulahal et al. 2007). When treated with a combined treatment which consisted of ultrasound at 40 kHz for 10 s, EDTA and/or in enzyme solutions (protease type XXII (36 U/mL), crystalline trypsin (7600 U/mL), amyloglucosidase (50 U/mL), papain (3 U/mL) and lysozyme (6000 U/mL)), 75% (±4%) of *Escherichia coli* and 100% (±15%) of *Staphylococcus aureus* biofilms were removed. This was more successful than when ultrasound was used alone. In the latter case the recovery rates were 49% (±5%) for *Escherichia coli* and 39 per cent (±5%) for *Staphylococcus aureus*.

Conclusion

Bacteria cause food-borne illnesses and thus the demand for novel technologies that enhance food safety has increased with time (Mukhopadhyay and Ramaswamy 2012). However, consumers also demand the use of methods that have a lower impact on the nutritional content and quality of the food. One of the technologies that can be used to meet these demands is ultrasound and assisted ultrasound. Through the application of assisted disinfection technologies, treated products could be free from harmful chemicals and safer both for consumers' health and the environment. New outbreaks of food-borne illness associated with consumption of contaminated food could be further controlled. Also, it is an environmentally friendly innovative approach with minimal chemical emissions, waste output, and water usage. Thus, further research should be carried out in order to optimize the processing variables of such technologies to achieve desirable results for producing safe and high quality food products when applying this novel technology at industrial scale.

Acknowledgments

This research was supported by a Marie Curie FP7-Reintegration-Grant within the 7th European Community Framework Programme under the project Development of novel Disinfection Technologies for Fresh Produce (DiTec), while it is partly funded by the Malta Government Scholarship Scheme through the post-graduate scheme scholarship awarded to DA.

Keywords: ultrasound, thermosonication, assisted ultrasound, manothermosonication, osmosonication, sonolysis, cavitation, micro-streaming

References

Adekunte, A., B. Tiwari, A. Scannell, P. Cullen and C. O'Donnell. 2010a. Modelling of yeast inactivation in sonicated tomato juice. Int. J. Food Microbiol. 137(1): 116–120.

Adekunte, A., V. Valdramidis, B. Tiwari, N. Slone, P. Cullen, C. Donnell and A. Scannell. 2010b. Resistance of *Cronobacter sakazaki* in reconstituted powdered infant formula during ultrasound at controlled temperatures: A quantitative approach on microbial responses. Int. J. Food Microbiol. 142(1): 53–59.

Alliger, H. 1975. Ultrasonic disruption. American Laboratory 10: 75–85.

Ananta, E., D. Voigt, M. Zenker, V. Heinz and D. Knorr. 2005. Cellular injuries upon exposure of *Escherichia coli* and *Lactobacillus rhamnosus* to high-intensity ultrasound. J. Appl. Microbiol. 99(2): 271–278.

Arroyo, C., G. Cebrián, R. Pagán and S. Condón. 2011. Synergistic combination of heat and ultrasonic waves under pressure for *Cronobacter sakazakii* inactivation in apple juice. Food Control 25(1): 342–348.

Awad, T., H. Moharram, O. Shaltout, D. Asker and M. Youssef. 2012. Applications of ultrasound in analysis, processing and quality control of food: A review. Food Res. Int. 48(2): 410–427.

Bartowsky, E. 2009. Bacterial spoilage of wine and approaches to minimize it. Lett. Appl. Microbiol. 48(2): 149–156.

Baumann, A.R., S.E. Martin and H. Feng. 2005. Power ultrasound treatment of listeria monocytogenes in apple cider. J. Food Prot. 68(11): 2333–2340.

Bermúdez-Aguirre, D., M.G. Corradini, R. Mawson and G.V. Barbosa-Cánovas. 2009. Modeling the inactivation of *Listeria innocua* in raw whole milk treated under thermo-sonication. Innov. Food Sci. Emerg. Technolo. 10(2): 172–178.

Blume, T., I. Martinez and U. Neis. 2002. Wastewater disinfection using ultrasound and UV light. TU Hamburg-Harburg Reports on Sanitary Engineering 35: 117–128.

Blume, T. and U. Neis. 2004. Improved wastewater disinfection by ultrasonic pre-treatment. Ultrason. Sonochem. 11(5): 333–336.

Brilhante São José, J.F. and M.C. Dantas Vanetti. 2012. Effect of ultrasound and commercial sanitizers in removing natural contaminants and *Salmonella enterica typhimurium* on cherry tomatoes. Food Control 24(1): 95–99.

Brown, J.C. and A. Salveson. 2006. Emerging Disinfection Technologies. Fl. Water Resour. J. pp. 4–9.

Cameron, M., L.D. McMaster and T.J. Britz. 2009. Impact of ultrasound on dairy spoilage microbes and milk components. Dairy Sci. Technol. 89(1): 83–98.

Cárcel, J., J. Garcia-Perez, J. Benedito and A. Mulet. 2012. Food process innovation through new technologies: Use of ultrasound. J. Food Eng. 110(2): 200–207.

Chandrapala, J., C. Oliver, S. Kentish and M. Ashokkumar. 2012. Ultrasonics in food processing—Food quality assurance and food safety. Trends Food Sci. Technol. 26(2): 88–98.

Char, C.D., E. Mitilinaki, S.N. Guerrero and S.M. Alzamora. 2010. Use of high-intensity ultrasound and UV-C light to inactivate some microorganisms in fruit juices. Food Bioprocess Tech. 3: 797–803.

Chemat, F. and M.K. Khan. 2011. Applications of ultrasound in food technology: Processing, preservation and extraction. Ultrason. Sonochem. 18(4): 813–835.

Dahl, E. 1976. Physicochemical aspects of disinfection of water by means of ultrasound and ozone. Water Res. 10(8): 677–684.

D'amico, D.J., T.M. Silk, J. Wu and M. Guo. 2006. Inactivation of microorganisms in milk and apple cider treated with ultrasound. J. Food Prot. 69(3): 556–563.

Demirdöven, A. and T. Baysal. 2008. The use of ultrasound and combined technologies in food preservation. Food Rev. Int. 25(1): 1–11.

Dolatowski, Z.J., J. Stadnik and D. Stasiak. 2007. Applications of ultrasound in food technology. Acta Sci. Pol. Technol. Aliment 6(3): 89–99.

Drakopoulou, S., S. Terzakis, M. Fountoulakis, D. Mantzavinos and T. Manios. 2009. Ultrasound-induced inactivation of gram-negative and gram-positive bacteria in secondary treated municipal wastewater. Ultrason. Sonochem. 16(5): 629–634.

Earnshaw, R., J. Appleyard and R. Hurst. 1995. Understanding physical inactivation processes: Combined preservation opportunities using heat, ultrasound and pressure. Int. J. Food Microbiol. 28(2): 197–219.

Entezari, M.H., S. Hagh Nazary and M. Haddad Khodaparast. 2004. The direct effect of ultrasound on the extraction of date syrup and its micro-organisms. Ultrason. Sonochem. 11(6): 379–384.

Feng, H., W. Yang and T. Hielscher. 2008. Power ultrasound. Food Sci. Technol. Int. 14(5): 433–436.

Ferrante, S., S. Guerrero and S.M. Alzamora. 2007. Combined use of ultrasound and natural antimicrobials to inactivate *Listeria monocytogenes* in orange juice. J. Food Prot. 70(8): 1850–1856.

Floros, J.D. and H. Liang. 1994. Acoustically assisted diffusion through membranes and biomaterials. Food Technol. 48(12): 79–84.

Gabriel, A.A. 2012. Microbial inactivation in cloudy apple juice by multi-frequency *Dynashock* power ultrasound. Ultrason. Sonochem. 19(2): 346–351.

Herceg, Z., A. Režek Jambrak, V. Lelas and S. Mededovic Thagard. 2012. The effect of high intensity ultrasound treatment on the amount of *Staphylococcus aureus* and *Escherichia coli* in milk. Food Technol. Biotechnol. 50(1): 46.

Huang, E., G. Mittal and M. Griffiths. 2006a. Inactivation of *Salmonella enteritidis* in liquid whole egg using combination treatments of pulsed electric field, high pressure and ultrasound. Biosystems Engineering 94(3): 403–413.

Huang, T.S., C. Xu, K. Walker, P. West, S. Zhang and J. Weese. 2006b. Decontamination efficacy of combined chlorine dioxide with ultrasonication on apples and lettuce. J. Food Sci. 71(4): M134–M139.

Hülsen, U. 1999. Alternative heat treatment processes. European Dairy Magazine 3: 20–24.

Jaeger, H., A. Schulz, N. Karapetkov and D. Knorr. 2009. Protective effect of milk constituents and sublethal injuries limiting process effectiveness during PEF inactivation of *Lb. rhamnosus*. Int. J. Food Microbiol. 134(1): 154–161.

Jiranek, V., P. Grbin, A. Yap, M. Barnes and D. Bates. 2008. High power ultrasonics as a novel tool offering new opportunities for managing wine microbiology. Biotechnol. Lett. 30(1): 1–6.

Johnson, L.L., R.V. Peterson and W.G. Pitt. 1998. Treatment of bacterial biofilms on polymeric biomaterials using antibiotics and ultrasound. Journal of Biomaterials Science, Polymer Edition 9(11): 1177–1185.

Joyce, E. and T.J. Mason. 2008. Sonication used as biocide: A review ultrasound a greener alternative to chemical biocides. Chemistry Today 26(6): 22–26.

Jyoti, K. and A. Pandit. 2004. Ozone and cavitation for water disinfection. Biochem. Eng. J. 18(1): 9–19.

Lee, B., S. Kermasha and B. Baker. 1989. Thermal, ultrasonic and ultraviolet inactivation of *Salmonella* in thin films of aqueous media and chocolate. Food Microbiol. 6(3): 143–152.
Lee, H., B. Zhou, W. Liang, H. Feng and S.E. Martin. 2009. Inactivation of *Escherichia coli* cells with sonication, manosonication, thermosonication, and manothermosonication: Microbial responses and kinetics modeling. J. Food Eng. 93(3): 354–364.
López-Malo, A., E. Palou, M. Jiménez-Fernández, S.M. Alzamora and S. Guerrero. 2005. Multifactorial fungal inactivation combining thermosonication and antimicrobials. J. Food Eng. 67(1): 87–93.
Mason, T., L. Paniwnyk and J. Lorimer. 1996. The uses of ultrasound in food technology. Ultrason. Sonochem. 3(3): S253–S260.
Mason, T.J. 1990. Sonochemistry: The use of ultrasound in chemistry. Royal Society of Chemistry, Cambridge, England.
McClements, D.J. 1995. Advances in the application of ultrasound in food analysis and processing. Trends Food Sci. Technol. 6(9): 293–299.
Mukhopadhyay, S. and R. Ramaswamy. 2012. Application of emerging technologies to control *Salmonella* in foods: A review. Food Res. Int. 45(2): 666–677.
Muñoz, A., I. Palgan, F. Noci, D. Cronin, D. Morgan, P. Whyte and J. Lyng. 2012. Combinations of selected non-thermal technologies and antimicrobials for microbial inactivation in a buffer system. Food Res. Int. 47(1): 100–105.
Naddeo, V., M. Landi, V. Belgiorno and R. Napoli. 2009. Wastewater disinfection by combination of ultrasound and ultraviolet irradiation. J. Hazard. Mater. 168(2): 925–929.
Oulahal, N., A. Martial-Gros, M. Bonneau and L. Blum. 2007. Removal of meat biofilms from surfaces by ultrasounds combined with enzymes and/or a chelating agent. Innov. Food Sci. Emerg. Technol. 8(2): 192–196.
Oulahal-Lagsir, N., A. Martial-Gros, E. Boistier, L. Blum and M. Bonneau. 2001. The development of an ultrasonic apparatus for the non-invasive and repeatable removal of fouling in food processing equipment. Lett. Appl. Microbiol. 30(1): 47–52.
Oulahal-Lagsir, N., A. Martial-Gros, M. Bonneau and L.J. Blum. 2003. "*Escherichia coli*-milk" biofilm removal from stainless steel surfaces: Synergism between ultrasonic waves and enzymes. Biofueling 19(3): 159–168.
Pagán, R., P. Manas, I. Alvarez and S. Condón. 1999. Resistance of *Listeria monocytogenes* to ultrasonic waves under pressure at sublethal (manosonication) and lethal (manothermosonication) temperatures. Food Microbiol. 16(2): 139–148.
Palacios, P., J. Burgos, L. Hoz, B. Sanz and J. Ordoñez. 1991. Study of substances released by ultrasonic treatment from *Bacillus stearothermophilus* spores. J. Appl. Microbiol. 71(5): 445–451.
Palgan, I., A. Muñoz, F. Noci, P. Whyte, D. Morgan, D. Cronin and J. Lyng. 2012. Effectiveness of combined pulsed electric field (PEF) and manothermosonication (MTS) for the control of *Listeria innocua* in a smoothie type beverage. Food Control 25(2): 621–625.
Patil, S., P. Bourke, B. Kelly, J.M. Frías and P.J. Cullen. 2009. The effects of acid adaptation on *Escherichia coli* inactivation using power ultrasound. Innov. Food Sci. Emerg. Technol. 10(4): 486–490.
Patist, A. and D. Bates. 2008. Ultrasonic innovations in the food industry: From the laboratory to commercial production. Innov. Food Sci. Emerg. Technol. 9(2): 147–154.
Piyasena, P., E. Mohareb and R. McKellar. 2003. Inactivation of microbes using ultrasound: A review. Int. J. Food Microbiol. 87(3): 207–216.
Raso, J., R. Pagan, S. Condon and F. Sala. 1998. Influence of temperature and pressure on the lethality of ultrasound. Appl. Environ. Microbiol. 64(2): 465–471.
Sagong, H.G., S.Y. Lee, P.S. Chang, S. Heu, S. Ryu, Y.J. Choi and D.H. Kang. 2011. Combined effect of ultrasound and organic acids to reduce *Escherichia coli* O157: H7, *Salmonella typhimurium*, and *Listeria monocytogenes* on organic fresh lettuce. Int. J. Food Microbiol. 145(1): 287–292.
Sams, A. and R. Feria. 2006. Microbial effects of ultrasonication of broiler drumstick skin. J. Food Sci. 56(1): 247–248.

Scherba, G., R. Weigel and W. O'brien. 1991. Quantitative assessment of the germicidal efficacy of ultrasonic energy. Appl. Environ. Microbiol. 57(7): 2079–2084.

Seymour, I., D. Burfoot, R. Smith, L. Cox and A. Lockwood. 2002. Ultrasound decontamination of minimally processed fruits and vegetables. Int. J. Food Sci. Tech. 37(5): 547–557.

Soria, A.C. and M. Villamiel. 2010. Effect of ultrasound on the technological properties and bioactivity of food: A review. Trends Food Sci. Technol. 21(7): 323–331.

Tiwari, B., K. Muthukumarappan, C. O'Donnell and P. Cullen. 2009. Inactivation kinetics of pectin methylesterase and cloud retention in sonicated orange juice. Innov. Food Sci. Emerg. Technol. 10(2): 166–171.

Ugarte-Romero, E., H. Feng, S.E. Martin, K.R. Cadwallader and S.J. Robinson. 2006. Inactivation of *Escherichia coli* with power ultrasound in apple cider. J. Food Sci. 71(2): E102–E108.

Ulusoy, B.H., H. Colak and H. Hampikyan. 2007. The use of ultrasonic waves in food technology. Res. J. Biol. Sci. 2(4): 491–497.

Villamiel, M. and P. de Jong. 2000. Inactivation of *Pseudomonas fluorescens* and *Streptococcus thermophilus* in trypticase soy broth and total bacteria in milk by continuous-flow ultrasonic treatment and conventional heating. J. Food Eng. 45(3): 171–179.

Williams, D. 2012. The physics of ultrasound. Anaesthesia and Intensive Care Medicine 13(6): 264–268.

Wong, E., F. Vaillant-Barka and E. Chaves-Olarte. 2012. Synergistic effect of sonication and high osmotic pressure enhances membrane damage and viability loss of *Salmonella* in orange juice. Food Res. Int. 45(2): 1072–1079.

Wordon, B., B. Mortimer and L. McMaster. 2011. Comparative real-time analysis of *Saccharomyces cerevisiae* cell viability, injury and death induced by ultrasound (20kHz) and heat for the application of hurdle technology. Food Res. Int. 47(2): 164–139.

Wrigley, D. and N. Llorca. 1992. Decrease of *Salmonella typhimurium* in skim milk and egg by heat and ultrasonic wave treatment. J. Food Prot. 55(9): 678–680

Yap, A., V. Jiranek, P.R. Grbin, M. Barnes and D. Bates. 2007. The application of high power ultrasonics to enhance winemaking processes and wine quality. Australian & New Zealand Wine Industry Journal 22(2): 44–48.

Zhou, B., H. Feng and Y. Luo. 2009. Ultrasound enhanced sanitizer efficacy in reduction of *Escherichia coli* O157: H7 population on spinach leaves. J. Food Sci. 74(6): M308–M313.

Zips, A., G. Schaule and H. Flemming. 1990. Ultrasound as a means of detaching biofilms. Biofouling 2(4): 323–333.

CHAPTER 5

Ozone: A Novel Microbial Inactivation Process

*S. Patil, P.J. Cullen and P. Bourke**

Introduction

Food preservation is a major concern for the food processing industry, where the issues of maintaining both product quality and microbiological safety are key. In tandem, consumer awareness of food safety and quality issues has risen. The principle culprits for food spoilage and food safety issues are microorganisms. Therefore, when developing a novel technology and/or progressing applications of existing technologies, a holistic approach to optimising quality characteristics whilst retaining safety is advised.

The efficacy of a preservation technology is often measured in terms of its efficacy against pathogenic and spoilage microorganisms. However, developing an understanding of microbial physiology and behaviour subject to a processing treatment, by means of investigating the mechanism of inactivation, should provide insights for optimizing application of a technology.

Ozone is a triatomic form of oxygen and is a potent disinfectant for water treatment, food processing and preservation purposes. The approval of ozone as an antimicrobial agent for food treatment, storage and processing by the US Food and Drug Administration (FDA) expanded interest in ozone applications. It is a powerful broad-spectrum antimicrobial agent active against bacteria, fungi, viruses, protozoa, and bacterial and fungal spores (Khadre et al. 2001). Ozone inactivates microorganisms through oxidization

School of Food Science and Environmental Health, Dublin Institute of Technology, Cathal Brugha Street, Dublin-1, Ireland.
Emails: sonalpatil81@gmail.com; patrick.j.cullen@dit.ie; paula.bourke@dit.ie
* Corresponding author

and residual ozone spontaneously decomposes to non-toxic products (i.e., oxygen), making it an environmentally friendly antimicrobial agent for use in the food industry (Kim et al. 1999b).

In this chapter, applications of ozone in food processing as well as approaches that can be employed for assessing the microbial inactivation or cell injury within a microbial population after exposure to inactivation treatment towards optimising application are described.

Production of Ozone

Ozone is a bluish gas with a pungent odour and strong oxidizing properties. Generally, the electrical (corona) discharge method is used for production of ozone at a commercial level. In this method, dried air or O_2 is passed between two high-voltage electrodes separated by a dielectric material, which is usually glass. When a voltage is supplied to the electrodes, a corona discharge forms between the two electrodes, and the O_2 in the discharge gap is converted to ozone. First, oxygen molecules (O_2) are split into oxygen atoms (O), subsequently the individual oxygen atoms combine with the remaining oxygen molecules to form ozone (O_3) (Muthukumarappan et al. 2008). Alternative methods that can be or are used for generation of ozone are described fully in the literature (Muthukumarappan et al. 2000, 2009, Sacco 2009).

Applications of Ozone in Food Processing

The oxidizing action of ozone for food application can be affected by environmental factors such as pH, temperature, relative humidity and organic compounds in the medium (Kim et al. 1999b). The influence of these factors on ozone efficiency is further discussed in the literature (Khadre et al. 2001, Patil and Bourke 2012).

The ozone demand of a food sample depends on the organic constituents present in that food which compete with microorganisms for ozone. Therefore, commodities with different chemical composition require different ozone dosage for effective sanitization. For example, ozone treatment of a meat sample will require higher doses of ozone than that required for fruits and vegetables (Kim et al. 2003). Thus for any required microbial inactivation, higher doses of ozone might be necessary which may lead to adverse effects on nutritional and quality parameters of treated food product. Ozone has been widely and successfully applied for decontamination of processing water in the food industries.

Chlorine has been used for decades for sanitation purposes, but has been found to produce toxic compounds. Ozone has also been applied for various sanitation applications in the food processing industry, such as in the meat industry and for detoxification of aflatoxins (Kim et al. 1999b, King and Prudente 2005). In the literature, a number of ozone applications for food processing have been described in detail (Kim et al. 2003, Guzel-Seydim et al. 2004, Tiwari et al. 2010).

Fruits and vegetables

Generally synthetic additives are used to preserve food or improve characteristics such as colour and flavour, however, consumers are concerned about usage of additives (Bruhn 2000). Contamination of fresh produce or minimally processed fresh-cut products comes from a number of sources such as postharvest handling, processing and microorganisms; natural contaminants for these products (Beuchat 1996). Chlorine based washing and decontamination procedures are used by a majority of fresh minimally processed fruit and vegetable manufacturers, but there are ongoing concerns with respect to the use of chlorine due to the risk of undesirable by-products upon reaction with organic matter due to the formation of toxic chlorine by-products such as tri-halomethanes and chloramines (Martín-Diana et al. 2007) and this may lead to new regulatory restrictions in the future (Barry-Ryan and Bourke 2012).

Ozone has been utilized as a strong antimicrobial agent for washing of fruits and vegetables (Karaca and Velioglu 2007) and has also shown to be beneficial with respect to extending storage life of fresh produce (Kim et al. 1999b, Skog and Chu 2001). Alothman et al. (2010) indicated that gaseous ozone treatment at a flow rate of 8±0.2 ml/s for up to 20 min would be useful for enhancing the nutritional properties of banana and pineapple fruits, but it would have a negative effect on the nutritional value of guava fruits. The results of their study indicated that ozone can be used in minimally processed fresh food industry to enhance the antioxidant status of fresh-cut fruits. Ozone has been reported to improve the quality and storage life of cold stored broccoli, cucumber and blackberries (Barth et al. 1995, Skog and Chu 2001). Tzortzakis et al. (2007) reported that ozone exposure did not affect quality related attributes of cold stored tomato fruit. Ozonated water washings (2 ppm) were reported to be more effective for reducing microbial load on red peppers, strawberries and watercress than simple water washings (Alexandre et al. 2011). Generally, the use of ozone for fruit or their products and vegetables processing is widely reported (Beuchat 1992, Sarig et al. 1996, Kim et al. 1999a, Song et al. 2000, Achen and Yousef 2001, Skog and Chu 2001, Palou et al. 2002, 2003, Garcia et al. 2003, Baur et al. 2004, Beltrán et al. 2005, 2005a, Zhang et al. 2005, Metzger et al. 2007, Akbas and Ozdemir 2008, Zorlugenç et al. 2008, Barboni et al. 2010, Gabler et al. 2010, Ozkan et al. 2011). However, ozone efficacy has been reported as dependent on various factors such as organic matter or composition of a food, ozone concentration, treatment time and temperature (Patil and Bourke 2012). The improper use of ozone can result in some deleterious effects on physiology and quality of products such as losses in sensory quality (Karaca and Velioglu 2007).

Poultry, seafood and meat

Ozone (gas or aqueous phase) has been used to inactivate microorganisms and decontaminate meat, eggs, poultry, fish, fruits, vegetables and dry foods (Fan et al. 2007). The application of ozone either for surface disinfection of meats (Sheldon and Brown 1986, Greer and Jones 1989, Rusch and Kraemer 1989, Pohlman et al. 2002, Castillo et al. 2003, Novak and Yuan 2003, 2004) or for preservation of meats (Chen et al. 1992) has been reported. The use of ozone for reconditioning recycled poultry chilling water was approved by the US Department of Agriculture in 1997 (Guzel-Seydim et al. 2004). Waldroup et al. (1993) investigated the use of ozone for recycling of poultry chill water and reported reductions in excess of 99 per cent for total aerobes, *Escherichia coli* or presumptive coliforms after ozonation. The effects of ozone have also been evaluated on poultry meat microflora (Yang and Chen 1979), where ozone washing in combination with refrigerated storage reduced microbial counts with an estimated extension of poultry shelf life by 2.4 days. Nieto et al. (1984) studied the effect of ozone on microbiological and sensory quality of poultry during refrigerated storage, where ozone exerted a strong bactericidal effect on coliforms, *E. coli* and *Salmonella* with no alteration in the sensory characteristics of the meat observed. The effect of ozone as an antimicrobial treatment or disinfectant for poultry carcasses has also been studied (Sheldon and Brown 1986, Fabrizio et al. 2002, Vadhanasin et al. 2004).

Rasanayagam (2006) reported that ozone concentrations of higher than 10,000 ppm reduced the population of inoculated *Salmonella* in beef and poultry by 99 per cent. Iacumin et al. (2012) investigated the use of ozonated air to limit mould growth on meat products. The gaseous ozone treatment prevented the growth of *Aspergillus ochraceus* and the presence of ochratoxin on the surface of sausages without adversely influencing ripening, physico-chemical parameters, peroxide value or sensory characters of the treated sausages. Coll Cárdenas et al. (2011) analyzed the effects of ozone treatment at refrigeration temperatures on microbial counts in culture media, beef samples and on beef quality parameters (surface colour and rancidity). Gaseous ozone (154×10^{-6} kg m^{-3}) treatment (3 or 24-hr at 0° and 4°C) applied in culture media inoculated with *E. coli* resulted in total inactivation of this organism. However, the treatment was less effective in beef tissue and quality characteristics such as surface colour and lipid peroxidation were undesirable.

Seafood is generally highly perishable and effective technologies to extend its shelf-life, in addition to refrigeration, are still required. A number of studies reported the application of ozone for seafood processing (Rice et al. 1982, Ravesi et al. 1988, Brooks and Pierce 1990, Kötters et al. 1997, Kim et al. 2000, Nash 2002, Gelman et al. 2005, Campos et al. 2006, Vaz-Velho et al. 2006, Pastoriza et al. 2008). Gonçalves (2009) summarized the effectiveness of ozone application for the seafood industry.

Davidson et al. (2011) reported use of ozone for an improved water quality environment within water recirculating aquaculture systems. Recently Crowe et al. (2012) investigated the application of ozone sprays (1 mg/L and 1.5 mg/L) to improve the microbial safety and quality of salmon fillets. Their study indicated that ozone spray at concentrations up to 1.5 mg/L were effective in reducing initial counts of aerobic bacterial populations and inoculated *Listeria innocua* counts without causing significant increase in lipid peroxidation levels in salmon fillets held under refrigerated storage at 4°C. Similarly, Bono and Badalucco (2012) reported that ozone in combination with modified atmosphere packaging was a promising technique to extend the shelf-life of striped red mullet maintaining both microbiological safety and quality. The combined treatment significantly reduced microbial spoilage; also no pro-oxidant effect on lipid degradation was detected.

Pesticides, herbicides, fungicides and mycotoxins

Ozone or ozonated water treatment has been reported to be effective for pesticide degradation on vegetables or on fresh or processed apples (Ong et al. 1996, Wu et al. 2007, Ikeura et al. 2011). An increased efficiency of pesticide degradation in natural waters was reported by the combined use of ozone and hydrogen peroxide (Chelme-Ayala et al. 2011).

Herbicides are commonly applied to control weeds in crops. An extensive application of these compounds has increased the concern over the presence of their residues on foods. Herbicides like 2, 4-dichlorophenoxyacetic acid (2, 4-D) are environmentally unfriendly due to their toxicity (Chu et al. 2004). Use of chlorinated oxidants for herbicide removal is not considered to be advisable, as it might form toxic by-products (Richardson 1983). Ozone application for this purpose is valuable as it is an environmentally-friendly agent which decomposes into oxygen. The efficacy of ozone alone or in combination with another process for degradation or removal of herbicide and fungicide was investigated (Xiong and Graham 1992, Hulsey et al. 1993, Beltrán et al. 1999, Javier Benitez et al. 2002, Acero et al. 2003, Drzewicz et al. 2004, Benitez et al. 2005, Hladik et al. 2005, Yu et al. 2006,

Gabler et al. 2010, Walse and Karaca 2011, Karaca et al. 2012). Karaca et al. (2012) investigated the effect of continuous ozone exposure (0.3μL/L) for degradation of some fungicides, which are commonly applied in table grape vineyards, during the cold storage of table grape berries. Ozone atmosphere during storage was reported to degrade fungicides such as fenhexamid, cyprodinil, or pyrimethanil more rapidly by comparison with storage in air atmosphere.

Mycotoxins, e.g., trichothecene are naturally occurring metabolites produced by fungi on a variety of cereal grains. They are linked with several diseases in animals and humans (Pittet 1998, Placinta et al. 1999). The effectiveness of ozone has also been reported for the degradation and detoxification of commonly occurring mycotoxins (McKenzie et al. 1997, Young et al. 2006, Inan et al. 2007, Zorlugenç et al. 2008, Beyer et al. 2009).

Microbiological Aspects of Ozone Applications in Food Processing

Bacteria

Food-borne pathogens pose a major threat to public health. Therefore, ensuring microbiological safety along with quality of a product is a necessity for any applied food processing technology.

The potential of ozone for inactivation of bacteria including food-borne pathogens has been studied (Restaino et al. 1995, Zhao and Cranston 1995, Achen and Yousef 2001, Weavers and Wickramanayake 2001, Han et al. 2002, Wang et al. 2004, Bialka and Demirci 2007a, 2007b, Selma et al. 2007, 2008, Cabo et al. 2009, Klockow and Keener 2009, Neal et al. 2012). Daş et al. (2006) studied the bactericidal effect of gaseous ozone treatment for surface decontamination of *Salmonella enteritidis* inoculated on tomatoes. Gaseous ozone treatment of 10 mg/l of for 5 and 15 min was effective for reducing low (approximately 3.0 $\log_{10}$ cfu/tomato) or high (approximately 7.0 $\log_{10}$ cfu/tomato) levels of *S. enteritidis* attached for 1 hr respectively. Complete inactivation of cells for high inoculum dose with a 4 hr attachment time was achieved with 20 min of ozone treatment and increasing ozone concentrations to 20 mg/l or 30 mg/l resulted in complete elimination of *S. enteritidis* cells in 15 or 5 min of treatment, respectively. Recently, Trinetta et al. (2011) evaluated the effectiveness of three non-thermal technologies, chlorine dioxide gas, ozone gas and e-beam irradiation for inactivation of *S. enterica* and *E. coli* O157:H7 on pre-inoculated tomato, lettuce and cantaloupe seeds. All treatments reduced initial load of pathogenic bacteria on seeds significantly ($p<0.05$). In particular, ozone gas treatment (4.3mg/l for 5min) consistently reduced pathogenic bacteria on seeds by 4.0 log CFU/g.

Combined application of ozone with other technologies/processes was also reported to be more effective for controlling pathogens than ozone treatments alone (Singh et al. 2002, Williams et al. 2004, 2005, Yuk et al. 2007, Vurma et al. 2009, Kim and Hung 2012). Singh et al. (2002) studied the efficacy of chlorine dioxide, ozone, and thyme essential oil for reducing *E. coli* O157:H7 on lettuce and baby carrots. In that study, sequential washing treatments, i.e., thyme oil followed by aqueous chlorine dioxide/ozonated water, or ozonated water/chlorine dioxide, were reported to be more effective in reducing bacterial load on lettuce and baby carrots compared to sanitizers evaluated alone. Similarly, Yuk et al. (2007) concluded combined treatment of ozone (3 ppm) and 1 per cent organic (citric) acid was efficient for significant reduction of tested bacteria on enoki mushroom. Kim and Hung (2012) reported combination of UV technology with ozone was more effective for reduction of *E. coli* O157:H7 on blueberries than the use of individual methods.

Yeasts and moulds

Food spoilage renders foods products unacceptable or unpalatable for human consumption. The range and predominant type of spoilage microorganism in any food product depends on a range of intrinsic and extrinsic parameters. Yeasts have the ability to grow at low pH, high sugar concentration and low water activity conditions as well as in products containing preservatives (Deak and Beuchat 1996). Yeasts have been responsible for the spoilage of a wide variety of foods including preserved, processed and also refrigerated food (dairy, meat products) (Fung and Liang 1990, Viljoen 2001, Cocolin et al. 2002). Watanabe et al. (2010) evaluated the effects of UV irradiation and ozone treatment for inactivation of yeast and mould in mineral water and reported that combining both technologies with heat treatment would achieve definite yeast inactivation.

Ozone has been reported to prevent fungal decay (Pérez et al. 1999, Palou et al. 2002) or reduction of fungi population on the surface of grains, figs (Kells et al. 2001, Öztekin et al. 2006, Raila et al. 2006, Wu et al. 2006, Zorlugenç et al. 2008). McDonough et al. (2011) evaluated the efficacy of ozone application to treat grain for insect pests, fungal contaminants, and mycotoxins. Ozone application (47,800 ppm) in a single pass through modified screw conveyor reduced the mould (*Aspergillus flavus*) counts by 96 per cent on the surface of corn kernels.

The efficacy of ozone for degradation or removal of aflatoxin has also been studied. Ozone effectiveness for degradation of aflatoxin on corn, in peanut industry, i.e., for peanut kernels and flour, also in figs has been reported (McKenzie et al. 1997, 1998, Prudente and King 2002, Proctor et al. 2004, Zorlugenç et al. 2008).

Viruses

Ozone was applied for several years for elimination of viruses (Burleson et al. 1975, Kim et al. 1980, Farooq and Akhlaque 1983, Vaughn et al. 1987, Shriniki et al. 1988, Herbold et al. 1989, Thurston-Enriquez et al. 2005, Lin and Wu 2006, Hudson et al. 2007, Sudhakar et al. 2007, Murray et al. 2008, Lénès et al. 2010). Tseng and Li (2008) investigated the inactivation of surface viruses by gaseous ozone. Effect of various parameters on viral inactivation such as ozone concentration, contact time, different capsid architecture of viruses, and relative humidity were studied. Ozone was effective for reducing the viral number between 1 and 3 logs on surfaces.

Recently, Hirneisen et al. (2011) studied ozone inactivation of human norovirus surrogates (feline calcivirus [FCV] and murine norovirus [MNV] on fresh produce (green onions and lettuce) and in sterile water. Ozone treatment (6.25 ppm) of fresh produce inoculated with norovirus surrogates was conducted at 0.5 to 10 min time intervals. Infectivity was measured by 50 per cent tissue culture infectious dose ($TCID_{50}$) and plaque assay for FCV and MNV respectively. Ozone was reported as effective for reduction of viral contamination on the surface of fresh produce.

Enzymes

Polyphenol oxidases (PPO) are a group of enzymes which are responsible for enzymatic browning of food, and were reported to be inactivated by ozone (Rico et al. 2006, Whangchai et al. 2006).

Additionally, bacterial cells generally respond to the oxidative stress exerted by ozone with antioxidant enzymes which play a role in protecting the cells. Whiteside and Hassan (1987) investigated the effects of ozone on biosynthesis of the antioxidant enzymes in *E. coli* to assess their role against ozone toxicity and increased activities of antioxidant enzymes in ozone exposed *E. coli* cultures was recorded. Ozone can stimulate the expression of antioxidant enzymes (Rahman et al. 1991, Rao et al. 1996). Lee et al. (2003) reported ozone inactivation of antioxidant enzymes including Cu/Zn superoxide dismutase (CuZnSOD), catalase and glutathione peroxidise. The primary target of ozone was CuZnSOD but presence of antioxidants like glutathione and ascorbate fully protect the enzyme from ozone unless cells were exposed to extreme oxidative stress which results in depletion of the antioxidants. Fisher et al. (2000) reported both catalase and superoxide dismutase were found to protect listerial cells from ozone attack, with superoxide dismutase being more important than catalase in this protection. Thus, one way to enhance the bactericidal effect of ozone is through inactivation of antioxidant enzymes.

Assessment of Microbial Inactivation Mechanism by Ozone

Estimating the microbial inactivation efficacy of an applied processing technology is critical to support application. In a majority of studies that investigated ozone efficacy, microbiological analysis was conducted using standard plate count techniques. However, understanding the basis or mechanism for that inactivation is also critical for informed optimisation of processing technologies. In this section, we describe techniques that were or could be applied for assessment of microbial inactivation mechanism by ozone.

Standard/conventional microbiological technique

In order to determine the cell viability of ozone treated cell populations using the standard plate count technique, aliquots of diluted or undiluted ozone treated samples can be plated on solid medium of interest such as tryptic soya agar (TSA). Following incubation under required conditions, inactivation kinetics are recorded in the form of colony-forming units per millilitre or gram (cfu/ml or cfu/g) depending on the type of product treated. Many studies applied this standard technique to assess microbial inactivation efficiency of ozone (Kim et al. 1999a, Williams et al. 2005, Fan et al. 2007, Patil et al. 2010, Trinetta et al. 2011, Choi et al. 2012, Crowe et al. 2012).

This technique efficiently determines the inactivation kinetics of culturable populations after ozone treatment. Other approaches that may be employed for better understanding of ozone action are described further.

Molecular methods and instrumental analysis

Measurement of Nucleic Acid Degradation. DNA is one of the targets of oxidative damage. Ozone is known to affect purines and pyrimidines in nucleic acid (Christensen and Giese 1954, Scott and Lesher 1963). Ozone produces single-strand breaks in DNA which, if unrepaired, cause extensive breakdown of DNA in *E. coli*, resulting in loss of cell viability (Hamelin et al. 1977, 1978). Ishizaki et al. (1987) reported that the plasmid DNA present in *E. coli* cells is converted from closed circular DNA to open circular DNA by ozone. The release of nucleic acid content or its disintegration after treatment can be measured as described below.

Spectroscopic Method. The amount of DNA and RNA released from cytoplasm can be estimated through measurement of absorbance at 260 nm (Chen and Cooper 2002). This method was used for measurement of bacterial cell membrane integrity after exposure to antibacterial compounds (Nobmann et al. 2010, Xing et al. 2009). Hence, a simple method to determine the release of nucleic acid following ozonation is measurement of absorbance at 260 nm. Previous studies employed this method for measuring nucleic acid leakage after ozone treatment of bacterial populations (Komanapalli and Lau 1996, Curtiellas et al. 2005, Patil et al. 2011). Thus absorbance measurement will indicate the damaged or compromised bacterial membrane after exposure to inactivation technique.

Agarose Gel Electrophoresis. DNA damage following treatment can be measured by performing agarose gel electrophoresis. This technique was used to demonstrate physical disintegration of DNA (Zirkle and Krieg 1996, Yoon et al. 2002). In recent times, plasma exposure of *E. coli* for different time intervals (0 to 180 sec) resulting in damaged genomic DNA was reported (Joshi et al. 2011). In that study, one of the methods employed for detecting DNA damage by plasma exposure was agarose gel electrophoresis. Cooper (2009) showed the degradation of plasmid DNA by exposure to dielectric barrier discharge (DBD) plasma using gel electrophoresis technique. Such determinations are important as bacteria such as *E. coli* can retain the integrity of their plasmids in the viable but non-culturable state (Byrd and Colwell 1990) and thus it is important to eliminate the presence of organism by disintegrating the presence of plasmid (Cooper 2009). Using a similar method Cooper (2009) also proved complete destruction of chromosomal DNA by DBD plasma only after 2 sec plasma treatment. A similar approach can be used to indicate DNA damage after application of ozone processing.

Polymerase chain reaction (PCR). Another method that can be employed for DNA damage detection is PCR (Fig. 1), in which amplification stops at the DNA damage site (Kumari et al. 2008). The highly conserved region of bacterial DNA, i.e., 16SrDNA, is a typical fingerprint of microorganism, used to detect and identify the presence of organism (Kolbert et al. 2004, Bae et al. 2008). Before and after ozone treatment, amplification of 16SrDNA can be done using PCR and agarose gel electrophoresis can be performed indicating damage profiles. Absence of a visible band indicates the possibility of degradation/damage of released DNA due to processing treatment (Bae et al. 2008, Cooper et al. 2010). Ozone is also known to produce DNA single- and double-strand breaks in *E. coli* which was previously determined by alkaline sucrose gradient sedimentation (Hamelin and Chung 1989).

Enzyme Linked Immunosorbent Assay (ELISA). Oxidative damage to DNA can be detected by measuring or determining 8-hydroxydeoxyguanosine (8-OHdG). It is a major marker of oxidative DNA damage in prokaryotic cells like *E. coli*. This oxidative DNA damage by-product is excreted during the repair process of damaged DNA *in vivo* by exonucleases (Cooper 2009).

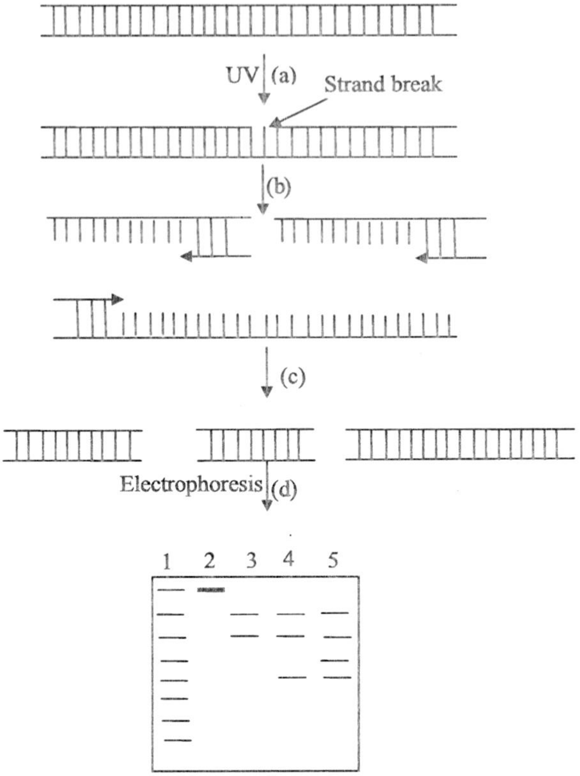

Figure 1. Schematic representation of the PCR application in DNA damage detection. Due to UV radiation or other agents, strand breakage occurs resulting in separate amplification of the single strand that is identified after electrophoresis: (a) Strand break by UV irradiation, (b) Denaturation and annealing, (c) Amplification and (d) Gel electrophoresis showing different bands- Lane 1- Marker, Lane 2- Control, Lane 3- 1 h, Lane 4- 2 h and Lane 5- 3 h of UV treatment. Adapted from Kumari et al. (2008).

Joshi et al. (2011) used enzyme linked immunosorbent assay (ELISA) technique to demonstrate oxidative DNA damage markers during plasma treatments. Determination or quantification of 8-OHdG in an unknown sample is done by comparing its absorption with that of a known standard curve (Joshi et al. 2011). Such measurements or detection would be an additional marker for damage exerted by ozone.

Comet Assay. This technique is used to detect single-strand break, double-strand break, oxidative DNA damage and single-strand break associated with incomplete excision repair sites caused by UV radiation, ultrasound, electromagnetic frequency radiation, etc. (Kumari et al. 2008). In this technique, the DNA tailing combined with the tail length represents the extent of DNA damage (Olive et al. 1990). Sastre et al. (2001) utilized the comet assay to detect UV-induced damage in DNA cells of *Rhodomonas* spp. (Cryptophyta) and indicated that the comet assay is more sensitive to UV radiation damage when used in conjunction with T4 endonuclease V. Dhawan et al. (2009) reviewed the use of comet assay to assess the DNA-damaging potential of chemicals and/or environmental conditions in different cell models. Liao et al. (2011) utilized the comet assay method to analyze DNA degradation in high pressure carbon dioxide (HPCD) treated *E. coli.* The schematic representation of the comet assay is shown in Fig. 2 (Adapted from Kumari et al. 2008).

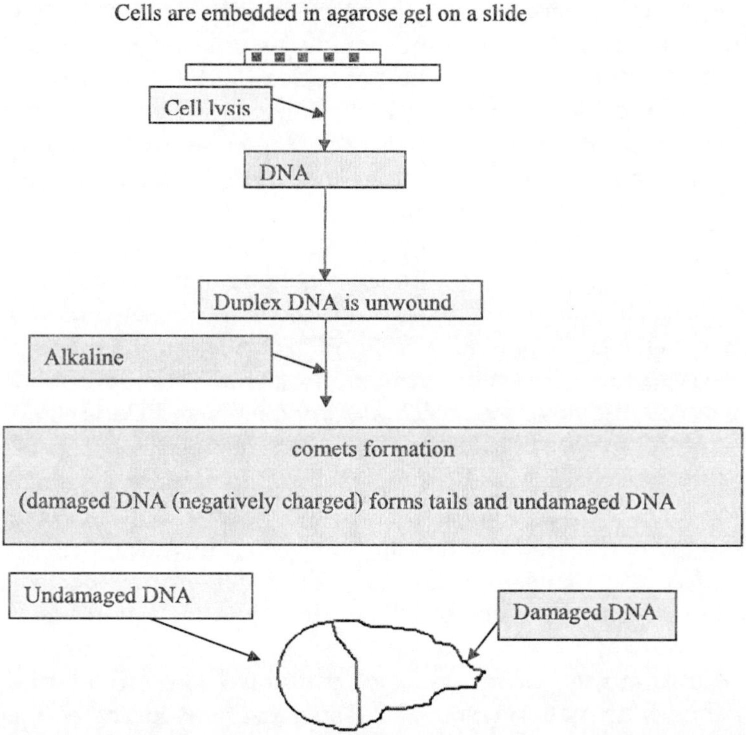

Figure 2. Schematic representation of the comet assay. Adapted from Kumari et al. 2008.

Gas Chromatography-Mass Spectrometry (GC-MS). GC-MS is used for the detection of wide range of oxidative DNA damage base products (Kumari et al. 2008). Fuciarelli et al. (1990) used GC-MS to investigate the radiation induced purine and pyrimidine products. However, this method can overestimate the level of oxidized bases due to the high or low temperature derivatization step of hydrolyzed DNA (Cadet et al. 2003) in the presence of air (Kumari et al. 2008).

Previously DNA damage profiles induced by different agents were determined by using endonucleases (Milligan et al. 1996, Rolda'n-Arjona and Sedgwick 1996). Kuipers and Lafleur (1998) studied the characterization of the DNA damage profiles due to γ-radiation induced water radicals. The endonuclease assay system was mainly employed for the detection of DNA damage.

High pressure liquid chromatography (HPLC) coupled with tandem mass spectrometry (MS/MS) detection technique is one of the more sensitive methods for the detection of oxidized DNA damage products allowing detection of DNA lesions at the femtomole (fmole) range (Ravanat 2005).

Enzymatic Activity Measurements. E. coli responds to oxidative stress by modifying the expression of many genes resulting in induction of proteins with antioxidant functions, e.g., superoxide dismutase and catalase. Ozone is known to stimulate the expression of antioxidant enzymes (Dubick and Keen 1983, Whiteside and Hassan 1987, Rahman et al. 1991) and the role of such enzymes provides information on the effect of ozone on specific enzymes, relating it to the possible generation of reactive species in the process. In *E. coli*, two redox-responsive transcription regulators (SoxR and OxyR) have been well defined. Both proteins have the remarkable ability of directly transducing oxidative signals to genetic regulation (Pomposiello and Demple 2001). Both regulators are induced in the presence of superoxide ($^{\bullet}O_2^-$) radicals (Greenberg et al. 1990) and activate various genes like *soxS* and *sod*, which in turn confer protection against these radicals through DNA repair or removal of the radicals (Pomposiello and Demple 2001). The DnaK and RpoS are two regulators of general stress genes, which, although not dedicated mechanisms of protection against oxidative radicals, have been previously shown to confer protection against them (Delaney 1990, Rockabrand et al. 1995, Loewen et al. 1998). Similar radicals are produced during ozone treatments, and therefore, these genes are expected to play an important role in protection of cells against the effects of this technology. Recently, Patil et al. (2011) investigated effect of oxidative stress of ozone treatment through responses of *E. coli* mutants lacking oxidative stress-related genes. Inactivation responses of such mutants indicated the role of enzymes in providing protection against ozone and a possible bactericidal mechanism of action of ozone exerted by production of ROS during the process.

Reaction of ozone and biomolecules results in the formation of reactive species which are responsible for the cytotoxic effect of ozone (Lee et al. 2003). Reaction of unsaturated fatty acids with ozone results in generation of aldehydes, hydrogen peroxide and organic radicals (Pryor and Church 1991, Pryor 1994).

Besides this, Takamoto et al. (1992) observed that ozone decreased enzyme activity in *E. coli* to a greater degree in the case of cytoplasmic beta-galactosidase than in case of the periplasmic alkaline phosphatase. Effect of ozone exposure on enzymatic activities was studied by Komanapalli et al. (1997). Inactivation of Glyceraldehyde-3-phosphate dehydrogenase and to lesser extents: pyruvate decarboxylase, phosphofructokinase 1 and NAD alcohol dehydrogenase was reported. Cho et al. (2010) studied ozone degradation kinetics of intracellular enzyme, β-D galactosidase and related it to the cell surface damage and subsequent change in cell wall permeability.

Lipid peroxidation. Lipid peroxidation is a process that occurs following oxidative stress in bacterial cells, which is known to be responsible for cellular damage (Yoon et al. 2002, Pérez et al. 2008) leading to loss of respiratory activity and cell death (Maness et al. 1999, Sokmen et al. 2001). During this process, reactive compounds such as malondialdehyde (MDA) and 4-hydroxynonenal are formed. A well-known method for detecting lipid peroxidation is Thiobarbituric Acid Reactive Substances (TBARS) assay. In this method, MDA forms a 1:2 adduct with thiobarbituric acid (TBA), and can be measured colorimetrically or fluorometrically using a MDA equivalent standard (Joshi et al. 2011). Joshi et al. (2011) observed lipid peroxidation after 60 sec of plasma treatment by both intact *E. coli* cells and membrane fractions. A significant accumulation of MDA following plasma treatment was reported. Similarly, Komanapalli et al. (1997) reported a significant time dependent increase in TBARS following ozone exposure of *E. coli* K-12 strains. Komanapalli and Lau (1998) observed a dose-dependent increase in the production of TBARS followed by ozone exposure of *E. coli* and *Candida albicans*. The amount of TBARS produced by *C. albicans* was greater than that produced by *E. coli* thus indicating the compromised membrane permeability of organisms. Cho et al. (2010) also observed the greatest amount of MDA formation by ozone treatment of *E. coli*. Murray et al. (2008) reported elevated lipid peroxidation levels, indicated by higher MDA concentrations after ozone treatment of viruses. This was related to the abundance of ROS which might be the cause of viral inactivation. Lipid peroxidation is one of the ways of ozone inactivation of viruses (Carbonneau et al. 1991, Dianzani 1993, Friedman and Stromberg 1993).

In biological membrane systems, lipid peroxidation products interact with DNA and proteins causing oxidative modifications (Joshi et al. 2011). Interaction of ozone with unsaturated lipids results in the production of free radicals or toxic intermediate products such as hydrogen peroxide and aldehydes (Pryor et al. 1991).

Protein Assay. Reaction of proteins or amino acid with ozone showed susceptibility to oxidation (Mudd et al. 1969). The amount of protein released during ozone disinfection process can be measured by simple absorbance measurement at 280 nm. In order to quantify the released protein concentration, the Bradford assay, a widely used method, can be utilized. The interaction of ozone primarily with either the lipid or protein component of biological membrane relating it to the characterization of early cellular events is important for better understanding of ozone toxicity (Komanapalli et al. 1997).

Cho et al. (2010) studied the amount of protein released after *E. coli* inactivation using several disinfectants including ozone and also determined the degradation of standard proteins (Fig. 3). After the treatment of *E. coli* with disinfectants, the supernatant was concentrated and analyzed using sodium dodecyl sulphate-polyacrylamide gel electrophoresis (SDS-PAGE) which showed several protein bands after ozone treatment. These results were similar to the protein assay with greater amounts of protein released after ozone treatment.

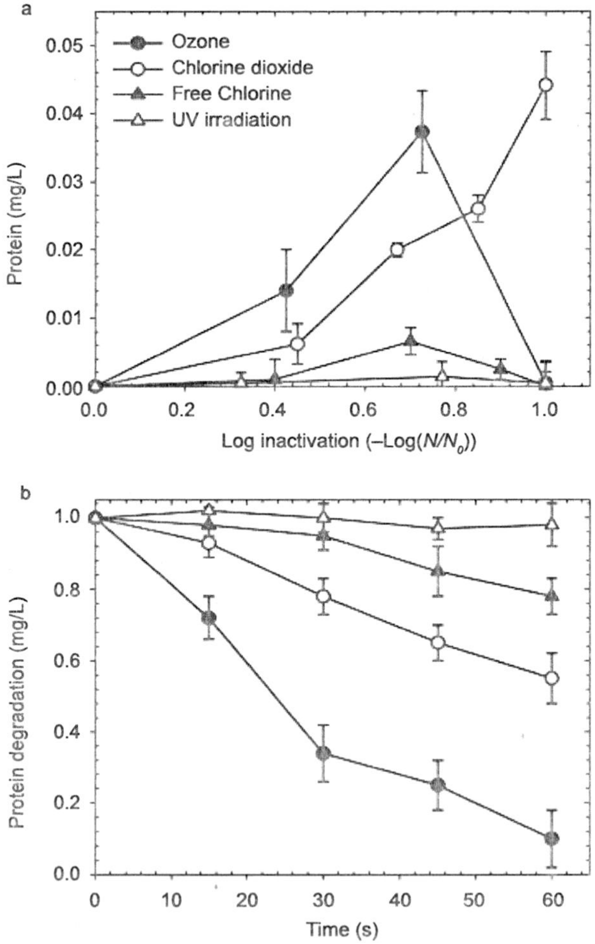

Figure 3. Bradford assay results: (a) Amount of protein released as a function of *E. coli* inactivation and (b) Degradation of standard protein by disinfectants (pH 7.1, 20°C, buffer condition). Adapted from Cho et al. (2010).

Determination of the release of cell constituents, e.g., protein, can indicate the cellular damage following the inactivation treatment.

Two-dimensional electrophoresis (2-DE) is another approach that can be used for detecting protein changes after exposure of cells to ozone treatment. Previously it has been used to demonstrate protein changes after exposure of bacteria to HPCD treatment (White et al. 2006, Kim et al. 2009).

Microscopic Techniques. In order to analyze the damage or alteration in cell structure of bacterial population after exposure to any bactericidal treatment, various microscopic techniques were employed.

Scanning electron microscopy (SEM) is one of the techniques used to image the damage or alteration in cell surface structure of bacterial population after treatments. Mahfoudh et al. (2010) utilized SEM to visualize eventual damage to the bacterial spore external structure after treatment with gaseous ozone. Following ozonation, severe alterations to the spore integrity were observed, but the morphology of a minority of spores was unaffected. Environmental SEM was used to visualize the effect of ozone exposure on xerophilic fungus, *Eurotium amstelodami* IS-SAB-01 (Antony-Babu and Singleton 2011). Diao et al. (2004) investigated the damage to the surface morphology and structure of the cells after various disinfection treatments including ozonation using SEM (Fig. 4).

Thanomsub et al. (2002) revealed ultrastructural changes of ozone treated gram positive and gram negative bacteria by SEM showing deformation, rough damage and surface destruction. Recently, Zhang et al. (2011) examined effect of ozone exposure on membrane permeability and ultrastructure in *Pseudomonas aeruginosa*. In that study, the authors observed that most of the cells were intact after ozone treatment with vesicles on the surface, which was determined by SEM. Thus, SEM was used to examine the cell structure after treatment which indicates the damage or alteration of the cell indicating the membrane permeability or cell lysis.

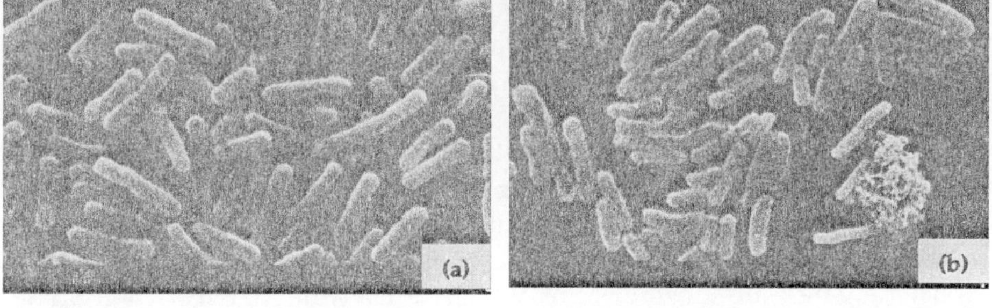

Figure 4. SEM photographs of *E. coli* cells in (a) fresh culture and after, (b) ozonation at 10 mg/l for 5 min. Adapted from Diao et al. (2004).

Transmission electron microscopy (TEM) is another method used to demonstrate bacterial cell surface damage or alteration after exposure to processing treatments. Khadre and Yousef (2001) demonstrated the damage to *Bacillus* spp. outer spore coat, the surface layer and also to some extent to the inner spore coat layer after ozone treatment using TEM. Cho et al. (2010) utilised TEM and demonstrated the damage to cell surfaces, inner components and morphological change in *E. coli* depending on the treatment with various disinfectants. A significant morphological change was apparent with ozone treated *E. coli*. Murray et al. (2008) demonstrated that the integrity of the viral envelope and protein coat was compromised by ozone treatment by performing electron microscopy. Zhang et al. (2011) examined the mechanism of ozone induced damage to cytoplasmic membrane and cell ultrastructure of *P. aeruginosa* using both SEM and TEM thus revealing the effect of ozone treatment on cell surface structure as well as effect on cytoplasmic contents.

Fluorescence microscopy is another technique used to indicate the live or dead population ratio after treatment with inactivation techniques. Nagayoshi et al. (2004) studied the effect of ozonated water on oral microflora and dental plaque. In that study, LIVE/DEAD® BacLight™ bacterial viability kit solution was used to indicate viable and dead population of bacteria following ozone treatment. Microscopic analysis revealed almost complete killing of *Streptococcus mutans* cells by ozonated water treatment. This approach could be useful for instantaneous detection of viable or dead bacterial population relating it to the treatment efficacy.

Fluorescent probes. A simple approach to distinguish between live and dead populations after ozone treatment uses fluorescent probes. Fluorescent probes can measure bacterial cell viability. Different probes detect metabolic or functional activity of a cell, such as propidium iodide, SYTO 9 indicating membrane integrity, fluorescein diacetate (FDA) for metabolic activity, dyes like 2,3-bis (2-methoxy-4-nitro-5-sulfophenyl)-5-[(phenylamino)carbonyl]-2H-tetrazo-lium hydroxide (XTT) for metabolic acitivity (Joshi et al. 2011), thiazole orange and propidium iodide to indicate compromised cell membranes and RNA and DNA damage (Fröhling et al. 2012). Fluorescent probes can target different cell structures and locations and are also available for cell components such as proteins, lipids, membranes, DNA, RNA. The fluorescence of probes or dyes is detected by using various instruments that include spectrofluorometers, flow cytometers, fluorescent microscopes, confocal microscopes and fluorescent microplate readers (Trevors 2003). Trevors (2003) described the use of fluorescent probes for membrane research in bacteria. There are specific probes which indicate the change in membrane potential after exposure of cells to disinfection treatment, e.g., Rhodamine 123 (Rh123), 3,3′-dihexyloxacarbocyanine; DiOC6(3). A slow response potential-sensitive probe DiSBAC2(3), i.e., (bis-(1,3-diethylthiobarbituric acid) trimethine oxonol) was used to measure the membrane potential of *E. coli* after plasma treatment in PBS (Joshi et al. 2011). A significant increase in fluorescence by this probe in proportion to the plasma energy applied was recorded. Joux and Lebaron (2000) summarised the different physiological target sites of fluorescent probes (Fig. 5).

Application of fluorescent probes can prove useful for analysis of how a processing technology is effecting a target microbial population with respect to retention of cell membrane integrity, or damage to cell components thus indicating the applicability or efficacy of the processing technology under a range of conditions.

The decomposition of ozone results in the generation of superoxide radicals, hydroperoxyl radicals and hydroxyl radicals (Adler and Hill 1950, Hoigné and Bader 1975). Joshi et al. (2011) utilised fluorescent probes for detection of reactive oxygen species (ROS) during plasma treatment. The presence of ROS in disinfection process can be identified by using specific fluorescent probes such as Amplex red, hydroethidine, 2-[6-(4V-Hydroxy) phenoxy-3H-xanthen-3-on-9-yl]benzoic acid (HPF) and 2-[6-(4V-amino) phenoxy-3H-xanthen-3-on-9-yl] benzoic acid (APF) (Gomes et al. 2005).

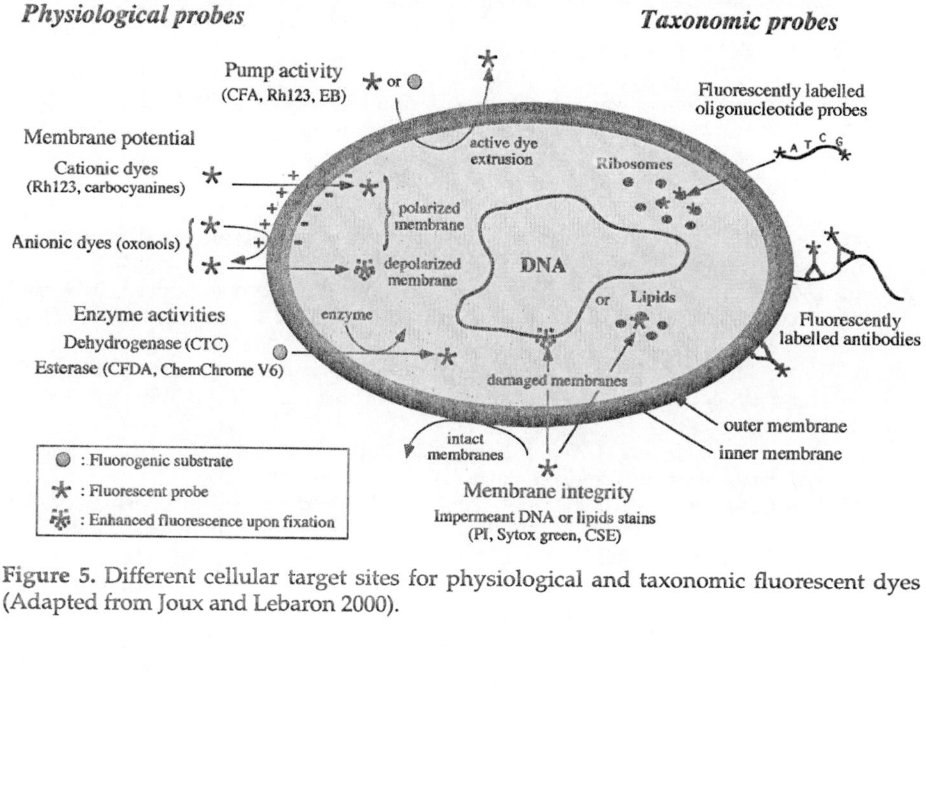

Figure 5. Different cellular target sites for physiological and taxonomic fluorescent dyes (Adapted from Joux and Lebaron 2000).

Conclusion

Ozone as a processing technology is effective for inactivation of a range of target microorganisms in different media including beverages, food or food products. The application of ozone is advantageous as it does not leave any harmful residues unlike chlorine. However, for efficient application of this technology, quality characteristic retention of treated product should be considered prior to its usage. To gain a deeper understanding of the mechanism of action of ozone, various strategies to determine the bactericidal action are briefly described. Utilising a molecular approach can provide insight to the various cellular targets of ozone action. Further exploring differential gene expression in microorganisms after ozone treatment would help to optimise the effective application of ozone treatment and the manipulation of critical control parameters.

Keywords: Ozone, Food processing, Microbial inactivation, Mechanism of action, Molecular methods, DNA damage, lipid peroxidation

References

Acero, J.L., F.J. Benitez, F.J. Real and C. Maya. 2003. Oxidation of acetamide herbicides in Natural Waters by Ozone and by the Combination of Ozone/Hydrogen Peroxide: Kinetic Study and Process Modeling. Ind. Eng. Chem. Res. 42(23): 5762–5769.

Achen, M. and A.E. Yousef. 2001. Efficacy of Ozone against *Escherichia coli* O157:H7 on Apples. J. Food Sci. 66(9): 1380–1384.

Adler, M.G. and G.R. Hill. 1950. The kinetics and mechanism of hydroxide iron catalyzed ozone decomposition in aqueous solutions. J. Am. Chem. Soc. 72: 1884–1886.

Akbas, M.Y. and M. Ozdemir. 2008. Application of gaseous ozone to control populations of *Escherichia coli, Bacillus cereus* and *Bacillus cereus* spores in dried figs. Food Microbiol. 25(2): 386–391.

Alexandre, E.M.C., D.M. Santos-Pedro, T.R.S. Brandão and C.L.M. Silva. 2011. Influence of aqueous ozone, blanching and combined treatments on microbial load of red bell peppers, strawberries and watercress. J. Food Eng. 105(2): 277–282.

Alothman, M., B. Kaur, A. Fazilah, R. Bhat and A.A. Karim. 2010. Ozone-induced changes of antioxidant capacity of fresh-cut tropical fruits. Innov. Food Sci. Emerg. Technol. 11(4): 666–671.

Antony-Babu, S. and I. Singleton. 2011. Effects of ozone exposure on the xerophilic fungus, *Eurotium amstelodami* IS-SAB-01, isolated from naan bread. Int. J. Food Microbiol. 144(3): 331–336.

Bae, E., J.W. Lee, B.H. Hwang, J. Yeo, J. Yoon, H.J. Cha and W. Choi. 2008. Photocatalytic bacterial inactivation by polyoxometalates. Chemosphere 72(2): 174–181.

Barboni, T., M. Cannac and N. Chiaramonti. 2010. Effect of cold storage and ozone treatment on physicochemical parameters, soluble sugars and organic acids in Actinidia deliciosa. Food Chem. 121(4): 946–951.

Barth, M.M., C. Zhou, J. Mercier and F.A. Payne. 1995. Ozone storage effects on anthocyanin content and fungal growth in blackberries. J. Food Sci. 60(6): 1286–1288.

Barry-Ryan, C. and P. Bourke. 2012. Essential oils for the treatment of fruits and vegetables. *In*: Vicente M. Gómez-López (ed.). Decontamination of fresh and minimally processed produce. John Wiley & Sons, Inc. pp. 225–246.

Baur, S., R. Klaiber, W.P. Hammes and R. Carle. 2004. Sensory and microbiological quality of shredded, packaged iceberg lettuce as affected by pre-washing procedures with chlorinated and ozonated water. Innov. Food Sci. Emer. Tec. 5(1): 45–55.

Beltrán, D., M.V. Selma, A. Marın and M.I. Gil. 2005. Ozonated water extends the shelf life of fresh-cut lettuce. J. Agri. Food Chem. 53: 5654–5663.

Beltrán, D., M.V. Selma, J.A. Tudela and M.I. Gil. 2005a. Effect of different sanitizers on microbial and sensory quality of fresh-cut potato strips stored under modified atmosphere or vacuum packaging. Postharvest Biol. Technol. 37: 37–46.

Beltrán, F., B. Acedo and J. Rivas. 1999. Use of Ozone to Remove Alachlor from Surface Water. B. Environ. Contam. Toxicol. 62(3): 324–329.

Benitez, F.J., J.L. Acero, F.J. Real and S. Roman. 2005. Oxidation of MCPA and 2,4-d by UV radiation, ozone, and the combinations UV/H_2O_2 and O_3/ H_2O_2. J. Environ. Sci. Heal. Part B 39(3): 393–409.

Beuchat, L.R. 1996. Pathogenic organisms associated with fresh produce. J. Food Prot. 59(2): 204–216.

Beuchat, L.R. 1992. Surface disinfection of raw produce. Dairy, Food Env. Sanitation 12: 6–9.

Beyer, M., I. Ferse, D. Mulac, E.U. Würthwein and H.U. Humpf. 2009. Structural elucidation of T-2 toxin thermal degradation products and investigations toward their occurrence in retail food. J. Agric. Food Chem. 57(5): 1867–1875.

Bialka, K.L. and A. Demirci. 2007a. Utilization of gaseous ozone for the decontamination of *Escherichia coli* O157:H7 and *Salmonella* on raspberries and strawberries. J. Food Prot. 70(5): 1093–1098.

Bialka, K.L. and A. Demirci. 2007b. Efficacy of aqueous ozone for the decontamination of *Escherichia coli* O157:H7 and *Salmonella* on raspberries and strawberries. J. Food Prot. 70(5): 1088–1092.

Bono, G. and C. Badalucco. 2012. Combining ozone and modified atmosphere packaging (MAP) to maximize shelf-life and quality of striped red mullet (*Mullus surmuletus*). LWT—Food Sci. Tec. 47(2): 500–504.

Brooks, G.M. and S.W. Pierce. 1990. Ozone applications for commercial catfish processing. Paper presented at 15th Annual Tropical and Subtropical Fisheries Technological Conference of the Americas, December 2-5, Orlando, Florida. Available at http://fshn.ifas.ufl.edu/seafood/sst/AnnPdf/15th_180.pdf (Accessed 10th May 2012).

Bruhn, C. 2000. Food labelling: consumer needs. *In*: J. Ralph Blanchfield (ed.). Food labelling. Woodhead Publishing Limited, Cambridge, pp. 5–13.

Burleson, G.R., T.M. Murray and M. Pollard. 1975. Inactivation of viruses and bacteria by ozone with and without sonication. Appl. Microbiol. 29: 340–344.

Byrd, J.J. and R.R. Colwell 1990. Maintenance of plasmids pBR322 and pUC8 in nonculturable *Escherichia coli* in the marine environment. Appl. Environ. Microbiol. 56(7): 2104–2107.

Cabo, M.L., J.J. Herrera, M.D. Crespo and L. Pastoriza. 2009. Comparison among the effectiveness of ozone, nisin and benzalkonium chloride for the elimination of planktonic cells and biofilms of *Staphylococcus aureus* CECT4459 on polypropylene. Food Control 20(5): 521–525.

Cadet, J., T. Douki, D. Gasparutto and J.-L. Ravanat. 2003. Oxidative damage to DNA: formation, measurement and biochemical features. Mutation Res. Fund. Mol. Mechanisms of Mutagenesis 531(1–2): 5–23.

Campos, C.A., V. Losada, Ó. Rodríguez, S.P. Aubourg and J. Barros-Velázquez. 2006. Evaluation of an ozone–slurry ice combined refrigeration system for the storage of farmed turbot (*Psetta maxima*). Food Chem. 97(2): 223–230.

Carbonneau, M.A., E. Peuchant, D. Sess, P. Canioni and M. Clerc. 1991. Free and bound malondialdehyde measured as thiobarbituric acid adduct by HPLC in serum and plasma. Clin. Chem. 37: 1423–1429.

Castillo, A., K.S. McKenzie, L.M. Lucia and G.R. Acuff. 2003. Ozone treatment for reduction of *Escherichia coli* O157:H7 and *Salmonella* serotype *typhimurium* on beef carcass surfaces. J. Food Protect. 66: 775–779.

Chelme-Ayala, P., M.G. El-Din, D.W. Smith and C.D. Adams. 2011. Oxidation kinetics of two pesticides in natural waters by ozonation and ozone combined with hydrogen peroxide. Water Res. 45(8): 2517–2526.

Chen, C.Z. and S.L. Cooper. 2002. Interactions between dendrimer biocides and bacterial membranes. Biomaterials 23(16): 3359–3368.

Chen, H.C., S.H. Huang, M.W. Moody and S.T. Jiang. 1992. Bacteriocisal and mutagenic effects of ozone on shrimp (Penaues-monodon) meat. J. Food Sci. 57(4): 923–927.

Cho, M., J. Kim, J.Y. Kim, J. Yoon and J.-H. Kim. 2010. Mechanisms of *Escherichia coli* inactivation by several disinfectants. Water Res. 44(11): 3410–3418.

Choi, M.-R., Q. Liu, S.-Y. Lee, S. Ryu and D.-H. Kang. 2012. Inactivation of *Escherichia coli* O157:H7, *Salmonella Typhimurium* and *Listeria monocytogenes* in apple juice with gaseous ozone. Food Microbiol. 32(1): 191–195.

Christensen, E. and A.C. Giese. 1954. Changes in absorption spectra of nucleic acids and their derivatives following exposure to ozone and ultraviolet radiations. Arch. Biochem. Biophys. 50: 209–216.

Chu, W., K.H. Chan and C.Y. Kwan. 2004. Modeling the ozonation of herbicide 2,4-D through a kinetic approach. Chemosphere 55(5): 647–652.

Cocolin, L., D. Aggio, M. Manzano, C. Cantoni and G. Comi. 2002. An application of PCR-DGGE analysis to profile the yeast populations in raw milk. Int. Dairy J. 12(5): 407–411.

Coll Cárdenas, F., S. Andrés, L. Giannuzzi and N. Zaritzky. 2011. Antimicrobial action and effects on beef quality attributes of a gaseous ozone treatment at refrigeration temperatures. Food Control 22(8): 1442–1447.

Cooper, M. 2009. Elucidation of levels of bacterial viability post non-equilibrium barrier discharge plasma treatment. PhD thesis. Drexel University.

Cooper, M., G. Fridman, A. Fridman and S.G. Joshi. 2010. Biological responses of *Bacillus stratosphericus* to floating electrode-dielectric barrier discharge plasma treatment. J. Appl. Microbiol. 109(6): 2039–2048.

Crowe, K.M., D. Skonberg, A. Bushway and S. Baxter. 2012. Application of ozone sprays as a strategy to improve the microbial safety and quality of salmon fillets. Food Control 25(2): 464–468.

Curtiellas, V., M. Gutierrez, E. Sanchez, I. Fernandez, C. Baluja, M. Bataller, S. Rodriguez and O. Ancheta. 2005. Characterization of *E. coli* lysis by ozone. IOA 17th World Ozone Congress-Strasbourg 2005. pp. I.3.1–I.3.8.

Daş, E., G.C. Gürakan and A. Bayındırlı. 2006. Effect of controlled atmosphere storage, modified atmosphere packaging and gaseous ozone treatment on the survival of *Salmonella Enteritidis* on cherry tomatoes. Food Microbiol. 23(5): 430–438.

Davidson, J., C. Good, C. Welsh and S. Summerfelt. 2011. The effects of ozone and water exchange rates on water quality and rainbow trout *Oncorhynchus mykiss* performance in replicated water recirculating systems. Aquacult. Eng. 44(3): 80–96.

Deak, T. and L.R. Beuchat. 1996. Handbook of food spoilage yeasts. CRC press, Boca Raton, FL.

Delaney, J.M. 1990. Requirement of the *Escherichia coli dnaK* gene for thermotolerance and protection against H_2O_2. J. Gen. Microbiol. 136: 2113–2118.

Dhawan, A., M. Bajpayee and D. Parmar. 2009. Comet assay: a reliable tool for the assessment of DNA damage in different models. Cell Biol. Toxicol. 25(1): 5–32.

Dianzani, M.U. 1993. Lipid peroxidation and cancer. Crit. Rev. Oncol. Hematol. 15: 125–147.

Diao, H.F., X.Y. Li, J.D. Gu, H.C. Shi and Z.M. Xie. 2004. Electron microscopic investigation of the bactericidal action of electrochemical disinfection in comparison with chlorination, ozonation and Fenton reaction. Process Biochem. 39(11): 1421–1426.

Drzewicz, P., M. Trojanowicz, R. Zona, S. Solar and P. Gehringer. 2004. Decomposition of 2,4-dichlorophenoxyacetic acid by ozonation, ionizing radiation as well as ozonation combined with ionizing radiation. Radiat. Phys. Chem. 69(4): 281–287.

Dubick, M.A. and C.L. Keen. 1983. Tissue trace elements and lung superoxide dismutase activity in mice exposed to ozone. Toxicol. Lett. 17: 355–360.

Fabrizio, K., R. Sharma, A. Demirci and C. Cutter. 2002. Comparison of electrolyzed oxidizing water with various antimicrobial interventions to reduce *Salmonella* species on poultry. Poultry Sci. 81(10): 1598–1605.

Fan, L., J. Song, K.B. McRae, B.A. Walker and D. Sharpe. 2007. Gaseous ozone treatment inactivates *Listeria innocua in vitro*. J. Appl. Microbiol. 103(6): 2657–2663.

Farooq, S. and S. Akhlaque. 1983. Comparative response of mixed cultures of bacteria and virus to ozonation. Water Res. 17: 809–812.

Fisher, C.W., D. Lee, B.A. Dodge, K.M. Hamman, J.B. Robbins and S.E. Martin. 2000. Influence of catalase and superoxide dismutase on ozone inactivation of *Listeria monocytogenes*. Appl. Env. Microbiol. 66(4): 1405–1409.

Friedman, L.I. and R.R. Stromberg. 1993. Viral inactivation and reduction in cellular blood products. Rev. Fr. Transfus. Hemobiol. 36: 83–91.

Fröhling, A., M. Baier, J. Ehlbeck, D. Knorr and O. Schlüter. 2012. Atmospheric pressure plasma treatment of *Listeria innocua* and *Escherichia coli* at polysaccharide surfaces: Inactivation kinetics and flow cytometric characterization. Innov. Food Scie. Emer. Technol. 13: 142–150.

Fuciarelli, A.F., B.J. Wegher, W.F. Blakely and M. Dizdaroglu. 1990. Yields of radiation-induced base products in DNA: Effects of DNA conformation and gassing conditions. Int. J. Rad. Biol. 58: 397–415.

Fung, D.Y.C., C. Liang and N.A. Cox. 1990. Critical review of isolation, detection, and identification of yeasts from meat products. Crit. Rev. Food Sci. Nutr. 29(5): 341–379.

Gabler, F.M., J.L. Smilanick, M.F. Mansour and H. Karaca. 2010. Influence of fumigation with high concentrations of ozone gas on postharvest gray mold and fungicide residues on table grapes. Postharvest Biol. Technol. 55(2): 85–90.

Garcia, A., J.R. Mount and P.M. Davidson. 2003. Ozone and Chlorine Treatment of Minimally Processed Lettuce, J. Food Sci. 68(9): 2747–2751.

Gelman, A., O. Sachs, Y. Khanin, V. Drabkin and L. Glatman. 2005. Effect of ozone pretreatment on fish storage life at low temperatures. J. Food Prot. 68(4): 778–784.

Gomes, A., E. Fernandes and J.L. Lima. 2005. Fluorescence probes used for detection of reactive oxygen species. J. Biochem. Biophys. Methods 65(2–3): 45–80.

Gonçalves, A.A. 2009. Ozone: an emerging technology for the seafood industry. Braz. Arch. Biol. Technol. 52: 1527–1539.

Greenberg, J.T., P. Monach, J.H. Chou, P. David Josephy and B. Demple. 1990. Positive control of a global antioxidant defense regulon activated by superoxide generating agents in *Escherichia coli*. Proc. Natl. Acad. Sci. USA 87: 6181–6185.

Greer, G.G. and S.D.M. Jones. 1989. Effects of ozone on beef carcass shrinkage, muscle quality and bacterial spoilage. Can. Inst. Food Sci. Technol. 22(2): 156–160.

Guzel-Seydim, Z.B., A.K. Greene and A.C. Seydim. 2004. Use of ozone in the food industry. LWT- Food Sci. Technol. 37(4): 453–460.

Hamelin, C., F. Sarhan and Y.S. Chung. 1977. Ozone induced DNA degradation in different DNA polymerase I mutants of *Escherichia coli* K12. Biochem. Biophys. Res. Commun. 77: 220–224.

Hamelin, C., F. Sarhan and Y.S. Chung. 1978. Induction of deoxyribonucleic acid degradation in *Escherichia coli* by ozone. Experentia 34: 1578–1579.

Hamelin, C. and Y. Sup Chung. 1989. Repair of ozone-induced DNA lesions in *Escherichia coli* B cells. Mutat. Res-Fund. Mol. M. 214(2): 253–255.

Han, Y., J.D. Floros, R.H. Linton, S.S. Nielsen and P.E. Nelson. 2002. Response surface modeling for the inactivation of *Escherichia coli* O157:H7 on green peppers (Capsicum annuum) by ozone gas treatment. J. Food Sci. 67(3): 1188–1193.

Herbold, K., B. Flehmig and K. Botzenhart.1989. Comparison of ozone inactivation, in flowing water, of hepatitis A virus, poliovirus 1, and indicator organisms. Appl. Environ. Microbiol. 55: 2949–2953.

Hirneisen, K.A., S.M. Markland and K.E. Kniel. 2011. Ozone Inactivation of norovirus surrogates on fresh produce. J. Food Prot. 74(5): 836–839.

Hladik, M.L., A.L. Roberts and E.J. Bouwer. 2005. Removal of neutral chloroacetamide herbicide degradates during simulated unit processes for drinking water treatment. Water Res. 39(20): 5033–5044.

Hoigné, J. and H. Bader. 1975. Ozonation of water: role of hydroxyl radicals as oxidizing intermediates. Sci. 190: 782–784.

Hudson, J.B., M. Sharma and M. Petric. 2007. Inactivation of Norovirus by ozone gas in conditions relevant to healthcare. J. Hosp. Infect. 66(1): 40–45.

Hulsey, R.A., B.W. Long, S.J. Randtke and C.D. Adams. 1993. Atrazine removal using ozone and GAC: a pilot plant study. Ozone: Sci. Eng. 15(3): 227–244.

Iacumin, L., M. Manzano and G. Comi. 2012. Prevention of *Aspergillus ochraceus* growth on and Ochratoxin a contamination of sausages using ozonated air. Food Microbiol. 29(2): 229–232.

Ikeura, H., F. Kobayashi and M. Tamaki. 2011. Removal of residual pesticide, fenitrothion, in vegetables by using ozone microbubbles generated by different methods. J. Food Eng. 103(3): 345–349.

Inan, F., M. Pala and I. Doymaz. 2007. Use of ozone in detoxification of aflatoxin B_1 in red pepper. J. Stored Prod. Res. 43(4): 425–429.

Ishizaki, K., K. Miura and N. Shinriki. 1987. Effect of ozone on plasmid DNA of *Escherichia coli in situ*. Water Res. 21: 823–827.

Javier Benitez, F., J.L. Acero and F.J. Real. 2002. Degradation of carbofuran by using ozone, UV radiation and advanced oxidation processes. J. Hazard. Mater. 89(1): 51–65.

Joshi, S.G., M. Cooper, A. Yost, M. Paff, U.K. Ercan, G. Fridman, G. Friedman, A. Fridman and A.D. Brooks. 2011. Nonthermal dielectric-barrier discharge plasma-induced inactivation involves oxidative DNA damage and membrane lipid peroxidation in *Escherichia coli*. 55(3): 1053–1062.

Joux, F. and P. Lebaron. 2000. Use of fluorescent probes to assess physiological functions of bacteria at single-cell level. Microbes Infect. 2(12): 1523–1535.

Karaca, H., S.S. Walse and J.L. Smilanick. 2012. Effect of continuous 0.3μL/L gaseous ozone exposure on fungicide residues on table grape berries. Postharvest Biol. Technol. 64(1): 154–159.

Karaca, H. and Y.S. Velioglu. 2007. Ozone Applications in Fruit and Vegetable Processing. Food Rev. Int. 23(1): 91–106.

Kells, S.A., L.J. Mason, D.E. Maier and C.P. Woloshuk. 2001. Efficacy and fumigation characteristics of ozone in stored maize. J. Stored Prod. Res. 37: 371–382.

Khadre, M.A., A.E. Yousef and J.G. Kim. 2001. Microbiological aspects of ozone applications in food: a review. JFS: Concise Reviews in Food Science 66(9): 1242–1252.

Khadre, M.A. and A.E. Yousef. 2001. Sporicidal action of ozone and hydrogen peroxide: a comparative study. Int. J. Food Microbiol. 71: 131–138.

Kim, C. and Y.C. Hung. 2012. Inactivation of *E. coli* O157:H7 on blueberries by electrolyzed water, ultraviolet light, and ozone. J. Food Sci. 77(4): M206–M211.

Kim, C.K., D.M. Gentile and O.J. Sproul. 1980. Mechanism of ozone inactivation of bacteriophage f2. Appl. Environ. Microbiol. 39: 210–218.

Kim, J.G., A.E. Yousef and G.W. Chism. 1999a. Use of ozone to inactivate microorganisms on lettuce. J. Food Safety 19: 17–34.

Kim, J.-G., A.E. Yousef and M.A. Khadre. 2003. Ozone and its current and future application in the food industry. Adv. Food Nutr. Res. 45: 167–218.

Kim, J.G., A.E. Yousef and S. Dave. 1999b. Application of ozone for enhancing the microbiological safety and quality of foods: A review. J. Food Prot. 62(9): 1071–1087.

Kim, S.R., H.T. Kim, H.J. Park, S. Kim, H.J. Choi, G.-S. Hwang, J.H. Yi, D.H. Ryu and K.H. Kim. 2009. Fatty acid profiling and proteomic analysis of *Salmonella enterica* serotype *Typhimurium* inactivated with supercritical carbon dioxide. Int. J. Food Microbiol. 134(3): 190–195.

Kim, T.J., J.L. Silva, R.S. Chamul and T.C. Chen. 2000. Influence of ozone, hydrogen peroxide, or salt on microbial profile, TBARS and color of channel catfish fillets. J. Food Sci. 65(7): 1210–1213.

King, J.M. and Jr. A.D. Prudente. 2005. Chemical detoxification of aflatoxins in food and feeds. *In:* H.K. Abbas (ed.). Aflatoxin and Food Safety. CRC press, Boca Raton, FL, pp. 543–554.

Klockow, P.A. and K.M. Keener. 2009. Safety and quality assessment of packaged spinach treated with a novel ozone-generation system. LWT: Food Sci. Technol. 42(6): 1047–1053.

Kolbert, C.P., P.N. Rys, M. Hopkins, D.T. Lynch, J.J. Germer, C.E. O'Sullivan, A. Trampuz and R. Patel. 2004. 16S ribosomal DNA sequence analysis for identification of bacteria in a clinical microbiology laboratory. *In*: Molecular Microbiology: Diagnostic Principles and Practice. D.H. Persing, F.C. Tenover, J. Versalovic, Y.-W. Tang, E.R. Unger, D.A. Relman and T.J. White (eds.). American Society for Microbiology, Washington, DC, pp. 361–378.

Komanapalli, I.R. and B.H.S. Lau. 1996. Ozone-induced damage of *Escherichia coli* K-12. Appl. Microbiol. Biotechnol. 46: 610–614.

Komanapalli, I.R. and B.H.S. Lau. 1998. Inactivation of bacteriophage λ, *Escherichia coli*, and *Candida albicans* by ozone. Appl. Microbiol. Biotechnol. 49(6): 766–769.

Komanapalli, I.R., J.B. Mudd and B.H.S. Lau. 1997. Effect of ozone on metabolic activities of *Escherichia coli* K-12. Toxicol. Lett. 90: 61–66.

Kötters, J., A. Prahst, B. Skura, H. Rosenthal, E.A. Black and J. Rodigues-Lopez. 1997. Observations and experiments on extending shelf-life of 'rockfish' (*Sebastes* spp.) products with ozone. J. Appl. Icthiology 13(1): 1–8.

Kuipers, G.K. and M.V.M. Lafleur. 1998. Characterization of DNA damage induced by gamma-radiationderived water radicals, using DNA repair enzymes. Int. J. Radiat. Biol. 74(4): 511–519.

Kumari, S., R.P. Rastogi, K.L. Singh, S.P. Singh and R.P. Sinha. 2008. DNA damage: detection strategies. EXCLI J. 7: 44–62.

Lee, Y.-K., S. Mok Kim and S. Han. 2003. Ozone-induced inactivation of antioxidant enzymes. Biochimie 85(10): 947–952.

Lénès, D., N. Deboosere, F. Ménard-Szczebara, J. Jossent, V. Alexandre, C. Machinal and M. Vialette. 2010. Assessment of the removal and inactivation of influenza viruses H5N1 and H1N1 by drinking water treatment. 44(8): 2473–2486.

Liao, H., F. Zhang, X. Hu and X. Liao. 2011. Effects of high-pressure carbon dioxide on proteins and DNA in *Escherichia coli*. Microbiol. 157(3): 709–720.

Lin, Y.C. and S.C. Wu. 2006. Effects of ozone exposure on inactivation of intra- and extracellular enetrovirus 71. Antiviral Res. 70: 147–153.

Loewen, P., B. Hu, J. Strutinsky and R. Sparling. 1998. Regulation in the rpoS regulon of *Escherichia coli*. Can. J. Microbiol. 44: 707–717.

Mahfoudh, A., J. Barbeau, M. Moisan, A. Leduc and J. Séguin. 2010. Biocidal action of ozone-treated polystyrene surfaces on vegetative and sporulated bacteria. Appl. Surface Sci. 256(10): 3063–3072.

Maness, P.C., S. Smolinski, D.M. Blake, Z. Huang, E.J. Wolfrum and W.A. Jacoby. 1999. Bactericidal activity of photocatalytic TiO_2 reaction: toward an understanding of its killing mechanism. Appl. Environ. Microbiol. 65(9): 4094–4098.

Martín-Diana, Ana B., D. Rico, C. Barry-Ryan, J.M. Frías, Gary T.M. Henehan and J.M. Barat. 2007. Efficacy of steamer jet-injection as alternative to chlorine in fresh-cut lettuce. Postharvest Biol. Technol. 45(1): 97–107.

McDonough, M.X., C.A. Campabadal, L.J. Mason, D.E. Maier, A. Denvir and C. Woloshuk. 2011. Ozone application in a modified screw conveyor to treat grain for insect pests, fungal contaminants, and mycotoxins. J. Stored Prod. Res. 47(3): 249–254.

McKenzie, K.S., A.B. Sarr, K. Mayura, R.H. Bailey, D.R. Miller, T.D. Rogers, W.P. Norred, K.A. Voss, R.D. Plattner, L.F. Kubena and T.D. Phillips. 1997. Oxidative degradation and detoxification of mycotoxins using a novel source of ozone. Food Chem. Toxicol. 35(8): 807–820.

McKenzie, K.S., L.F. Kubena, A.J. Denvir, T.D. Rogers, G.D. Hitchems, R.H. Bailey, R.B. Harvey, S.A. Buckley and T.D. Phillips. 1998. Aflatoxicosis in turkey poults is prevented by treatment of naturally contaminated corn with ozone generated by electrolysis. Poultry Sci. 77: 1094–1102.

Metzger, C., J.D. Barnes, I. Singleton and P. Andrews. 2007. Effect of low level ozone enrichment on the quality and condition of citrus fruit under semi-commercial conditions. *In:* Proceedings of the IOA Conference and Exhibition, Valencia, Spain, 29–31 October 2007. Available at: http://www.chimia.it/pdf/Paper%205.11.pdf. pp. 5.11-1–7.

Milligan, J.R., J.Y. Ng, C.C. Wu, J.A. Aguilera, J.F. Ward, Y.W. Kow, S.S. Wallace and R.P. Cunningham. 1996. Methylperoxyl radicals as intermediates in the damage to DNA irradiated in aqueous dimethyl sulfoxide with gamma rays. Radiat. Res. 146(4): 436–443.

Mudd, J.B., R. Leavitt, A. Ongun and T.T. McManus. 1969. Reactions of ozone with amino acids and proteins. Atmos. Environ. 3: 669–682.

Murray, B.K., S. Ohmine, D.P. Tomer, K.J. Jensen, F. Brent Johnson, J.J. Kirsi, R.A. Robison and K.L. O'Neill. 2008. Virion disruption by ozone-mediated reactive oxygen species. J. Virol. Methods 153: 74–77.

Muthukumarappan, K., F. Halaweish and A.S. Naidu. 2000. Ozone. *In:* A.S. Naidu (ed.). Natural Food Anti-Microbial Systems. CRC Press, Boca Raton, FL, pp. 783–800.

Muthukumarappan, K., C.P. O'Donnell and P.J. Cullen. 2008. Ozone utilization. Enc. Agri. Food Biol. Eng. Taylor and Francis Group, New York, USA (DOI: 10.1081/E-EAFE-120044004).

Muthukumarappan, M., C.P. O'Donnell and P.J. Cullen. 2009. Ozone treatment of food materials. *In:* S. Jun and J.M. Irudayaraj (eds.). Food Processing Operations Modeling Design and Analysis. CRC Press Taylor and Francis Group, Boca Raton, Florida, pp. 263–280.

Nagayoshi, M., T. Fukuizumi, C. Kitamura, J. Yano, M. Terashita and T. Nishihara. 2004. Efficacy of ozone on survival and permeability of oral microorganisms. Oral Microbiol. Immunol. 19(4): 240–246.

Nash, B. 2002. Ozone effective in preserving seafood freshness. Marine Extension News, North Carolina Sea Grant, Spring.

Neal, J.A., M. Marquez-Gonzalez, E. Cabrera-Diaz, L.M. Lucia, C.A. O'Bryan, P.G. Crandall, S.C. Ricke and A. Castillo. 2012. Comparison of multiple chemical sanitizers for reducing *Salmonella* and *Escherichia coli O157:H7* on spinach *(Spinacia oleracea)* leaves. Food Res. Int. 45(2): 1123–1128.

Nieto, J.C., F. Jiménez-Colmenero and M.C. Peláez. 1984. Effect of ozone on bacterial flora in poultry during refrigerated storage. Int. J. Refrig. 7(6): 389–392.

Nobmann, P., P. Bourke, J. Dunne and G. Henehan. 2010. *In vitro* antimicrobial activity and mechanism of action of novel carbohydrate fatty acid derivatives against *Staphylococcus aureus* and MRSA. J. Appl. Microbiol. 108(6): 2152–2161.

Novak, J.S. and J.T.C. Yuan. 2003. Viability of *Clostridium perfringens*, *Escherichia coli*, and *Listeria monocytogenes* surviving mild heat or aqueous ozone treatment on beef followed by heat, alkali, or salt Stress. J. Food Protect. 66(3): 382–389.

Novak, J.S. and J.T.C. Yuan. 2004. The fate of Clostridium perfringens spores exposed to ozone and/or mild heat pretreatment on beef surfaces followed by modified atmosphere packaging. Food Microbiol. 21(6): 667–673.

Olive, P.L., J.P. Bana'th and R.E. Durand. 1990. Detection of etoposide resistance by measuring DNA damage in individual Chinese hamster cells. J. Natl. Cancer Inst. 82: 779–783.

Ong, K.C., J.N. Cash, M.J. Zabik, M. Siddiq and A.L. Jones. 1996. Chlorine and ozone washes for pesticide removal from apples and processed apple sauce. Food Chem. 55: 153–160.

Ozkan, R., J.L. Smilanick and O.A. Karabulut. 2011. Toxicity of ozone gas to conidia of *Penicillium digitatum*, *Penicillium italicum*, and *Botrytis cinerea* and control of gray mold on table grapes. Postharvest Biol. Technol. 60(1): 47–51.

Öztekin, S., B. Zorlugenç and F.K.l. Zorlugenç. 2006. Effects of ozone treatment on microflora of dried figs. J. Food Eng. 75(3): 396–399.

Palou, L., C.H. Crisosto, J.L. Smilanick, J.E. Adaskaveg and J.P. Zoffoli. 2002. Effects of continuous 0.3 ppm ozone exposure on decay development and physiological responses of peaches and table grapes in cold storage. Postharvest Biol. Technol. 24(1): 39–48.

Palou, L., J.L. Smilanick, C.H. Crisosto, M. Mansour and P. Plaza. 2003. Ozone gas penetration and control of the sporulation of *Penicillium digitatum* and *Penicillium italicum* within commercial packages of oranges during cold storage. Crop Protection 22(9): 1131–1134.

Pastoriza, L., M. Bernárdez, G. Sampedro, M.L. Cabo and J.J.R. Herrera. 2008. Use of sterile and ozonized water as a strategy to stabilize the quality of stored refrigerated fresh fish. Food Control 19(8): 772–780.

Patil, S. and P. Bourke. 2012. Ozone processing of fluid foods. *In*: P.J. Cullen, B.K. Tiwari and V.P. Valdramidis (eds.). Novel thermal and non-thermal technologies for fluid foods. Elsevier Inc. pp. 225–261.

Patil, S., V.P. Valdramidis, K.A.G. Karatzas, P.J. Cullen and P. Bourke. 2011. Assessing the microbial oxidative stress mechanism of ozone treatment through the responses of *Escherichia coli* mutants. 111(1): 136–144.

Patil, S., V.P. Valdramidis, P.J. Cullen, J. Frias and P. Bourke. 2010. Inactivation of *Escherichia coli* by ozone treatment of apple juice at different pH levels. Food Microbiol. 27(6): 835–840.

Pérez, A.G., C. Sanz, J.J. Ríos, R. Olías and J.M. Olías. 1999. Effects of ozone treatment on postharvest strawberry quality. J. Agri. Food Chem. 47(4): 1652–1656.

Pérez, J.M., F.A. Arenas, G.A. Pradenas, J.M. Sandoval and C.C. Vásquez. 2008. *Escherichia coli* YqhD exhibits aldehyde reductase activity and protects from the harmful effect of lipid peroxidation-derived aldehydes. J. Biochem. 283(12): 7346–7353.

Pittet, A. 1998. Natural occurrence on mycotoxins in foods and feeds-an updated review. Revue de Médecine Vétérinaire 149: 479–492.

Placinta, C.M., J.P.F. D'Mello and A.M.C. Macdonald. 1999. A review of worldwide contamination of cereal grains and animal feed with *Fusarium* mycotoxins. Anim. Feed Sci. Technol. 78: 21–37.

Pohlman, F.W., M.R. Stivarius, K.S. McElyea, Z.B. Johnson and M.G. Johnson. 2002. The effects of ozone, chlorine dioxide, cetylpyridinium chloride and trisodium phosphate as multiple antimicrobial interventions on microbiological, instrumental color, and sensory color and odor characteristics of ground beef. Meat Sci. 61(3): 307–313.

Pomposiello, P.J. and B. Demple. 2001. Redox-operated genetic switches: the SoxR and OxyR transcription factors. Trends Biotechnol. 19(3): 109–114.

Proctor, A.D., M. Ahmedna, J.V. Kumar and I. Goktepe. 2004. Degradation of aflatoxins in peanut kernels/flour by gaseous ozonation and mild heat treatment. Food Additives and Contaminants 21(8): 786–793.

Prudente, A.D. and J.M. King. 2002. Efficacy and safety evaluation of ozonation to degrade aflatoxin in corn. J. Food Sci. 67(8): 2866–2872.

Pryor, W.A., B. Das and B.F. Church. 1991. The ozonation of unsaturated fatty acids: Aldehydes and hydrogen peroxides as products and possible mediators of ozone toxicity. Chem. Res. Toxicol. 4: 341–348.

Pryor, W.A. 1994. Mechanisms of radical formation from reactions of ozone with target molecules in the lung. Free Radic. Biol. Med. 17: 451–465.

Rahman, I., L.B. Clerch and D. Massaro. 1991. Rat lung antioxidant enzyme induction by ozone. Am. J. Physiol. 260: 412–418.

Raila, A., A. Lugauskas, D. Stenonavicius, M. Railiene, A. Stenonavicius and E. Zvicevicius. 2006. Application of ozone for reduction of mycological infection in wheat grain. Ann. Agr. Env. Med. 13: 287–294.

Rao, M.V., G. Paliyath and D.P. Ormrod. 1996. Ultraviolet-B- and ozone-induced biochemical changes in antioxidant enzymes of *Arabidopsis thaliana*. Plant Physiol. 110: 125–136.

Rasanayagam, V. 2006. Optimization ozone application to increase the selective reactivity. Division 15c Bioengineering poster session, IFT annual meeting.

Ravanat, J.C. 2005. Measuring oxidized DNA lesions as biomarkers of oxidative stress: An analytical challenge. FABAD J. Pharm. Sci. 30: 100–113.

Ravesi, E.M., J.J. Licciardello and L.D. Racicot. 1988. Ozone treatments of fresh atlantic cod, Gadus morhua. Marine Fisheries Rev. 49(4): 37–42.

Restaino, L., E.W. Frampton, J.B. Hemphill and P. Palnikar. 1995. Efficacy of ozonated water against various food-related microorganisms. Appl. Environ. Microbiol. 61(9): 3471–3475.

Rice, R.G., J.W. Farguhar and L.J. Bollyky. 1982. Review of the applications of ozone for increasing storage times of perishable foods. Ozone Sci. Eng. 4: 147–163.

Richardson, M.L. 1983. Dyes—The Aquatic Environment and the Mess made by Metabolites. Coloration Technol. 99(7–8): 198–200.

Rico, D., A.B. Martin-Diana, J.M. Frias, T.M.H. Gary and C. Barry-Ryan. 2006. Effect of ozone and calcium lactate treatments on browning and texture properties of fresh-cut lettuce. J. Sci. Food Agric. 86(13): 2179–2188.

Rockabrand, D., T. Arthur, G. Korinek, K. Livers and P. Blum. 1995. An essential role for the *Escherichia coli* DnaK protein in starvation-induced thermotolerance, H_2O_2 resistance, and reductive division. J. Bacteriol. 177(13): 3695–703.

Roldân-Arjona, T. and B. Sedgwick. 1996. DNA base damage induced by ionizing radiation recognized by *Escherichia coli* UvrABC nuclease but not Nth or Fpg proteins. Mol. Carcinogenesis 16: 188–196.

Rusch, A. and J. Kraemer. 1989. Influence of instruments for elimination of microorganisms on surface bacterial contamination of fresh meat and on airborne microorganisms in cold stores with increased. Relative Humidity. Arch. Lebensmittelhyg 40: 61–65.

Sacco, A. 2009. Introduction to ozone generation techniques—corona, UV and electrochemistry in http://ezinearticles.com/?Introduction-to-Ozone-Generation-Techniques—Corona,-UV-and-Electrochemistry&id=1686734 (Accessed on 15th May 2012).

Sarig, P., T. Tahavi, Y. Zutkhi, S. Yannai, N. Lisker and R. Arie. 1996. Ozone for control of postharvest decay of table grapes caused by Rhizopus stolonifer. J. Physiol. Molecular Plant Pathol. 48(6): 403–415.

Sastre, M.P., M. Vernet and S. Steinert. 2001. Single-cell Gel/Comet Assay Applied to the Analysis of UV Radiation–induced DNA Damage in *Rhodomonas* sp. (Cryptophyta). Photochem. Photobiol. 74(1): 55–60.

Scott, D.B.M. and E.C. Lesher. 1963. Effect of ozone on survival and permeability of *Escherichia coli*. J. Bacteriol. 85: 567–576.

Selma, M.V., A.M. Ibáñez, A. Allende, M. Cantwell and T. Suslow. 2008. Effect of gaseous ozone and hot water on microbial and sensory quality of cantaloupe and potential transference of *Escherichia coli* O157:H7 during cutting. Food Microbiol. 25(1): 162–168.

Selma, M.V., D. Beltran, A. Allende, E. Chacon-Vera and M. Gil. 2007. Elimination by ozone of *Shigella sonnei* in shredded lettuce and water. Food Microbiol. 24: 492–499.

Sheldon, B.W. and A.L. Brown. 1986. Efficacy of ozone as a disinfectant for poultry carcasses and chill water. J. Food Sci. 51(2): 305–309.

Shriniki, N., K. Ishizaki, T. Yoshizaki, K. Miura and T. Ueda. 1988. Mechanism of inactivation of tobacco mosaic virus with ozone. Water Res. 22(7): 933–938.

Singh, N., R.K. Singh, A.K. Bhunia and R.L. Stroshine. 2002. Efficacy of chlorine dioxide, ozone, and thyme essential oil or a sequential washing in killing *Escherichia coli* O157:H7 on lettuce and baby carrots. LWT- Food Sci. Technol. 35(8): 720–729.

Skog, L.J. and C.L. Chu. 2001. Effect of ozone on qualities of fruits and vegetables in cold storage. Can. J. Plant Sci. 81(4): 773–778.

Sokmen, M., F. Candan and Z. Sumer. 2001. Disinfection of *E. coli* by the Ag-TiO2/UVsystem: lipidperoxidation. J. Photoch. Photobio. A-Chemistry 143(2–3): 241–244.

Song, J., L. Fan, P.D. Hildebrand and C.F. Forney. 2000. Biological effects of corona discharge on onions in a commercial storage facility. Hort. Technol. 10: 608–612.

Sudhakar, N., D. Nagendra-Prasad, N. Mohan and K. Murugesan. 2007. A bench-scale, cost effective and simple method to elicit *Lycopersicon esculentum* cv. PKM1 (tomato) plants against Cucumber mosaic virus attack using ozone-mediated inactivated Cucumber mosaic virus inoculum. J. Virol. Methods 146(1–2): 165–171.

Takamoto, Y., H. Maeba and M. Kamimura. 1992. Changes in survival rates of enzyme activities in *Escherichia coli* with ozone. Appl. Microbiol. Biotechnol. 37: 393–395.

Thanomsub, B., V. Anupunpisit, S. Chanphetch, T. Watcharachaipong, R. Poonkhum and C. Srisukonth. 2002. Effects of ozone treatment on cell growth and ultrastructural changes in bacteria. J. Gen. Appl. Microbiol. 48: 193–199.

Thurston-Enriquez, J.A., C.N. Haasb, J. Jacangeloc and C.P. Gerba. 2005. Inactivation of enteric adenovirus and feline calicivirus by ozone. Water Res. 39: 3650–3656.

Tiwari, B.K., C.S. Brennan, T. Curran, E. Gallagher, P.J. Cullen and C.P. O' Donnell. 2010. Application of ozone in grain processing. J. Cereal Sci. 51(3): 248–255.

Trevors, J.T. 2003. Fluorescent probes for bacterial cytoplasmic membrane research. J. Biochem. Biophys. Methods 57(2): 87–103.

Trinetta, V., N. Vaidya, R. Linton and M. Morgan. 2011. A comparative study on the effectiveness of chlorine dioxide gas, ozone gas and e-beam irradiation treatments for inactivation of pathogens inoculated onto tomato, cantaloupe and lettuce seeds. Int. J. Food Microbiol. 146(2): 203–206.

Tseng, C. and C. Li. 2008. Inactivation of surface viruses by gaseous ozone. J. Environ. Health 70(10): 56–62.

Tzortzakis, N., I. Singleton and J. Barnes. 2007. Deployment of low-level ozone-enrichment for the preservation of chilled fresh produce. Postharvest Biol. Technol. 43(2): 261–270.

Vadhanasin, S., A. Bangtrakulnonth and T. Chidkrau. 2004. Critical control points for monitoring salmonellae reduction in Thai commercial frozen broiler processing. J. Food Prot. 67(7): 1480–1483.

Vaughn, J.M., Y.S. Chem, K. Lindburg and D. Morales. 1987. Inactivation of human and simian rotaviruses by ozone. Appl. Env. Microbiol. 53: 2218–2221.

Vaz-Velho, M., M. Silva, J. Pessoa and P. Gibbs. 2006. Inactivation by ozone of *Listeria innocua* on salmon-trout during cold-smoke processing. Food Control 17(8): 609–616.

Viljoen, B.C. 2001. The interaction between yeasts and bacteria in dairy environments. Int. J. Food Microbiol. 69(1–2): 37–44.

Vurma, M., R.B. Pandit, S.K. Sastry and A.E. Yousef. 2009. Inactivation of *Escherichia coli* O157:H7 and natural microbiota on spinach leaves using gaseous ozone during vacuum cooling and simulated transportation. J. Food Prot. 72(7): 1538–1546.

Waldroup, A.L., R.H. Hierholzer, M.J. Forsythe and Miller. 1993. Recycling of poultry chill water using ozone. J. Appl. Poultry Sci. Res. 2: 330–336.

Walse, S.S. and H. Karaca. 2011. Remediation of fungicide residues on fresh produce by use of gaseous ozone. Environ. Sci. Technol. 45(16): 6961–6969.

Wang, H., H. Feng and Y. Luo. 2004. Microbial reduction and storage quality of fresh-cut cilantro washed with acidic electrolyzed water and aqueous ozone. Food Res. Int. 37(10): 949–956.

Watanabe, M., H. Masaki, T. Mori, T. Tsuchiya, H. Konuma, Y. Hara-Kudo and K. Takatori. 2010. Inactivation effects of UV irradiation and ozone treatment on the yeast and the mold in mineral water. J. Food Prot. 73(8): 1537–1542.

Weavers, L.K. and G.B. Wickramanayake. 2001. Disinfection and sterilization using ozone. *In:* S.S. Block (ed.). Disinfection, Sterilization and Preservation (5th ed.). Lippincott Williams & Wilkins, Philadelphia, pp. 205–214.

Whangchai, K., K. Saengnil and J. Uthaibutra. 2006. Effect of ozone in combination with some organic acids on the control of postharvest decay and pericarp browning of longan fruit. Crop Protection 25: 821–825.

White, A., D. Burns and T.W. Christensen. 2006. Effective terminal sterilization using supercritical carbon dioxide. J. Biotechnol. 123: 504–515.

Whiteside, C. and H.M. Hassan. 1987. Induction and inactivation of catalase and superoxide dismutase of Escherichia coli by ozone. Arch. Biochem. Biophys. 257(2): 464–471.

Williams, R., S. Sumner and D. Golden. 2004. Survival of *Escherichia coli* O157:H7 and *Salmonella* in apple cider and orange juice as affected by ozone and treatment temperature. J. Food Prot. 67(11): 2381–2386.

Williams, R., S. Sumner and D. Golden. 2005. Inactivation of *Escherichia coli* O157:H7 and *Salmonella* in apple cider and orange juice treated with combinations of ozone, dimethyl dicarbonate, and hydrogen peroxide. J. Food Sci. 70(4): 197-201.

Wu, J., H. Doan and M.A. Cuenca. 2006. Investigation of gaseous ozone as an antifungal fumigant for stored wheat. J. Chem. Technol. Biotechnol. 81: 1288–1293.

Wu, J., T. Luan, C. Lan, T.W. Hung Lo and G.Y.S. Chan. 2007. Removal of residual pesticides on vegetable using ozonated water. Food Control 18(5): 466–472.

Xing, K., X.G. Chen, M. Kong, C.S. Liu, D.S. Cha and H.J. Park. 2009. Effect of oleoyl-chitosan nanoparticles as a novel antibacterial dispersion system on viability, membrane permeability and cell morphology of *Escherichia coli* and *Staphylococcus aureus*. Carbohydrate Polymers 76(1): 17–22.

Xiong, F. and N.J.D. Graham. 1992. Rate constants for herbicide degradation by ozone. Ozone: Sci. Eng. 14(4): 283–301.

Yang, P.P.W. and T.C. Chen. 1979. Effects of ozone treatment on microflora of poultry meat. J. Food Process. Pres. 3: 177–185.

Yoon, S.J., J.E. Park, J.H. Yang and J.W. Park. 2002. OxyR regulon controls lipid peroxidation-mediated oxidative stress in *Escherichia coli*. J. Biochem. Mol. Biol. 35: 297–301.

Young, J.C., H. Zhu and T. Zhou. 2006. Degradation of trichothecene mycotoxins by aqueous ozone. Food Chem. Toxicol. 44(3): 417–424.

Yu, Y.-H., J. Ma and Y.-J. Hou. 2006. Degradation of 2,4-dichlorophenoxyacetic acid in water by ozone-hydrogen peroxide process. J. Environ. Sci. (China) 18(6): 1043–1049.

Yuk, H.-G., M.-Y. Yoo, J.-W. Yoon, D.L. Marshall and D.H. Oh. 2007. Effect of combined ozone and organic acid treatment for control of *Escherichia coli* O157:H7 and *Listeria monocytogenes* on enoki mushroom. Food Control 18(5): 548–553.

Zhang, L., Z. Lu, Z. Yu and X. Gao. 2005. Preservation fresh-cut celery by treatment of ozonated water. Food Control 16: 279–283.

Zhang, Y.Q., Q.P. Wu, J.M. Zhang and X.H. Yang. 2011. Effects of ozone on membrane permeability and ultrastructure in *Pseudomonas aeruginosa*. J. Appl. Microbiol. 111(4): 1006–1015.

Zhao, J. and P.M. Cranston. 1995. Microbial decontamination of black pepper by ozone and the effect of the treatment on volatile oil constituents of the spice. J. Sci. Food Agri. 68(1): 11–18.

Zirkle, R.E. and N.R. Krieg. 1996. Development of a method based on alkaline gel electrophoresis for estimation of oxidative damage to DNA in *Escherichia coli*. J. Appl. Bacteriol. 81: 133–138.

Zorlugenç, B., F. Kıroğlu Zorlugenç, S. Öztekin and I.B. Evliya. 2008. The influence of gaseous ozone and ozonated water on microbial flora and degradation of aflatoxin B1 in dried figs. Food Chem. Toxicol. 46(12): 3593–3597.

CHAPTER 6

Nonthermal Plasma Technology for Decontamination of Foods

N.N. Misra,[1] *Lu Han,*[1] *B.K. Tiwari,*[2] *Paula Bourke*[1] and *P.J. Cullen*[1,*]

Introduction

The food industry is currently experiencing a rising consumer demand for safer, nutritious and appealing convenience products. At the same time, new risks are encountered due to the changing characteristics of the relevant micro-organisms, changing production methodologies, changes in the environment and the ecology, and an increase in the global trade of food stuffs (Havelaar et al. 2010, Misra et al. 2011b). Recently, in October 2011, fresh, whole cantaloupes from Colorado-based Jensen farms in the USA were linked to a multistate outbreak of *Listeria monocytogenes* infection (listeriosis) (FDA 2012). In yet another incident in Europe, *Escherichia coli* O104:H4 bacteria caused a serious outbreak of foodborne illness focussing in northern Germany, in May through June 2011 (Kennedy 2011). The outbreak was originally thought to have been caused by an enterohemorrhagic (EHEC) strain of *E. coli*, but it was later shown to have been caused by the novel enteroaggregative *E. coli* (EAEC) strain that had acquired the genes to produce Shiga toxins. Thus, consumer demands, evolving risks

[1] School of Food Science & Environmental Health, Dublin Institute of Technology, Dublin 1, Ireland.
[2] Department of Food Biosciences, Teagasc Food Research Centre, Dublin, Ireland.
* Corresponding author

and the shortcomings of the existing technologies trigger the need for the development of novel and nonthermal approaches to food processing. Further, to address issues of chemical contamination, many processors are seeking to minimize or avoid the use of conventional preservatives and chemical antimicrobials (Misra et al. 2011a).

Before the recent ban on its use, ethylene oxide (EtO) was the gas of choice for the decontamination of foods and associated surfaces. The disadvantages of EtO are toxic residues and the long storage time required for venting the toxic vapours. Irradiation has been constantly shown to be effective in inactivating human pathogens on produce. However poor consumer acceptance has prevented widespread adoption of this technology. It is no exaggeration to state that there is currently no ideal method to achieve acceptable decontamination at ambient temperature. This has provided the impetus to develop novel strategies for decontamination of foods.

Within this scenario, nonthermal plasma (NTP) technology offers a great opportunity for food decontamination. Atmospheric pressure NTP in air or within various gas mixtures generated by electrical discharges, present considerable interest for a wide range of environmental, bio-medical and industrial applications, such as air pollution control, waste water cleaning, bio-decontamination and sterilization, material and surface treatment, electromagnetic wave shielding, carbon beneficiation and nanotube growth, and element analysis (Machala et al. 2007). The fact that NTP can be generated using a variety of energy sources, in various configurations and using relatively low power, makes them a powerful adaptive tool for pasteurization/sterilization of foods and decontamination of biological materials. The source of NTP does not need to come in direct contact with the target to produce its effect, hence minimizing the risk of any possible contamination from the plasma generator. However, the interaction and effects of NTP on inherent biological functions separate to decontamination effects need to be considered for appropriate adoption of this technology to more sensitive systems.

This chapter introduces the reader to NTP, its physics and chemistry, followed by a discussion on the effects of reactive plasma species on the microbial cells and the proven and potential applications to decontaminate foods. Emphasis has been placed on understanding the mode of action of individual plasma species on micro-organisms, cell functionality and the effect of process variables on the efficacy of decontamination.

Fundamentals of NTP Technology

Plasmas are conductive assemblies of charged particles, neutrals and fields that exhibit a collective effect. When a gas is energized to such an extent that the molecules of the gas break down to release free electrons, radicals, positive and negative ions, and quanta of electromagnetic radiation, while some molecules may still remain neutral (incomplete ionization), then this state of gas is known as plasma. Plasma is often referred to as the fourth state of matter according to a scheme expressing an increase in the energy level from solid to liquid to gas and ultimately to plasma (Fig. 1). Thus, any source of energy which can ionise a gas to plasma may be employed for generation of plasma.

Two classes of plasma, namely *thermal* and *nonthermal plasma* (NTP) can be distinguished on the basis of conditions in which they are generated. The temperature of any plasma is dictated by the mean energies of the plasma species (neutral and charged) and their relevant degrees of freedom (translational, rotational, vibrational and those associated with electronic excitation). Electrons are usually first to receive energy from an electric field and then distribute it among other components of plasma. In electrical discharge plasmas, electrons accumulate the energy from the electric discharges following collisions and then transfer a small fraction of this energy to the heavy particles. The reason that only a small fraction of energy is transferred is that electrons are much lighter than the heavy particles. Thus, NTPs are characterized by an electron temperature much above that of gas (macroscopic temperature) and consequently do not present a local thermodynamic equilibrium.

The behaviour of plasma is largely determined by the plasma chemistry. Plasma chemistry can be very complex with a large number of different constituent species (Fig. 2) at any given point of time. For example in the

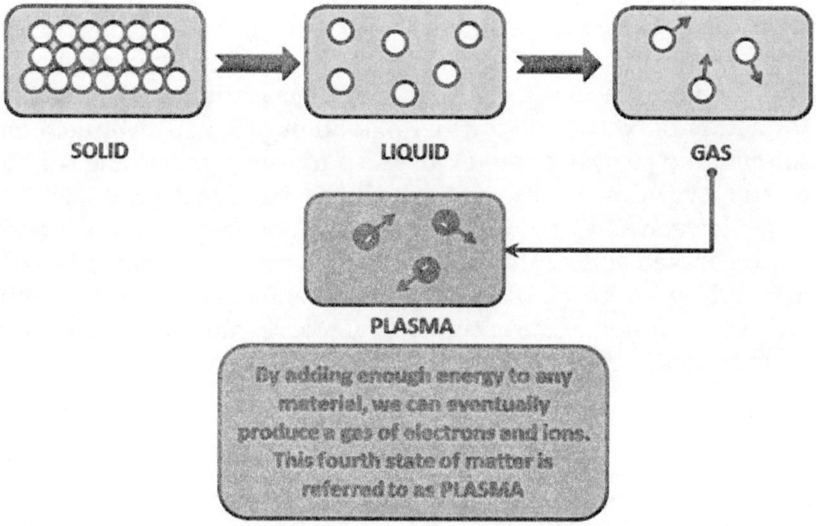

Figure 1. The four states of matter.

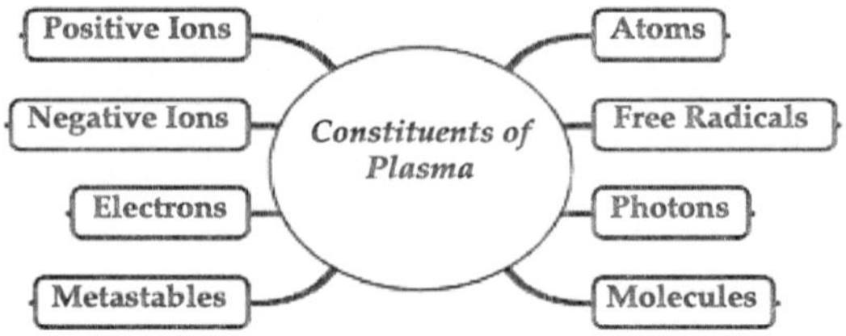

Figure 2. Constituents of plasma.

plasma generated in the air, the chemistry is believed to include more than 75 species and almost 500 reactions (Gordillo-Vázquez 2008). In air, chemical reactions are mainly initiated by the impact of electrons with oxygen and nitrogen. It is a general belief that dissociation is much easier than ionization and this is the reason for higher radical densities than ion densities in plasma. However, the reality is that dissociation is typically only a factor of 3–10 more frequent than ionization and that the radical densities are higher due to slower transport rather than more effective production (Schram 2002).

The power input to the discharge, working gas and mode of operation of a plasma system has a major impact on the species that are formed and their extent of formation. For example, ozone generated by corona discharges has been extensively studied by various theoretical and experimental investigations (Mennad et al. 2010). Gases commonly employed for atmospheric pressure plasma operations include helium, argon, sulphur hexafluoride, nitrogen/CO_2 mixtures and air. The selection of operating gas would largely dictate the process cost. In NTPs, oxygen molecules are first dissociated by electrons to form oxygen atoms which subsequently react with O_2 to form O_3. In fact, mechanisms for O_3 generation in oxygen plasmas can involve more than 50 chemical reactions and at least 10 species (Lee et al. 2004). Of greatest practical interest are reactions leading to the destruction and formation of NO and its further oxidation products. At much longer times in a humid air system there can be a build-up of ozone and the higher oxides and hydrides of nitrogen, which are the "final products" of the discharge (Herron and Green 2001). The reactions leading to the formation of various charged and neutral species in a humid air discharge have been summarized in (Sieck et al. 2000) and (Herron and Green 2001) respectively. A detailed description of the antimicrobial roles evaluated to date, of some of the prominent molecular species will be presented later in this chapter.

Plasma Sources

Plasma can be formed in any neutral gas by providing sufficient energy capable of causing ionization of the gas. Thus the source of energy is very high temperatures from flames or arcs in case of thermal plasmas and electrical energy or electromagnetic waves in case of most NTPs. The technologies used to obtain NTP vary from the use of radiofrequency waves, to capacitive or inductive coupling methods or even dielectric barrier discharges. The ability to generate nonthermal plasma discharges at atmospheric pressure makes the NTP decontamination process easier and less expensive (Kim et al. 2011). Plasma generation at atmospheric pressure is of interest, both technically and commercially to the food industry because it does not require extreme processing conditions (Misra et al. 2011b) that may prove detrimental to other quality and nutritional characteristics. Typical examples of plasma generation at atmospheric pressure include the corona discharge, dielectric barrier discharges (DBD), One Atmosphere Uniform Glow Discharge Plasma (OAUGDP), gliding arc discharge, radio-frequency plasmas (RFP) and the microwave plasmas. These are described below.

Microwave plasmas

These plasma systems are based on high frequency (GHz range) microwave discharges in the gas. Microwave plasma technology permits dense non-equilibrium plasmas to be generated at pressures up to 0.26–0.39 torr and at power levels reaching 1 MW. The key advantage of high-frequency fields is a stabilizing effect on the plasma and generation of streamerless plasmas in molecular gases (Bárdos and Baránková 2010).

Radiofrequency plasmas

These are capacitive discharges produced more often in noble gases such as helium or argon by the application of RF power (e.g., 13.56 MHz) to one powered electrode, with a grounded electrode held at a small separation. Low frequency RF (<100 kHz) electric field is also used to generate glow discharge between two metal electrodes. A practical example of glow discharges includes fluorescent lamps.

Corona discharges

Corona refers to a weak luminous discharge, which is generated at atmospheric pressure near sharp points, edges, or thin wires using very large electric fields (see Fig. 3).

Coronas are thus inherently non-uniform discharges that develop in the high field region near the sharp electrode spreading out towards the planar electrode. This phenomenon of local breakdown is called corona discharge. Corona discharges have been used in ozone generators and particle precipitators.

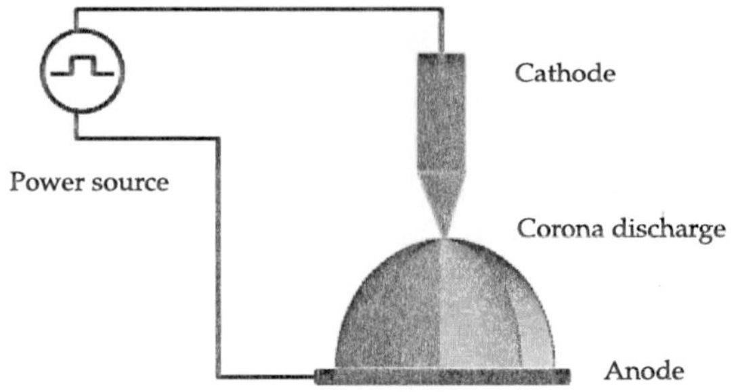

Figure 3. Schematic of corona discharge.

Dielectric barrier discharge (DBD)

In DBD, plasma is generated between two electrodes, separated by one or more dielectric barriers (see Fig. 4). The main characteristic of DBDs is the presence of a dielectric material between the discharge gap and one of the two electrodes. When the potential across the gap reaches the breakdown voltage the dielectric acts as a stabilizing material leading to the formation of a large number of micro-discharges. Thus, the dielectric material inhibits the glow-to-arc transition in a DBD. DBD plasmas in air, helium and several gas mixtures have been employed for in-package ionization and treatment of fresh produce in recent times. In-package ionization is a dry, nonthermal process which can extend shelf-life of packaged foods by reducing spoilage organisms inside the package (Klockow and Keener 2009). Depending on the gas mixture, dielectric surface properties and operating conditions, vastly different modes ranging from filamentary to completely diffuse barrier discharges, can be achieved (Kogelschatz 2002).

Despite the various technologies to generate NTP, in general, from an application standpoint, these can be classified into three modes of operation: electrode contact, direct treatment and remote treatment (see Fig. 5) (Niemira and Sites 2008). A detailed discussion on various cold plasma sources has been provided in Bárdos and Baránková (2008, 2010).

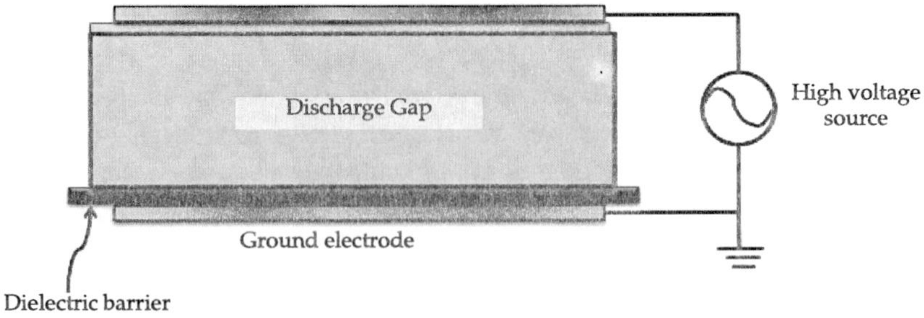

Figure 4. Dielectric barrier discharge (DBD) with a single dielectric barrier configuration.

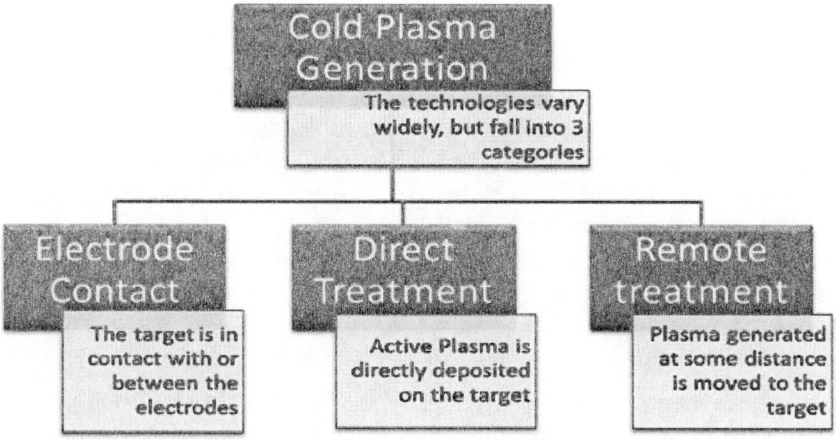

Figure 5. Categorization of NTP technologies based on modes of operation.

Effect of Plasma on Microbial Cells

Role of plasma species

NTP produced in a gas discharge results in a cocktail of active molecules that continually react (often very rapidly) with other molecules and particles present. The anti-microbial agents that could be generated in a gas discharge include reactive oxygen species (ROS), ultraviolet (UV) radiation, energetic ions, and charged particles. Among the ROS; ozone, atomic oxygen, superoxide, peroxide, and hydroxyl radicals are some of the important radicals when the plasma is generated in air. The combined action of all these species could have a synergistic effect on inactivation of the microbial cells and spores. The mechanism of microbial destruction by the gas discharge is undoubtedly complex and heterogeneous in nature. The radical production by dissociation can be roughly related to the ion production by ionization, which can in turn be related to the dissipated power in the source (Schram 2002). However, there could be a myriad of uncharacterised transient species formed in a discharge, with a very low rate constant for the reactions. Reported methods employed for reactive species (RS) characterization in plasma include broadband UV–Visible optical emission spectroscopy, as outlined in Chiper et al. (2004, 2011); electron spin resonance (ESR) spectroscopy for free radical identification, as in Yang et al. (2009); low pressure mass spectrometry of plasma plumes, for example in Yamamoto et al. (2002); laser-induced resonance fluorescence (reviewed in Amorim et al. (2000)); and short-term gas analyser tubes, as in Tamazawa et al. (2006) and Keener et al. (2012). Currently, no well accepted or completely reliable and accurate method exists to quantitatively characterize all these species within plasma. Thus, it is very challenging to relate or explain inactivation patterns associated with active species with their effects on microbial cells. Moreover, the process of determining cell death is not always straightforward in microbiological studies. Inactivation or loss of viability occurs at a point when vital cellular components suffer certain levels of irreversible damage. The following section focuses on the effect of various species in air plasma on the vital molecules in a cell, structural elements of microbial cells and the resulting effects on cell dysfunction and inactivation.

Ozone

Ozone is an unstable molecule which rapidly decays into oxygen with a time constant that is dependent upon temperature and the presence of oxidizable compounds. At molecular level, ozone reactions are selective and usually restricted to unsaturated aromatic and aliphatic compounds in the microbial cell. Thus, the double bonds of unsaturated lipids are particularly vulnerable to ozone attack (Guzel-Seydim et al. 2004). Microbial enzymes and spores are particularly abundant in sulfhydryl groups and ozone oxidizes these groups, thereby inactivating them. When the double bonds of unsaturated lipids and the sulfhydryl groups of enzymes are oxidized by ozone, disruption of normal cellular activity including cell permeability and rapid death ensues. The action of ozone on amino acids and peptides is significant, especially at neutral and basic pH. Ozone can attack the nitrogen atom or the alkyl group of an amino acid or both (Khadre et al. 2001). Ozone reacts quickly with nucleic acid bases, especially thymine, guanine, and uracil. Reaction of ozone with the nucleotides releases carbohydrate and phosphate ions. Ozone generated in plasma can also react with polysaccharides, leading to breakage of glycosidic bonds and formation of aliphatic acids and aldehydes (Bablon et al. 1991). However, reaction with polysaccharides is usually slow. In spite of these identified effects, there is little agreement among researchers regarding the relative sensitivity of different microorganisms to ozone. Ozone has the ability to open the circular plasmid DNA and reduce its transforming ability, producing single- and double-strand breaks in plasmid DNA and decreasing transcription activity (Di Mascio et al. 1989, Hamelin 1985, Victorin 1992).

UV photons

The production of UV photons of different wavelengths has been proposed to be involved in dimerizing the thymine bases of DNA of microbial cells, including that of spores (Munakata et al. 1991). DNA has an absorption spectrum with a maximum in the region 260–265 nm. The role of UV photons in bacterial death when they are submitted to a plasma treatment has been discussed elsewhere in detail (Boudam et al. 2006). By exclusion of reactive particles and spectral fractions of UV radiation from access to the spores, Roth et al. (2010) revealed that UV-C radiation was the most effective inactivation agent in plasma. Ultraviolet (UV) photons play a less important role in atmospheric pressure glow discharge (APGD) because they are easily absorbed by gas atoms and molecules at atmospheric pressure (Vleugels et al. 2005). Moreover, in bacteria, the various DNA repair mechanisms are generally very efficient and rapid, meaning that cell death only occurs when the UV photon hits are so numerous that the bacterial repair mechanisms are overwhelmed (Gaunt et al. 2006).

Carbon monoxide

Reports on the use of carbon monoxide (CO) for food-borne pathogen inactivation are scarce in the literature. The use of CO for colour retention in fresh beef was studied by El-Badawi et al. (1964), who showed that a mixture of 2% CO and 98% air significantly preserved the colour of fresh beef for 15 days. In the context of packaged meat, CO binds strongly to the meat pigment myoglobin to form stable carboxymyoglobin which has a cherry red colour (Jayasingh et al. 2001). Thus, it can be hypothesised that plasma treatment of red meats may provide an added advantage of enhanced colour retention (which is a very important quality parameter) and extension of shelf-life.

NOx species

Nitric oxide and superoxide in a plasma combine to form peroxynitrite and possibly several other oxidants. Peroxynitrite is a potent oxidant capable of inflicting damage through oxidation of sulfhydryl groups and nitration of tyrosine residues in cellular proteins. Nitrous acid is a much more active species than nitrite, but present in very low concentrations at neutral pH (the pKa of nitrous acid is 3.4). Nitrous acid and N-nitroso compounds can diazotise and deaminate amino groups in, for example, nucleotides (Cammack et al. 1999). The inhibition of respiration has long been considered a possible mechanism of the bacteriostatic action of nitrite, at least for aerobic organisms (Davidson and Juneja 1990). Chemistry of DNA damage from nitric oxide and peroxynitrite has been reviewed by Burney et al. (1999).

Hydrogen peroxide

Hydrogen peroxide is a bacteriostatic rather than bactericidal agent, dramatically increasing the mean generation time of bacteria at micromolar concentrations (Hyslop et al. 1995). Hydrogen peroxide has been shown to remove protein from spore coats in *B. cereus* and *C. bifermentans* (Russell 1982). Setlow and Setlow (1993) have proposed that hydrogen peroxide, or possibly the free hydroxyl radicals resulting from its degradation, gain access to the core of spores of certain *B. subtilis* mutants and kill these spores at least in part by DNA damage. While the mechanism by which hydrogen peroxide kills spores is not fully understood, inactivation of vegetative bacteria involves DNA damage, much of which is DNA backbone cleavage caused by hydroxyl radicals generated from the hydrogen peroxide (Imlay and Linn 1988). The oxidative decarboxylation of pyruvic acid by peroxide *in vivo* has also been considered to explain the toxicity of peroxide against *Escherichia coli* (Académie des and Centre national de la recherche 1835).

Singlet and atomic oxygen

Atomic oxygen is potentially a very effective sterilizing agent. Recent results reveal that the combined impact of H or O atoms and argon ions resulted in very effective etching and perforation of the bacterial spore coat that leads to inactivation (Raballand et al. 2008). Etching refers to an intense bombardment of the microbial cells by the radicals, which provoke surface lesions that the living cells cannot repair with sufficient speed. It has been seen from SEM images that the bacterial spore's morphology is significantly changed in the case of oxygen plasma sterilization (Singh et al. 2009).

The reactive oxygen species (ROS) act on the unsaturated fatty acids of the lipid bilayer of the cell membrane, thereby impeding the transport of biomolecules across it. Membrane lipids are assumed to be more significantly affected by the reactive oxygen species (ROS) due to their location along the surface of bacterial cell, which allows them to be bombarded by these strong oxidizing agents (Montie et al. 2002).

Other free radicals and charged particles

The presence of water vapour in a discharge feed gas results in the formation of $OH^{\bullet}$ by electron impact dissociation of H_2O and by reactions of electronically excited oxygen atoms and nitrogen molecules (Laroussi et al. 2004). Hydroxyl radicals are one of the most strongly oxidative species produced in water by plasmas and are also the building blocks of the formed H_2O_2 which is considered to be an important agent in the chemical reactivity of plasmas in and, in contact with water (Bruggeman and Schram

2010). When moist foods (such as fresh cut fruits and vegetables) are placed in a plasma discharge, the moisture on the surface of produce is likely to dissociate and yield hydroxyl radical.

Effect on cell constituents and function

It has been confirmed that NTP causes acidification of lipid films (Helmke et al. 2009), though lipid peroxidation in cells remains a point of controversy (Leduc et al. 2010). Oxidation of lipid in cell membranes of microbial cells increases membrane permeability, decreases fluidity and the products emitted react with DNA (Marnett 2002). Lipid peroxidation is initiated when a reactive species with sufficient reactivity to abstract a hydrogen atom from a methylene group attacks a poly-unsaturated fatty acid (PUFA). The abstraction of the hydrogen atom leaves an unpaired electron on the carbon atom, and this carbon atom most likely reacts with O_2 to form a peroxyl radical ($ROO^{\bullet}$), which propagates the reaction thereafter (Leduc et al. 2010).

The reactive oxygen species (ROS) generated from various plasma sources can oxidize the cell membrane and cause the leakage of cytoplasm (Xingmin et al. 2011). Transmission electron micrographs (TEM) of plasma treated cells, as shown in Fig. 6 reveal severe cytoplasmic deformations and leakage of bacterial chromosome (Hong et al. 2009). Laroussi et al. (2003) and Mendis et al. (2002) have attributed cell wall rupture of plasma exposed bacterial cells to electrostatic forces due to accumulation of charges at the outer surface of cell membranes. SEM images demonstrate that plasma-treated cell surface were altered or dehydrated as cells exhibited a shrunken morphology but visually no cell lysis was observed (Ziuzina et al. 2013). It

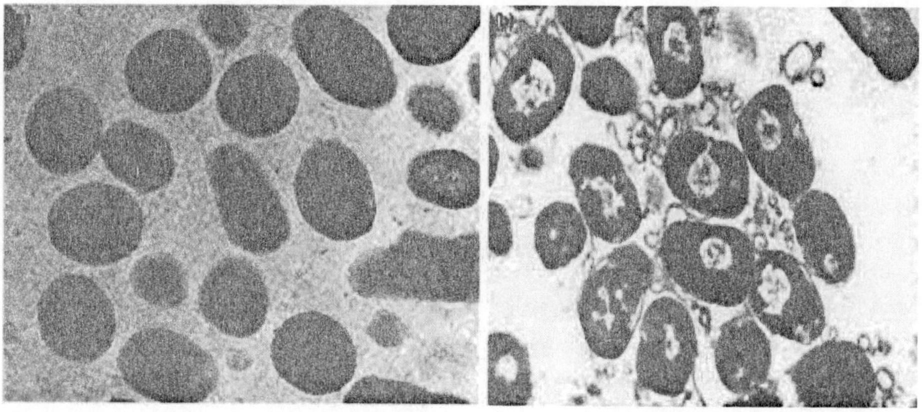

Figure 6. The Transmission Electron Microscopy (TEM) of *Escherichia coli* (X4400). (a) Control cells. (b) Cells after 2 min of plasma treatment at 75 W. Adapted from Hong et al. 2009, with permission.

was previously reported that ozone-treated *E. coli* cells remain intact and cell lysis was not the major mechanism of bacterial inactivation (Patil et al. 2011). The possible bactericidal action of plasma was proposed as likely due to the diffusion of plasma-generated reactive species through the cell membrane into the cell where they react and possibly damage proteins and nucleic acids leading to cell death (Ziuzina et al. 2013). On the other hand, Helmke et al. (2011) report that scanning electron microscopy (SEM) images of treated bacterial cells show no alteration of cell walls, while fluorescence microscopy reveals lethal damage to cell membranes. The ability of NTP from a DBD to fragment naked DNA as demonstrated by Leduc et al. (2010) and direct degradation of bacterial DNA (unlike indirect mechanism of UV light) as observed by Helmke et al. (2011) are notable in recent literature. Further investigation is warranted to get clearer insights into the sites and mode of action of NTP generated species on cellular components.

DBD plasma has been found to inactivate bacterial endotoxin quickly and effectively by Xingmin et al. (2011) who observed a very high inactivation effect, such that the detection sensitivity was less than 0.005 endotoxin unit/mL for *E. coli* endotoxin after 60 s treatment. The authors considered that ROSs oxidizing lipoid may be the main reason for the inactivation of lipopolysaccharide (LPS) which is the major chemical constituent of bacterial endotoxins.

The details of interaction of the different plasma agents with the various components of bacterial or other microbial cells or spores are currently very limited. However, by varying the process parameters involved in plasma generation, a multitude of mechanisms can be actuated which may act individually or synergistically (Misra et al. 2011b).

Variables Governing the Microbial Efficacy of NTP

The fact that reactive species in plasma and their concentration under different scenarios of generator, working gas, discharging method, and so on may differ, explains the varying degree of inactivation effect on microorganisms (Moisan et al. 2001) and thus the importance of considering the inactivation kinetics. Given the diversity of plasma generation approaches and systems, there is a growing body of information regarding the effect of process variables on efficacy of cold plasma decontamination. Salient findings of research studies are presented in brief in the following section and their practical effects with regard to foods are highlighted where appropriate (see also applications to food processing).

Effect of microbial cell type, cell loading and growth stage

Gram positive and negative bacteria, spores, viruses, bacteriophages and fungi have all been studied to assess their susceptibility to nonthermal plasma sterilisation. Bacteria, spores and fungi can be inactivated relatively easily by NTP. Electron micrographs and studies of the inhibition of bacteriophage infectivity suggest that the primary injury is to the cell envelope. The plasma cycle denatures isolated bacterial proteins and inactivates bacteriophages, but it was reported as having no effect on isolated DNA and bacterial proteins within exposed bacteria (Venezia et al. 2008).

The cell characteristics of bacteria and bacterial spores dictate their susceptibility to plasma sterilization process. Baier et al. (1992) mentioned that gram-negative bacteria are more resistant to plasma processing, because of their extra proteins and lipopolysaccharide walls. The mode of action of observed lysis in gram-negative bacteria (e.g., *E. coli*) has been shown to result from the subsequent disruption of its outer cell membrane due to electrostatic forces (Laroussi et al. 2003). This phenomenon may not occur with gram-positive bacterium (e.g., *Bacillus subtilis*) (no sign of morphological change), because of its different membrane structure.

The inverse relation between initial bacterial concentration and the rate of inactivation is well-established. For example, Fernandez et al. (2012) observed the highest D-value at the highest cell concentrations for *S. typhimurium*. The lower efficacy of NTP at higher cell concentrations can be explained on the basis of insufficient biocide plasma molecules present per cell to cause death and the possible production of proteins capable of quenching the active species of plasma by cultivable cells (Kamgang-Youbi et al. 2008).

Bacterial endospores are more resistant to traditional methods of sterilization, such as heating, chlorination and UV radiation. For example, *Bacillus subtilis* forms a tough protective endospore, with several proteinaceous outer layers and cortex peptidoglycan with a compact genetic core. This allows the organism to tolerate extreme conditions of high temperature, strong acidic/alkaline environments.

The bacterial growth phase and growth temperature usually play a minor role in the efficiency of inactivation by NTP. For example, the maximum difference in inactivation observed between stationary and logarithmic phase differs only by ~1 log cycle for *E. coli* (Deng et al. 2007, Yu et al. 2006) and *S. typhimurium* (Fernandez et al. 2013) treated with NTP.

Effect of gas composition

Because of the cost, the most popular gases for NTP sterilization are argon, helium and a mixture of nitrogen and oxygen. Helium (noble gases, in general) was traditionally the preferred gas for reasons including high thermal conductivity (for heat removal), rich ultraviolet emission spectrum and a lower operating discharge voltage at atmospheric pressure (Konesky 2009). With the presence of oxygen, the most aggressive of the active chemical species hydroxyl radicals (OH) and ozone (O_3) are always created. Water in NTP discharging procedure, either as humidity in gas or as liquid in system, will result in the abundance of hydroxyl radicals (OH), hydrogen peroxide H_2O_2 and hydronium ions H_3O^+. Depending on their sensitivities to the plasma constituent's, different groups of micro-organisms exhibit different inactivation kinetics and changes to cellular constituents.

Effect of power input and exposure conditions

Major factors which determine the ionization potential, and therefore the density and reactivity of the resultant plasma, are operating pressure of the system (partial vacuum compared with atmospheric pressure), input power and the gas composition. Inactivation effects of NTP vary with the voltage and frequency applied. An increase in inactivation with increasing voltage and frequency is due to increased energy input and energy density (Deng et al. 2007).

The inactivation effect of plasma also depends on the contact between plasma and bioindicators (Purevdorj et al. 2002). Fridman et al. (2007) investigated the influence of direct and indirect exposure to non-thermal DBD plasma on bacterial cells. Bacterial cells directly in contact with the plasma field were found to inactivate much faster than those exposed indirectly through after-glow or jet. This enhanced effect is believed to be an outcome of action of charged particles in direct exposure. When samples are placed at a distance from the plasma volume, meaning remotely exposed, the short-lived charged particles do not play any significant role since they recombine before reaching the sample.

Effect of food type

For solid foods, the efficiency of inactivation by NTP is related to food surface structures. The topographic features of food surface may allow bacterial cell to obscure thereby escaping the action of active plasma species. Examples of such features include stomata of lettuce, the convolutions of strawberry surfaces and the walls of the eukaryotic cells of potato tissue

(Fernandez et al. 2013). The intrinsic features of food surfaces including waxy cuticles, or other components could also contribute to lower degrees decontamination by NTP. In relation to the effect of food type, the importance of any protective effect of surface structures may be related to the microbial ecology present and the nature of that microbial community as well as NTP exposure mode and time.

Duration of exposure

The microbial inactivation efficiency of plasma is dependent on the duration of treatment. Higher treatment times, in general, lead to greater inactivation. This may be valid for any type of cold plasma source, however, practical limits exist for the duration of treatment for foods and other biological materials when considering changes in quality and degradation of sensitive components. Thus, it becomes evident that optimizing factors to yield shorter treatment times for inactivation may prove beneficial for application of a given cold plasma source.

Applications to Food Processing

NTP has been proved effective for reducing microbial population on the surface of materials, such as glass, metals, fabrics, and agar, with wide range of microorganisms including vegetative forms, spores, fungi, viruses and parasites, etc. Although most of the previous work on bacterial inactivation has been conducted with bacteria deposited on the surface of abiotic materials such as membrane filters and glass slides, data on the disinfection of foodstuffs is steadily accumulating (Noriega et al. 2011). Nonthermal plasma technology can have many potential applications in the food industry including the dry disinfection of food surfaces (like meat, poultry, fish and freshly harvested horticultural produce), granular and particulate foods (dried milk, herbs and spices) and sprouted seeds.

Treatment of raw fruits and vegetables

Microorganisms can occur on raw and minimally processed produce at populations ranging from 10^3 to 10^9 CFU/g. Although washing produce in water may remove some soil and other debris, it cannot be relied upon to completely remove microorganisms and may result in cross-contamination of food preparation surfaces (Nguyen-the and Carlin 1994). Chlorine wash is the most prevalent method for the preliminary decontamination of fresh produce at an industrial scale, primarily due to its relatively low cost, efficacy, availability and ease of operation (Connolly et al. 2010). Commercially, only around 100 to 200 ppm chlorine is used for sanitation purposes. However, the chlorine derivatives are not considered safe compounds since they react with organic material to form reactive by-products such as chloramines and trihalomethanes. In European countries, such as the Netherlands, Sweden, Germany and Belgium, the use of chlorine on minimally processed vegetable products is prohibited (Brilhante São José and Dantas Vanetti 2012, Rico et al. 2007). Unlike chlorine, which selectively destroys certain enzymes in bacteria, ozone (the prominent anti-microbial species formed in most NTP with oxygen as a component) acts as a general protoplasmic oxidant (Khadre et al. 2001, Sykes 1965).

Niemira and Sites (2008) significantly reduced the viable populations of *Salmonella* and *E. coli* 0157:H7 inoculated on apple surfaces using cold plasma generated from a gliding arc. A 2.9 to 3.7 $\log_{10}$ CFU/ml and 3.4 to 3.6 $\log_{10}$ CFU/ml reduction of *Salmonella* Stanley and *E. coli*, respectively was observed. The authors reported the highest flow rate of the air (discharge medium) i.e., 40 litres/min, to be most effective. Recent important findings related to plasma based inactivation of microflora associated with fresh fruits and vegetables have been summarized in Table 1.

Decontamination of seeds and nuts

The main entry of mycotoxins into the human and animal food chain is from agricultural products such as cereal grains and oilseeds or products derived from these (Jiménez et al. 1991). Low-pressure cold plasma with sulphur hexafluoride (SF_6) treatment for 10 min has been found to result in approximately a 5-log decrease in fungal population on the surface of hazelnuts, peanuts, and pistachio nuts (Basaran et al. 2008). In addition, with similar treatments for 20 min using air plasma a 50% reduction in total aflatoxins (AFB1, AFB2, AFG1, and AFG2) has been reported (Basaran et al. 2008).

Decontamination of tomato seeds, wheat (*Triticum durum*), bean, chickpea, soybean, barley, oat, rye, lentil (*Lens culinaris*) and corn contaminated with *Aspergillus parasiticus* 798 and *Penicillium spp.* to less than 1 per cent of initial count for treatment times ranging from 30 s to 30 min have also been reported in literature (Selcuk et al. 2008) (Fig. 7). These effects were obtained using a custom designed batch type low pressure cold plasma (LPCP) prototype unit operating under vacuum, in air and SF_6 gases. It is worth noting that seeds remain viable post-treatment.

Deng et al. (2007) have reported effective reduction of *E. coli* on almond by almost 5-log after 30-sec treatment at 30 kV and 2000 Hz using a DBD plasma. Enhanced inactivation of *E. coli* inoculated on almonds was associated with increasing voltage and frequency input to the DBD

Table 1. Recent studies in the area of atmospheric nonthermal plasma for inactivation of microorganisms associated with fresh fruits and vegetables.

Organism	Plasma conditions	Treatment surface/ medium	Salient result	References
Escherichia coli, *Saccharomyces cerevisiae*, *Pantoea agglomerans*, *Gluconacetobacter liquefaciens*	Cold atmospheric plasma generated by an AC voltage (variable12 kV and 16kV)	Pericarps of mangoes and melons	*S. cerevisiae* was the most resistant amongst all test organisms. An increase in the applied voltage led to more efficient production of reactive plasma species (oxygen atoms) which was attributed for better inactivation	Perni et al. (2008)
Escherichia coli O157:H7, *Salmonella* sp., *Listeria monocytogenes*	One atmosphere uniform glow discharge plasma (OAUGDP) operated at 9 kV power and 6 kHz frequency	Apples, Cantaloupe and Lettuce	Inactivation was observed in all the cases. Extent of log reduction varied with the organisms	Critzer et al. (2007)
Escherichia coli O157:H7 *Salmonella* Stanley	Gliding Arc plasma	On agar plates and inoculated onto surfaces of Golden Delicious apples	Bacterial inactivation was shown to be a function of flow rate and duration of exposure	Niemira and Sites (2008)
Escherichia coli O157:H7 6460	Dielectric barrier discharge operated at 40 W and 60 Hz with 12 kV of potential between the electrodes.	Whole spinach leaves	A reduction of 3–5 $\log_{10}$ CFU/leaf after 24 h of storage in package	Klockow and Keener (2009)

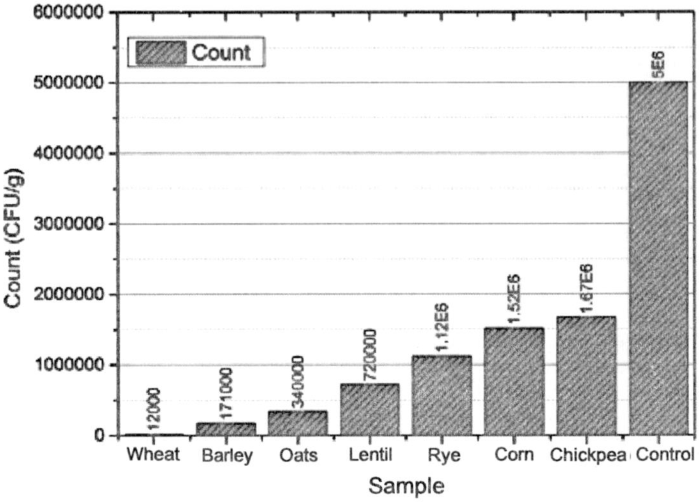

Figure 7. Survival curves of SF6 low pressure cold plasma exposed *Aspergillus* spp. on various seed surface. Data adapted from Selcuk et al. (2008).

system in this research. Niemira (2012) also reported a decrease in the count of *Salmonella* and *Escherichia coli* O157:H7 on almonds treated using a jet plasma operating with air and nitrogen feed gases.

Surface decontamination of eggs

Salmonella spp. is a well-recognized hazard for egg consumers and a need for alternative methods of decontaminations has been highlighted by Davies and Breslin (2003). Vertical transmission of *Salmonella* and *Campylobacter* from infected breeders to their offspring through contamination of the eggs is one possibility of how pathogens spread throughout flocks and/ or are present inside eggs (Newell and Fearnley 2003, Snelling et al. 2005). *Salmonella* has been shown to be present in eggs (Berchieri et al. 2001) and as such contaminated eggs are also a major food safety risk. In *Campylobacter* spp. this has been shown but is not thought to be the major route of transmission (Doyle and Erickson 2006).

Several nonthermal approaches for the surface decontamination of eggs, such as pulsed light technology (Hierro et al. 2009), ozone, UV radiation (Fuhrmann et al. 2010, Rodriguez-Romoand and Yousef 2005) and electrolyzed water (Cao et al. 2009) have already been researched. These attempts are in extension to the failure of conventional techniques in efficient cleaning of the surface of eggs. Ragni et al. (2010) reported efficacy of resistive barrier discharge (RBD) plasma for decontamination of shell egg surfaces. They identified a maximum reduction of 2.2–2.5

Log CFU/eggshell in *Salmonella enteritidis* levels following a 60–90 min of treatment at 35% RH. Efficacy was enhanced at a higher RH level of 65%, where maximum declines of 3.8 and 4.5 Log CFU/eggshell were achieved after 90 min of exposure. Similar observations were made for *Salmonella typhimurium* with an overall reduction of 3.5 Log CFU/eggshell, after 90 min treatment. The enhanced effects of increased RH on the efficiency of the treatments have been explained on the basis of the presence of oxygen species as detected in the discharge emission spectra. These results are at par with the UV and ozone treatments of eggs reported earlier (Rodriguez-Romoand and Yousef 2005).

Donner and Keener (2011) evaluated the use of dielectric barrier discharges for in-package ozonation to reduce *Salmonella enteritidis* on raw, shell eggs. After a 24-hour storage period, they observed 3 $\log_{10}$ reductions. Their results demonstrate effectiveness of in-package ozonation treatment from dielectric barrier discharge plasma for reducing *Salmonella enteritidis* on raw, shell eggs without significant effects on measured egg quality over time.

Decontamination of meat, fish and poultry

Surface contamination of ready to eat (RTE) meat may occur during processing since *Listeria monocytogenes* can persist in the processing environment and be transferred to the product from equipment such as slicers (Bērziņš et al. 2010). In products where the contamination is primarily localized to the surface, the prevalence and level of *L. monocytogenes* may be lowered via surface decontamination. Rød et al. (2012) investigated the application of cold atmospheric pressure plasma for decontamination of a sliced ready-to-eat (RTE) meat product (bresaola) inoculated with *Listeria innocua*. This group treated the inoculated samples at 15.5, 31, and 62 W for 2–60 s inside sealed linear-low-density-polyethylene bags containing 30 per cent oxygen and 70 per cent argon. The treatments resulted in a reduction of *L. innocua* ranging from 0.8±0.4 to 1.6±0.5 log cfu/g with no significant effects of time and intensity while multiple treatments at 15.5 and 62 W of 20 s with a 10 min interval increased reduction of *L. innocua* with increasing number of treatments.

Moon et al. (2009) treated pork skin using a plasma source powered by a 13.56 MHz radio frequency (RF) supply through an impedance matching network powered between 20 and 150 W. Figure 8 depicts the effects of plasma and UV lamp on *E. coli* sterilization. The sterilization efficiency of plasma (1 min treatment time) against *E. coli* inoculated on pork surface was comparable with that of a conventional UV sterilizer (30 sec treatment).

Kim et al. (2011) studied the effect of atmospheric pressure plasma (at 75, 100, and 125 W of input power for 60 and 90 s) on the inactivation

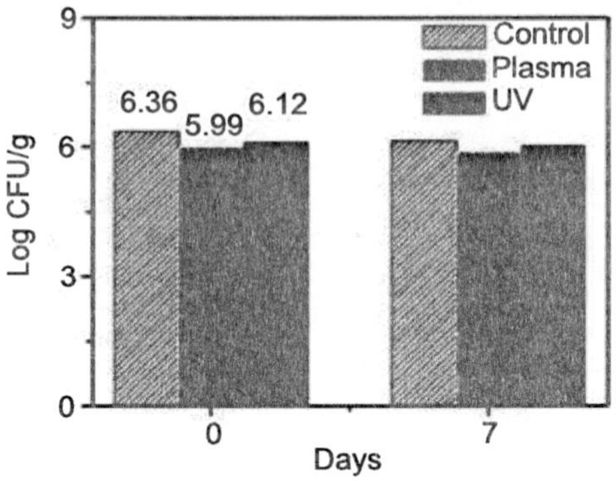

Figure 8. Sterilization efficiency of *E. coli* inoculated on pork surface by plasma and conventional ultraviolet sterilizer. Adapted from Moon et al. (2009).

of *Listeria monocytogenes* (KCTC 3596), *Escherichia coli* (KCTC 1682), and *Salmonella typhimurium* (KCTC 1925) inoculated onto sliced bacon. They used two gas compositions; helium (10 l/min) and a mixture of helium and oxygen (10 l/min and 10 sccm, respectively) for the plasma generation. The authors reported that plasma with helium only reduced inoculated pathogens by about 1–2 $\log_{10}$ cycles, while the helium/oxygen gas mixture exhibited microbial reduction of about 2–3 $\log_{10}$. Based on log reduction data, microscopic and colorimetric studies, Kim et al. (2011) concluded that APP treatment was effective for the inactivation of the three pathogens used in their study. Lee et al. (2011) also reported a reduction of *L. monocytogenes* inoculated onto ham by 1.94 to 6.52 log units for 2 min treatments with APP jets of He, He + O_2, N_2, or N_2 + O_2 feed gases.

In a study of cold plasma treatment of cold smoked salmon, Chiper et al. (2011) employed a DBD operating at atmospheric pressure in 100 per cent argon and (93 per cent argon+ 7 per cent CO_2) gas mixture with a potential difference of 13 kV (peak to peak) across the electrodes. They observed ~3 $\log_{10}$ CFU reduction after 60–120s treatment for *Photobacterium phosphoreum* and an insignificant effect on the inactivation of *Listeria monocytogenes* or *Lactobacillus sakei*.

Poultry carcasses are commonly contaminated with enteric pathogens such as *Salmonella* sp., *Campylobacter* sp. and *Listeria monocytogenes* (Murphy et al. 2004). NTP treatment of cooked chicken breast using jet plasma was investigated by Lee et al. (2011), who observed 1.37 to 4.73 log reduction of *L. monocytogenes* after 2 min treatment with He, N_2, and their mixtures with O_2 gas. The lethality of the plasma jets were reported to be lower for cooked chicken samples than for bacteria grown on agar plates. Noriega et al. (2011) employed a cold atmospheric plasma pen (CAP-Pen) for decontamination of chicken and have reported > 3 log reductions of *L. innocua* on membrane filters after 10s treatment, 1 log reduction on chicken skin after 8 min treatment, and > 3 log reductions treatment on chicken muscle after 4 min under optimal conditions. Very recently Dirks et al. (2012) demonstrated the feasibility of NTP DBD as an intervention to help reduce foodborne pathogens *Salmonella enterica* and *Campylobacter jejuni* on the surface of raw poultry (boneless skinless chicken breast and chicken thigh with skin).

Decontamination of packaging materials

Most regulatory guidelines specify microbiological requirements for food packaging and in many cases the packaging process is an important control point in hazard analysis critical control point (HACCP) (Mittendorfer et al. 2002). NTP has the potential to replace or complement popular chemical based sterilization methods for food packaging materials in industries. From a food industry point of view, current commercial applications of cold plasma are primarily limited to the packaging industry and include sterilization of anti-fouling and printable surfaces and permeability reduction of polymers for carbon dioxide and oxygen (Basaran et al. 2008, Schneider et al. 2005). Packaging materials such as plastic bottles, lids and films can be rapidly sterilised using cold NTP without adversely affecting their properties or leaving any residues.

Plasma deposition of heat-sensitive materials such as vitamins, antioxidants and antimicrobials into the packaging material may be sought as potential alternative in the emerging field of antimicrobial and active packaging (Misra et al. 2011b). Fernández-Gutierrez et al. (2010) have demonstrated the application of cold gas plasma for deposition of vanillin film over red delicious apples. The antimicrobial nature of vanillin against bacteria, yeasts and fungi has been proved in literature (Cerrutti and Alzamora 1996, Fitzgerald et al. 2004).

NTP processing of fruit juices

Changes in consumer preferences towards healthier and minimally processed beverages has led to increased research interest in non-thermal processing technologies in order to manufacture safe and 'fresh-like' products, with enhanced retention of organoleptic quality (Palgan et al. 2011). Recent studies indicate that atmospheric NTP technology has potential significance for commercial level orange juice treatments, considering the fact that only a 5 log reduction is required for pasteurization of liquid foods. Shi et al. (2011) employed a DBD to effectively kill *Staphylococcus aureus*, *Escherichia coli* and *Candida albicans* inoculated in orange juice by >5 $\log_{10}$ after 12, 8, and 25 s treatments respectively at peak voltage of 30 kV. This research group demonstrated the ability of NTP to extend the shelf life of freshly squeezed orange juice. Prior to this, Montenegro et al. (2002) employed direct current corona discharges obtained using a pulsating 0–15 kV DC power supply to generate NTP in liquid media. They reported a reduction in the number of *Escherichia coli* O157:H7 in apple juice by more than 5 log CFU/mL in 40 s. It is a well-recognized fact that the resultant ionization that occurs during plasma discharge activity allows substantial levels of ozone to be formed and to dissolve in liquids (Espie et al. 2001).

Processing of milk and milk products

Milk and milk products have long been associated with infectious diseases including campylobacteriosis, salmonellosis, yersinosis, listeriosis, tuberculosis, brucellosis, staphylococcal enterotoxin poisoning, streptococcal poisoning and *Escherichia coli* O157:H7 infection (Gurol et al. 2012, Sampedro et al. 2005). In this respect, nonthermal food processing has played an important role in improving food quality by maintaining its microbiological safety and quality profile while only minimally affecting its sensory and health-related properties (Sobrino-López and Martín-Belloso 2010). Pulsed electric field (PEF) processing has already been extensively researched and to some extent commercialized for treatment of milk. The first attempt to decontaminate milk using cold plasma generated from a corona discharge (9 kV of AC power supply) has been made by Gurol et al. (2012). This group has reported *E. coli* count reduction from 7.78 log CFU/ml to 3.63 log CFU/ml after 20 min treatment in whole milk and an average 54% reduction after 3 min treatment irrespective of fat content. Thus, it appears that cold plasma generated under controlled conditions has potential to decontaminate milk.

Milk products including sliced cheese are often associated with foodborne listeriosis. Recently the effects of atmospheric NTP on the inactivation of 3-strain cocktail of *L. monocytogenes* in sliced cheese have been examined by Song et al. (2009). This group reported ~6 log reduction in the viable population of *L. monocytogenes* after 120s treatment at 125W power using radiofrequency plasma (13.65 MHz) operating in helium gas. Further, the microbial log reduction was enhanced with increased input power and plasma exposure time.

Conclusions

Nonthermal plasma decontamination of foods is a complex and interdisciplinary area involving amalgamation of core concepts from the areas of plasma physics, electronics, chemistry, microbiology and food science. Ample work has already been done to establish sufficiently high microbial population reductions of commercial significance on various inert materials, model foods and some real food systems. Several energetic species simultaneously bombard the food surface during the plasma treatment and it is still a matter of debate which of them is/are chiefly responsible for surface decontamination. This in turn is due to the complexity of the plasma chemistry and the combined effect of different plasma species. It is important to note that in-spite of the high log reductions published in scientific literature, the mechanism of action of plasma is not completely unravelled. Additionally, many of the antimicrobial studies are on controlled monoculture populations which do not mimic the microbial complexity often presented in real food systems. In particular, the respective roles of ions and photons are still unresolved, with both capable of independent microbial inactivation. Despite the fact that direct comparison of results reported in literature are not possible because of different NTP systems employed for different studies, NTP may generally be considered to be a candidate method for ensuring food safety during processing (Lee et al. 2006). In order to fully exploit the benefits of NTP technology, the changes in the chemical constituents, consideration of complex microbial ecology, food structure effects as well as the effects of NTP on quality and nutritional characteristics of foods require deeper investigation.

Keywords: Nonthermal plasma, decontamination, Food, Pathogen, Microwave, Radio-frequency, Corona, Dielectric barrier discharge.

References

Académie des, s., s. Centre national de la recherche. 1835. Comptes rendus hebdomadaires des séances de l'Académie des sciences. Comptes rendus hebdomadaires des séances de l'Académie des sciences.

Amorim, J., G. Baravian and J. Jolly. 2000. Laser-induced resonance fluorescence as a diagnostic technique in non-thermal equilibrium plasmas. J. Phys. D: Appl. Phys. 33(9): R51–R65.

Bablon, G., W.D. Bellamy, M.-M. Bourbigot, F.B. Daniel, M. Dore, F. Erb, G. Gordon, B. Langlais, A. Laplanche, B. Legube, G. Martin, W.J. Masschelein, G. Pacey and D.A.C.V. Reckhow. 1991. Fundamental aspects. *In*: B. Langlais, D.A. Reckhow and D.R. Brank (eds.). Ozone in Water Treatment: Application and Engineering. Lewis Publishers, Inc., Chelsea, Mich., USA, pp. 11–132.

Baier, R., J. Carter, S. Sorensen, A. Meyer, B. McGowan and S. Kasprzak. 1992. Radiofrequency gas plasma (glow discharge) disinfection of dental operative instruments, including handpieces. J. Oral implantol. 18(3): 236.

Bárdos, L. and H. Baránková. 2008. Plasma processes at atmospheric and low pressures. Vacuum 83(3): 522–527.

Bárdos, L. and H. Baránková. 2010. Cold atmospheric plasma: Sources, processes, and applications. Thin Solid Films 518(23): 6705–6713.

Basaran, P., N. Basaran-Akgul and L. Oksuz. 2008. Elimination of Aspergillus parasiticus from nut surface with low pressure cold plasma (LPCP) treatment. Food Microbiol. 25(4): 626–632.

Berchieri Jr., A., P. Wigley, K. Page, C. Murphy and P. Barrow. 2001. Further studies on vertical transmission and persistence of *Salmonella enterica* serovar Enteritidis phage type 4 in chickens. Avian pathol. 30(4): 297–310.

Bērziņš, A., S. Hellström, I. Siliņš and H. Korkeala. 2010. Contamination patterns of Listeria monocytogenes in cold-smoked pork processing. J. Food Protect. 73(11): 2103.

Boudam, M., M. Moisan, B. Saoudi, C. Popovici, N. Gherardi and F. Massines. 2006. Bacterial spore inactivation by atmospheric-pressure plasmas in the presence or absence of UV photons as obtained with the same gas mixture. J. Phys. D: Appl. Phys. 39: 3494.

Brilhante São José, J.F. and M.C. Dantas Vanetti. 2012. Effect of ultrasound and commercial sanitizers in removing natural contaminants and Salmonella enterica Typhimurium on cherry tomatoes. Food Control 24(1–2): 95–99.

Bruggeman, P. and D.C. Schram. 2010. On OH production in water containing atmospheric pressure plasmas. Plasma Sources Sci. Technol. 19(4): 045025.

Burney, S., J.L. Caulfield, J.C. Niles, J.S. Wishnok and S.R. Tannenbaum. 1999. The chemistry of DNA damage from nitric oxide and peroxynitrite. Mutat. Res. 424(1–2): 37–49.

Cammack, R., C. Joannou, X.Y. Cui, C. Torres Martinez, S.R. Maraj and M.N. Hughes. 1999. Nitrite and nitrosyl compounds in food preservation. Biochimica et Biophysica Acta (BBA)-Bioenergetics 1411(2–3): 475–488.

Cao, W., Z.W. Zhu, Z.X. Shi, C.Y. Wang and B.M. Li. 2009. Efficiency of slightly acidic electrolyzed water for inactivation of Salmonella enteritidis and its contaminated shell eggs. Int. J. Food. Microbiol. 130(2): 88–93.

Cerrutti, P. and S.M. Alzamora. 1996. Inhibitory effects of vanillin on some food spoilage yeasts in laboratory media and fruit purees. Int. J. Food. Microbiol. 29(2): 379–386.

Chiper, A.S., V. Anița, C. Agheorghiesei, V. Pohoața, M. Anița and G. Popa. 2004. Spectroscopic Diagnostics for a DBD Plasma in He/Air and He/N2 Gas Mixtures. Plasma Processes Polym. 1(1): 57–62.

Chiper, A.S., W. Chen, O. Mejlholm, P. Dalgaard and E. Stamate. 2011. Atmospheric pressure plasma produced inside a closed package by a dielectric barrier discharge in Ar/CO(2) for bacterial inactivation of biological samples. Plasma Sources Sci. Technol. 20(2): 10.

Connolly, J., J.P. Mosnier, V.P. Valdramidis, E. Byrne, P.J. Cullen and K.M. Keener. 2010. Diagnostic of reactive species in a dielectric barrier discharge air/helium plasma for the treatment of a pre-packed food simulant. Paper # P4.317. 37th European Physical Society Conference on Plasma Physics. Dublin, Ireland.

Critzer, F., K. Kelly-Wintenberg, S. South and D. Golden. 2007. Atmospheric plasma inactivation of foodborne pathogens on fresh produce surfaces. J. Food Protect. 70(10): 2290.

Davidson, P.M. and V.K. Juneja. 1990. Antimicrobial agents. *In*: A.L. Branen, P.M. Davidson, S., S. (eds.). Food Additives. Marcel Dekker, New York and Basel, pp. 97–100.

Davies, R. and M. Breslin. 2003. Investigations into possible alternative decontamination methods for Salmonella enteritidis on the surface of table eggs. J. Vet Med. B. 50(1): 38–41.

Deng, S., R. Ruan, C.K. Mok, G. Huang, X. Lin and P. Chen. 2007. Inactivation of *Escherichia coli* on almonds using nonthermal plasma. J. Food Sci. 72(2): M62–66.

Di Mascio, P., H. Wefers, H.P. Do-Thi, M.V.M. Lafleur and H. Sies. 1989. Singlet molecular oxygen causes loss of biological activity in plasmid and bacteriophage DNA and induces single-strand breaks. Biochimica et Biophysica Acta (BBA)-Gene Structure and Expression 1007(2): 151–157.

Dirks, B.P., D. Dobrynin, G. Fridman, Y. Mukhin, A. Fridman and J.J. Quinlan. 2012. Treatment of Raw Poultry with Nonthermal Dielectric Barrier Discharge Plasma To Reduce *Campylobacter jejuni* and *Salmonella enterica*. J. Food Protect. 75(1): 22–28.

Donner, A. and K.M. Keener. 2011. Investigation of in-package ionisation. Journal of Purdue Undergraduate Research 1: 10–15.

Doyle, M. and M. Erickson. 2006. Reducing the carriage of foodborne pathogens in livestock and poultry. Poultry Science 85(6): 960–973.

El-Badawi, A., R. Cain, C. Samuels and A. Anglemeier. 1964. Color and pigment stability of packaged refrigerated beef. Food Technol. 18(5): 159–163.

Espie, S., L. Marsili, S.J. MacGregor and J.G. Anderson. 2001. Investigation of dissolved ozone production using plasma discharge in liquid. Pulsed power plasma science, Digest of Technical Papers 1: 616–619.

FDA. 2012. Information on the Recalled Jensen Farms Whole Cantaloupes. U.S. Food and Drug Administration, USA.

Fernández-Gutierrez, S.A., P.D. Pedrow, M.J. Pitts and J. Powers. 2010. Cold Atmospheric-Pressure Plasmas Applied to Active Packaging of Apples. Plasma Science, IEEE Transactions on 38(4): 957–965.

Fernandez, A., E. Noriega and A. Thompson. 2013. Inactivation of *Salmonella enterica* serovar Typhimurium on fresh produce by cold atmospheric gas plasma technology. Food Microbiol. 33(1): 24–29.

Fernandez, A., N. Shearer, D.R. Wilson and A. Thompson. 2012. Effect of microbial loading on the efficiency of cold atmospheric gas plasma inactivation of *Salmonella enterica* serovar Typhimurium. Int. J. Food. Microbiol. 152(3): 175–180.

Fitzgerald, D., M. Stratford, M. Gasson, J. Ueckert, A. Bosand and A. Narbad. 2004. Mode of antimicrobial action of vanillin against *Escherichia coli, Lactobacillus plantarum* and *Listeria innocua*. J. Appl. Microbiol. 97(1): 104–113.

Fridman, G., A.D. Brooks, M. Balasubramanian, A. Fridman, A. Gutsol, V.N. Vasilets, H. Ayan and G. Friedman. 2007. Comparison of Direct and Indirect Effects of Non-Thermal Atmospheric-Pressure Plasma on Bacteria. Plasma Processes Polym. 4(4): 370–375.

Fuhrmann, H., N. Rupp, A. Büchner and P. Braun. 2010. The effect of gaseous ozone treatment on egg components. J. Sci. Food Agr. 90(4): 593–598.

Gaunt, L.F., C.B. Beggs and G.E. Georghiou. 2006. Bactericidal action of the reactive species produced by gas-discharge nonthermal plasma at atmospheric pressure: a review. Plasma Science, IEEE Transactions on 34(4): 1257–1269.

Gordillo-Vázquez, F.J. 2008. Air plasma kinetics under the influence of sprites. J. Phys. D: Appl. Phys. 41(23): 234016.

Gurol, C., F.Y. Ekinci, N. Aslan and M. Korachi. 2012. Low temperature plasma for decontamination of *E. coli* in milk. Int. J. Food. Microbiol. 157(1): 1–5.

Guzel-Seydim, Z.B., A.K. Greene and A.C. Seydim. 2004. Use of ozone in the food industry. Lebensm-Wiss-Technol. 37(4): 453–460.

Hamelin, C. 1985. Production of single-and double-strand breaks in plasmid DNA by ozone. Int. J. Radiat. Oncol. 11(2): 253–257.

Havelaar, A.H., S. Brul, A. de Jong, R. de Jonge, M.H. Zwietering and B.H. ter Kuile. 2010. Future challenges to microbial food safety. Int. J. Food. Microbiol. 139(1): S79–S94.

Helmke, A., D. Hoffmeister, F. Berge, S. Emmert, P. Laspe, N. Mertens, W. Viol and K.D. Weltmann. 2011. Physical and Microbiological Characterisation of *Staphylococcus epidermidis* inactivation by dielectric barrier discharge Plasma. Plasma Processes Polym. 8(4): 278–286.

Helmke, A., D. Hoffmeister, N. Mertens, S. Emmert, J. Schuette and W. Viol. 2009. The acidification of lipid film surfaces by non-thermal DBD at atmospheric pressure in air. New J. Phys. 11(11): 115025.

Herron, J.T. and D.S. Green. 2001. Chemical kinetics database and predictive schemes for nonthermal humid air plasma chemistry. Part II. Neutral species reactions. Plasma Chem. Plasma P. 21(3): 459–481.

Hierro, E., S. Manzano, J.A. Ordóñez, L. de la Hoz and M. Fernández. 2009. Inactivation of *Salmonella enterica* serovar Enteritidis on shell eggs by pulsed light technology. Int. J. Food. Microbiol. 135(2): 125–130.

Hong, Y., J. Kang, H. Lee, H. Uhm, E. Moon and Y. Park. 2009. Sterilization effect of atmospheric plasma on *Escherichia coli* and *Bacillus subtilis* endospores. Lett. Appl. Microbiol. 48(1): 33–37.

Hyslop, P.A., D.B. Hinshaw, I.U. Scraufstatter, C.G. Cochrane, S. Kunz and K. Vosbeck. 1995. Hydrogen peroxide as a potent bacteriostatic antibiotic: implications for host defense. Free Radical Bio. Med. 19(1): 31–37.

Imlay, J.A. and S. Linn. 1988. DNA damage and oxygen radical toxicity. Science 240(4857): 1302–9.

Jayasingh, P., D.P. Cornforth, C.E. Carpenter and D. Whittier. 2001. Evaluation of carbon monoxide treatment in modified atmosphere packaging or vacuum packaging to increase color stability of fresh beef. Meat Sci. 59(3): 317–324.

Jiménez, M., R. Mateo, A. Querol, T. Huerta and E. Hernández. 1991. Mycotoxins and mycotoxigenic moulds in nuts and sunflower seeds for human consumption. Mycopathologia 115(2): 121–127.

Kamgang-Youbi, G., J.M. Herry, J.L. Brisset, M.N. Bellon-Fontaine, A. Doubla and M. Naïtali. 2008. Impact on disinfection efficiency of cell load and of planktonic/adherent/detached state: case of *Hafnia alvei* inactivation by Plasma Activated Water. Appl. Microbiol. Biot. 81(3): 449–457.

Keener, K.M., J. Jensen, V. Valdramidis, E. Byrne, J. Connolly, J. Mosnier and P. Cullen. 2012. Decontamination of Bacillus subtilis spores in a sealed package using a non-thermal plasma system. *In*: Z. Machala, K. Hensen and Y. Akishev (eds.). Plasma for Bio-Decontamination, Medicine and Food Security, Springer, Netherlands, pp. 445–455.

Kennedy, M. 2011. *E. coli* outbreak: WHO says bacterium is a new strain. Guardian, UK.

Khadre, M.A., A.E. Yousef and J.G. Kim. 2001. Microbiological aspects of ozone applications in food: A review. J. Food Sci. 66(9): 1242–1252.

Kim, B., H. Yun, S. Jung, Y. Jung, H. Jung, W. Choe and C. Jo. 2011. Effect of atmospheric pressure plasma on inactivation of pathogens inoculated onto bacon using two different gas compositions. Food Microbiol. 28(1): 9–13.

Klockow, P.A. and K.M. Keener. 2009. Safety and quality assessment of packaged spinach treated with a novel ozone-generation system. LWT—Food Sci. Technol. 42(6): 1047–1053.

Kogelschatz, U. 2002. Filamentary, patterned, and diffuse barrier discharges. Plasma Science, IEEE Transactions on 30(4): 1400–1408.

Konesky, G. 2009. Dwell time considerations for large area Cold Plasma decontamination. *SPIE* 7304, Chemical, Biological, Radiological, Nuclear, and Explosives (CBRNE) Sensing X, 73040N. http://dx.doi.org/10.1117/12.818323.

Laroussi, M., D. Mendis and M. Rosenberg. 2003. Plasma interaction with microbes. New J. Phys. 5: 41.

Laroussi, M., K.H. Schoenbach, U. Kogelschatz, R.J. Vidmar, S. Kuo, M. Schmidt, J.F. Behnke, K. Yukimura and E. Stoffels. 2004. Applications of atmospheric pressure air plasma. *In*: K.H. Becker, U. Kogelschatz, K.H. Schoenbach and R.J. Barker (eds.). Non-Equilibrium Air Plasma at Atmospheric Pressure. IOP, Bristol, UK, pp. 537–671.

Leduc, M., D. Guay, S. Coulombe and R.L. Leask. 2010. Effects of non-thermal plasmas on DNA and mammalian cells. Plasma Processes Polym. 7(11): 899–909.

Lee, H.J., H. Jung, W. Choe, J.S. Ham, J.H., Lee and C. Jo. 2011. Inactivation of *Listeria monocytogenes* on agar and processed meat surfaces by atmospheric pressure plasma jets. Food Microbiol. 28(8): 1468–1471.

Lee, H.M., M.B. Chang and T.C. Wei. 2004. Kinetic modeling of ozone generation via dielectric barrier discharges. Ozone: Sci. Eng. 26(6): 551–562.

Lee, K., K. Paek, W.T. Ju and Y. Lee. 2006. Sterilization of bacteria, yeast, and bacterial endospores by atmospheric-pressure cold plasma using helium and oxygen. J. Microbiol. 44(3): 269–275.

Machala, Z., M. Janda, K. Hensel, I. Jedlovský, L. Leštinská, V. Foltin, V. Martišovitš and M. Morvová. 2007. Emission spectroscopy of atmospheric pressure plasmas for bio-medical and environmental applications. J. Mol. Spectrosc. 243(2): 194–201.

Marnett, L.J. 2002. Oxy radicals, lipid peroxidation and DNA damage. Toxicology 181: 219–222.

Mendis, D., M. Rosenberg and F. Azam. 2002. A note on the possible electrostatic disruption of bacteria. Plasma Science, IEEE Transactions on 28(4): 1304–1306.

Mennad, B., Z. Harrache, D. Amir Aid and A. Belasri. 2010. Theoretical investigation of ozone production in negative corona discharge. Curr. Appl. Phys. 10(6): 1391–1401.

Misra, N.N., S.U. Kadam and S.K. Pankaj. 2011a. An overview of nonthermal technologies in food processing. Indian Food Industry 30(5–6): 45–52.

Misra, N.N., B.K. Tiwari, K.S.M.S. Raghavarao and P.J. Cullen. 2011b. Nonthermal plasma inactivation of food-borne pathogens. Food Eng. Rev. 3(3–4): 159–170.

Mittendorfer, J., H. Bierbaumer, F. Gratzl and E. Kellauer. 2002. Decontamination of food packaging using electron beam—status and prospects. Radiat. Phys. Chem. 63(3): 833–836.

Moisan, M., J. Barbeau, S. Moreau, J. Pelletier, M. Tabrizian and L.H. Yahia. 2001. Low-temperature sterilization using gas plasmas: a review of the experiments and an analysis of the inactivation mechanisms. Int. J. Pharm. 226(1–2): 1–21.

Montenegro, J., R. Ruan, H. Ma and P. Chen. 2002. Inactivation of *E. coli* O157 : H7 using a pulsed nonthermal plasma system. J. Food Sci. 67(2): 646–648.

Montie, T.C., K. Kelly-Wintenberg and J.R. Roth. 2002. An overview of research using the one atmosphere uniform glow discharge plasma (OAUGDP) for sterilization of surfaces and materials. Plasma Science, IEEE Transactions 28(1): 41–50.

Moon, S.Y., D.B. Kim, B. Gweon, W. Choe, H.P. Song and C. Jo. 2009. Feasibility study of the sterilization of pork and human skin surfaces by atmospheric pressure plasmas. Thin Solid Films 517(14): 4272–4275.

Munakata, N., M. Saito and K. Hieda. 1991. Inactivation action spectra of *Bacillus subtilis* spores in extended ultraviolet wavelengths (50–300 nm) obtained with synchrotron radiation. Photochem. Photobiol. 54(5): 761.

Murphy, R., T. Osaili, L. Duncan and J. Marcy. 2004. Thermal inactivation of *Salmonella* and *Listeria monocytogenes* in ground chicken thigh/leg meat and skin. Poultry Science 83(7): 1218–1225.

Newell, D. and C. Fearnley. 2003. Sources of *Campylobacter* colonization in broiler chickens. Appl. Environ. Microbiol. 69(8): 4343–4351.

Nguyen-the, C. and F. Carlin. 1994. The microbiology of minimally processed fresh fruits and vegetables. Crit. Rev. Food Sci. Nutr. 34(4): 371–401.

Niemira, B.A. 2012. Cold Plasma Reduction of *Salmonella* and *Escherichia coli* O157:H7 on almonds using ambient pressure gases. J. Food Sci. 77(3): M171–M175.

Niemira, B.A. and J. Sites 2008. Cold plasma inactivates *Salmonella* Stanley and *Escherichia coli* O157: H7 inoculated on golden delicious apples. J. Food Prot. 71(7): 1357–1365.

Noriega, E., G. Shama, A. Laca, M. Díaz and M.G. Kong. 2011. Cold Atmospheric Gas Plasma Disinfection of Chicken Meat and Chicken Skin contaminated with *Listeria innocua*. Food Microbiol. 28(7): 1293–1300.

Palgan, I., I.M. Caminiti, A. Muñoz, F. Noci, P. Whyte, D.J. Morgan, D.A. Cronin and J.G. Lyng. 2011. Effectiveness of High Intensity Light Pulses (HILP) treatments for the control of *Escherichia coli* and *Listeria innocua* in apple juice, orange juice and milk. Food Microbiol. 28(1): 14–20.

Patil, S., V.P. Valdramidis, K.A.G. Karatzas, P.J. Cullen and P. Bourke. 2011. Assessing the microbial oxidative stress mechanism of ozone treatment through the responses of *Escherichia coli* mutants. J. Appl. Microbiol. 111(1): 136–144.

Perni, S., D.W. Liu, G. Shama and M.G. Kong. 2008. Cold atmospheric plasma decontamination of the pericarps of fruit. J. Food Prot. 71(2): 302–308.

Purevdorj, D., N. Igura, I. Hayakawa and O. Ariyada. 2002. Inactivation of *Escherichia coli* by microwave induced low temperature argon plasma treatments. J. Food Eng. 53(4): 341–346.

Raballand, V., J. Benedikt, J. Wunderlich and A.V. Keudell. 2008. Inactivation of *Bacillus atrophaeus* and of *Aspergillus niger* using beams of argon ions, of oxygen molecules and of oxygen atoms. J. Phys. D: Appl. Phys. 41(11): 115207.

Ragni, L., A. Berardinelli, L. Vannini, C. Montanari, F. Sirri, M.E. Guerzoni and A. Guarnieri. 2010. Non-thermal atmospheric gas plasma device for surface decontamination of shell eggs. J. Food Eng. 100(1): 125–132.

Richard A. Venezia, P., M.S. Michael Orrico, P. Edward Houston, P. Shu=Min Yin and P. Yelena Y. Naumova. 2008. Lethal activity of nonthermal plasma sterilization against microorganisms •. Infect. Control Hosp. Epidemiol. 29(5): 430–436.

Rico, D., A. Martin-Diana, J. Barat and C. Barry-Ryan. 2007. Extending and measuring the quality of fresh-cut fruit and vegetables: a review. Trends Food Sci. Tech. 18(7): 373–386.

Rød, S.K., F. Hansen, F. Leipold and S. Knøchel. 2012. Cold atmospheric pressure plasma treatment of ready-to-eat meat: Inactivation of *Listeria innocua* and changes in product quality. Food Microbiol. 30(1): 233–238.

Rodriguez-Romoand, L.A. and A.E. Yousef. 2005. Inactivation of *Salmonella enterica* serovar Enteritidis on shell eggs by ozone and UV radiation. J. Food Prot. 68(4): 711–717.

Roth, S., J. Feichtinger and C. Hertel. 2010. Characterization of *Bacillus subtilis* spore inactivation in low pressure, low temperature gas plasma sterilization processes. J. Appl. Microbiol. 108(2): 521–531.

Russell, A.D. 1982. The Destruction of Bacterial Spores. Academic Press London, New York.

Sampedro, F., M. Rodrigo, A. Martinez, D. Rodrigo and G. Barbosa-Cánovas. 2005. Quality and safety aspects of PEF application in milk and milk products. Crit. Rev. Food Sci. Nutr. 45(1): 25–47.

Schneider, J., K. Baumgärtner, J. Feichtinger, J. Krüger, P. Muranyi, A. Schulz, M. Walker, J. Wunderlich and U. Schumacher. 2005. Investigation of the practicability of low-pressure microwave plasmas in the sterilisation of food packaging materials at industrial level. Surf. Coat. Tech. 200(1–4): 962–966.

Schram, D.C. 2002. Plasma processing and chemistry. Pure Appl. Chem. 74(3): 369–380.

Selcuk, M., L. Oksuz and P. Basaran. 2008. Decontamination of grains and legumes infected with *Aspergillus* spp. and *Penicillum* spp. by cold plasma treatment. Bioresour. Technol. 99(11): 5104–5109.

Setlow, B. and P. Setlow. 1993. Binding of small, acid-soluble spore proteins to DNA plays a significant role in the resistance of *Bacillus subtilis* spores to hydrogen peroxide. Appl. Environ. Microbiol. 59(10): 3418–3423.

Shi, X.-M., G.-J. Zhang, X.L. Wu, Y.X. Li, Y. Ma and X.J. Shao. 2011. Effect of low-temperature plasma on microorganism inactivation and quality of freshly squeezed orange juice. IEEE Transactions on Plasma Science 39(7): 1591–1597.

Sieck, L.W., J.T. Heron and D.S. Green. 2000. Chemical Kinetics Database and Predictive Schemes for Humid Air Plasma Chemistry. Part I: Positive Ion–Molecule Reactions. Plasma Chem. Plasma P. 20(2): 235–258.

Singh, M., A. Ogino and M. Nagatsu. 2009. Inactivation factors of spore-forming bacteria using low-pressure microwave plasmas in an N_2 and O_2 gas mixture. New J. Phys. 11: 115027.

Snelling, W., J. McKenna, D. Lecky and J. Dooley. 2005. Survival of *Campylobacter jejuni* in waterborne protozoa. Appl. Environ. Microbiol. 71(9): 5560–5571.

Sobrino-López, A. and O. Martín-Belloso. 2010. Review: potential of high-intensity pulsed electric field technology for milk processing. Food Eng. Rev. 2(1): 17–27.

Song, H.P., B. Kim, J.H. Choe, S. Jung, S.Y. Moon, W. Choe and C. Jo. 2009. Evaluation of atmospheric pressure plasma to improve the safety of sliced cheese and ham inoculated by 3-strain cocktail *Listeria monocytogenes*. Food Microbiol. 26(4): 432–436.

Sykes, G. 1965. Disinfection and Sterilisation (2nd ed). E. & F.N. Spon., Ltd., London.

Tamazawa, K., Y. Tamazawa and H. Shimauchi. 2006. Evaluation of Plasma Sterilization Mechanism and H_2O_2 Residues. International Association of Dental Research (IADR), Brisbane.

Victorin, K. 1992. Review of the genotoxicity of ozone. Mutat. Res. Rev. Genet 277(3): 221–238.

Vleugels, M., G. Shama, X.T. Deng, E. Greenacre, T. Brocklehurst and M.G. Kong. 2005. Atmospheric plasma inactivation of biofilm-forming bacteria for food safety control. Plasma Science, IEEE Transactions on 33(2): 824–828.

Xingmin, X., L. Yaxi, Z. Guanjun, M.A. Yue and S. Xianjun. 2011. Experimental study on inactivation of bacterial endotoxin by using dielectric barrier discharge. Plasma Sci. Technol. 13(6): 651–655.

Yamamoto, M., M. Nishioka and M. Sadakata. 2002. Sterilization by H_2O_2 droplets under corona discharge. J. Electrostat. 56: 173–187.

Yang, L., J. Chen, J. Gao and Y. Guo. 2009. Plasma sterilization using the RF glow discharge. Appl. Surf. Sci. 255(22): 8960–8964.

Yu, H., S. Perni, J.J. Shi, D.Z. Wang, M.G. Kong and G. Shama. 2006. Effects of cell surface loading and phase of growth in cold atmospheric gas plasma inactivation of *Escherichia coli* K12. J. Appl. Microbiol. 101(6): 1323–1330.

Ziuzina, D., S. Patil, P.J. Cullen, K.M. Keener and P. Bourke. 2013. Atmospheric cold plasma inactivation of *Escherichia coli* in liquid media inside a sealed package. J. Appl. Microbiol. 114(3): 778–787.

CHAPTER 7

Novel Approach to Control Pathogenic and Harmful Microorganisms in Nonthermal Way: Photosensitization

Zivile Luksiene

Introduction

> *Photodynamic therapy is an entirely new treatment modality and its development can be likened to that of the discovery of antibiotics. This is just the beginning, and its possible uses are only limited by the imagination.*
>
> J.S. McCaughan, *Drugs & Aging*, 1999; 15: 49–68.

It was earlier supposed that the discovery of antibiotics would have been a powerful tool to control pathogens and infectious diseases. Unfortunately, after 80 years we must state, that the fight against microbes is still continuing and remains as one of the permanent challenges before mankind (Wainwright 2004). The emergence of multi-antibiotic-resistant pathogens has enhanced the risk of infectious diseases, making it necessary for microbial safety of foods and preservation postharvest (Brovko 2010).

Vilnius University, Inst. Applied Sciences, Sauletekio 9, LT-10223, Vilnius, Lithuania.
Email: Zivile.Luksiene@tmi.vu.lt

Despite tremendous progress in technology and biomedical science, the number of reported food-borne diseases continues to rise. Health experts estimate that every year food-borne illnesses in the USA alone cost 5–6 billion US dollars in direct medical expenses and lost productivity. More than 1.5 billion cases of food-borne illness and more than 3 million deaths from these diseases are estimated every year (WHO 2007).

About 1.0 million people are infected with non-typhoidal salmonellae every year, resulting in 19,336 hospitalizations and 378 deaths in the USA yearly (Scallan et al. 2011). The European Union reported 138,469 cases of human salmonellosis in 2008 (ECDC 2010). Human listeriosis infections are mostly food-borne (98% of the total cases) and account for about 2500 cases of illness and approximately $200 million in monetary loss in the US annually. Listeriosis is considered to be among the most frequent causes of food-borne death, with a mortality rate of 20–30% (Mead et al. 1999). Certain strains of *Escherichia coli* are a major cause of foodborne illness. In May 2011, an *E. coli* O104:H4 bacteria outbreak in Germany affected many as 3228 people and 35 deaths were reported on 13 June 2011 and a few more cases were reported in several countries including Switzerland, Poland, the Netherlands Sweden, Denmark, the United Kingdom, Canada and the USA (Nillian et al. 2011). Today, approximately 1 billion dollars each year is spent on 5000 antimicrobial products registered with the US Environmental Protection Agency (US EPA 2004). So far, antibacterial cleaners are insufficient and aid mostly in the development of resistant bacteria.

Traditional thermal food preservation technologies are effective antimicrobials, but usually they induce plenty of uncontrolled chemical reactions in the food matrix and subsequently reduce the nutritional quality (Leistner and Gould 2002). Thus, considerable efforts are underway to develop novel strategies to find effective decontamination treatments. The emerging food safety technologies including chemical treatments, ultraviolet light, ionizing irradiation, and high pressure, are still at a research stage. Being effective decontamination tools, most of them are not effective against bacterial spores, biofilms or viruses and make different undesirable changes in the food matrix; for instance, altering the structure of proteins and polysaccharides cause changes in the texture, physical appearance, functionality of food, thus affecting the organoleptic properties. Application of natural compounds, such as essential oils, chitosan, nisin or lysozyme is hampered due to the impact on organoleptic properties of foods (Manas and Pagan 2005).

In addition, some of these processes also have certain disadvantages which include high costs of application, requirements for specialized equipment, extended processing time, degradation of exposed materials, microbial resistance and lower efficiency. For instance, irradiation is known as one of the best nonthermal sterilization methods to destroy pathogenic and spoilage microorganisms. However, it needs special facilities, highly trained personnel and is not acceptable to consumers. Atmospheric pressure plasma (APP) treatment is effective for the inactivation of *L. monocytogenes* and is applicable for disposable food containers. However, the type of material is crucial and appropriate treatment conditions should be considered for achieving satisfactory inactivation levels (Luksiene and Zukauskas 2009).

Inevitably, a new approach to inactivate pathogenic and harmful microorganisms in different fields, including medicine, food manufacturing and safety in cost-effective and environmentally friendly way is highly needed. To this end, modern biophotonic technology based on photosensitization is successfully used to cure cancer and infectious diseases (photodynamic therapy) (Jori et al. 2006). Experimental studies conducted over the last decade confirm that photosensitization using plant-based photosensitizers or photoactive food additives might serve as a promising tool to decontaminate food and food-related surfaces (Luksiene and Zukauskas 2009). This inevitably poses a question: What is it and how does it work?

This treatment is based on the application of a photosensitizer, which is preferentially accumulated in the microbial cells. The subsequent illumination with visible light, in the presence of oxygen, specifically produces cell damage that inactivates the microorganisms. It is known that the main cytotoxic agent produced after absorption of light by the photosensitizer is singlet oxygen ($^1O_2$). As a result, many cytotoxic reactions are triggered in the cell which disrupt cell membranes, inactivate enzymes and damage DNA (Luksiene 2005, Jori et al. 2006). One of the most important advantages of photosensitization in comparison with other antibacterial tools is the absence of any bacterial resistance to this treatment (Tavares et al. 2010), as the process is free radical-mediated (Polio et al. 2002). Moreover, the cytotoxic action is local, as the photosensitizer is accumulated only in micro-organisms and its diffusion distance is less than 20 nm (Moan 1990). It allows us to achieve an extensive decrease in the population of pathogens with minimal damage or thermal effects on the surrounding matrix.

Basic Concepts of Antibacterial Action of Photosensitization

History of photosensitization

The era of photosensitization was initiated by Raab in 1900. He observed the death of *Paramecium caudatum* after light exposure in the presence of acridine orange. Subsequently Von Tappeiner and Jesionek described the use of topical eosin and visible light for the treatment of skin tumors. The use of photodynamic cancer treatment is based on the pioneering work of Dougherty (Malik et al. 2010).

The interest in photosensitization as an effective tool to eradicate pathogenic microorganisms can be traced back to before the age of chemotherapy. Ehrlich was the first, to introduce the idea of the "magic bullet" and laid the foundations of modern chemotherapy: if a living microorganism accumulates vital stain and can be afterwards selectively detected, it should be possible to destroy the stained microbe after the illumination (Wainwright 1998).

The development of antimicrobial photosensitization-based treatment was stopped following the introduction of antibiotics. However, in recent years, attention has been drawn towards research into alternative antimicrobial approaches. Photosensitization can be a potential alternative to antimicrobials or antibiotics, since the mechanism of its action is different (Jori et al. 2006). This technique has been shown to be effective *in vitro* against resistant bacteria, yeasts, viruses and parasites (Luksiene et al. 2005a, Brovko et al. 2009). Currently photosensitization is used for disinfection of blood and treating local infections (Huang et al. 2010). The chief advantages of antimicrobial action of photosensitization are as follows:

1. Treatment efficiency is independent of the antibiotic resistance pattern of the strain.
2. It inactivates pathogen population by up to 6 orders of magnitude without any harmful effect on surrounding.
3. There is no mutagenicity or bacterial resistance to photosensitization which is not the case with commonly used sanitizers.
4. It is a low-cost, easy to maintain, environmentally friendly and effective antimicrobial treatment (Jori et al. 2006).

Photophysical and photochemical processes: role of oxygen

Photosensitization is a result of the combined action of two indispensable and non-toxic components, photosensitizer and light, in the presence of oxygen, thus it is necessary to describe how they interact.

Absorption of light by a photosensitizer, results in the excitation of molecules from the ground state (S_0) to the excited singlet state (S_1). The molecules can relax to the ground state by fluorescence or by internal conversion, whereby the energy is lost as heat to the surroundings. The third way to relax is intersystem crossing, when the excitation transfers from the S_1 state to the lower excited triplet T_1 state with a longer lifetime (Fig.1).

Relaxation from the T_1 state results in either phosphorescence or induction of two types of photooxidative reactions. In Type I pathway the excited photosensitizer in T_1 can react directly with the substrate that

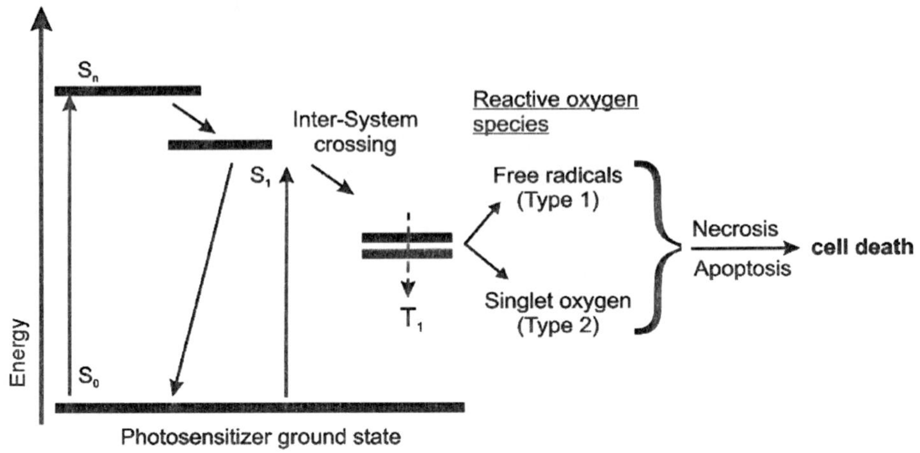

Figure 1. Jablonski scheme.

is located in its close vicinity and transfer a proton or electron to form superoxide anion or hydroxyl radical. These free radicals further react with molecular oxygen to produce reactive oxygen species (ROS).

In Type II reaction the excited photosensitizer in T_1 can transfer its excitation energy directly to ground state oxygen ($^3O_2$). As a result, highly reactive singlet oxygen is formed (Ergaieg et al. 2008). It is worth nothing that the destruction of a cell in this case is strictly localized due to a very short half-life of $^1O_2$ (<40ns) and consequently short diffusion distance (20 nm) (Moan 1990).

ROS, together with singlet oxygen produced in Type II reaction, are oxidizing agents that can directly react with the surrounding biomolecules, such as proteins and nucleic acids, causing their damage and leading to cell death (Plaetzer et al. 2009). Mostly, cell injuries induced by photosensitization include disruption of cell membrane, inactivation of different enzymes and damage of DNA (Luksiene 2005). Thus, just molecules and structures that are in close proximity to the localization of photosensitizer are affected and destructed.

Photosensitizers (photoactive dyes)

A great deal of work has been carried out to evaluate the correlation between antimicrobial efficiency and structure of the photosensitizer. As a rule, the photosensitizers are usually based on tetrapyrrole nucleus including porphyrins, chlorines, bacteriochlorins, phthalocyanines and texaphyrins. These molecules have low toxicity, can form long-lived triplet excited states and have high affinity to the life-essential molecules. Several lines of evidence indicate that physico-chemical properties of a photosensitizer have potential impact on the efficacy of photosensitization. Lipophilicity (logP), ionization (pK_a), light-absorption characteristics and the efficiency of singlet oxygen production (Φ_Δ) must be included in a putative photoantimicrobial profile (Redi et al. 2002). Moreover, neutral or anionic photosensitizers are ineffective against gram (–) bacteria (Nitzan and Pechatnikov 2011).

Some bacteria, being organisms with characteristic high metabolic turnover, are known to produce significant amounts of endogenous porphyrins from 5-aminolevulinic acid (ALA) which itself is not photoactive. Thus, it seems reasonable to exploit cell metabolism for the production of endogenous photosensitizers using a well-known precursor (ALA) (Figs. 2 and 3).

ALA is a naturally occurring metabolite, the first intermediate in heme biosynthetic pathway which is highly conserved across eukariotic as well as prokaryotic cells. Application of exogenous ALA to bacterial cell induces the production and accumulation of sufficient amounts of endogenous photosensitizers protoporphyrin IX (PpIX), uroporphyrin and coproporphyrin. Thus, photosensitization could be based on the endogenous synthesis of porphyrin-type photosensitizers by δ-aminolevulinic acid (ALA) applied exogenously. It was postulated that the presence of endogenous porphyrins within the cell with no need to penetrate any cell barriers would result in effective photodestruction of the strains that can produce high amounts of endogenous porphyrins. Talking of it, some specificity in this case was observed: for instance, gram (+) bacteria *B. cereus* is producing endogenous porphyrin 10 times more than gram (–) *S. enterica* (Luksiene et al. 2009).

There are, however, many examples of natural photosensitizers. For instance, psoralen derivatives have been used in Asia for millennia for the treatment of various skin disorders. Similarly, traditional Chinese medicine used the extract of *Hypocrella bambusae* containing antiviral hypocrellin. Anionic photosensitizer Rose Bengal is a biological stain effective in producing the triplet excited state T_1 and exhibiting significant antibacterial properties. Cationic photosensitizer Malachite Green has been used for the

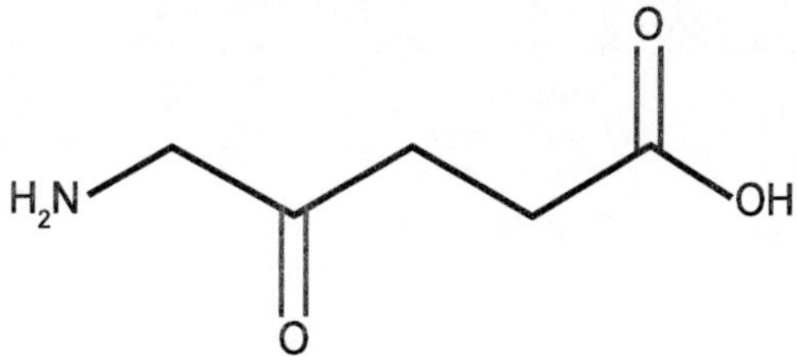

Figure 2. Chemical formula of 5–aminolevulinic acid.

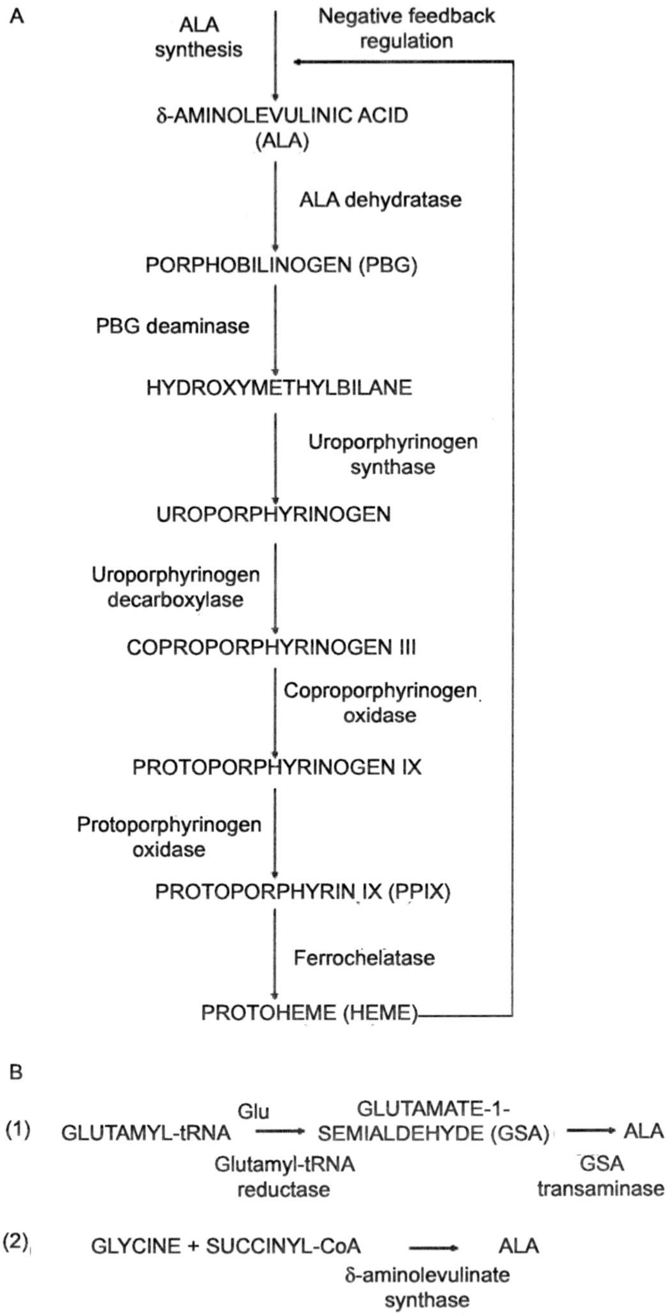

Figure 3. Synthesis of endogenous porphyrins from exogenously applied 5-aminolevulinic acid in procaryotic(1) and nonplant eukaryotic(2) cells. Part A shows the most conservative biosynthetic pathaways. Part B-shows the pathways which changed (adapted from Harris and Pierpoint 2011).

treatment of parasitic, fungal and bacterial infections in fish. Negatively-charged Phloxine B is approved for use in food products as a colorant, has negligible toxicity and was used for detoxication and disinfection of waste water (Brovko et al. 2009). Methylene blue also demonstrated high photosensitizing activity against bacteria and fungi (Street et al. 2009). As a rule, photosensitizers, derived from vital stains, are known to be non-toxic in much higher concentrations than those required for effective pathogen killing.

Hypericin (policyclic phenanthroperylenedione) is a plant pigment which has been founded in *Hipericum perforatum* (also called Saint John´s wort) and well-known for its antidepressant, antiviral and antitumor effects. The *H. perforatum*-derived products are available as tea, tinctures, juices, and oily macerates (Gaedcke 2003) and are often used both for therapeutic purposes and as a flavoring in the preparation of food and alcoholic beverages (Mašković et al. 2011). Phosensitizing properties of hypericin have been recognized since 1930 and widely investigated for further development of hypericin-based photodynamic cancer therapy. This natural pigment is described as 'one of the most powerful photodynamic agents in nature' (Fig.4).

It has been convincingly shown that hypericin exerted high singlet oxygen generation, high fluorescence yield and strong absorption at the range (600–700 nm). It is furthermore important to note that hypericin is devoid of toxic or genotoxic effects, and is an effective photosensitizer against bacteria and yeasts (Luksiene and de Witte 2002, Bernal et al. 2011, Kairyte et al. 2012).

Unlike natural chlorophyll, chlorophyllin is a water-soluble food additive (E141) and is used as food colorant in dietary supplements and in cosmetics (Lopez-Carballo et al. 2008) (Fig. 5).

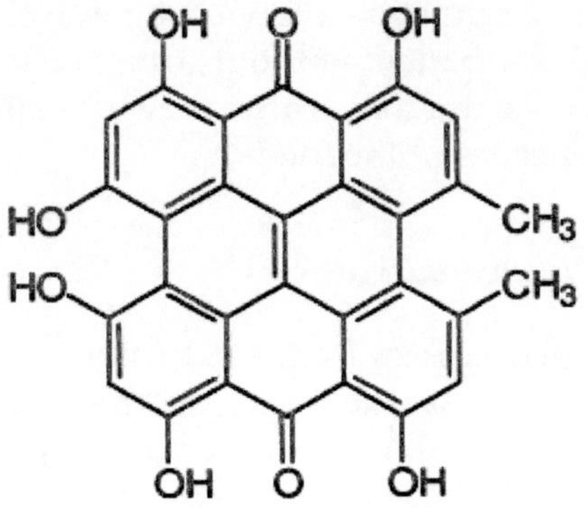

Figure 4. Chemical structure of hypericin.

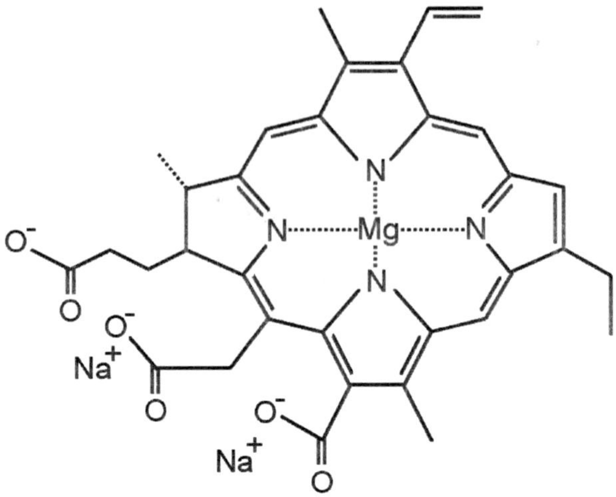

Figure 5. Chemical structure of chlorophyllin Na salt.

The conjugation of porphyrins with biopolymers (chitosan) or their incorporation into polymeric films (Funes et al. 2009) resulted in a further enhancement of the increased targeting of pathogen and enhanced photokilling effect. This can be explained by the enhanced interaction of the positively-charged polymer-dye conjugates with negatively charged bacterial cell wall which probably bring photosensitizer closer to the target cell facilitating photodestruction.

To summarize, chemical purity, capability to accumulate in a microorganism in a strategically important localization inside it, high killing efficiency and a lack of mutagenicity or genotoxicity are desirable features of an ideal photosensitizer (Luksiene and Zukauskas 2009). To use it for food preservation, additional features are essential. For instance, a desirable photosensitizer must be a low-cost, natural compound, may be food additive or food component without strong colour, taste and flavour, working at very low concentrations with no negative effect on the nutritional and organoleptic properties of the foods.

Light sources for photosensitization

The wavelength of light necessary for the induction of lethal photobiological reaction in a microorganism depends on the structure and electron absorption spectrum of the photosensitizer (Nowis et al. 2005). Common porphyrin and chlorine derivatives have a characteristic absorption band between 400 and 430 nm (Soret band) and smaller absorption bands (Q-bands) above 600 nm. The wavelength also determines the penetration depth of light into tissue: 400–500 nm light penetrates by about 300–400 μm (surface treatment), whereas 600–700 nm does by about 50–200% more (deeper treatment).

Initially, photosensitization was performed using conventional gas discharge and incandescent lamps equipped with colour-glass filters for narrowing the spectrum. More recent applications involve incoherent sources of light specifically designed for photosensitization needs: metal halide lamp, which emits in the range of 600–800 nm, short-arc xenon lamp (400–1200 nm), as well as narrow-band UV lamp (407–420 nm) (Nitzan et al. 2004).

Alternative light sources for activation of photosensitizers are light-emitting diodes (LEDs), which are in between lasers and conventional lamps in view of spectral properties and radiation pattern (Brancaleon and Moseley 2002). With the rapid development of LEDs, these devices achieved maturity sufficient for their application in life sciences (Zukauskas et al. 2002). The principle of operation of LEDs offers unsurpassed radiant efficiency. Besides, LEDs feature numerous advantages over conventional sources of light, such as low driving voltage, robustness, shock and vibration resistance, the absence of hazardous agents (mercury), compactness, light-weightness, flexibility in assembling into arrays of various forms, narrow-band emission, and the absence of unwanted spectral components. This makes them attractive for using in photosensitizing luminaries that can be safe, portable, battery driven, free of thermal side effect, and low-maintenance.

The present choice of the LED emission wavelengths covers a wide range from about 250 nm to about 7 μm. For surface treatment, deep blue and near UV InGaN LEDs, which emit at 380–450 nm, can be used. UV LEDs have a line width in the range of 10–15 nm. The output power of LEDs ranges from ~1 mW to 1 W. Differently from many lamps, LEDs usually feature no abrupt failure. Instead, the output flux gradually decreases with time. Advanced manufacturers provide LEDs with the output maintaining 70% of the initial value after 50,000 to 100,000 hours of exploitation (Fig. 6).

Although various light sources can be used for photosensitization, the sufficient intensity of delivered light must be in the range 0.5–100 mW/cm^2. Higher light intensities can induce thermal effects which actually are not desirable. The time of exposure depends on light intensity and varies from seconds to minutes. According to the World Meteorological Organization, the intensity of direct sunlight is > 12 mW/cm^2 (Anon 2008). It means that the illumination required for food-processing areas (220–540 lux) would be sufficient to photosensitize harmful and pathogenic microorganisms (Brovko 2010).

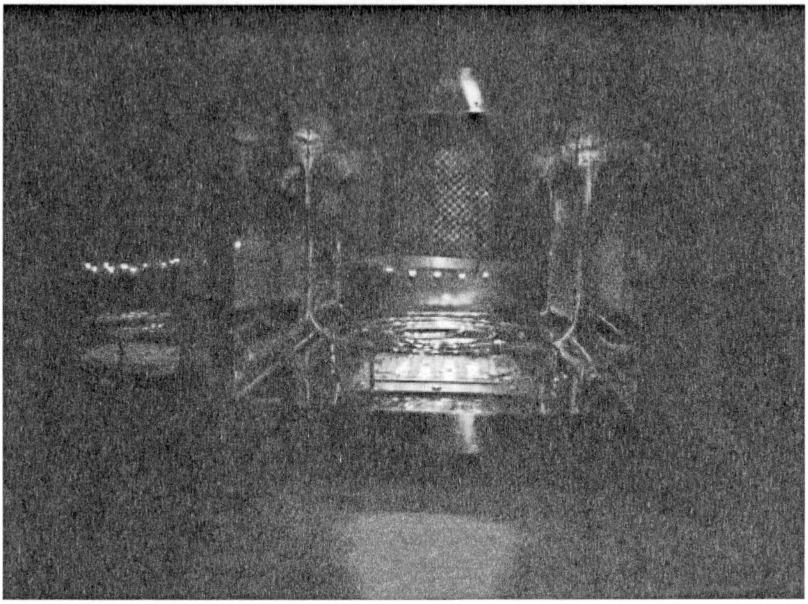

Figure 6. LED-based light source prototype for decontamination of fruits and vegetables constructed in Vilnius University.

Antimicrobial efficiency of photosensitization

General antibacterial efficiency of photosensitization consists of three stages:

1. Preferential accumulation of exogenous photosensitizer in the bacterial cell.
2. Illumination of bacteria at an approximate light wavelength that induces generation of highly reactive singlet oxygen $^1O_2$.
3. Reaction of $^1O_2$ with surrounding biomolecules and production of lethal cell damage and death (Type II) (Fig. 7).

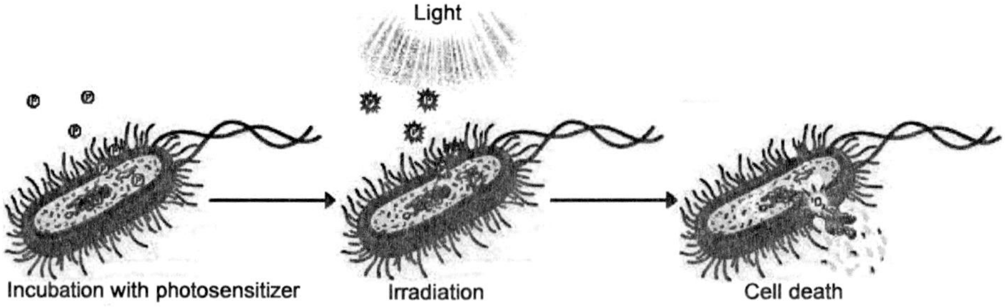

Figure 7. Mechanism of destructive action of photosensitization in the cell: P -photosensitizer, $^1O_2$–singlet reactive oxygen.

When endogenous porphyrins are used for bacterial inactivation, Type I reactions occur in the cell and inactivation process can be described in following way:

1. Incubation of bacteria with ALA.
2. Illumination of bacteria at an approximate light wavelength that excite endogenous photosensitizers.
3. Transfer of electrons to molecular oxygen generating superoxide radical anion.
4. Eventually conversion of superoxide radical to hydroxyl free radical and singlet oxygen occurs that induce bacterial death (Nitzan and Pechatnikov 2011).

A lot of work has been done on the ALA-based photoinactivation of microorganisms which induce infectious diseases (Luksiene 2005, Harris and Pierpoint 2011). Moreover, ALA-based photosensitization was used to inactivate main food pathogens. According to the obtained data, fast and significant inactivation of *Bacillus cereus* (6.5 log cycles), *Listeria monocytogenes* (4 log cycles), *Salmonella typhimurium* (6 log cycles) was achieved after this treatment (Buchovec et al. 2009, Le Marc et al. 2009, Luksiene et al. 2009).

Chlorophyllin, being food additive (E140), is photoactive and after exposure to λ=400 nm light, exhibited photosensitizing properties. Thermotolerant and thermoresistant strains of *L. monocytogenes* were inactivated by Chl-based photosensitization by 7 log cycles. Results show that *B. cereus* after Chl-based photosensitization was inactivated by 7 log cycles as well. The comparison of ALA- and Chl-based photoinactivation of *B. cereus* indicates some advantages of the later treatment. First, Chl as photosensitizer "requires" comparatively shorter incubation time (from 20 to 2 min). Second, it is effective at 4 times reduced illumination time (from 20 to 5 min). Third, Chl "works" at extremely low concentrations ($7.5 \cdot 10^{-8}$ M), whereas ALA "working" concentration was much higher ($7.5 \cdot 10^{-3}$ M) (Luksiene et al. 2009).

A plethora of bacteria and fungi are known to form spores as a strategy to survive for a longer period in unfavorable conditions. The extremely high resistance of spores to chemical and physical antimicrobials is explained by their multilayer impermeable chemical structure. Some spore-forming bacteria produce special DNA-protecting proteins. Strong hypochlorite solutions, chlorine dioxide, or ionizing radiation (which are not always compatible with food processing environment) are mostly used to combat spores. More than 90% of packaging contamination is composed of aerobic, spore-forming bacteria (Pirttijarvi et al. 1996). Due to very high resistance of bacterial spores to UV (7–50 times more resistant than vegetative cells) (Nicholson et al. 2000), germicidal lamps are insufficient to ensure sterilization of surfaces, including packaging materials. In some cases hydrogen peroxide solution (35% v/v) is sprayed onto the surface of packaging and subsequently removed by a stream of hot sterile air. Hence, there is a recognized risk of chemical food contamination by residues of hydrogen peroxide, as relatively high concentration of this compound is employed. Data indicated that the *Bacilus* spores are susceptible to ALA-based and Chl-based photosensitization. A 3.1 log cycle reduction in spore population was observed after ALA-based photosensitization (Buchovec et al. 2009, Luksiene et al. 2009, Luksiene et al. 2010a). Chl-based photosensitization diminished *Bacillus* spore population by 4 log cycles. The higher level of spore inactivation (5 log cycles) required a higher Chl concentration (Luksiene et al. 2010a).

Microbial biofilms are sometimes 1000-fold more resistant to antibiotics than single cells (Maisch 2007, Biel 2010). They can be defined as a structured community of cells enclosed in a self-produced polymeric matrix and adherent to the surface. Bacteria living in the biofilm have significantly different properties from planctonic ones, as the highly dense and protected environment of the film enables them to cooperate, communicate and interact in the other way. One of the benefits of this bacterial environment is increased (up to 1000-fold) resistance to detergents and antibiotics (Biel 2010). In general, it is difficult to remove biofilms from the surface by conventional cleaning methods as their resistance to antimicrobials is very high (Wirtanen et al. 2000). According to Soucos et al. (2003), *Streptococcus mutants, S. sobrinus, S. sanquinis* biofilms can be destroyed by 95–99% chlorine (e6)-based photosensitization.

The ALA-based photoinactivation of *Listeria* biofilms by 3.1 log cycles indicates that this treatment has the potential to combat biofilms (Buchovec et al. 2010a,b). Moreover, *Listeria* biofilms are also susceptible to Chl-based photosensitization. Actually, biofilms are less susceptible to photosensitization in comparison with vegetative cells, thus a higher Chl concentration ($1.5{\cdot}10^{-4}$ M) had to be used to inactivate *L. monocytogenes* biofilms by 4.5 log cycles. In a nutshell, the efficiency of photoinactivation depends on the used photosensitizer concentration and illumination time. Thus, there is always a possibility to increase pathogen inactivation, if necessary. This finding is particularly important since biofilms are usually resistant to commonly employed antibacterial agents. Hence, photosensitization could represent an innovative approach in sterilization of packaging contaminated by bacterial biofilms (Luksiene et al. 2010b).

Many reports have been published on the susceptibility of yeasts and microfungi to photosensitization (Donelly et al. 2008, Bernal et al. 2011). Data have indicated that yeast *Saccharomyces cerevisiae* as well as micromycetes *Ulocladium oudemansii, Trichotecium roseum* and *Aspergillus flavus* might be inactivated by photosensitization (Luksiene et al. 1989, Luksiene et al. 2004a, b). Moreover, the inhibition of spore germination was further observed in *Aureobasidium* sp., *Rhodotorula* sp., *Penicillium stoloniferum, Aspergillus fumigatus, Aureobasidium pullulans, Ulocladium chartarum, Alternaria alternata, Rhizopus oryzea, Fusarium avenaceum* and *Acremonium strictum* (Luksiene et al. 2005a,b). It seems that a plethora of harmful micromycetes that destroys food might be inactivated by photosensitization, a method that is completely safe, reproducible, nonmutagenic, noncarcinogenic, environmentally and human friendly.

Mechanism of microbial inactivation

The successful implementation of any novel technology relies on understanding the mechanism of microbial inactivation. As shown in Fig. 6, interaction of the photosensitizer with bacteria is the main prerequisite for its effective photo-inactivation. This interaction strongly depends on the physiological state of the bacteria: in the exponential growth phase, bacteria exhibit better accumulation of the photosensitizer than the corresponding cells in the lag phase. Thus, two types of cell-photosensitizer interaction exist:

1. Photosensitizer forms a tight complex with the bacterial cell wall surface receptors. The main oxidative damage in this case is localized in the cell wall. Mostly disintegration and leakage of intracellular material are the main injuries which lead to the cell death.
2. Photosensitizer is transported inside the cell where it associates with essential for surviving macromolecules. In this case photosensitization disrupts cell metabolism which subsequently causes cell death.

As mentioned before, there was significant difference in susceptibility to photosensitization between gram (+) and gram (–) bacteria.

The resistance of gram (–) bacteria to photosensitization when negatively-charged photosensitizers are used is attributed to highly complex multilayered structure of their cell wall (Fig. 8). As a rule, neutral or anionic photoactive dyes might efficiently bind and subsequently, after the illumination, inactivate gram (+) bacteria. This might be easily explained by the fact that gram (+) bacteria have a cytoplasmic membrane surrounded by a relatively porous layer of peptidoglycan (15–80 nm) and lipoteichoic acid traversing this wall and allowing the photosensitizer to cross it. Macromolecules with molecular weight of 30.000–60.000 Da (glycopeptides, polysaccharides) can readily diffuse to the inner plasma membrane. Taking into account that molecular weight of most photosensitizers does not exceed 1,500–1,800 Da, the outer wall of gram (+) bacteria is not a permeability barrier. As a result, positively-charged photosensitizers interact with negatively charged teichoic and lipoteichoic acid.

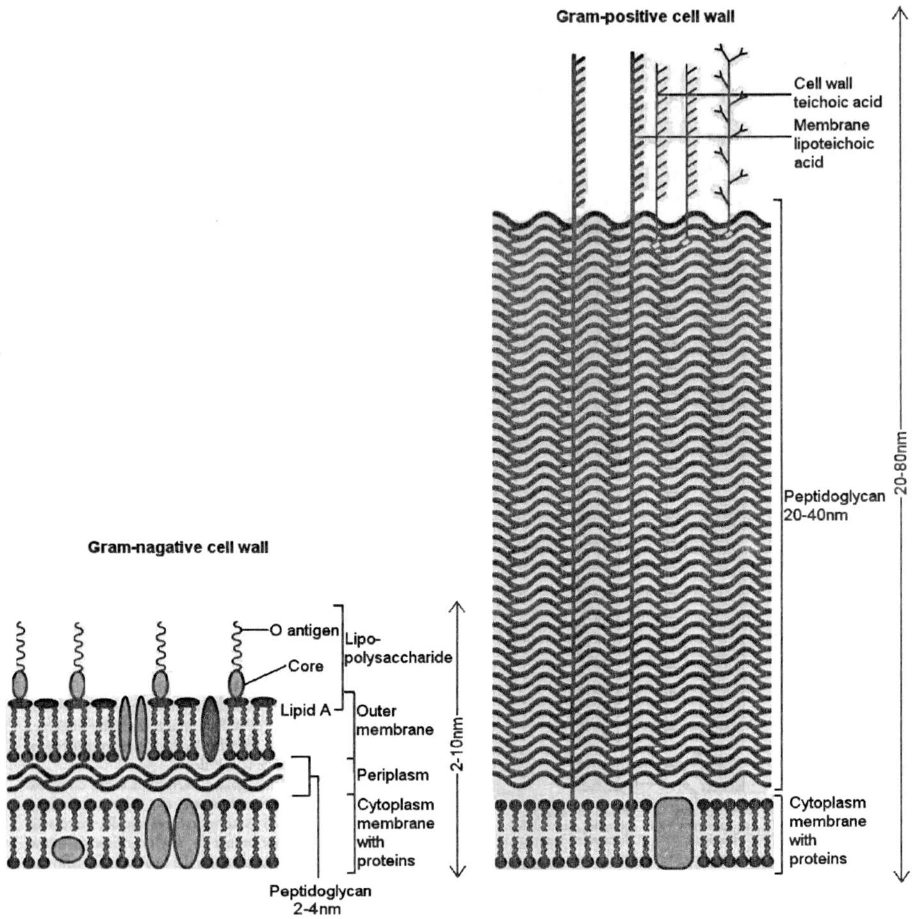

Figure 8. Cell wall structure of gram (–) and gram(+) bacteria.

The cell envelope of gram (–) bacteria consists of an inner cytoplasmic membrane and an outer membrane which are separated by the very dense peptidoglycan-containing periplasm. The outer membrane contains phospholipids, pore-forming proteins, lipopolisaccharide trimers and lipoproteins. It gives the outer surface a quasi-continuum of densely packed negative charges. Such a highly organized system inhibits the penetration of several compounds: even hydrophilic 600–700 Da molecules can diffuse through the porin channel (Nikaido 1994). The total thickness of such cell wall is much lower (2–5 nm). Despite it, the very dense cell wall of gram (–) bacteria form a physical and functional barrier to communicate with the surrounding. So, just positively-charged photosensitizers can bind the negatively-charged cell surface.

As a rule, for gram (+) bacteria and yeasts the photosensitizer accumulates in the cell wall. After irradiation by visible light, reactive oxygen species induce rapid disintegration of the cell wall. Chlorophyllin-based photosensitization induced surface blebbing in *L. monocytogenes*, but no significant injuries were observed in *S. enterica* (Fig. 9).

Usually reactive ROS interacts with unsaturated fatty acids, amino acid residues, such as cystein, histidine, tryptophan, nucleic acid bases of DNA, particularly guanine and thymidine (Girotti 2001, Milanesio et al. 2001). Breaks in both single- and double-stranded DNA have been detected in both gram (+) and gram (–) bacteria after photosensitization (Bertoloni et al. 2000).

A very attractive feature, peculiar to photosensitization as antimicrobial treatment, is the possibility of reactive oxygen species to destroy a lot of secreted virulence factors. For instance, Komerik et al. (2000), showed that LPS from *E. coli* and proteases of *P. aeruginosa* were inactivated after exposure to red light and Toluidine blue O.

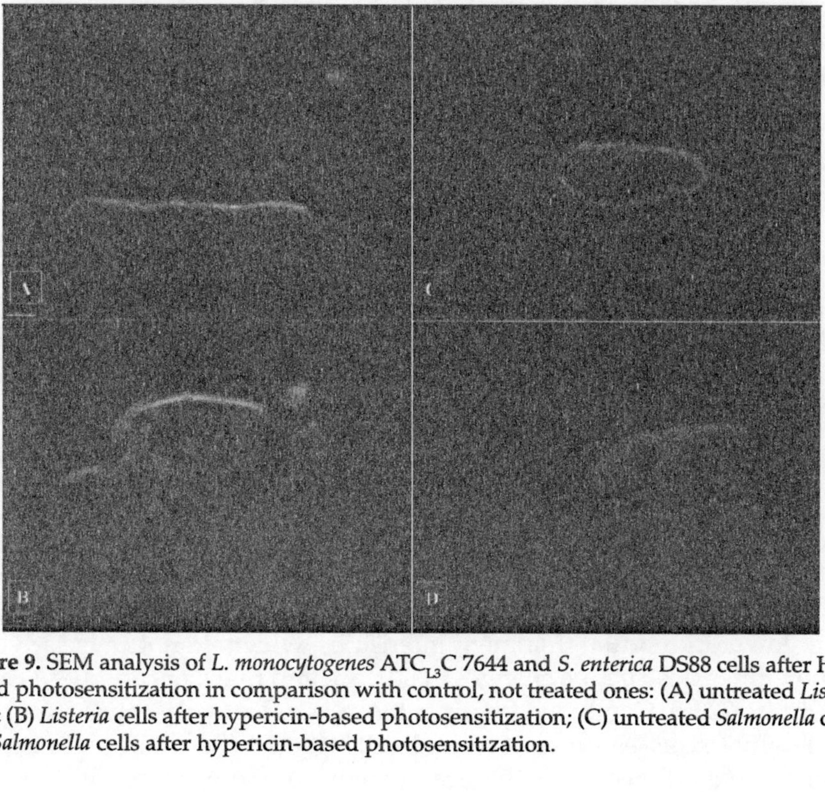

Figure 9. SEM analysis of *L. monocytogenes* ATC$_{L3}$C 7644 and *S. enterica* DS88 cells after Hyp-based photosensitization in comparison with control, not treated ones: (A) untreated *Listeria* cells; (B) *Listeria* cells after hypericin-based photosensitization; (C) untreated *Salmonella* cells; (D) *Salmonella* cells after hypericin-based photosensitization.

Novel approaches to increase the susceptibility of Gram (–) bacteria to photosensitization

Photosensitization-based inactivation failed to kill gram (–) bacteria when neutral or anionic photosensitizers have been used. To overcome this problem, polycationic agent polymixin B nanopeptide (PMNP) as non-toxic bacterial membrane structure disorganizing agent was applied. It enables the penetration of negatively-charged photosensitizers into gram (–) bacteria. After illumination complete inhibition of bacterial growth was observed (Nitzan and Pechatnikov 2011).

One of the most effective approaches to kill gram (–) bacteria is the use of cationic photosensitizers. The blue dyes known as phenothiazinium salts, such as toluidine blue O and methylene blue have been studied for many years. Tegos et al. (2008) demonstrated effective inactivation of gram (–) bacteria using phenotiazinium salts. The photoinactivation of *S. enterica* C1058 using cationic photoactive dye Malachite green, resulted in 2,3 log cycles reduction (Brovko et al. 2009). George et al. (2009) has found that the uptake of cationic photosensitizers by the cell is mediated by electrostatic interactions and self-promoted uptake pathways. After accumulation in the cell, cationic sensitizers usually tend to bind DNA molecules.

Photosensitization of gram (–) bacteria by endogenously produced porphyrins is another effective approach to their eradication. Enhanced production of endogenous porphyrins by ALA has been observed in such food pathogens as *S. enterica, B. cereus,* and *L. monocytogenes* (Buchovec et al. 2009, Buchovec et al. 2010a,b).

Chitosan, a natural cationic linear polysaccharide can be a mediator between negatively charged bacterial cell wall and negatively charged photosensitizer. Thus, photoactivated chlorophyllin-chitozan conjugates were able to inactivate *S. enterica* by 6 log cycles *in vitro* (Buchovec et al. 2012). Moreover, chitozan-chlorophyllin conjugates can form "active" antimicrobial films and serve as a carrier of wide range of food additives, including various antimicrobials (Aider 2010).

In order to increase the inactivation of gram (–) bacteria it seems promising to combine photosensitization with high power pulsed light, which in 2000 was approved by the USA Food and Drug Administration as a safe technology for food surface decontamination (FDA 2000). It is well known that combination of several antimicrobial agents (concept of hurdle technologies) can enhance the microbial inactivation and allow the use of lower individual treatment intensities, which do not cause adverse or sometimes destructive changes of surface and matrix properties (Wood and Bruhn 2000). Combined treatment is more effective because a number of sublethal stresses on a microbial cell "press" the microorganism to expend energy to overcome the stressful environment, potentially leading to metabolic exhaustion and death (Leistner and Gould 2002). High power pulsed light is a non-thermal food preservation technology that proposes decontamination of surfaces by intense and short duration (microseconds) pulses of broad spectrum light (200–1000 nm). It offers effective inactivation of pathogens by destructive action on their DNA, is mercury-free, has limited energy costs and has a short exposure time (Elmnasser et al. 2007). When photosensitization and HPPL treatments were combined, a remarkable reduction of viable count (6-7 log cycles) of gram (–) *Salmonella* and gram (+) *Listeria* was observed (Kairyte et al. 2012). Thus, these results confirm that combination of photosensitization and high power pulsed light might be an effective tool to combat both gram-positive and gram-negative bacteria in an effective and uniform way.

Interesting results were obtained combining cocktails of ALA and negatively-charged chlorophyllin to inactivate both gram (–) and gram (+) bacteria (Luksiene et al. 2013, *in press*). In this case both *S. enterica* and *L. monocytogenes* were inactivated by 7 log cycles.

Decontamination of Packaging Surface from Pathogens by Photosensitization

Food-borne pathogens are generally eliminated by conventional treatments during food processing. However, during post-processing, contamination can occur via working and packaging surfaces, equipment and environment. It has been determined that contamination with *B. cereus* and its highly resistant spores can occur via food packaging (Pirttijarvi et al. 1996, Nicholson et al. 2000).

In general, the current methods of surface decontamination include chemical washing, heat treatment and irradiation. Each of these methods has proven to be effective for certain situations. Meanwhile, all of them have specific disadvantages. For instance, chemical washing is not environmental friendly and has been associated with occupational and operational hazards. For instance, the fresh-cut industry commonly uses NaOCl and acids as disinfection agents, but the by-products such as trihalometanes and chloramines are potentially harmful for humans (Leistner and Gould 2002). Heat treatment in autoclaves is not practical for large objects. Ionizing irradiation equipment is rather expensive, needs a lot of space and requires highly qualified personnel. Ultraviolet light (UV) has been used for decontamination of food-related surfaces. However, the high microbial resistance to this treatment, the long exposure time, the enormous light intensities required and the degradation of exposed materials including food itself, limit the wide application of this technique (Guerrero-Beltran and Barbosa-Canovas 2004). Thus, for reasons of cost, portability, environmental impact and safety, the existing methods have limited practicality and attractiveness (Luksiene and Zukauskas 2009).

Considerable efforts are underway to find effective surface decontamination treatments. Much work has been carried out on the photoinactivation of food pathogens using endogenous photosensitizers produced by microorganisms metabolizing colourless and odourless ALA to endogenous porphyrins (Buchovec et al. 2009, Buchovec et al. 2010a,b, Le Marc et al. 2009, Luksiene et al. 2009). Decontamination of packaging material from *B. cereus* adhered to the surface of packaging material (polyolefine, a mixture of polyethylene/polypropylene) by photosensitization seems promising. More than 4 log cycles reduction was achieved after ALA-based photosensitization. In addition, the inactivation of *Listeria* cells attached on the surface of packaging after ALA-based photosensitization can reach 2.3–3.7 log cycles. It was shown that ALA-based photosensitization and following inactivation of pathogens depended on the total light dose (illumination time multiplied on light intensity) as well as on the amount of endogenous porphyrins (which depends on initial ALA concentration and time of incubation with it). So far, ALA is a comparatively expensive compound for wide use in food industry. Moreover, synthesis of endogenous porphyrins from ALA in the bacteria requires additional time (20–60 min) which increases the duration and the costs of antibacterial technology (Luksiene and Zukauskas 2009).

Inactivation of *L. monocytogenes* on the surface of packaging material by Chl-based photosensitization indicates that this treatment can totally clean *Listeria* from surfaces. Decontamination of packaging material from thermoresistant *L. monocytogenes* 56 Ly cells adhered to the surface reveals that a higher Na-Chl concentration ($1.5 \cdot 10^{-4}$ M) was necessary to reduce the bacterial population from 4 down to 0 log cycles. Inactivation of *B. cereus* on the surface of packaging material by Chl-based photosensitization was also remarkable, since 5 log cycles reduction was succeeded (Luksiene et al. 2010a,b).

Photosensitization was compared with conventional surface treatments such as washing with sterile water or Na-hypochlorite. It was shown that washing with water can diminish *L. monocytogenes* population on the surface of the packaging material by less than 1 log cycle Washing with 100 ppm Na-hypochlorite was more effective and reduced the pathogen population about 1.2 log cycle When a higher Na-hypochlorite concentration was used (200 ppm), the reduction of bacterial population on the surface of packaging was 1.8 log cycles. Meanwhile, Chl-based photosensitization seems to be more effective as it reduced *Listeria* and *Bacillus* population by 4.5 and 5 log cycles respectively. In the case of inactivation of thermoresistant *L. monocytogenes* 56 Ly strain on the surface of packaging material, it was found that total elimination of bacteria from the surface required higher Chl concentration (Luksiene et al. 2010a,b).

Physico-chemical analysis of polyolefine packaging material indicated that there were no significant physico-chemical modifications in the packaging surface structure and properties after the treatment by light, photosensitizer alone, as well as after Na-Chl-based photosensitization (Luksiene and Paskeviciute 2011b). More specifically, chemical analysis of polyolefine packaging material treated by photosensitization was conducted to monitor any possible modifications of the composition material. Results obtained by IR analysis indicated that no modification occurred in the polyolefine after incubation with Chl and illumination with light Chl-based photosensitization treatment. In addition, no specific migration of the photoactive compound was found as migration was below the detection limit (0.05 mg/dm^2). Chromatographic fingerprint analysis indicated negligible variations of the followed molecules compared to the overall migration results. Therefore, there was no significant modification of the chemical composition of the polyolefine packaging material treated by photosensitization. The overall migrations of molecules treated by photosensitization polyolefine packaging material were the highest but the uncertainty of 2 mg/dm^2 could be considered as negligible. The physical analysis was focused mainly on the modification of the packaging material structure. The surface of the treated polyolefine was observed by scanning electronic microscope (SEM) and compared with the control. Modifications of the surface, appearances of ruptures were looked for. No differences were observed between the control and treated polyolefine. Investigation of possible changes in mechanistic polyolefine properties after photosensitization by traction and puncture tests indicated no changes as well. Finally, data obtained on the permeability of carbon dioxide in treated samples indicated that photosensitization did not affect the porosity of the packaging which keeps its barrier properties toward the food (Luksiene and Paskeviciute 2011b).

Photosensitization for Microbial Control of Fruits and Vegetables

Fresh produce is becoming more popular all over the world. Adequate fresh fruit and vegetables consumption is a necessary component for healthy diet and due to the changes in dietary habits, the tendency of consumption of raw vegetables has increased (Nillian et al. 2011).

However, raw fruit and vegetables have been known as vehicles for transmission of human diseases and is the second leading cause of food-borne illnesses, which cost for instance the U.S. economy $ 6.9 billion of loss in productivity and medical expenses (ERS 2005). The last *E. coli* O104:H4 bacteria outbreak was in Germany in May 2011. A number of 3228 cases and 35 deaths were reported on 13 June 2011 and a handful of cases was also reported in several countries including Switzerland, Poland, the Netherlands Sweden, Denmark, the United Kingdom, Canada and the USA. All affected people had been in Germany shortly before getting ill. German officials issued that outbreak was suspected when they identified some organic cucumber, bean sprouts and others raw vegetables contaminated by *E.* coli O104:H4. Obviously, there is a need to develop novel processing technologies that are more effective and do not diminish the organoleptic properties and nutritional value of the treated produce.

During the past decade the emphasis in post-harvest fruit protection has shifted from using chemicals to various alternative techniques including biological control (Sharma et al. 2009) and physical methods such as controlled atmosphere (Zheng et al. 2008, Sandhya 2010) or irradiation (Dionisio et al. 2009).

Decontamination of strawberries by photosensitization

Strawberries (*Fragaria ananassa* Duch.) are a popular and nutritious fruit worldwide. It has been reported that strawberries contain the highest antioxidant capacity among twelve fruits analyzed (Kähkönen et al. 2001). The main contributors to antioxidant activity are phenolic compounds which prevent cancer, cardiovascular diseases, and age-induced oxidative stress (Olsson et al. 2004). Anthocyanins are the most abundant flavonoids (phenolics) and are present in high levels in mature strawberries (Cortenunsi et al. 2005). Short ripening and senescence periods of strawberries, susceptibility to mechanical injury, contamination during storage with fungi and bacteria reduce significantly their shelf life. Strawberries spoilage losses can be as high as 40% (Satin 1996).

Several outbreaks associated with consumption of strawberries have been reported (FDA 1999). A variety of pathogenic bacteria such as *L. monocytogenes*, *Salmonella* spp. as well as pathogenic *E. coli* strains may be present on fresh fruits (Knudsen et al. 2001, Johannessen et al. 2002).

The most widely known postharvest treatments to reduce microbial spoilage are low temperature and modified atmosphere packaging (Nielsen and Leufven 2008). However, it has been reported that these technologies are not effective enough and can have a negative impact on the quality of strawberries (Ayala-Zavala et al. 2007). For instance, carbon dioxide treatment reduced anthocyanins content and changed internal fruit colour (Gil et al. 1997). Washing alone or conventional sanitizers have been shown to have limited efficacy at removing spoilage and pathogenic bacteria from the surface (Yuk et al. 2006). Irradiation of strawberries with 2–3 kGy was found to suppress fungi on stored berries and doubled their shelf life, but unfortunately irradiation gave rise to changes in strawberry texture, cell wall composition and colour (Yu et al. 1995). Ionising radiation has been approved for decontamination of food in the USA but is strongly discouraged in the EU as consumers mostly prefer non-irradiated products (Neyssen 2000).

Marquenie et al. (2003) studied the combined effect of three physical methods—high power pulsed light, heat, and UV-C illumination on strawberry decontamination from fungus *Botrytis cinerea*. The results revealed that combined treatment of all three techniques reduced the size of *B. cinerea* mycelia and did not affect fruit firmness. This technique also prolonged the disease-free period, increasing the shelf-life by 1–2 days. Allende et al. (2007) determined the effect of UV-C light, gaseous O_3, super atmospheric O_2 and CO_2-enriched atmospheres applied in combination on the shelf-life of strawberries. This combination reduced the growth of yeast and microfungi by 1 log for up to 5 days, and prolonged the storage period respectively.

Photosensitization-based treatment decontaminated strawberries from inoculated *Listeria* by 1.8 log cycles. Population of mesophillic microorganisms naturally distributed on the surface of strawberries was also reduced by 1.7 log cycles. It is important to note that spoilage yeasts and microfungi showed slightly lower susceptibility to photosensitization and their population was reduced by only 0.9 log cycles. The overall reduction of this microbial contamination prolonged the disease-free period of berries by two days. The irregularity and the different light reflecting properties of the illuminated strawberry surfaces can possibly account for the lower antimicrobial efficiency of the treated food matrix (Luksiene and Paskeviciute 2011a). However, the obtained antibacterial efficiency can be significantly enhanced by usage of more powerful light sources (Light emitting diodes-LED) or increase of illumination time.

The distinguishing feature of photosensitization is its non-thermal action. It should be noted that there was no significant temperature increase on the surface of strawberries which would be detrimental to the quality of fruit after photosensitization treatment. During 20 min of treatment the temperature on the surface of strawberries changed from 20 to 25°C. It must be emphasized that to find a physical antimicrobial technique without thermal effects is a complicated task. For instance, high power pulsed light is an effective disinfection method, but temperature increase up to 80°C on the surface of berry is unacceptable (Bialka and Demirci 2008).

The beneficial properties of fruits and vegetables have been partially attributed to the presence of antioxidant compounds (Tulipani et al. 2008). Antioxidants can scavenge free radicals and reactive oxygen species which usually induce toxic processes in the living cell including oxidative damage to proteins and DNA, membrane lipid oxidation, enzyme inactivation, and gene mutation that may finally lead to cancer genesis or other oxidative cardiovascular or inflammatory diseases (Eberhardt et al. 2000). Therefore, it might be possible that photosensitization being an effective antimicrobial treatment modality can affect the nutritional quality of strawberries. Data indicated that photosensitization induced significant increase in total antioxidant activity (19%) in strawberries, which could be associated with enhanced cellular capacity to detoxify reactive oxygen species (Luksiene and Paskeviciute 2011a). However, the mechanisms of these effects have not been thoroughly elucidated so far. In fact, plant cells usually keep the ROS level under tight control by production or activation of scavenging enzymes (Bailly 2004).

Recently, phenolic compounds such as flavonoids (anthocyanins), flavanols (catechins) and flavonols (quercetin) have caught the attention as potential antioxidants. They are capable of simulating carcinogen-detoxifying enzymes and counteract inflammatory processes (Parr and Bolwell 2000). Due to the fact that phenolic compounds as effective antimicrobials and UV screens are accumulated in the epidermal tissue, the question arises, does photosensitization affect their content in the strawberry fruit? According to Luksiene and Paskeviciute (2011a), the photosensitization did not affect the level of soluble phenolics, while storage for only 24 hours at 6°C diminished their amount from 1750 to 1400 mg/100g.

As anthocyanins prevent cardiovascular and other disorders (Zafra-Stone et al. 2007), to maintain their concentration in strawberries is important. A lot of factors such as pH, light, oxygen, enzymes, and high temperature can induce anthocyanin degradation (Wang and Xu 2007). For instance Tiwari et al. (2009) studied the effect of ozone on strawberry juice anthocyanins and found that ozone induced significant reductions in their content (98.2%). Luksiene and Paskeviciute (2011a) found that the amount of anthocyanins after Chl-based photosensitization did not change and remained very close to the initial content (140µg/g), while slight decrease was observed even after storage for 24 hours at 6°C (121 µg/g). Gil et al. (1997) studied the effect of carbon dioxide treatment on the amount of anthocyanins and other polyphenols in strawberries. Their results revealed that anthocyanins content of CO_2-treated fruit was reduced compared to air-stored fruit. Odriozola-Serrano et al. (2008) also found that high-intensity pulsed electric field (HIPEF) affected strawberry nutritional quality. The loss of phenolic content of HIPEF and thermally-processed strawberry juices after 56 days storage at 4°C was in the range of 21.5–24.1 mg/100 ml.

Colour is one of the important quality indicators in fresh strawberry appearance and greatly contributes to fruit quality. The bright red colour of strawberry fruit is due to the presence of anthocyanin pigments in the fruit epidermis and cortex (Nunes et al. 2005). In addition such factors as co-pigmentation, pH and anthocyanin metabolism may play a significant role in the expression of colour in strawberries. The most effective co-pigments flavonols are located in external tissue. To determine, whether photosensitization had any effect on the colour of strawberry, the fruits were analyzed immediately after treatment. According to the obtained data, the colour of strawberry surface was not markedly affected by the photosensitization (Luksiene and Paskeviciute 2011a). Other studies performed on decontamination of strawberries using carbon dioxide treatment indicated that fruit surface colour did not change, but remarkable changes were observed in internal flesh colour (Gil et al. 1997).

Preliminary data were obtained on the taste of treated berries. Twenty-one volunteers tasted treated strawberries and compared them with control ones. Eighteen out of the 21 volunteers found no sweetness or firmness changes in the control and treated berries. It can be easily explained as visible light alone is not producing any effects on fruits and vegetables, and Chl is a well known food additive and food component (Luksiene and Paskeviciute 2011a).

Decontamination of sprouts by photosensitization

The areas of organic production is expanding rapidly throughout the world as the demand for organic products is growing. Organic cereals of good quality are widely used both in product processing and expansion of organic crop areas. Photosensitization being environmentally friendly, effective and free from chemical substances, can be used for decontamination of germinated sprouts.

Prevailing genus of micromicetes *Alternaria* (frequency of occurrence 100%) was reduced from the surface of wheat by 75% after photosensitzation treatment. Such harmful micromycetes as *Penicillium* genus (frequency of occurrence 62.5%) and *Mucor hiemalis* (frequency of occurrence 37.5%) were effectively reduced from the surface of wheat as well (Luksiene et al. 2005b, Luksiene et al. 2006).

Considering the application of ALA-based photosensitization to decontaminate sprouts, the question arises how does ALA interact with food matrix. Data indicated that ALA stimulates the growth of wheat seedlings and roots without impairing the vigor of germination and the viability of the seeds. Moreover, 5-ALA increases the rate of photosynthesis and the activities of antioxidant enzymes, which can be associated with enhanced cellular capacity to detoxify reactive oxygen species (Luksiene et al. 2007). Moreover, ALA is an essential precursor of such tetrapyrole compounds as vitamin B_{12} and hemes, which serve as prosthetic groups of respiratory enzymes and chlorophyll in plants (Granick 1961). Appropriate ALA concentrations have promotive effects on the growth rates and photosynthesis. Crop yields were enhanced by the application of ALA at the leafstage of rice, barley, potato and garlic (Tanaka et al. 1992). Foliar application of ALA (hundred milligram per milliliter) on date palm has promoting effects on the fruit weight, volume and sugar content (Hotta et al. 2004).

Photosensitization Treatment of Edible Biopolymer-based Films and Coatings

Although food packaging materials have traditionally been based on non-renewable materials, increasing attention is being given to sustainability and the replacement of non-renewable resources with renewable ones. Hence, alternative packaging materials based on biological materials have been developed. Edible biopolymers are typically used as films and coatings, and can be consumed along with foods. The polysaccharide-based biopolymers include starch, cellulose derivatives, seaweed extracts (alginates and carrageenan), and chitosan. The protein-based biopolymers include whey protein isolate, collagen, gelatin, corn zein, wheat gluten, and soy protein isolate (Chen and Neetoo 2010).

It is well-known that chitonan (CHS) itself is an antimicrobial agent (Tayel et al. 2010). Antibacterial activity of CHS was assessed for a wide range of gram (–) and gram (+) bacteria (No et al. 2002). The actual mechanism of its antimicrobial activity is not yet fully understood, but gram(–) bacteria are less susceptible to its action. However, photoactivated CHS-chlorophyllin conjugates reduced gram (–) *S. enterica* population by 6·07 log cycles. Thus, CHS can potentiate the Chl-based photosensitization efficacy against gram (–) food pathogen *S. enterica*. It might be, that synergetic antibacterial effect of chitosan and chlorophyllin rely on the complexation of these two compounds (Fig. 10). Soluble conjugates of CHS- Chl possess positive charge, and it facilitate the attachment of them on the cell wall.

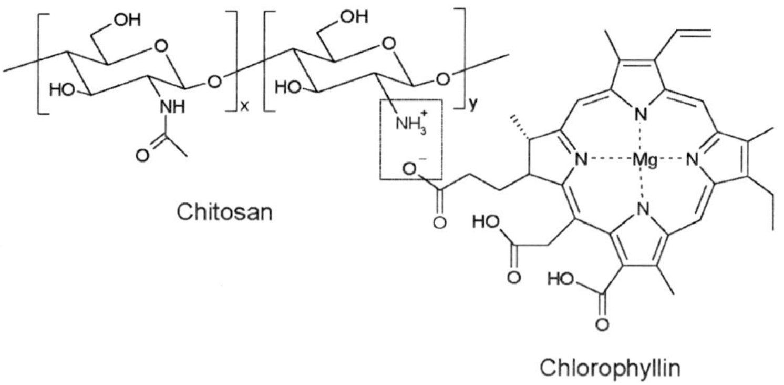

Figure 10. Hypothetical formula of Chl-CHS conjugate.

Thus, chitosan serves as a carrier of Chl, increasing its concentration at a close vicinity to the cells. High local concentration of chlorophyllin around cells could be a crucial factor significantly enhancing antibacterial activity of photoactivated Chl-CHS complex (Buchovec et al. 2012). These experimental data can serve as background for further development of "active" edible antibacterial biofilms.

Lopez-Carballo et al. (2008) studied the phototoxic effect of Chl-incorporated gelatin-based edible films against *Staphylococcus aureus* and *L. monocytogenes*. Results revealed that photoactivated Chl immobilized in gelatin film had a strong bactericidal effect on the viability of *S. aureus*, which its population decreased by 4 log cycles after Chl-based photosensitization treatment. Similar results were obtained when stationary phase cultures of *L. monocytogenes*, *E. coli* and *S. enterica* were used.

Photosensitization Treatment of Photoactive Antimicrobial Packaging

Active packaging refers to the incorporation of certain additives into packaging systems with the aim of extending product quality and shelf-life. Active packaging includes additives or 'freshness enhancers' that can participate in a host of packaging applications and by so doing, enhance the preservation function of the primary packaging system. The following methods are most commonly used to incorporate antimicrobials into packaging materials: (1) incorporation of antimicrobial agents directly into polymers; (2) coating or adsorbing antimicrobials onto polymer surfaces; (3) immobilization of antimicrobials to polymers via ionic or covalent linkages; and (4) use of polymers that are inherently antimicrobial. Antimicrobial packaging is an emerging and exciting area of food technology, which can confer many preservation benefits on a wide range of foods. Post-pasteurization/prepackaging contamination of ready-to-eat (RTE) foods by foodborne pathogen *L. monocytogenes* is a significant problem currently faced by the food industry. Recent research has shown great promise for the use of antimicrobial packaging films to control *L. monocytogenes* in various RTE foods such as frankfurters and smoked salmon as well as to reduce the risk of surface contamination of spoilage microorganisms (Chen and Neetoo 2010). Incorporation of chlorophyllin into packaging material to make antibacterial photoactive packaging looks promising (Lopez-Carballo et al. 2008).

Photosensitization in Comparison with other Emerging Nonthermal Technologies

Traditional thermal food preservation technologies provide safe foods, but there is a great loss of food quality, associated with the induction of thousands of uncontrolled chemical reactions which lead to undesirable protein denaturation, non-enzymatic browning, loss of vitamins and volatile flavour compounds (Gould 2001).

Most of the developing novel technologies (high hydrostatic pressure, pulsed electric field, pulsed light, ultrasound, plasma technology and biopreservatives) are still at a basic level of research. High hydrostatic pressure (HHP) is an effective antimicrobial tool but the implementation of it is expensive, thus, its use is limited to high value products (Senorans et al. 2003). High intensity pulsed electric field (PEF) failed to inactivate bacterial spores. Moreover, the main problems in the development of PEF technology deal with the generation of high electric field intensities and design of technique with optimal parameters (Devlieghere et al. 2004). The main disadvantage of high power pulsed light is that light cannot penetrate deeper into the food matrix and can sterilize it just superficially (Gomez-Lopez et al. 2007). Treatment of food with ionizing radiation is effective and legal in more than 40 countries. Nevertheless, this technology remains not approved in Europe mainly due to the suspicions of consumers (Gould 2001). A plethora of natural antimicrobial compounds derived from plants are known as antimicrobials (phytoalexins, phenolic compounds). The main disadvantage of these antimicrobials is its strong flavour, high working concentration and relatively low antimicrobial efficiency (Leistner and Gould 2002).

Photosensitization, in comparison with the above mentioned emerging nonthermal techniques is environmentally and human friendly, low-cost, effective and easy to maintain. It induces lethal effects in vegetative cells, spores and biofilms without any mutagenicity. No microbial resistance to this treatment was observed, whereas very high resistance to high hydrostatic pressure and moderate resistance to ionizing radiation as well as pulsed electric field were identified (Manas and Pagan 2005). Moreover, photosensitization does not require high intensities (can be used as hurdle), thus the preservation of sensory, nutritional and functional properties of foods is higher. The main limitation of this technique is its superficial action.

Conclusions

Multi-antibiotic resistance of pathogens is a rapidly growing and alarming phenomenon. Hence, the discovery of novel, cost–effective and human friendly technologies to inactivate harmful and pathogenic microorganisms seems to be imperative. To this end, photosensitization as really effective technique against a range of food-borne bacterial pathogens, yeasts and microfungi, should encourage its use in a wider arena, supplementing the existing antimicrobial armamentarium. Photosensitization of bacteria has repetitively been shown to be independent of the antibiotic resistance-spectrum, it induces loss of viral infectivity, and is notmutagenic or genotoxic. It is important to note that no bacterial resistance to this treatment was detected. Therewith photosensitizer-biopolymer conjugates way serves for the development of antimicrobial bio-destructive "active" packaging, edible biofilms or "self-cleaning" surfaces. In our opinion, this phenomenon opens a new and prospective avenue for the development of effective, human and environmentally-friendly antimicrobial treatment. Its proper application for the treatment of foods, packaging, processing equipment and industrial environment might be really useful to increase microbial food control. Nevertheless, this is just the beginning. Taking into account the infinity and complexity of food matrices, microbial ecology of food, interaction of food matrix, shelf-life and quality of foods after photosensitization treatment, further deep research is needed. Whilst it is not suggested that photosensitization will solve all problems of antimicrobial issues, improvements may be obtained by developing concept of hurdle technologies, combining photosensitization with accepted technologies.

Acknowledgements

This work was partially supported by European Commission (FP6, STREP project HighQRTE). Author thanks Kristina Kairyte for technical support.

Keywords: photosensitization, inactivation, food pathogens, photooxidative reactions, photosensitizers, aminolevulinic acid, hypericin, chlorophyllin

References

Aider, M. 2010. Chitosan application for active bio-based films production and potential in the food industry: Review. LWT-Food Sci. Technol. 43: 837–842.

Allende, A., A. Marin, B. Buendia, F. Tomas-Barberan and M.I. Gil. 2007. Impact of combined postharvest treatments (UV-C light, gaseous O_3, superatmospheric O_2 and high CO_2) on health promoting compounds and shelf-life of strawberries. Postharv. Biol. Technol. 46: 201–211.

Anon. 2008. Measurement of sunshine duration. *In*: CIMO Guide. World Meteorological Organization, Chapter 8.

Ayala-Zavala, F.J., W.Y. Shiow, W.Y. Chien and G.A. Gonzalez-Aguilar. 2007. High oxygen treatment increases antioxidant capacity and postharvest life of strawberry fruit. Food Technol. Biotech. 45: 166–173.

Bailly, C. 2004. Active oxygen species and antioxidants in seed biology. Seed Sci. Res. 14: 93–107.

Bernal, C., J. Rodrigues, A. Guimaraes, A. Ribeiro, K. Oliveira, H. Imasato and J. Perussi. 2011. Selective photoinactivation of *C. albicans* and *C. dubliniensis* with hypericin. Laser Phys. 21: 245–249.

Bertoloni, G., F.M. Lauro, G. Cortella and M. Merchat. 2000. Photosensitizing activity of hematoporphyrin on *Staphylococcus aureus* cells. Biochim. Biophys. Acta 1475: 169–174.

Bialka, K.L. and A. Demirci. 2008. Efficacy of pulsed UV-light for the decontamination of *Escherichia coli* O157: H7 and *Salmonella* spp. on raspberries and strawberries. J. Food Sci. 73: 201–207.

Biel, M.A. 2010. Photodynamic therapy of bacterial and fungal biofilm infections. *In:* C.J. Gomer [ed.]. Photodynamic Therapy: Methods and Protocols. Methods Mol. Biol. 635: 175–194.

Brancaleon, L. and H. Moseley. 2002. Laser and non-laser light sources for photodynamic therapy. Laser. Med. Sci. 17: 173–186.

Brovko, L. 2010. Photodynamic treatment: a new effective alternative for surface sanitation. Adv. Food Nutr. Res. 61: 119–147.

Brovko, L., A. Meyer, A. Tiwana, W. Chen, H. Liu, C. Filipe and M.W. Griffiths. 2009. Photodynamic treatment—A novel method for sanitation of food handling and food processing surfaces. J. Food Protect. 72: 1020–1024.

Buchovec, I., Z. Vaitonis and Z. Luksiene. 2009. Novel approach to control *Salmonella enterica* by modern biophotonic technology: photosensitization. J. Appl. Microbiol. 106: 748–754.

Buchovec, I., E. Paskeviciute and Z. Luksiene. 2010a. Photodynamic inactivation of food pathogen *Listeria monocytogenes* by non-thermal treatment. Food Technol. Biotechnol. 48: 207–213.

Buchovec, I., E. Paskeviciute and Z. Luksiene. 2010b. Photosensitization-based inactivation of food pathogen *Listeria monocytogenes in vitro* and on the surface of packaging material. J. Photochem. Photobiol. B 99: 9–14.

Buchovec, I., R. Kokstaite, V. Pamedytyte, R. Makuska and Z. Luksiene. 2012. Antimicrobial efficiency of photoactivated chlorophyllin-chitosan complex against Gram (–) bacteria *Salmonella enterica*. Chem. Technol. 3: 54–57.

Chen, H. and H. Neetoo. 2010. Antimicrobial packaging. Encyclopedia of Biotechnol. in Agric. Food. 1: 43–46.

Cordenunsi, B., M. Genovese, J. Nascimento, N. Hassimotto, R. Dos Santos and F. Lajolo. 2005. Effects of temperature on the chemical composition and antioxidant activity of three strawberry cultivars. Food Chem. 91: 113–121.

Devlieghere, F., L. Vermeiren and J. Debevere. 2004. New preservation technologies. Possibilities and limitations. Int. Dairy J. 14: 273–285.

Dionisio, A.P., R.T. Gomes and M. Oetterer. 2009. Ionizing Radiation Effects on Food Vitamins —A Review. Braz. Arch. Biol. Tech. 52: 1267–1278.

Donnelly, R.F., P.A. McCarron and M.M. Tunney. 2008. Antifungal photodynamic therapy. Microbiol. Res. 163: 1–12.

Eberhardt, M.V., C.Y. Lee and R.H. Liu. 2000. Antioxidant activity of fresh apples. Nature 405: 903–904.

[ECDC] European Centre for Disease Prevention and Control. 2010. Annual Epidemiological report on communicable diseases in Europe. Prevention and Control. Available at: http://ecdc.europa.eu/en/press/news/Lists/News/ECDC_DispForm.aspx?List=32e43ee8-e230-4424-a783-85742124029a&ID=402.

Economic Research Service. 2005. Foodborne illness cost calculator. Available at: http://www.cspinet.org/foodsafety/outbreak.alert.pdf. Accessed 22 August 2007.

Elmnasser, N., S. Guillou, F. Leroi, A. Orange, A. Bakhrouf and M. Federighi. 2007. Pulsed-light system as novel food decontamination technology: A review. Can. J. Microbiol. 53: 813–821.

Ergaieg, K., M. Chevanne, J. Cillard and R. Seux. 2008. Involvement of both Type I and Type II mechanisms in Gram(+) and Gram(–) bacteria photosensitization by a meso-substituted cationic porphyrin. Sol. energy 82: 1107–1117.

[FDA] Food and Drug Administration. 2000. Food and Drug Administration issues approval for pulsed UV light in the production, processing and handling of food. *Code 21CFR179.41*, 438.

Food and Drug Administration. 1999. FDA survey of imported fresh produce. Available from: http:// www.cfsan.fda.gov/dms/prodsir6/html. Accessed September 13, 2004.

Funnes, M.D., D.A. Caminos, M.G. Alvarez, F. Fungo, L.A. Otero and E.N. Durantini. 2009. Photodynamic properties and photodynamic action of electrochemically generated porphyrin polymeric films. Environ. Sci. Technol. 43: 902–908.

Gaedcke, F. 2003. St. John's wort herb extracts: manufacturing, standardization and characterization. *In:* E. Ernst [ed.]. The genus Hypericum (Medicinal and Aromatic Plants - Industrial Profiles). Taylor and Francis Group, London, UK, pp. 49–64.

George, S., M.R. Hamblin and A. Kishen. 2009. Uptake pathways of anionic and cationic photosensitizers into bacteria. Photochem. Photobio. Sc. 8: 788–795.

Gil, M.I., D.M. Holcroft and A.A. Kader. 1997. Changes in strawberry anthocyanins and other polyphenols in response to carbon dioxide treatments. J. Agr. Food Chem. 45: 1662–1667.

Girotti, A.W. 2001. Photosensitized oxidation of membrane lipids: reaction pathways, cytotoxic effects, and cytoprotective mechanisms. J. Photochem. Photobiol. 63: 103–113.

Gomez-Lopez, V.M., P. Ragaert, J. Devebere and F. Devlieghere. 2007. Pulsed light for food decontamination. Trends Food Sci. Tech. 18: 464–473.

Gould, G.W. 2001. New processing technologies: an overview. P. Nutr. Soc. 60: 463–474.

Granick, S. 1961. Magnesium protoporphyrin monoester and protoporphyrin monomethyl ester in chlorophyll biosynthesis. J. Biol. Chem. 236: 1168–1172.

Guerrero-Beltran, J.A. and G.V. Barbosa-Canova. 2004. Advantages and limitations on processing foods by UV light. Food Sci. Technol. Int. 10: 137–147.

Harris, F. and L. Pierpoint. 2011. Photodynamic Therapy based on 5-aminolevulinic acid and its use as an antimicrobial agent. Med. Res. Rev. Wiley Periodicals 1–36.

Hotta, Y., T. Tanaka, H. Tanaka, Y. Takeuchi, M. Konnai and H. Al-Khateeb. 2002. Promotive effects of 5-aminolevulinic acid on the yield of several crops. Plant Growth Regul. 2: 109–114.

Huang, L., T. Dai and M.R. Hamblin. 2010. Antimicrobial photodynamic inactivation and photodynamic therapy for infections. Methods Mol. Biol. 635: 155–73.

Johannessen, G.S., S. Loncarevic and H. Kruse. 2002. Bacteriological analysis of fresh produce in Norway. Int. J. Food Microbiol. 77: 199–204.

Jori, G., C. Fabris, M. Soncin, S. Ferro, O. Coppelotti, D. Dei, L. Fantetti, G. Chiti and G. Roncucci. 2006. Photodynamic therapy in the treatment of microbial infections: basic principles and perspective applications. Laser. Surg. Med. 38: 468–481.

Kähkönen, M.P., A.I. Hopia and M. Heinonen. 2001. Berry phenolics and their antioxidant activity. J. Agr. Food Chem. 49: 4076–4082.

Kairyte, K., S. Lapinskas, V. Gudelis and Z. Luksiene. 2012. Effective inactivation of food pathogens *Listeria monocytogenes* and *Salmonella enterica* by combined treatment of hypericin-based photosensitization and high power pulsed light. J. Appl. Microbiol. 112: 1144–1151.

Knudsen, D.M., S.A. Yamamoto and L.J. Harris. 2001. Survival of *Salmonella* spp. and *Escherichia coli* 0157:H7 on fresh and frozen strawberries. J. Food Protect. 64: 1438–1488.

Komerik, N., M. Wilson and S. Poole. 2000. The effect of photodynamic action on two virulence factors of gram-negative bacteria. Photochem. Photobiol. 72: 676–680.

Le Marc, Y., I. Buchovec, J. Baranyi and Z. Luksiene. 2009. Modeling the photosensitization-based inactivation of *Bacillus cereus*. J. Appl. Microbiol. 107: 1006–1011.

Leistner, L. and G.W. Gould. 2002. Hurdle Technologies: Combination Treatments for Food Stability, Safety and Quality. Kluwner Academic Premium Publishers, New York. USA.

Lopez-Carballo, G., P. Hernandez-Munoz, R. Gavara and M.J. Ocio. 2008. Photoactivated chlorophyllin-based gelatin films and coatings to prevent microbial contamination of food products. Int. J. Food Microbiol. 126: 65–70.

Luksiene, Z., I. Bagdoniene, V. Smilgevicius, A. Piskarskas and B. Juodka. 1989. Action of sensitizers and light on biomolecules and cells: inactivation of *S. cerevissiae* growth after laser and porphyrin treatment. Biology 27: 161–169.

Luksiene, Z. and P. de Witte. 2002. Hypericin-based photodynamic therapy: comparative antiproliferative activity and uptake studies in murine Ehrlich ascite carcinoma. J. Photosci. 9: 524–526.

Luksiene, Z., D. Peciulyte and A. Lukauskas. 2004a. Photodynamic inactivation of harmful and pathogenic microorganisms. Vet. Medicine and Zootechnics 26: 58–60.

Luksiene, Z., D. Peciulyte and A. Lugauskas. 2004b. Inactivation of fungi *in vitro* by photosensitization: Preliminary results. Ann. Agr. Env. Med. 11: 215–220.

Luksiene, Z. 2005. New approach to inactivate harmful and pathogenic microorganisms: photosensitization. Food Technol. Biotech. 43: 411–418.

Luksiene, Z., D. Peciulyte, S. Jurkoniene and R. Puras. 2005a. Inactivation of possible fungal food contaminants by photosensitization. Food Technol. Biotech. 43: 1–8.

Luksiene, Z., V. Rutkoviene and D. Sileikiene. 2005b. Biophotonic technology for decontamination of organic cereals from fungi. State-of-the-art and problems of agricultural science and education jubilee scientific conference with international participation. Agricultural University of Plovdiv, Scientific Works 1: 125–130.

Luksiene, Z., H. Danilcenko, Z. Taraseviciene and V. Baliukoniene. 2006. Several experimental evidences to develop biophotonic technology for inactivation for decontamination of germinated seeds. Chem. Technol. 4: 64–68.

Luksiene, Z., H. Danilcenko, Z. Taraseviciene, Z. Anusevicius, A. Maroziene and H. Nivinskas. 2007. New approach to the decontamination of germinated wheat from microfungi: effects of aminolevulinic acid. Int. J. Food Microbiol. 116: 153–158.

Luksiene, Z. and A. Zukauskas. 2009. Prospects of photosensitization in control of pathogenic and harmful microorganisms. J. Appl. Microbiol. 107: 1415–1424.

Luksiene, Z., I. Buchovec and E. Paskeviciute. 2009. Inactivation of food pathogen *Bacillus cereus* by photosensitization. J. Appl. Microbiol. 107: 2037–2046.

Luksiene, Z., I. Buchovec and E. Paskeviciute. 2010a. Inactivation of *Bacillus cereus* by Na-chlorophyllin-based photosensitization on the surface of packaging. J. Appl. Microbiol. 109: 1540–1548.

Luksiene, Z., I. Buchovec and E. Paskeviciute. 2010b. Inactivation of several strains of *Listeria monocytogenes* attached to the surface of packaging material by Na-Chlorophyllin based photosensitization. J. Photochem. Photobiol. B 101: 326–331.

Luksiene, Z. and E. Paskeviciute. 2011a. Novel approach to the microbial decontamination of strawberries: chlorophyllin-based photosensitization. J. Appl. Microbiol. 110: 1274–1283.

Luksiene, Z. and E. Paskeviciute. 2011b. Novel approach to decontaminate food packaging from pathogens in non-thermal and not-chemical way: chlorophyllin-based photosensitization. J. Food Eng. 101: 326–331.

Luksiene, Z., R. Kokstaite, P. Katauskis and V. Skakauskas. 2013. Novel approach for effective and uniform inactivation of Gram-positive and Gram-negative bacteria treated by photosensitization. Food Technol. Biotechnol., accepted.

Maisch, T. 2007. Antimicrobial photodynamic therapy: useful in the future. Laser. Med. Sci. 22: 83–91.

Malik, R., A. Manocha and D. Suresh. 2010. Photodynamic therapy—A strategic review. Indian J. Dent. Res. 21: 285–291.

Manas, P. and R. Pagan. 2005. Microbial inactivation by new technologies of food preservation. J. Appl. Microbiol. 98: 1387–1399.

Marquenie, D., C.W. Michiels, J.F. Impe, E. Schrevens and B.N. Nicolaï. 2003. Pulsed white light in combination with UV-C and heat to reduce storage rot of strawberry. Postharvest Biol. Tec. 28: 455–461.

Mašković, P.Z., J.D. Mladenović, M.S. Cvijović, G. Aćamović-Đoković, S.R. Solujić and M.M. Radojković. 2011. Phenolic content, antioxidant and antifungal activities of acetonic, ethanolic and petroleum ether extracts of *Hypericum perforatum* L. Hem. Ind. 65: 159–164.

McCaughan, J.S. 1999. Photodynamic therapy: a review. Drugs & Aging 15: 49–68.

Mead, P.S., L. Slutsker, V. Dietz, L.F. McCaig, J.S. Bresee, C. Shapiro, M. Griffin and V. Tauxe. 1999. Food-related illness and death in the United States. Emerg. Infect. Dis. 5: 607–625.

Milanesio, M.E., M.G. Alvarez, E. Yslas, C.D. Borsarelli, J.J. Silber, V. Rivarola and E.N. Durantini. 2001. Photodynamic studies of metallo 5, 10, 15, 20-tetrakis (4-methoxyphenyl) porphyrin: photochemical characterization and biological consequences in a human carcinoma cell line. Photochem. Photobiol. 74: 14–21.

Moan, J. 1990. On the diffusion lenght of singlet oxygen in cells and tissues. J. Photochem. Photobiol. B 6: 343–344.

Neyssen, P.J.G. 2000. Practical implications of developments in legislation on food irradiation in the European Union. Radiat. Phys. Chem. 57: 215–217.

Nicholson, W.L., N. Munakata, G. Horneck, H.J. Melosh and P. Setlow. 2000. Resistance of *Bacillus* endospores to extreme terrestrial and extraterrestrial environments. Microbiol. Mol. Biol. R. 64: 548–572.

Nielsen, T. and A. Leufven. 2008. The effect of modified atmosphere packaging on the quality of Honeoye and Korona strawberries. Food Chem. 107: 1053–1063.

Nikaido, H. 1994. Prevention of drug access to bacterial targets: permeability barrier and active efflux. Science 264: 362–368.

Nillian, E., Ch. Ching, P. Fung, T. Robin, U. Anyi, T. Zainazor Tuan Chilek, S. Radu and M. Nishibuchi. 2011. Simultaneous Detection of *Salmonella* spp., *Salmonella* Enteritidis and *Salmonella* Typhimurium in Raw Salad Vegetables and Vegetarian Burger Patties. Food Nutr. Sc. 2: 1077–1081.

Nitzan, Y. and I. Pechatnikov. 2011. Approaches to kill gram-negative bacteria by photosensitized processes. *In:* M.R. Hamblin and G. Jori [eds.]. Photodynamic inactivation of microbial pathogens: medical and environmental applications. Royal Society of Chemistry, UK, 45–65.

Nitzan, Y., M. Salmon-Divon, E. Shporen and Z. Malik. 2004. ALA induced photodynamic effects on Gram positive and negative bacteria. Photochem. Photobiol. Sci. 3: 430–439.

No, H.K., N.Y. Park, S.H. Lee and S. Meyers. 2002. Antibacterial activity of chitosans and chitosan oligomers with different molecular weights. Int. J. Food Microbiol. 74: 65–72.

Nowis, D., M. Makowski, T. Stoklosa, M. Legat, T. Issat and J. Golab. 2005. Direct tumor damage mechanisms of photodynamic therapy. Acta Biochim. Pol. 52: 339–352.

Nunes, M.C.N., J.K. Brecht, A.M. Morais and S.A. Sargent. 1995. Physical and chemical quality characteristics of strawberries after storage are reduced by a short delay to cooling. Postharvest Biol. Tec. 6(1): 17–28.

Odriozola-Serrano, I., R. Soliva-Fortuny and O. Martín-Belloso. 2008. Phenolic acids, flavonoids, vitamin C and antioxidant capacity of strawberry juices processed by high-intensity pulsed electric fields or heat treatments. Eur. Food Res. Technol. 228: 239–248.

Olsson, M.E., J. Ekvall, K.E. Gustavsson, J. Nilsson, D. Pillai, I. Sjoholm, U. Svensson, B. Akesson and M.G.L. Nyman. 2004. Antioxidants, low molecular weight carbohydrates, and total antioxidant capacity in strawberry (*Fragaria × ananassa*): effects of cultivar, ripening, and storage. J. Agr. Food Chem. 52: 2490–2498.

Parr, A.J. and G.P. Bolwell. 2000. Phenols in the plant and in man. The potential for possible nutritional enhancement of the diet by modifying the phenols content or profile. J. Sci. Food Agr. 80: 985–1012.

Pirttijarvi, T.S.M., T.H. Graeffe and M.S. Salkinoja-Salonen. 1996. Bacterial contaminants in liquid packaging boards: assessment of potential for food spoilage. J. Appl. Bacteriol. 81: 445–458.

Plaetzer, K., B. Krammer, J. Berlanda, F. Berr and T. Kiesslich. 2009. Photophysics and photochemistry of photodynamic therapy: fundamental aspects. Laser. Med. Sci. 24: 259–268.

Polio, L., G. Valduga, G. Jori and E. Reddi. 2002. Low density lipoprotein receptors in the uptake of tumour photosensitizers by human and rat transformed fibroblasts. Int. J. Biochem. Cell B. 34: 10–23.

Reddi, E., M. Ceccon, G. Valduga, G. Jori, J. Bommer, F. Elisei, L. Latterini and U. Mazzucato. 2002. Photophysical properties and antibacterial activity of meso-substituted cationic porphyrins. Photochem. Photobiol. 75: 462–470.

Sandhya. 2010. Modified atmosphere packaging of fresh produce: current status and future needs. LWT-Food Sci. Technol. 43: 381–392.

Satin, M. 1996. The prevention of food losses after harvesting. *In:* Food Irradiation/2nd edition. Technomic, Lancaster, USA, pp. 81–94.

Scallan, E., R.M. Hoekstra, F.J. Angulo, R.V. Tauxe, M.A. Widdowson, S.L. Roy, J.L. Jones and P.M. Griffin. 2011. Foodborne illness acquired in the United States—major pathogens. Emerg. Infect. Dis. 17: 7–15.

Senorans, F.J., E. Ibanez and A. Cifuentes. 2003. New trends in food processing. Crit. Rev. Food Sci. 43: 507–526.

Sharma, R.R., D. Singh and R. Singh. 2009. Biological control of postharvest diseases of fruits and vegetables by microbial antagonists: A review. Biol. Control 50: 205–221.

Street, C.N., A. Gibbs, L. Pedigo, D. Andersen and N.G. Loebel. 2009. *In vitro* photodynamic eradication of Pseudomonas aeroginosa in planktonic and biofilm culture. Photochem. Photobiol. 85: 137–143.

Tanaka, Y., A. Tanaka and H. Tsuji. 1992. Stabilization of apoproteins of light-harvesting chlorophyll-a/b protein complex by feeding 5-aminolevulinic acid under intermittent illumination. Plant Physiol. Bioch. 30: 365–370.

Tavares, A., C. Carvalho, M. Faustino, M. Neves, J. Tome, A. Tome, J. Cavaleiro, A. Cunha, N. Gomes, E. Alves and A. Almeda. 2010. Antimicrobial photodynamic therapy: study of bacterial recovery viability and potential development of resistance after treatment. Mar. Drugs 8: 91–105.

Tayel, A.A., S. Moussa, K. Opwis, D. Knittel, E. Schollmeyer and A. Nickisch-Hartfiel. 2010. Inhibition of microbial pathogens by fungal chitosan. Int. J. Biol. Macromol. 47: 10–14.

Tegos, G.P., K. Masago, F. Aziz, A. Higginbotham, F.R. Stermitz and M.R. Hamblin. 2008. Inhibitors of bacterial multidrug efflux pumps potentiate antimicrobial photoinactivation. Antimicrob. Agents Ch. 52: 3202–3209.

Tiwari, B.K., C.P. O'Donnell, A. Patras, N. Brunton and P.J. Cullen. 2009. Effect of ozone processing on anthocyanins and ascorbic acid degradation of strawberry juice. Food Chem. 113: 1119–1126.

Tulipani, S., B. Mezzetti, F. Capocasa, S. Bompadre, J. Beekwilder, C.H. Ric De Vos, E. Capanoglu, A. Bovy and M. Battino. 2008. Antioxidants, phenolic compounds, and nutritional quality of different strawberry genotypes. J. Agr. Food Chem. 56: 696–704.

U.S. Environmental Protection Agency. 2004. Antimicrobial pesticide products. Available at: http://www.epa.gov/opp00001/factsheets/antimic.htm (accessed 22 December 2008).

Wainwright, M. 1998. Photodynamic antimicrobial chemotherapy. J. Antimicrob. Chemoth. 42: 13–28.

Wainwright, M. 2004. Photoinactivation of viruses. Photoch. Photobio. Sci. 3: 406–411.

Wang, W.D. and S.Y. Xu. 2007. Degradation kinetics of anthocyanins in blackberry juice and concentrate. J. Food Eng. 82: 271–275.

Wirtanen, G., M. Saarela and T. Mattila-Sandholm. 2000. Biofilms-impact on hygiene in food industries. *In:* J. Bryers [ed.]. Biofilms II: Process analysis and application, Wiley-Liss, New York, USA, 327–372.

[WHO] World Health Organization. 2007. Food safety and food borne illness. Available at: http://www.who.int/mediacentre/factsheets/fs237/en/ (last accessed 08.12.10).

Wood, O.B. and C.M. Bruhn. 2000. Position of the American dietetic association: food irradiation. J. Am. Diet. Assoc. 100: 246–253.

Yu, L., C.A. Reitmeier, M.L. Gleason, G.R. Nonnecke, D.G. Olson and R.J. Gladon. 1995. Quality of electron beam irradiated strawberries. J. Food Sci. 60: 1084–1087.

Yuk, H., J. Bartz and K. Schneider. 2006. The effectiveness of sanitizer treatment in inactivation of *Salmonella* spp. from bell pepper, cucumber, and strawberry. J. Food Sci. 71: 95–99.

Zafra-Stone, S., T. Yasmin, M. Bagchi, A. Chatterjee, J.A. Vinson and D. Bagchi. 2007. Berry anthocyanins as novel antioxidants in human health and disease prevention. Mol. Nutr. Food Res. 51: 675–683.

Zheng, Y., Z. Yang and X. Chen. 2008. Effect of high oxygen atmospheres on fruit decay and quality. Food Control 19: 470–474.

Zukauskas, A., M.S. Shur and R. Gaska. 2002. Introduction to Solid-State Lighting. Wiley, New York. USA.

CHAPTER 8

Encapsulation and Delivery of Antimicrobial Compounds

*Atul Kumar Singh, Preetam Sarkar, Srinivas Janaswamy, Yuan Yao and Arun Kumar Bhunia**

Introduction

Antimicrobials are the oldest and widespread defense molecules used by many forms of life from prokaryotes to humans. They include ribosomally synthesized antimicrobial peptides (positively charged +2 to +9) of 12–100 amino acids, mainly produced by bacteria and animals (Jenssen et al. 2006, Li et al. 2012b), and secondary metabolites such as antibiotics, essential oils and polyphenols from bacteria, fungi and plants (Bassole and Juliani 2012, Cotter et al. 2005, Jenssen et al. 2006, Sadaka et al. 2013). Secondary metabolites produced by bacteria, fungi and plants are small molecules. Though they are inessential for the normal growth and development of organisms, they serve as: (1) competitive molecules against other bacteria, fungi, amoebae, insects, and large animals; (2) metal transporting agents; (3) symbiosis agents between microbes and plants, nematodes, insects, and higher animals; (4) pheromones; (5) differentiation effectors; and (6) in some cases, stimulant and/or inhibitor of spore formation or germinations (Demain and Fang 2000). In food applications, antimicrobial compounds have gained considerable attention due to their unique pharmacological

Department of Food Science, 745 Agriculture Mall Drive, PurdueUniversity, West Lafayette, Indiana-47907, USA.
Emails: aksingh@purdue.edu; psarkar@purdue.edu; janaswam@purdue.edu; yao1@purdue.edu
* Corresponding author: bhunia@purdue.edu

properties. For example, microbial growth is one of the primary reasons why food loses its organoleptic attributes (flavor, texture and color) and become unsafe for consumption. Packaging of food under modified atmosphere coupled with storage at low temperature could extend the shelf-file and prevent growth of pathogens; however, these processes alone are not sufficient to reduce foodborne outbreaks and resulting health concerns. In 2005, about 1.8 million deaths were reported from diarrheal diseases worldwide, and the impact is much worse in industrialized countries with around 30 per cent people suffering from foodborne diseases every year (WHO 2007). In the United States, foodborne pathogens cause about 48 million illnesses, 128,000 hospitalizations and 3,000 deaths annually (Scallan et al. 2011). The annual costs are estimated to be $72 billion resulting from health care expenses and production loss. Therefore, measures to mitigate microbial contamination are of economic and social significance, and several secondary metabolites and antibacterial peptides readily serve the required purpose. However, their direct addition to food systems is restricted due reasons such as: (i) loss of activity; (ii) modulation of organoleptic property; (iii) poor solubility and stability; and (iv) food matrix affecting the antimicrobial activity. In order to overcome these challenges and deliver antimicrobials in safe and sustained manner, several delivery systems are being adapted (Donsi et al. 2012, Parveen et al. 2012). This chapter provides an overview of antimicrobial compounds and their delivery systems with an emphasis to food application.

Classes/Properties of Antimicrobial Compounds

Moderation of spoilage and pathogenic microorganism has always been a challenge to microbiologist, which led to the discovery and development of antimicrobial compounds such as antibiotics, antimicrobial peptides and, organic and phytochemicals since early twentieth century. Extensive use of antibiotics increases antibiotic resistance and virulence in bacterial pathogens (Beceiro et al. 2013, Rennie 2012). Such issues, compelled food and regulatory agency to look for viable alternatives such as antimicrobial peptides (AMPs), plant derived essential oils and phytochemicals that are safe and have least chance of developing antimicrobial resistance (Bassole and Juliani 2012, Guani-Guerra et al. 2010, Nishie et al. 2012, Papetti 2012, Sadaka et al. 2013). Antimicrobial compounds can be classified into six categories based on following criteria: (1) *biosynthesis:* ribosomally synthesized or produced as primary or secondary metabolite; (2) *biological source:* produced by bacteria, animals, plants or invertebrates; (3) *biological functions:* antibacterial, antifungal, antiparasital or insecticidal; (4) *molecular properties:* compounds on the basis of charge, hydrophobicity and size; (5) *structure and composition:* a chemical agent or biomolecule having different topology; and (6) *molecular targets:* extracellular or intracellular. A representative scheme elucidating classification of antimicrobial compounds based on source of production is described in Fig. 1. Most of the antimicrobial compounds produced by bacteria and plant have found their applications in food systems as bio-preservatives toward extending the shelf-life of food products, whereas those from vertebrates and invertebrates are being used by the host for immunity or immunological modulation as therapeutic agents. A variety of antimicrobials produced by bacteria, plant and animals are specifically described in the following sections.

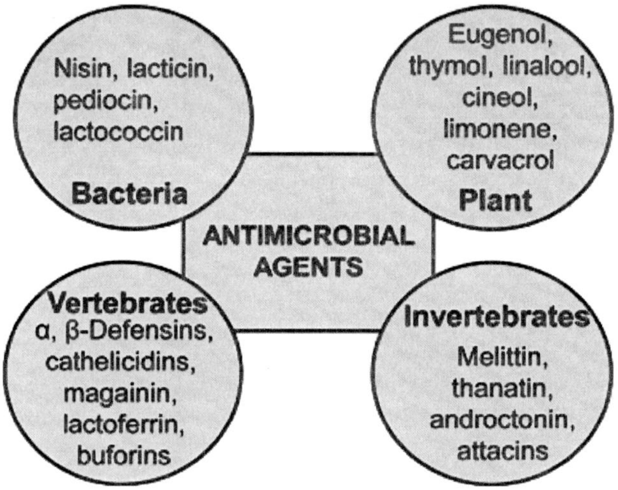

Figure 1. Schematic representation of antimicrobial compound classification based on its source of production. Data partly adapted from (Bassole and Juliani 2012, Davidson et al. 2013, Jenssen et al. 2006, Sadaka et al. 2013, Tiwari et al. 2009).

Bacterial antimicrobials

Bacteria produce many antimicrobials with inhibitory effect on other bacteria and can be used in controlling spoilage or pathogenic microorganisms. The antimicrobials are mainly produced by gram-positive bacteria that include lactic acid bacteria (LAB), *Bacillus* and *Streptomyces* (Brownlee et al. 1952, Davidson et al. 2013, Tiwari et al. 2009). LABs are Generally Recognized as Safe (GRAS) and have significant contribution in the preparation of healthy and functional foods, and during fermentation they produce a gamut of antimicrobial metabolites such as organic acid, diacetyl, acetoin, hydrogen peroxide, antibiotics and bacteriocins to increase the shelf-life and organoleptic value of food products (Topisirovic et al. 2006). The low pH built by the acidic environment as well as the action of other fermented byproducts inhibits the growth of many spoilage and pathogenic bacteria and helps food products acquire a unique aroma, flavor and texture due to various physiological attributes of LAB (Coolbear et al. 2008, Leroy and De Vuyst 2004). In the food industry, LABs are primarily used as starter cultures to instigate the fermentation and increase the shelf-life of fermented foods through innate production of bacteriocins and other antimicrobial compounds (Cotter et al. 2005, Drider et al. 2006, O'Sullivan et al. 2002).

Bacteria produce myriads of proteinaceous and non-proteinaceous antimicrobial compounds. Proteinaceous antimicrobial compounds include mainly ribosomally synthesized bacteriocins such as, nisin, pediocin, lacticin, plantaracin, produced by LAB; bacitracin produced by *Bacillus subtilis*; and polymyxin produced by *Paenibacillus polymyxa* and other secreted AMPs. Non-proteinaceous antimicrobial compounds include reuterin (β-hydroxypropionaldehyde) produced by *Lactobacillus reuteri*; propionic acid produced by propionic acid bacteria, lactic acid produced by LAB; and natamycin produced by *Streptomyces natalensis*. One of the non-proteinaceous antimicrobial, termed as Micrograd™. Micrograd™, is a United States Food and Drug Administration (FDA) approved fermentate of *Propionibacterium freudenreichii* subsp. *shermanii*, for cottage cheese manufacture, where propionic acid is the main antimicrobial component. Among the bacteriocins produced by LABs, nisin is the only FDA approved bacteriocin that can be used as a food preservative at acidic pH. Bacteriocins are mostly active against Gram-positive bacteria that include *Listeria monocytogenes*, *Bacillus cereus*, *Staphylococcus aureus*, *Lactobacillus* spp. and *Streptococcus* spp. (Cotter et al. 2005, Davidson et al. 2013, Ross et al. 2002, Singh and Ramesh 2008, Tiwari et al. 2009).

Plant antimicrobials

Essential oils are important members of plant-derived antimicrobial compounds. They are extracted through steam distillation of plant materials such as flowers, buds, seeds and roots. These phytochemicals are volatile and contain a complex mixture of biosynthetic components of plants secondary metabolites such as, thymol and carvacrol in thyme oil; eugenol in basil and clove oil; linalool in coriander and lavender oil and; limonene in lemon oil and lemon grass (Burt 2004, Sadaka et al. 2013). Phytochemicals demonstrate antimicrobial activity against both gram-positive and gram-negative bacteria, including organisms that are of concern for safety and quality such as *Listeria*, *Salmonella*, and O157 STEC. Composition, structure as well as functional groups of these compounds appear to play an important role in bestowing the antibacterial activity. The primary functionality of essential oil compounds include perturbation of bacterial cell wall and membrane that affects electrochemical gradient across the membrane, denatures the cell wall and cytoplasmic proteins, and leads to oxidative damage of bacterial cells. Such multifaceted action of plant antimicrobials impairs bacteria to develop any resistance for essential oils (Burt 2004, Sadaka et al. 2013).

Animal antimicrobials

Animal-derived antimicrobial compounds can be from vertebrate (mammals, fish, avian and amphibians) and invertebrate (insects and crustaceans) sources. In the food industry, antimicrobial compounds secreted from vertebrates have found significant impact due to their high-specificity and least cytotoxicity (Davidson et al. 2013, Li et al. 2012b, Tiwari et al. 2009). Some examples include, chitosan, a polymer of glucosamine and *N*-acetylglucosamine from the exoskeleton of crustaceans; conalbumin/ ovotransferrin, an iron-chelating protein found in egg-white; lactoferrin, an iron-chelating protein in milk, colostrum, polymorphonuclear leuckocytes and other physiological fluids; lactoperoxidase, a glycoprotein found naturally in raw milk, colostrum, saliva and other biological secretions; and lysozyme, a *N*-acetylmuramide glycanhydrolyase enzyme that degrades cell walls of bacteria and exists in egg-white, tears and other biological secretions.

Animal-derived antimicrobials exhibit broad-spectrum inhibitory activity against Gram-positive and Gram-negative bacteria and fungi. Chitosan inhibits lactic acid bacteria, *Listeria monocytogenes*, *Salmonella* Typhimurium, *Byssochlamys* spp. and *Zygosaccharomyces bailii*. Antimicrobial compound Conalbumin or ovotransferrin inhibits *E. coli* O157:H7, *B. subtilis*, *B. stearothermophilus*, *Carnobacterium viridans*, *L. monocytogenes*, and *Candida albicans*. Milk protein, lactoferrin is effective against *E. coli* O157:H7, enteropathogenic *E. coli* (EPEC), LAB (including *Lactobacillus* and *Pediococcus*), *L. monocytogenes*, *Staphylococcus aureus*, *Shigella*, *Salmonella*, *Yersinia enterocolitica* and *Candia* sp. Antimicrobial enzyme, lactoperoxidase inhibits *E. coli* O157:H7, *L. monocytogenes*, *S. aureus*, *B. cereus*, *Campylobacter jejuni*, *Salmonella*, *Y. enterocolitica*, *Aspergillus*, *Byssochlamys fulva*, *Candida*, *Mucor*, and *Rhodotorula*, whereas lysozyme works well against *Alicyclobacillus*, *Bacillus coagulans*, *B. stearothermophilus*, *B. cereus*, *Clostridium tyrobutyricum*, *Clostridium thermosaccharolyticum* (*Thermoanaerobacterium thermosaccharolyticum*), *Clostridium botulinum*, *Lactobacillus* spp., *L. monocytogenes*, *Micrococcus*, *C. jejuni*, *E. coli*, *Pseudomonas aeruginosa*, *Pseudomonas fluorescens*, *S.* Enteritidis, *Y. enterocolitica*, *Aspergillus*, *Candida*, *Fusarium graminearum*, *Paecilomyces*, *Penicillium*, *Saccharomyces*, and *Sporothrix* (Davidson et al. 2013, Tiwari et al. 2009).

Several reviews highlight the advantages of AMPs and essential oils (Juneja et al. 2012, Mills et al. 2011), especially in active packaging of food products, where antimicrobial usage elucidates synergistic effect in minimizing the microbial contaminations (Davidson et al. 2013, Imran et al. 2010, Kim et al. 2012). However, their delivery in food systems is still a problem of concern due to poor solubility, cytotoxicity and rapid degradation. Such predicaments can be overcome by designing and fabrication of suitable delivery systems. For example, liposomes, natural and synthetic polymers are found to be effective towards neutraceuticals, functional food components and drugs (Coelho et al. 2010, McClements et al. 2009).

Delivery Systems

Alterations in desirable functionalities and bioavailability of bioactive materials caused by environmental and chemical factors can be restricted by using suitable delivery or carrier systems. In the food system, food matrix is one of the environmental factors that can impact the availability of micronutrient and other bioactive compounds (Turgeon and Rioux 2011). Delivery systems can be made up of naturally occurring polymers and polysaccharides, synthetic polymer or nanomaterials to tailor encapsulation and delivery of antimicrobials, drugs or bioactive molecules (Coelho et al. 2010, Janaswamy and Youngren 2012, Janaswamy et al. 2013, McClements et al. 2009, Pillai and Panchagnula 2001). These delivery systems present exciting opportunity for food and pharmaceutical technologists to enhance bioavailability, stability, sustained activity and shelf-life of food and drug through encapsulation and controlled release. The subsequent section describes the different delivery systems, especially for food application.

Polymers

Attributes of efficient delivery systems or carriers include GRAS status, biocompatible with no side-effects, inexpensive, and stability during processing and storage. Furthermore, delivery system should be able to (1) preserve the functionality of the embedded compounds until the time of delivery, and (2) deliver the preserved form effectively at the target site with predictable release profiles. In this regard, several biodegradable polymers are explored as favorable alternatives (Ranade 1990a, Ranade 1990b). For example, polyesters and polyanhydrides are being engineered as drug carriers as well as to develop intelligent vehicles that respond to physical, chemical and biological stimuli (Fredenberg et al. 2011, Schmaljohann 2006). These biodegradable polymers also serve as basic ingredients in implants, biosensors and bioadhesives. They can be tailored to incorporate labile groups such as ester, anhydride, carbonate, amide in their backbone, which further enhances their utility, efficacy and applicability. The chemical structure, composition and molecular weight of biopolymers and its distribution, morphology (amorphous, semicrystalline and microstructure), and mechanism of degradation (enzymes and hydrolysis) are some of the influencing factors for its selection as delivery system. Overall, their intrinsic properties such as hydrophilicity and lubricity coupled with their ability to endure physical strength, durability and degradability, render favorable opportunities not only for the design and delivery of several functional ingredients but also as carriers of food antimicrobials. A list of biopolymers and encapsulated antimicrobials are highlighted in Table 1.

Table 1. Biopolymer carriers used in the delivery and encapsulation of antimicrobial or therapeutic agents.

S. no.	Polymer carrier	Monomer	Encapsulated molecule	Reference
1.	Hydrogel	Dendrtic amine	None	(Cuggino et al. 2013)
2.	Phytoglycogen Octinyl succinate (PGOS)	D-glucose	Nisin	(Bi et al. 2011b)
3.	Polysaccharide	Iota-carageenan	Eugenol	(Janaswamy et al. 2013)
4.	Milk fat globule membrane (MFGM)	Lipid	α-tocopherol and lycopene	(Bezelgues et al. 2009)
5.	Poly (DL-lactide-co-glycolide) (PLGA)	DL-lactide-co-glycolide	trans-Cinnamaldehyde and Eugenol	(Gomes et al. 2011)
6.	Soyprotein	Glycinin and β-conglycinin	Rifampicin	(Caillard et al. 2009, Petruccelli and Anon 1995)
7.	Corn zein protein	Prolamines	Nisin and thymol	(Xiao et al. 2011)

Polysaccharides

Polysaccharides are ubiquitous and a plethora of renewable resources constitute as feasible ingredients for their extraction and utility. They exhibit a wide range of chemical structures and physiological functions (Whistler and BeMiller 1993). Their intrinsic non-toxic nature readily extends their utility in food, pharmaceutical and medicinal applications as thickeners, viscosifiers, and gelling agents. One of their important utilities in the pharmaceutical sector is excipients. Their primary role is to provide the necessary weight, consistency and volume for the correct administration of the active pharmaceutical ingredient (API). Starch, cellulose, agar, alginate, guar gum, xanthan, and pectin are some known examples of excipients (Beneke et al. 2009). Polysaccharides are also utilized in the design and development of drug delivery carriers (Hamman 2006), which includes release of drugs in response to external stimuli, and targeted delivery to a specific region in the gastrointestinal (GI) tract. Polysaccharides are resistant to digestion, and hence retain their integrity and remain intact in the stomach and small intestine. However, upon reaching the colon, they are degraded by bacterial polysaccharidases, and this property suits well for colon-targeted drug formulations.

Loading active ingredients in polysaccharide fibers is another way of antimicrobial delivery, and electrospun fibers of cellulose acetate phthalate (Huang et al. 2012) and cellulose acetate (Phiriyawirut and Phaechamud 2012) are successful examples of polysaccharide based delivery systems. Moreover, carriers with stable network structure are advantageous than those with non-rigid structures for retaining the functional state of the encapsulated molecules and effective delivery at the target site. More recently, it has been demonstrated that well-oriented and crystalline polysaccharide fibers, especially iota-carrageenan, can be utilized to encapsulate nutraceuticals (Janaswamy and Youngren 2012) and drugs (Janaswamy et al. 2013) that provides protection from melting and controlled release of candidate molecule. The iota-carrageenan network is quite stable because of well-organized and cation-mediated inter-helical interactions with pockets of free space (8–15 Å) between pairs of carrageenan helices that entraps small molecules through van der Waals, ionic and hydrogen bonding interactions (Fig. 2). These fibers are effective toward encasing antimicrobial components, essential oils, and sustaining antimicrobial activity for prolonged durations (unpublished results).

Nanoparticles

Nanoparticles are small microscopic particles in the form of nanopowder, nanocluster or nanocrystal, where at least one dimension is less than 100 nm, and physiochemical properties vary with the size and dimension of the nanoparticles. These particles, being in nanometer range, offer unique physicochemical properties such as increased surface area to volume ratio and high reactivity. Consequently, over the past few decades, nanoparticles have emerged as novel carriers of active ingredients in food, pharmaceutical and medical applications. In the case of drugs, pharmacokinetics and therapeutic index can be enormously improved by using nanocarriers that also provide an elegant opportunity for targeted delivery, e.g., treating intracellular infections. In this regard, a wide variety of nanoparticle-based drug delivery systems are being investigated, and liposomes (Torchilin 2005), polymeric nanoparticles (Soppimath et al. 2001), solid lipid nanoparticles (Muller et al. 2000) and dendrimers (Svenson and Tomalia 2005) are highly noticeable nanoparticle-based delivery systems. Briefly, liposomes are spherical lipid vesicles based on amphiphilic lipid molecules and possess bilayered membrane structures. In the liposome

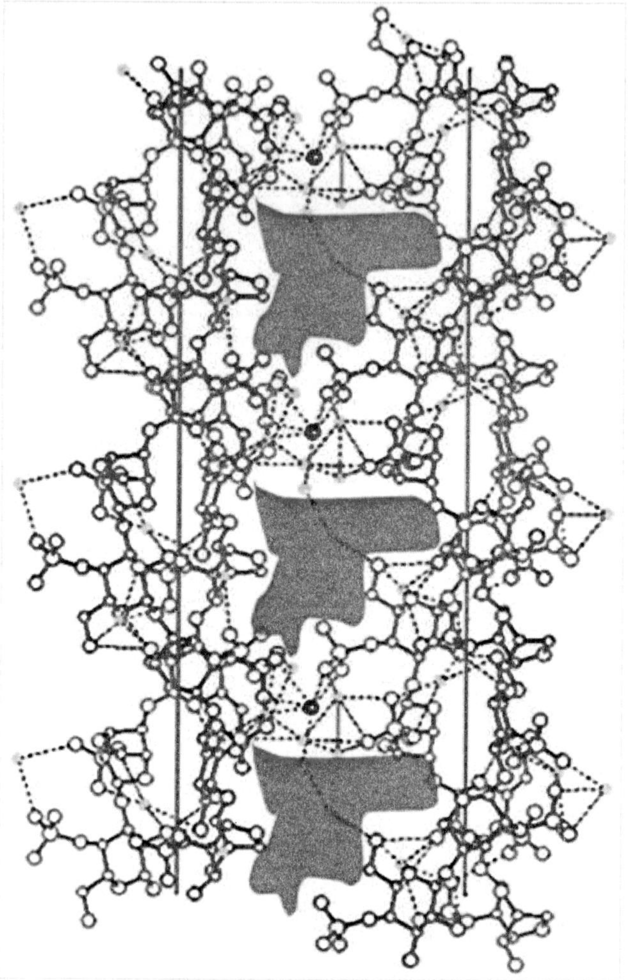

Figure 2. Association of a pair of iota-carrageenan double helices (one chain in red and other blue) in the presence of calcium ions (black circles) and water molecules (green circles). The broken lines represent hydrogen bonds. The green blocks highlight the water pockets suitable for encapsulation.

Color image of this figure appears in the color plate section at the end of the book.

structure, a lipid bilayer membrane surrounding an aqueous cavity uniquely enables them to carry both hydrophobic and hydrophilic compounds. For example, 0.39 g of nisin encapsulated in 100% phosphotidylcholine (PC) and PC-cholesterol (7:3 ratio) resulted in >2 log CFU/ml reduction of *Listeria monocytogenes* in the treatment compared to free nisin (Were et al. 2004). On the other hand, diblock copolymers, mostly biodegradable, containing hydrophilic and hydrophobic segments and liner polymers such as polyalkyl acrylates and polymethyl methacrylate are known examples of polymeric nanoparticles. Solid lipid nanoparticles are derived from lipids that are in solid phase (e.g., fatty acids, triglycerides and steroids) at room temperature and act as surfactants for emulsification. Finally, dendrimers are highly ordered, branched globular macromolecules and their highly branched characteristics, provides larger surface area to size ratio, leading to increased reactivity in actual delivery scenario. In addition, inorganic nanoparticles such as zinc oxide (ZNO) (Padmavathy and Vijayaraghavan 2008), metal titanate (Wang et al. 2008) and silver particles (Rai et al. 2009) also found to inhibit the growth of broad range of bacteria. In the next section, we shall discuss the protection and delivery of nisin and thymol in detail.

Exemplary Cases: Protection and Delivery of Nisin and Thymol

Nisin

Nisin is a positively charged, lantibiotic, antimicrobial peptide extracted from the cells of *Lactococcus lactis* subsp. *lactis* (Lactic acid bacteria group). It consists of 34 amino acids residues and is characterized within the lantibiotic family as it contains certain modified amino acids such as lanthionine and methylanthionine (Bonev et al. 2000, Breukink et al. 2000, Breukink et al. 1997, Liu and Hansen 1990). Nisin consists of 2 different variants, nisin-A and nisin-Z. They differ by one amino acid residue at position 27, where histidine is present in nisin-A and asparagine is present in nisin-Z (Laridi et al. 2003). Nisin has been found to be highly active against Gram-positive bacteria, but not against Gram-negative ones. Nisin kills bacterial cells by pore formation (Breukink and de Kruijff 2006). It has been extensively used as a biopreservative in the food processing industry for the past fifty years. The stability of nisin depends on various factors such as pH, temperature and food constituents namely enzymes (proteases), glutathione, titanium dioxide and sodium metabisulphite (Delves-Broughton 2005). All of these factors adversely affect nisin stability thus reducing its activity over time. Therefore, there is a need to complex nisin with a delivery system that will be enable to protect it against such depletion factors and maintain nisin concentration above minimum inhibitory concentration (MIC) against target bacteria.

Carbohydrate nanomaterials have been used to protect and deliver nisin (Bi et al. 2011b). Several types of nanoparticles were prepared and used such as, phytoglycogen succinate (PG-S); phytoglycogen β-dextrin succinate (PGB-S); phytoglycogen octenyl succinate (PG-OS); and phytoglycogen β-dextrin octenyl succinate (PGB-OS). The nisin preparations were stored in a BHI-agar deep-well model system (at 4°C) and the residual nisin activity was studied using an agar diffusion assay against *Listeria monocytogenes*

V7. Nanoparticle complexed nisin exhibited prolonged antilisterial activity compared with free nisin in the model system over extended period of time. In case of modified nanoparticles, octenyl succinylation led to greater retention of nisin activity compared to succinylation. This was primarily due to the effect of both hydrophobic and electrostatic interactions between PG-OS and nisin. PG-OS was also used to prepare oil-in-water emulsions to protect and deliver nisin (Bi et al. 2011b). PG-OS stabilized emulsion showed significant retention of nisin activity till 50 days. In contrast, waxy corn starch octenyl succinate (WCS-OS) and tween-20 stabilized emulsion retained nisin activity till 20–30 and 10–20 days, respectively. Free nisin on the other hand was depleted rapidly and lost its activity by 10 days of storage (Bi et al. 2011a).

Nisin has also been encapsulated using liposomes, and its antimicrobial efficacy was tested against *Listeria innocua*, *Lactococcus* spp., and *Lactobacillus casei* subsp. *casei*. The study demonstrated that about 90% of initial nisin activity could be recovered in encapsulated form, whereas only 12% could be recovered when nisin was in free form (Benech et al. 2003), which underscores the benefit of encapsulation. In a similar study, nisin Z complexed with liposomes were added to cheese milk and the antibacterial efficacy was observed against *Listeria innocua*. After cheese production step, the bacterial count showed 3 log reduction with encapsulated form compared to a 1.5 log reduction with free nisin (Benech et al. 2003).

Nisin has been incorporated within different types of biopolymer based films to enhance its antimicrobial efficacy against bacteria and fungi (Basch et al. 2012, Cao-Hoang et al. 2010, Pintado et al. 2010). Nisin was complexed within sodium caseinate-based films at the concentration of 1000 IU/cm^2 and its antimicrobial efficacy was studied in a cheese model. The study demonstrated 1.1 log reduction of *Listeria innocua* when compared with free nisin after 7 days (Cao-Hoang et al. 2010). Nisin along with natamycin and malic acid were incorporated within whey protein-based films and tested against *Pseudomonas aeruginosa*, *Listeria monocytogenes*, *Penicillium commune* and *Penicillium chrysogenum*. The study suggested that complexing nisin within whey protein films provides a strong hurdle against organisms obtained from cheese (Pintado et al. 2010). Nisin and potassium sorbate alone or in combination were also incorporated within starch-hydroxypropylmethylcellulose (starch-HPMC) based films and tested against *Listeria innocua* and *Zygosaccharomyces bailii*. The study confirmed that the use of starch-HPMC films containing nisin and potassium sorbate resulted in greater antimicrobial efficacy when compared with the antimicrobials used alone (Basch et al. 2012).

Thymol

Thymol (2-isopropyl-5-methylphenol) is a major monoterpene phenol obtained from thyme (*Thymus vulgaris*) essential oil. Thymol binds with bacterial proteins and unsaturated fatty acids, disintegrates the cell membrane and enhances cell permeability, and leads to the loss of cell viability (Burt 2004, Burt and Reinders 2003). As an antimicrobial compound, thymol needs to be protected as it can bind with hydrophobic food components such as lipids and proteins. Binding with food matrix reduces its availability against foodborne pathogens (Choi et al. 2010, Shah et al. 2012). Thymol has limited water solubility; therefore a sufficient concentration could be difficult to achieve in the aqueous phase, which may affect its antimicrobial properties (Guarda et al. 2011, Liolios et al. 2009, Shah et al. 2012, Weiss 2009). Additionally, thymol has a strong, medicinal aroma, which can create sensory defects in food at higher concentrations. Evidently, there is a need for encapsulation and rational delivery of thymol as an antimicrobial component in aqueous and food systems.

Thymol encapsulated within whey protein isolate-maltodextrin (WPI-MD) nanocapsules was evaluated for antimicrobial activity against *Listeria monocytogenes* Scott A and 101, and *Escherichia coli* O157:H7 ATCC 43889 and 43894 (Shah et al. 2012). Apple cider and 2 per cent reduced-fat milk were used as the model foods. At pH 5.5 in apple cider, WPI-MD nano-dispersed thymol at 500 μg/ml concentration was found to be bacteriostatic against both *Listeria monocytogenes* and *Escherichia coli* till 48 h. Whereas, when nano-dispersed thymol was tested in a low-fat milk system, 4500 μg/ml of thymol was found to be inhibitory for both the pathogens (Shah et al. 2012).

In thymol-chitosan complex, five different water-soluble chitosan based on their degree of deacetylation were complexed with thymol. Thymol and chitosan complex formed bimolecular micelles, and the chitosan complexed thymol exhibited stronger antibacterial activity against *Staphylococcus aureus* and *Bacillus subtilis* than free thymol (Hu et al. 2009). In another case, zein-sodium caseinate nanoparticle based antimicrobial films were developed to load thymol as the antimicrobial agent (Li et al. 2012a). These antimicrobial films were tested against *Escherichia coli* and *Salmonella*, and the inhibitory effect was evaluated using agar diffusion assay, where the zone of inhibition ranged between 15.89 and 18.81 mm for *E. coli* and from 16.02 to 18.12 mm for *Salmonella*. In the following section, challenges faced with different delivery systems are described briefly.

Challenges with Different Delivery Systems

Controlled release of active compounds towards retarding microbial growth and enhancing food quality is an active research area and being pursued by several researchers all over the world. Over the past several years, this field has gained enormous progress with the design and development of novel sustained-release antimicrobial formulations, especially with the availability of several biodegradable and natural materials as well as synthetic polymers. Among all the delivery systems, nanoparticles based vehicles are gaining considerable attention in recent times mainly due to their ultra-small sizes and long circulation characteristics. However, there are still some challenges to overcome, such as optimizing the shapes and sizes, directing them to the target site, loading capacity as well as production costs. Most importantly, lack of complete understanding of their impact on the functional properties of food and antimicrobials, and biocompatibility, impedes their wholesome utility in food systems and clinical applications. These shortcomings can be addressed, to some extent, by utilizing carriers containing porous structures comprising biopolymers and inorganic materials. For example, oriented fibers based on edible carbohydrate polymers and polysaccharides featuring stable network structure appear to be a facile route for delivering antimicrobial compounds. However, size restrictions of encapsulated molecules and associated loading efficiencies are some limiting factors. Challenges and limitation with delivery systems is still a developing field and provides a future scope for detailed review. The next section provides short elaboration about the assessment of toxicity and safety of engineered nanomaterials.

Assessment of Toxicity and Safety of Engineered Nanomaterials

Assessing toxicity and safety of engineered nanomaterials (ENMs) before their application in food system is a parameter of paramount importance. Tier-based assessing system has been systematically reviewed in the recently published article (Cockburn et al. 2012). ENMs can enter into the consumer directly through food, water, cosmetics or drugs; and its retention in the consumer is highly dependent on the route of ENM intake. It was reported that ENM uptake through food are not absorbed into gut lymphatics and does not lead to systematic distribution of ENM to organs and ENM are eliminated in the feces. In contrast, ENM administered intravenously were retained at least for a week with >70 per cent accumulation in the liver (Nel et al. 2006). A review on nanotechnology and health safety has also highlighted the various important parameters for the toxicity and risk assessment of nanostructured materials on human health (Singh and Nalwa 2007).

With the emphasis on the development of ENM for food application in the field of nanoscience and nanotechnology, various guidelines have been set in defining and assessing the toxicity of nanomaterials (Cockburn et al. 2012). The fate of nanomaterials in the body depends on its biokinetics, which involves absorption and metabolism/excretion of nanoparticle. Absorbance of particulate material occurs across the gastrointestinal tract (mainly in small intestine) through transcytosis. Uptake of large nanoparticles (20–100 nm) and small micro particles (100–500 nm) occurs through M-cells of the Peyer's Patches. Coockburn et al. (2012) in their review has described that metabolism/excretion of particulates, including ENMs initiates through endosomal/lysosomal degradation attempt. Toxicity of nanomaterials is a developing field of nanoscience/nanotechnology, and a detailed insight on nanotoxicity relies on the worldwide ongoing research outcome.

Future Perspectives and Conclusions

Food products are perishable, purchased everyday and consumed by people worldwide, which could be the primary route of spreading foodborne pathogens to humans. After consuming contaminated food, several individuals become ill and are hospitalized with some fatalities. Thus, food safety and foodborne illnesses are growing public health concern worldwide. Measures to enhance the shelf-life of food products and alleviate microbial contamination during preparation, storage and distribution are of economic and social significance. Traditional food preservation techniques such as salt addition and refrigeration alone cannot ensure food safety. Besides, consumption of excess salt is discouraged since it may lead to health problems such as high blood pressure and cardiovascular disease (WHO 2002). These predicaments coupled with the ever-growing demands for minimally processed, easily prepared, and ready-to-eat foods warrant novel methods for maintaining food safety. Food preservation techniques utilizing pulsed electric (Calderon-Miranda et al. 1999), and magnetic fields (San Martín et al. 2001), and irradiation (Mainville et al. 2001) are actively pursued as alternatives. In recent years, antimicrobial compounds composed of green chemicals or having GRAS status are of particular interest and are gaining popularity in food applications. In this regard, direct surcharge of antimicrobial compounds to food and/or packing materials appears to be effective. However, they lose the activity quickly due to instability and conditions encountered during processing, storage and transportation of foods. Thus, encapsulation and delivery through GRAS carriers is a promising approach and certainly holds credibility in preventing and/or delaying microbial proliferation that can lead to improved shelf-life and safety of foods and reduction in food spoilage.

Acknowledgements

Research in the authors laboratories was funded by the National Science Foundation (NSF) project number 0932586, Agricultural Research Service of the US Department of Agriculture (USDA) project number 1935-42000-072-02G, Whistler Center for Carbohydrate Research (WCCR) and the Center for Food Safety Engineering at Purdue University.

Keywords: Antimicrobials, bacteriocins, biopolymers, delivery system, encapsulation, lactic acid bacteria, lactoferrin, nanoparticles, nisin, polysaccharides, thymol, toxicity and safety of engineered nanomaterials

References

Basch, C.Y., R.J. Jagus and S.K. Flores. 2012. Physical and antimicrobial properties of tapioca starch-HPMC edible films incorporated with nisin and/or potassium sorbate. Food and Bioprocess Technology 1–10.

Bassole, I.H. and H.R. Juliani. 2012. Essential oils in combination and their antimicrobial properties. Molecules 17(4): 3989–4006.

Beceiro, A., M. Tomas and G. Bou. 2013. Antimicrobial resistance and virulence: a successful or deleterious association in the bacterial world? Clin. Microbiol. Rev. 26(2): 185–230.

Benech, R.-O., E. Kheadr, C. Lacroix and I. Fliss. 2003. Impact of Nisin Producing Culture and Liposome-encapsulated Nisin on Ripening of *Lactobacillus* added-Cheddar Cheese. Journal of Dairy Science 86(6): 1895–1909.

Beneke, C.E., A.M. Viljoen and J.H. Hamman. 2009. Polymeric plant-derived excipients in drug delivery. Molecules 14(7): 2602–2620.

Bezelgues, J.B., F. Morgan, G. Palomo, L. Crosset-Perrotin and P. Ducret. 2009. Short communication: Milk fat globule membrane as a potential delivery system for liposoluble nutrients. J. Dairy Sci. 92(6): 2524–2528.

Bi, L., L. Yang, A.K. Bhunia and Y. Yao. 2011a. Carbohydrate nanoparticle-mediated colloidal assembly for prolonged efficacy of bacteriocin against food pathogen. Biotechnology and Bioengineering 108(7): 1529–1536.

Bi, L., L. Yang, G. Narsimhan, A.K. Bhunia and Y. Yao. 2011b. Designing carbohydrate nanoparticles for prolonged efficacy of antimicrobial peptide. Journal of Controlled Release 150(2): 150–156.

Bonev, B.B., W.C. Chan, B.W. Bycroft, G.C. Roberts and A. Watts. 2000. Interaction of the lantibiotic nisin with mixed lipid bilayers: a 31P and 2H NMR study. Biochemistry 39(37): 11425–11433.

Breukink, E. and B. de Kruijff. 2006. Lipid II as a target for antibiotics. Nature Reviews Drug Discovery 5(4): 321–332.

Breukink, E., P. Ganz, B. de Kruijff and J. Seelig. 2000. Binding of Nisin Z to bilayer vesicles as determined with isothermal titration calorimetry. Biochemistry 39(33): 10247–10254.

Breukink, E., C. van Kraaij, R.A. Demel, R.J. Siezen, O.P. Kuipers and B. de Kruijff. 1997. The C-terminal region of nisin is responsible for the initial interaction of nisin with the target membrane. Biochemistry 36(23): 6968–6976.

Brownlee, G., S. Bushby and E.I. Short. 1952. The chemotherapy and pharmacology of the polymyxins. British Journal of Pharmacology and Chemotherapy 7(1): 170–188.

Burt, S. 2004. Essential oils: their antibacterial properties and potential applications in foods—a review. International Journal of Food Microbiology 94(3): 223–253.

Burt, S.A. and R.D. Reinders. 2003. Antibacterial activity of selected plant essential oils against *Escherichia coli* O157: H7. Letters in Applied Microbiology 36(3): 162–167.

Caillard, R., A. Petit and M. Subirade. 2009. Design and evaluation of succinylated soy protein tablets as delayed drug delivery systems. Int. J. Biol. Macromol. 45(4): 414–420.
Calderon-Miranda, M.L., G.V. Barbosa-Canovas and B.G. Swanson. 1999. Inactivation of Listeria innocua in skim milk by pulsed electric fields and nisin. International Journal of Food Microbiology 51(1): 19–30.
Cao-Hoang, L., L. Grégoire, A. Chaine and Y .Waché. 2010. Importance and efficiency of in-depth antimicrobial activity for the control of listeria development with nisin-incorporated sodium caseinate films. Food Control 21(9): 1227–1233.
Choi, K.-O., J. Ryu, H.-S. Kwak and S. Ko. 2010. Spray-dried conjugated linoleic acid encapsulated with Maillard reaction products of whey proteins and maltodextrin. Food Science and Biotechnology 19(4): 957–965.
Cockburn, A., R. Bradford, N. Buck, A. Constable, G. Edwards, B. Haber, P. Hepburn, J. Howlett, F. Kampers, C. Klein et al. 2012. Approaches to the safety assessment of engineered nanomaterials (ENM) in food. Food Chem. Toxicol. 50(6): 2224–2242.
Coelho, J.F., P.C. Ferreira, P. Alves, R. Cordeiro, A.C. Fonseca, J.R. Gois and M.H. Gil. 2010. Drug delivery systems: Advanced technologies potentially applicable in personalized treatments. EPMA J. 1(1): 164–209.
Coolbear, T., V. Crow, J. Harnett, S. Harvey, R. Holland and F. Martley. 2008. Developments in cheese microbiology in New Zealand—Use of starter and non-starter lactic acid bacteria and their enzymes in determining flavour. International Dairy Journal 18(7): 705–713.
Cotter, P.D., C. Hill and R.P. Ross. 2005. Bacteriocins: Developing innate immunity for food. Nature Reviews Microbiology 3(10): 777–788.
Cuggino, J.C., G. Charles, G. Gatti, M.C. Strumia and C.I. Alvarez Igarzabal. 2013. New hydrogel obtained from a novel dendritic monomer as a promising candidate for biomedical applications. J. Biomed. Mater. Res. A.
Davidson, P.M., F.J. Critzer and T.M. Taylor. 2013. Naturally occurring antimicrobials for minimally processed foods. Annu. Rev. Food Sci. Technol. 4: 163–190.
Delves-Broughton, J. 2005. Nisin as a food preservative. Food Australia 57(12): 525–527.
Demain, A.L. and A. Fang. 2000. The natural functions of secondary metabolites. Adv. Biochem. Eng. Biotechnol. 69: 1–39.
Donsi, F., M. Annunziata, M. Vincensi and G. Ferrari. 2012. Design of nanoemulsion-based delivery systems of natural antimicrobials: effect of the emulsifier. J. Biotechnol. 159(4): 342–350.
Drider, D., G. Fimland, Y. Hechard, L.M. McMullen and H. Prevost. 2006. The continuing story of class IIa bacteriocins. Microbiology and Molecular Biology Reviews 70(2): 564-+.
Fredenberg, S., M. Wahlgren, M. Reslow and A. Axelsson. 2011. The mechanisms of drug release in poly(lactic-co-glycolic acid)-based drug delivery systems—a review. Int. J. Pharm. 415(1–2): 34–52.
Gomes, C., R.G. Moreira and E. Castell-Perez. 2011. Poly (DL-lactide-co-glycolide) (PLGA) nanoparticles with entrapped trans-cinnamaldehyde and eugenol for antimicrobial delivery applications. J. Food Sci. 76(2): N16–24.
Guani-Guerra, E., T. Santos-Mendoza, S.O. Lugo-Reyes and L.M. Teran. 2010. Antimicrobial peptides: General overview and clinical implications in human health and disease. Clinical Immunology 135(1): 1–11.
Guarda, A., J.F. Rubilar, J. Miltz and M.J. Galotto. 2011. The antimicrobial activity of microencapsulated thymol and carvacrol. International Journal of Food Microbiology 146(2): 144–150.
Hamman, J.H.T., C. 2006. Functional excipients. Chemistry Today Magazine 24(5): 6.
Hu, Y., Y. Du, X. Wang and T. Feng. 2009. Self-aggregation of water-soluble chitosan and solubilization of thymol as an antimicrobial agent. Journal of Biomedical Materials Research Part A 90(3): 874–881.
Huang, C., S.J. Soenen, E. van Gulck, G. Vanham, J. Rejman, S. Van Calenbergh, C. Vervaet, T. Coenye, H. Verstraelen, M. Temmerman et al. 2012. Electrospun cellulose acetate phthalate fibers for semen induced anti-HIV vaginal drug delivery. Biomaterials 33(3): 962–969.

Imran, M., A.M. Revol-Junelles, A. Martyn, E.A. Tehrany, M. Jacquot, M. Linder and S. Desobry. 2010. Active food packaging evolution: transformation from micro- to nanotechnology. Crit. Rev. Food Sci. Nutr. 50(9): 799–821.

Janaswamy, S., K.L. Gill, O.H. Campanella and R. Pinal. 2013. Organized polysaccharide fibers as stable drug carriers. Carbohydrate Polymers 94(1): 209–215.

Janaswamy, S. and S.R. Youngren. 2012. Hydrocolloid-based nutraceutical delivery systems. Food Funct. 3(5): 503–507.

Jenssen, H., P. Hamill and R.E. Hancock. 2006. Peptide antimicrobial agents. Clin. Microbiol. Rev. 19(3): 491–511.

Juneja, V.K., H.P. Dwivedi and X.H. Yan. 2012. Novel Natural Food Antimicrobials. Annual Review of Food Science and Technology 3(3): 381–403.

Kim, J., J. Bang, L.R. Beuchat, H. Kim and J.H. Ryu. 2012. Controlled fermentation of kimchi using naturally occurring antimicrobial agents. Food Microbiol. 32(1): 20–31.

Laridi, R., E. Kheadr, R.-O. Benech, J. Vuillemard, C. Lacroix and I. Fliss. 2003. Liposome encapsulated nisin Z: optimization, stability and release during milk fermentation. International Dairy Journal 13(4): 325–336.

Leroy, F. and L. De Vuyst. 2004. Lactic acid bacteria as functional starter cultures for the food fermentation industry. Trends in Food Science & Technology 15(2): 67–78.

Li, K.-K., S.-W. Yin, X.-Q. Yang, C.-H. Tang and Z.-H. Wei. 2012a. Fabrication and Characterization of novel antimicrobial films derived from thymol-loaded zein–sodium caseinate (SC) Nanoparticles. Journal of Agricultural and Food Chemistry 60(46): 11592–11600.

Li, Y., Q. Xiang, Q. Zhang, Y. Huang and Z. Su. 2012b. Overview on the recent study of antimicrobial peptides: origins, functions, relative mechanisms and application. Peptides 37(2): 207–215.

Liolios, C.C., O. Gortzi, S. Lalas, J. Tsaknis and I. Chinou. 2009. Liposomal incorporation of carvacrol and thymol isolated from the essential oil of *Origanum dictamnus* L. and *in vitro* antimicrobial activity. Food Chemistry 112(1): 77–83.

Liu, W. and J.N. Hansen. 1990. Some chemical and physical properties of nisin, a small-protein antibiotic produced by Lactococcus lactis. Applied and Environmental Microbiology 56(8): 2551–2558.

Mainville, I., D. Montpetit, N. Durand and E.R. Farnworth. 2001. Deactivating the bacteria and yeast in kefir using heat treatment, irradiation and high pressure. International Dairy Journal 11(1): 45–49.

McClements, D.J., E.A. Decker, Y. Park and J. Weiss. 2009. Structural design principles for delivery of bioactive components in nutraceuticals and functional foods. Crit. Rev. Food Sci. Nutr. 49(6): 577–606.

Mills, S., C. Stanton, C. Hill and R.P. Ross. 2011. New developments and applications of bacteriocins and peptides in foods. Annu. Rev. Food Sci. Technol. 2: 299–329.

Muller, R.H., K. Mader and S. Gohla. 2000. Solid lipid nanoparticles (SLN) for controlled drug delivery—a review of the state of the art. Eur. J. Pharm. Biopharm. 50(1): 161–177.

Nel, A., T. Xia, L. Madler and N. Li. 2006. Toxic potential of materials at the nanolevel. Science 311(5761): 622–627.

Nishie, M., J.I. Nagao and K. Sonomoto. 2012. Antibacterial peptides "Bacteriocins": An overview of their diverse characteristics and applications. Biocontrol Science 17(1): 1–16.

O'Sullivan, L., R.P. Ross and C. Hill. 2002. Potential of bacteriocin-producing lactic acid bacteria for improvements in food safety and quality. Biochimie 84(5–6): 593–604.

Padmavathy, N and R. Vijayaraghavan. 2008. Enhanced bioactivity of ZnO nanoparticles—an antimicrobial study. Science and Technology of Advanced Materials 9(3).

Papetti, A. 2012. Isolation and characterization of antimicrobial food components. Current Opinion in Biotechnology 23(2): 168–173.

Parveen, S., R. Misra and S.K. Sahoo. 2012. Nanoparticles: a boon to drug delivery, therapeutics, diagnostics and imaging. Nanomedicine 8(2): 147–166.

Petruccelli, S. and M.C. Anon. 1995. Soy protein isolate components and their interactions. Journal of Agricultural and Food Chemistry 43(7): 1762–1767.

Phiriyawirut, M. and T. Phaechamud. 2012. Gallic acid-loaded cellulose acetate electrospun nanofibers: thermal properties, mechanical properties, and drug release behavior. Open J. Polymer. Chem. 2: 21–29.

Pillai, O. and R. Panchagnula. 2001. Polymers in drug delivery. Current Opinion in Chemical Biology 5(4): 447–451.

Pintado, C.M., M.A. Ferreira and I. Sousa. 2010. Control of pathogenic and spoilage microorganisms from cheese surface by whey protein films containing malic acid, nisin and natamycin. Food Control 21(3): 240–246.

Rai, M., A. Yadav and A. Gade. 2009. Silver nanoparticles as a new generation of antimicrobials. Biotechnology Advances 27(1): 76–83.

Ranade, V.V. 1990a. Drug delivery systems: 3A. Role of polymers in drug delivery. J. Clin. Pharmacol. 30(1): 10–23.

Ranade, V.V. 1990b. Drug delivery systems: 3B. Role of polymers in drug delivery. J. Clin. Pharmacol. 30(2): 107–120.

Rennie, R.P. 2012. Current and future challneges in the development of antimicrobial agents. "Antibiotic Resistance, Handbook of Experimental Pharmacology" A.R.M. Coates (ed.). Berlin Heidelberg: Springer-Verlag.

Ross, R.P., S. Morgan and C. Hill. 2002. Preservation and fermentation: past, present and future. International Journal of Food Microbiology 79(1–2): 3–16.

Sadaka, F., C. Nguimjeu, C. Brachais, I. Vroman, I. Tighzert and J. Couvercelle. 2013. Review on antimicrobial packaging containing essential oils and their active biomolecules. Innovative Food Science & Emerging Technologies DOI 10.1016/j.ifset.2013.01.004.

San Martin, M.F., F.M. Harte, H. Lelieveld, G.V. Barbosa-Cánovas and B.G. Swanson. 2001. Inactivation effect of an 18-T pulsed magnetic field combined with other technologies on *Escherichia coli*. Innovative Food Science & Emerging Technologies 2(4):273–277.

Scallan, E., R.M. Hoekstra, F.J. Angulo, R.V. Tauxe, M.A. Widdowson, S.L. Roy, J.L. Jones and P.M. Griffin. 2011. Foodborne illness acquired in the United States—major pathogens. Emerging Infectious Diseases 17(1): 7–15.

Schmaljohann, D. 2006. Thermo- and pH-responsive polymers in drug delivery. Adv. Drug Deliv. Rev. 58(15): 1655–1670.

Shah, B., P.M. Davidson and Q. Zhong. 2012. Nanocapsular Dispersion of Thymol for Enhanced dispersibility and increased antimicrobial effectiveness against *Escherichia coli* O157: H7 and Listeria monocytogenes in Model Food Systems. Applied and Environmental Microbiology 78(23): 8448–8453.

Singh, A.K. and A. Ramesh. 2008. Succession of dominant and antagonistic lactic acid bacteria in fermented cucumber: insights from a PCR-based approach. Food Microbiol. 25(2): 278–287.

Singh, S. and H.S. Nalwa. 2007. Nanotechnology and health safety—toxicity and risk assessments of nanostructured materials on human health. J. Nanosci. Nanotechnol. 7(9): 3048–3070.

Soppimath, K.S., T.M. Aminabhavi, A.R. Kulkarni and W.E. Rudzinski. 2001. Biodegradable polymeric nanoparticles as drug delivery devices. Journal of Controlled Release 70(1–2): 1–20.

Svenson, S. and D.A. Tomalia. 2005. Dendrimers in biomedical applications—reflections on the field. Adv. Drug Deliv. Rev. 57(15): 2106–2129.

Tiwari, B.K., V.P. Valdramidis, C.P. O'Donnell, K. Muthukumarappan, P. Bourke and P.J. Cullen. 2009. Application of natural antimicrobials for food preservation. J. Agric. Food Chem. 57(14): 5987–6000.

Topisirovic, L., M. Kojic, D. Fira, N. Golic, I. Strahinic and J. Lozo. 2006. Potential of lactic acid bacteria isolated from specific natural niches in food production and preservation. International Journal of Food Microbiology 112(3): 230–235.

Torchilin, V.P. 2005. Recent advances with liposomes as pharmaceutical carriers. Nature Reviews Drug Discovery 4(2): 145–160.

Turgeon, S.L. and L.E. Rioux. 2011. Food matrix impact on macronutrients nutritional properties. Food Hydrocolloid 25(8): 1915–1924.

Wang, Y.M., G.J. Du, H. Liu, D. Liu, S.B. Qin, N. Wang, C.G. Hu, X.T. Tao, J. Jiao, J.Y. Wang et al. 2008. Nanostructured sheets of Ti-O nanobelts for gas sensing and antibacterial applications. Advanced Functional Materials 18(7): 1131–1137.

Weiss, J.G.S., P.M. Davidson and D.J. McClements. 2009. Nanostructured encapsulation systems: food antimicrobials. *In*: B.C. Gustavo, M. Alan, L. David, S. Walter, B. Ken and C. Paul (eds.). Global Issues in Food Science and Technology. Academic Press, San Diego pp. 425–479.

Were, L.M., B. Bruce, P.M. Davidson and J. Weiss. 2004. Encapsulation of nisin and lysozyme in liposomes enhances efficacy against *Listeria monocytogenes*. J. Food Prot. 67(5): 922–927.

Whistler, R.L. and J. BeMiller. 1993. Polysaccharides and their derivatives. Industrial gums.

WHO. 2002. World health report 2002: Reducing risks, promoting healthy life; Geneva, 2002.

WHO. 2007. Food Safety and foodborne illness; Geneva.

Xiao, D., P.M. Davidson and Q. Zhong. 2011. Spray-dried zein capsules with coencapsulated nisin and thymol as antimicrobial delivery system for enhanced antilisterial properties. J. Agric. Food Chem. 59(13): 7393–7404.

CHAPTER 9

Application of Edible Films and Coatings on Food

Anastasia E. Kapetanakou, Stavros G. Manios and *Panagiotis N. Skandamis**

Introduction

Packaging as the final stage of food production chain consists one of the main factors for ensuring quality and safety of the final product. Given that foods are dynamic systems, the improvement of classic packaging techniques (e.g., modified atmospheres) or developing of new ones is a continuous challenge for the food industry. In the last decade, among the most challenging packaging techniques for research, was active packaging. Specifically, active packaging is a system that involves interactions between food and packaging atmosphere, packaging material, or certain (bio-) active package components, in order to maintain or increase the nutritional, quality, and safety attributes of packaged food (Coma 2008). Oxygen scavengers, carbon dioxide emitters/absorbers, moisture and ethylene absorbers, flavor releasing/absorbing systems, and antimicrobial agents are among the most distinctive types of active packaging. Antimicrobial agents may either be released through evaporation in the headspace (i.e., essential oils) or migrate into the food through diffusion and partitioning (i.e., nisin, organic acids). Migration could take place by adding directly the antimicrobial compound onto the surface of the food product or by incorporating it into edible carrier materials such as films or coatings (Fig. 1).

Laboratory of Food Quality Control & Hygiene, Department of Food Science & Technology, Agricultural University of Athens, Iera Odos 75, 118 55, Athens Greece.
Email: pskan@aua.gr
* Corresponding author

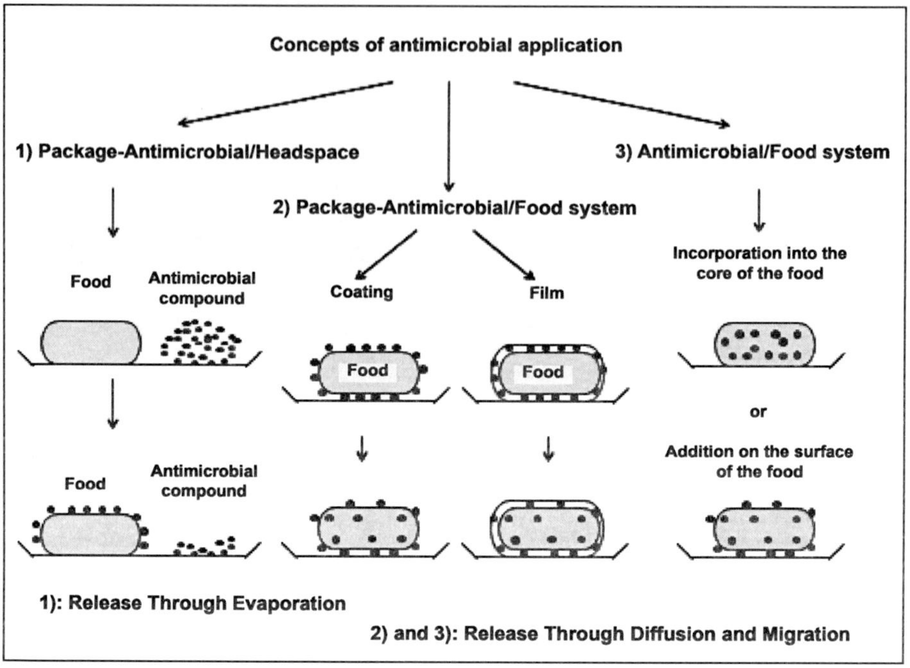

Figure 1. Presentation of the potential antimicrobial packaging concepts and their mode of action (based on Coma 2008). *Blue points correspond to antimicrobial compound.

Color image of this figure appears in the color plate section at the end of the book.

Edible films and coatings have taken such an attention, nowadays, due to their ability to provide a barrier to moisture, oxygen (O_2), carbon dioxide (CO_2), and lipids between the food and the surrounding atmosphere (Bourtoom and Chinnan 2008). In contrast to conventional synthetic packaging material (i.e., plastic), the application of edible films and coatings reduces packaging waste, is environmentally friendly and low-priced (Suppakul et al. 2003, Han and Gennadios 2005). They can also improve the mechanical handling properties or structural integrity of food product (Aydinli et al. 2004). To summarize, edible films and coatings constitute an alternative, environmental friendly and cost-effective technology that may enhance food quality (i.e., prevention of physical damage, enhancement of color, and total appearance) and safety (i.e., inhibition of microbial growth) (Campos et al. 2011).

To differentiate films from coatings, films are defined as a dried, thin layer of biopolymer separately added on food product, while coatings involve the formation of films directly on the surface of the food product (Krochta 2002). However, both are characterized as an extensively interacting polymer network that possesses a three-dimensional gel-type structure. Although coatings are applied directly on the surface of foods and their removal is possible, they are regarded as part of the final product (Han 2002). Given the above, edible coatings should affect the food product by a way that positively enhances the sensory or mechanical properties of the coated food such as color, odor, taste, flavor, and texture (Valencia-Chamorro et al. 2011). On the contrary, edible films can be applied as covers, wraps, or separation layers on food products, e.g., ham slices, beef patties, thereby being easily removed from the food product applied, providing the selection to the consumer not to consume them (Oussalah et al. 2007, Emiroğlu et al. 2010).

Rojas-Graü et al. (2009) summarized the regulatory status for edible films and coatings and addressed some points that should be taken into account, regarding their commercialization and the acceptability by the consumers. Different regulations exist for food additives between countries, i.e., in contrast to European directions, in the USA organic acids and their salts are considered as GRAS (Generally Recognized As Safe). The appropriate labeling is another concern, since most of biopolymers used for the production of edible films and coatings are classified as food allergens such as peanut protein, fish gelatin, and milk proteins (i.e., lactose intolerance) (Rojas-Graü et al. 2009). European directions and USA regulations classify edible films and coatings as food products, food ingredients, food additives, food contact substances, or food packaging materials (EU Reg. No 1935/2004, FDA 2006, EU Reg. No 450/2009). However, since they constitute a part of the final food product, they should serve all regulations required for food ingredients (Guilbert and Gontard 1995). Moreover, all biopolymers and any functional additive in the film or coating should be food-grade non-toxic materials, and all process facilities should meet high standards of hygiene (Han 2002).

Types of Edible Films and Coatings

Edible films and coatings are classified into: (1) hydrocolloids, (2) lipids, and (3) composites of the aforementioned components. Hydrocolloids are subcategorized as polysaccharides and proteins. Lipids include waxes, oils, and resins, while composites contain both hydrocolloids and lipids components (Baldwin 1999) (Fig. 2). Moreover, the selection of biopolymer and application type (coating or film) is strongly dependent on its desired function and applied food.

Each biopolymer has a different preparation procedure; however in all cases plasticizers are required for polysaccharides or proteins, while emulsifiers on lipids. Plasticizers are low molecular weight agents that are incorporated into biopolymer solutions to decrease the intermolecular forces between polymer chains, which results in greater coating or film

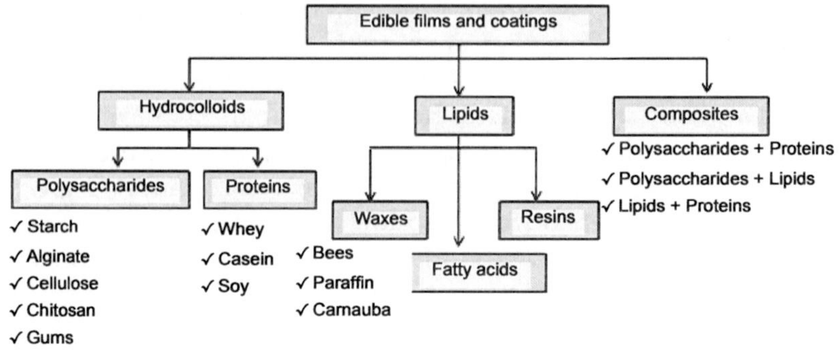

Figure 2. Classification of edible films and coatings applied in food according to their components.

flexibility, elongation, and toughness to prevent cracking (Barreto et al. 2003). However, at high concentrations, they may cause undesirable changes in barrier (e.g., increased gas permeability) or mechanical properties (e.g., decreased tensile strength) (Han and Gennadios 2005, Mali et al. 2006). The most commonly used plasticizers are glycerol, sucrose, and sorbitol. Emulsifiers are surface-active agents with amphiphilic nature that interact at the water-lipid interface reducing surface tension of water-lipid or water-air interfaces improving emulsion stability (Han and Gennadios 2005). The most commonly used emulsifiers include fatty acids, ethylene glycol or glycerol monostearate, and esters of fatty acids.

Hydrocolloids

Polysaccharides. Edible films and coatings produced by polysaccharides are transparent, cohesive, and homogeneous with adequate mechanical properties (Bourtoom and Chinnan 2008, Bourtoom 2009). To overcome their main disadvantage, their hydrophilic character (high water vapor permeability), alternative ways have been studied, i.e., using blend of biopolymers or addition of hydrophobic materials (Garcia et al. 2000, Anker et al. 2001). Regarding gas permeability, they may be selectively permeable to O_2 and CO_2 (Kester and Fennema 1986). Polysaccharide biopolymers include starch, cellulose, alginates, carrageenans, pectin, chitosan, agar, and gums (Han and Gennadios 2005).

Alginates

Alginates are being extracted from brown seaweeds of the *Phaephyceae* class. These polysaccharides are salts of alginic acid, a linear co-polymer of D-mannuronic and L-guluronic acid monomers (Sanderson 1981). Alginates react irreversibly and instantaneous with polyvalent metal cations (Ca^{+2} and

Ca^{+3}) and produce water insoluble polymers, which are impervious to oils and fats and have high water vapor permeability (Rhim 2004).

Different methods have been proposed for obtaining casted alginate types (coatings, films or beads). The first step consists of the preparation of alginate solution ranging from 0.5–3 per cent (w/v) concentration. The most commonly used plasticizer is glycerol (2–5 per cent w/v of alginate). To cast alginate coatings, food products were immersed into alginate solution. Subsequent immersion or spraying by $CaCl_2$ solution should take place (Pranoto et al. 2005a) (Fig. 3a). On the contrary, films are produced by drying a thin layer of alginate solution (for 20–24 hr at 25–40°C) and by adding $CaCl_2$ (2–20 per cent w/v) for 1–20 min (Pranoto et al. 2005a, Oussalah et al. 2006, 2007, Millette et al. 2007, Jiang et al. 2011a) (Fig. 3b).

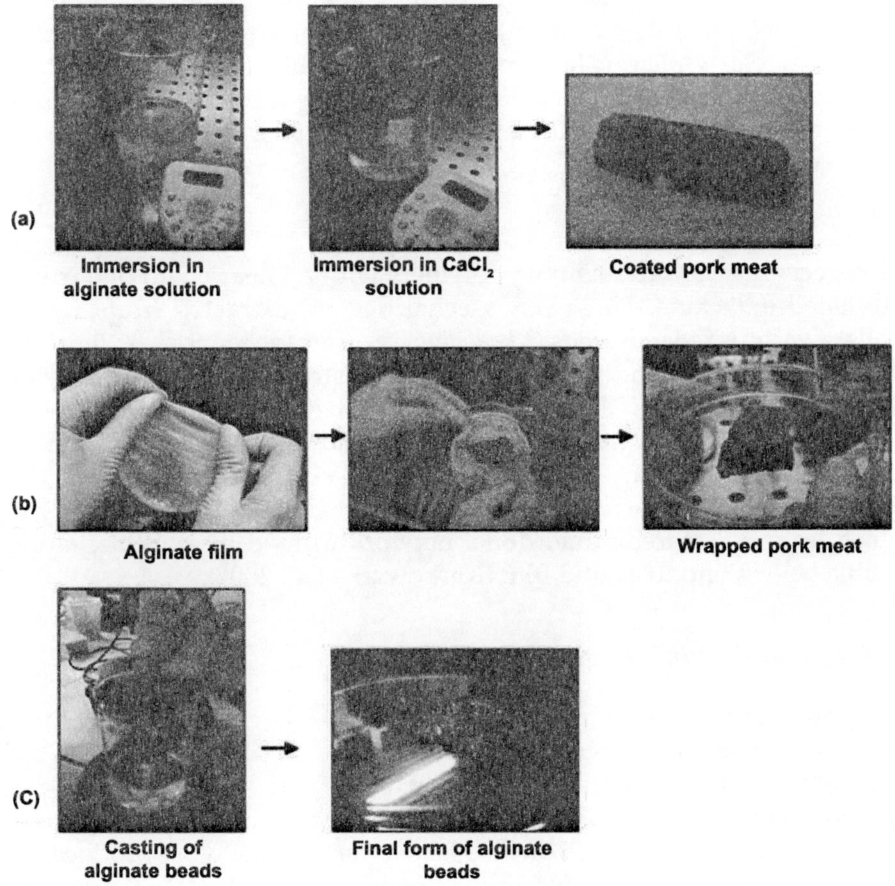

Figure 3. Lab process for casting and applying alginate based (a) coatings, (b) films, and (c) beads on pork meat.

Color image of this figure appears in the color plate section at the end of the book.

Drying process should be well-controlled, since this stage will determine film thickness and structural characteristics (Campos et al. 2011). Another application is adding alginate-beads inside the foodstuff, i.e., yogurt or placing them into the packaging without coming in contact with the product, thereby releasing activity of the bioactive compound through evaporation (Millette et al. 2007, Chan 2011). Beads are produced by drops of alginate solution in $CaCl_2$ using a 0.8–0.9 mm diameter syringe (Fig. 3c) (Millette et al. 2007, Nualkaekul et al. 2012).

Agar

Agar is a mixture of agarose and agaropectin which is derived from a variety of red seaweeds. It consists of a mixture of agarose (gelling fraction) and agaropectin (non-gelling fraction). It has the ability to form strong, thermal-reversible gels, while it is known for its hydrophilicity. Agar gel melts on heating (85°C) and resets on cooling (31–40°C). Agar coatings and films are clear and strong, but is insoluble in water under ambient conditions.

Carrageenans

Carrageenans are water-soluble polymers with a linear chain of partially sulphated galactans. These polysaccharides are extracted from the cell walls of various red seaweeds (*Rhodophyceae*). Carrageenans are extensively used in food, dairy, and pharmaceutical industry as gelling, emulsifying, and stabilizing agents. Their melting point is between 50 and 70°C, while gelatinization temperature is between 30 and 50°C, depending on the solution concentration. The main types are ι-, κ-, and λ-carrageenan (Cha and Chinnan 2004). Carrageenan film formation includes moderate drying which leads to a three-dimensional network formed by polysaccharide-double helices and to a solid film (Karbowiak et al. 2007).

Cellulose and cellulose derivatives

Cellulose is composed of linear chains of (1→4)-β-D-glucopyranosyl units and constitutes the structural polysaccharide of plants (Whistler and Daniel 1985). Native cellulose is a crystalline, cold water-insoluble, and high molecular weight polymer. Chemical substitution of some hydroxyl groups at positions 2, 3, and 6 on the glucosyl-units of cellulose produce two types of derivatives: (1) ionic (carboxymethylcellulose), and (2) non-ionic cellulose ethers (methylcellulose, hydroxypropyl cellulose, hydroxypropyl methylcellulose). Coatings and films produced by cellulose derivatives are tough, flexible, transparent, odorless, tasteless, resistant to fats and oils and highly biodegradable, while they are efficient O_2, CO_2, and aroma barriers, but with moderate resistance to water vapor (Villalobos et al. 2005, Vargas et al. 2008).

Cellulose and cellulose derivative solutions (1–6% w/v) are prepared by dissolution in distilled water with or without ethanol. The dispersion should take place under heating at temperatures close to 65–85°C for 10 min to 2 hr. The casting of coatings may be carried out either by immersing the food in biopolymer solution and drying under certain conditions (i.e., at 25°C for 2–3 hr), while the production of films takes place by drying a thin layer of solution at 20–35°C for 24–48 hr and applying it on the surface of food product (Imran et al. 2010, Sánchez-González et al. 2011, Sayanjali et al. 2011).

Chitosan

Chitosan is a natural carbohydrate polymer derived by deacetylation of chitin [poly-β-(1→4)-N-acetyl-D-glucos- amine] which is a major component of the shells of crustacean, i.e., crab, shrimp, and crawfish (Aider 2010). Films from aqueous chitosan are stable, clear, tough, flexible and good O_2 barriers (Kaplan et al. 1993). In spite of the selectivity of chitosan-based coatings and films on gas permeability, their main disadvantage is their high water vapor permeability. Chitosan is considered as a promising antibacterial and antifungal biopolymer due to its ability to be applied alone without the need to harbour any other antimicrobial compound (Zivanovic et al. 2005, Ziani et al. 2009). Recently, chitosan has been approved for use as food contact material in Europe (EU Reg. No 749/2012). Chitosan-based coatings and films have high antimicrobial activity, when the average molecular weight of chitosan is less than 10 kDa and/or at low pH values. Moreover, chitosan solubility and charge density which allows the appearance of the antimicrobial effect is directly affected by the degree of deacetylation (Aider 2010). Preparation procedure is described diagrammatically in Fig. 4.

Gums

Gums are categorized as: (1) exudate gums (natural exudates of trees—gum arabic), (2) extractive or seed gums (locust bean and guar gums), and (3) microbial fermentation gums (xanthan and gellan gums) (Williams and Phillips 2000). Guar gum is used as stabilizer and enhances water binding and viscosity building. Xanthan gum is easily dispersed in water, while the least viscous is guar gum. Gums have been also studied in blends with lipids (excellent water vapor barrier properties) and starches (Soares et al. 2005, Flores et al. 2010).

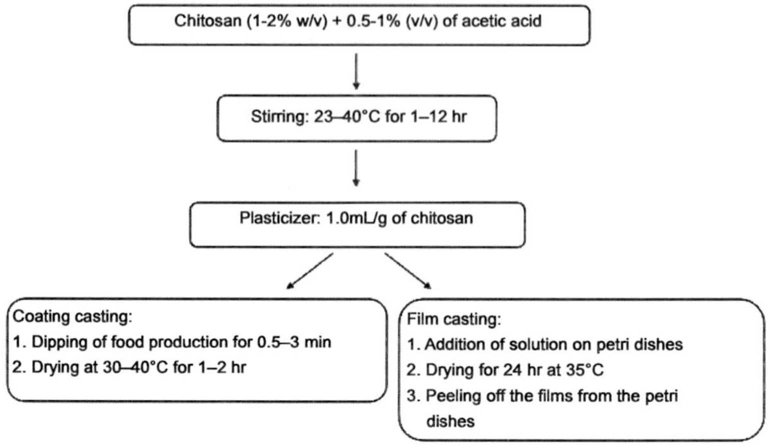

Figure 4. Preparation of chitosan solution and casting of films or coatings (Jeon et al. 2002, Sánchez-González et al. 2010, Jiang et al. 2011b, Moreira et al. 2011a,b).

Pectins

Pectins are biopolymers mainly composed of (1→4) α-D-galactopyranosyluronic acid units naturally esterified with methanol. It is a purified polysaccharide product obtained by aqueous extraction of some edible plant material, i.e., citrus fruits or apples. Pectin-based coatings have been studied for their ability to retard lipid migration and moisture loss, and to improve appearance during food handling and preservation (Maftoonazad and Ramaswamy 2008, Moalemiyan et al. 2011).

It is recommended that pectin (2.5–3% w/v) should be rehydrated in distilled water for 12–18 hr at 20°C (Maftoonazad and Ramaswamym 2008, Moalemiyan et al. 2011, Ravishankar et al. 2012). Sorbitol has been studied as a plasticizer at concentrations 45–50% (w/v). Heating is usually taking place at 80°C for 1 hr. Coatings may be casted by immersing the food in the solution (1 min at 20°C) and dried under cold-air draft for 3–4 hr (Maftoonazad and Ramaswamym 2008, Moalemiyan et al. 2011).

Starch

Starch is a polymeric carbohydrate composed of anhydroglucose units, which contains two types of glucose polymers: amylose, a linear chain molecule and amylopectin, a branched polymer of glucose (Rodriguez et al. 2006). Specifically, amylose is responsible for the excellent film-forming capacity of starches, rendering to strong, isotropic, odorless, tasteless, and colorless films (Campos et al. 2011, Jiménez et al. 2012). The main advantages of those edible films and coatings are low cost, biodegradability and good mechanical properties. They are hydrophilic, while they can exhibit selective permeability to O_2 and CO_2 (Pagella et al. 2002). Moreover, starch-based films and coatings constitute excellent lipid barriers, retarding lipid oxidation, while they have high water solubility and brittleness (Campos et al. 2011).

According to literature, the concentration of starch (derived from, i.e., sweet potato, rice, tapioca) solutions applied ranges from 1–5 per cent (w/v), while propylene glycol, sucrose, sorbitol, and glycerol as plasticizers have been tested at concentrations ranging from 15 to 50 per cent (w/v) (Bourtoom and Chinnan 2008, Chen et al. 2009, Ehivet et al. 2011). Starch suspension should take place under continuous agitation and heating at 40–100°C for 15–30 min to allow full gelatinization. Factors such as heating temperature (close to gelatinization temperature of starch), gelatinization time (increased) and drying rate (low) generally result in a more solid network and smaller water vapor permeability (Flores et al. 2007). The coatings may be produced either by immersing in or spraying the food with starch solution, while films are being casted by evaporation of water from a thin layer of starch solution at 25°C for *ca.* 72 hr (Chen et al. 2009, Ehivet et al. 2011, Garcia et al. 2012, Mehyar et al. 2014).

Proteins. Protein-based films and coatings have an impressive gas barrier and mechanical properties; however their poor water vapor resistance constitutes their main drawback (Campos et al. 2011). The limited resistance of protein films to water vapor transmission is attributed to the inherent hydrophilicity of proteins and the addition of hydrophilic plasticizers. Improvement of protein films could be attained by modifying protein properties and combining them with hydrophobic materials or biopolymers, e.g., waxes (Bourtoom 2009). Various types of proteins have been used as edible films and coatings, such as gelatin, casein, whey protein, and soy protein.

Gelatin

Gelatin is obtained by controlled hydrolysis from the fibrous insoluble protein, collagen, which is widely found as the major constituent of skin, bones and connective tissue (Bourtoom and Chinnan 2008). At approximately 35°C (melting point), gelatin forms an aqueous solution which becomes a physical, thermo-reversible gel on cooling, depending on the concentration. According to Guilbert et al. (1996), gelatin-based films and coatings should be prepared by 20–30 per cent gelatin-water solutions and added 10–30 per cent plasticizer (glycerin or sorbitol).

Milk protein

Casein and whey proteins represent the main milk protein fractions, *ca.* 80 per cent and 20 per cent (w/w), respectively. Campos et al. (2011) summarized the advantages of milk-protein-based films or coatings in their high nutritional added value, good taste, excellent gas barrier ability and good mechanical properties.

Whey proteins are by-products of cheese-making process which remain in the milk serum after coagulation of the caseins. Whey protein isolates (WPI), the purest form of whey proteins, can produce transparent films or coatings with good mechanical strength and are excellent O_2, lipid, and aroma barriers; however they have poor moisture barrier and tensile properties (Gounga et al. 2008, Dragich and Krochta 2010, Ramos et al. 2012). Preparation procedure is described in Fig. 5.

Caseins are quite soluble in water and the range of studied concentrations is 2.5–7.5 per cent (w/v) (Chen 2002, Schou et al. 2005, Kristo et al. 2008, Caprioli et al. 2009, Gialamas et al. 2010). Mechanism of casein-based coatings and films formation takes place by hydrogen bonds, electrostatic interactions and most probably hydrophobic forces (McHugh and Krochta 1994). Sodium caseinate-based coatings and films (by acid precipitation of casein) are also extensively used and present thermoplastic and film-forming properties (Longares et al. 2005). Casein and caseinate-based coatings and films are transparent, flavorless, flexible and highly soluble and have good mechanical properties (Schou et al. 2005).

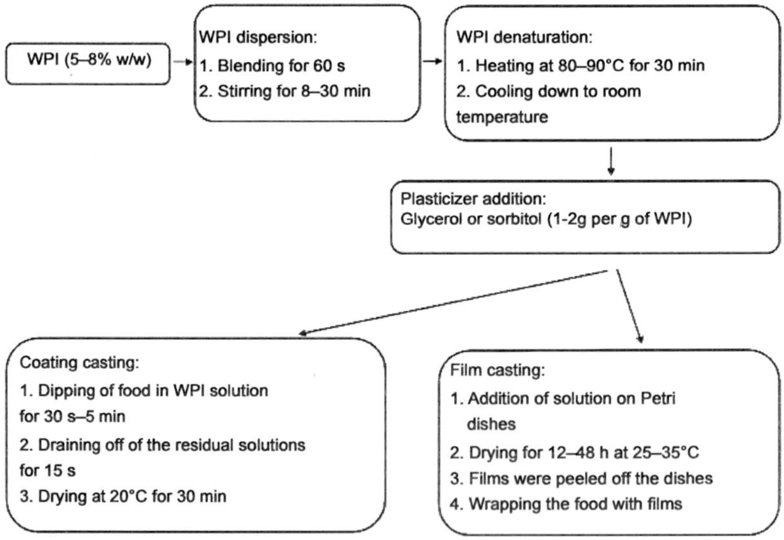

Figure 5. Preparation of whey protein solution and casting of edible films and coatings in foods (Seydim and Sarikus 2006, Reinoso et al. 2008, Gounga et al. 2008, Rodriguez-Turienzo et al. 2011).

Soy protein isolates

Soy protein-based films can be produced in two ways: (1) surface film formation on heated soymilk or (2) film formation from solutions of soy protein isolates (SPI). Although most soy proteins are insoluble in water, they present high solubility in dilute neutral salt solutions (Bourtoom and Chinnan 2008).

To form films from both soymilk and SPI, two steps are required: (1) heating of the biopolymer solution which leads to disruption of protein structure, cleaving native disulfide bonds and exposing sulfhydryl groups and hydrophobic groups, and (2) drying of certain volumes of the biopolymer solution to form new disulfide, hydrophobic and hydrogen bonds (Gennadios et al. 1994, Emiroğlu et al. 2010).

Lipids. Lipids consist of wax, oils and resins. For several decades, waxes (carnauba wax, beeswax and paraffin wax) and oils (mineral and vegetable oil) have been commercially used as protective coatings for fresh fruits and vegetables (Baldwin 1994). Lipids-based coatings and films are good moisture barriers (low polarity), while their main disadvantages include the formation of cracks, poor adherence to the food, lack of homogeneity, and waxy taste. Therefore it is recommended to be used as blends with proteins or polysaccharides to provide coatings or films with mechanical strength and better water barrier characteristics (Karbowiak et al. 2007, Maftoonazad et al. 2007, Bourtoom 2009).

Waxes

Waxes belong to the non-polar lipid class, while they are highly hydrophobic. Carnauba, beeswax, paraffin, and candelilla are among the most commonly tested wax coatings in fresh fruits and vegetables. Water vapor and oxygen permeability of the aforementioned waxes are mainly dependent on their chemical composition and their crystal type. Waxing of apples is a method that is being used by several retail markets, since may improve their appearance (i.e., look more glossy) and retard moisture loss. Candelilla wax is recognized by the FDA as a GRAS substance for application on foods (Saucedo-Pompa et al. 2009).

Waxes are applied on the whole or fresh-cut fruits and vegetables (except for leafy greens) by dipping or spraying, while casting of coating is carried out by drying (16–18 hr at 21°C) (Kenney and Beuchat 2002, Saucedo-Pompa et al. 2009, Chiumarelli and Hubinger 2012). Both applying methods should be very well standardized, since they may cause variations in coating thickness and subsequently in their gas barrier and mechanical properties.

Resins

Resins are soluble in alcohols and in alkaline solutions. Resins and their derivatives are widely used as coatings of citrus, mandarins and other fruits (Hagenmaier 2000, 2002). Fruits with resin-based coatings indicate lower internal O_2 and higher CO_2 than waxed fruits; however they have a similar glossy appearance. Some types of resins, i.e., shellac are not classified as GRAS substance, while others, such as oleoresins may be applied safely as food additives (Bourtoom and Chinnan 2008).

Composites of biopolymers. Edible films and coatings may also consist of a blend of polysaccharides, protein, and/or lipids. The possible combinations of biopolymers composites are: (1) proteins and polysaccharides, (2) proteins and lipids, and (3) polysaccharides and lipids. The new generation of blended biopolymers includes polysaccharides, i.e., alginates or pectin, mixed with purees of fruits (Rojas-Graü et al. 2007, Sothornvit and Pitak 2007). Coatings or films composed of one biopolymer either act as good gas and/or aroma barriers or have good mechanical properties, while coatings and films produced by composites of biopolymers usually have the best properties (mechanical or barrier) from each film or coating type.

Campos et al. (2011) suggest that the production of biopolymers composites may be achieved *via* blending, extruding, or laminating, with blending being the easiest and most effective way to achieve homogenous solutions. Composite of biopolymers can be produced either as bi-layer, or stable emulsions. In bi-layer composite biopolymers, the lipid forms a second layer over the polysaccharide or protein layer, while in emulsion the lipid is dispersed and immobilized in the supporting matrix of protein or polysaccharide (Pérez-Gago and Krochta 2005).

Antimicrobial Edible Films and Coatings

Edible films and coatings have been used as carriers of antioxidant, flavoring, and coloring agents; however, in the last decade, the incorporation of antimicrobial substances has also been investigated as a mean to develop an improved active form of films and coatings (Coma et al. 2002). Such techniques may control microbial contamination by reducing the growth rate and maximum growth population and/ or extending the lag-phase of the target microorganism or even by inactivating microorganisms. Migration and effectiveness of each antimicrobial can be affected by film or coating-forming conditions and by film composition.

Potential antimicrobial compounds for the above purpose are divided into: (1) natural and (2) chemical. Antimicrobial selection should be decided considering the sensitivity of the targeted microorganism and any possible interactions among the antimicrobial compound, the biopolymer, and the others food components present (Campos et al. 2011). The antimicrobial effect of edible films and coatings is differential and many times is limited in food products compared to laboratory media; therefore the review of *in vitro* and *in situ* studies is necessary.

In vitro studies

Natural antimicrobials. A common practice in the area of antimicrobial edible packaging is to select antimicrobials from natural sources, using GRAS compounds. Natural antimicrobials are easily available, biodegradable and have limited side effects and/or toxicity (Campos et al. 2011). Essential oils, nisin, and lactoperoxidase system are the most commonly applied as described below. However, in many cases chemical preservatives are also used taking into account the legislation levels.

Essential oils

Essential oils (EOs) are aromatic oily liquids obtained from individual or integrated plant material by applying steam distillation. Given their high concentration of terpenes, terpenoids, and phenolic compounds, EOs may have antioxidant and antimicrobial effect. Hydrophobicity of EOs, gives them the ability to pass through cell membranes and even enter mitochondria, disturbing the internal structures and rendering the membranes more permeable (Burt 2004).

The commercial application of EOs in foods is yet limited due to the impact of their flavor on the sensory properties of foods, the variability of their composition, and their moderate activity in foods at sensory acceptable levels due to interactions with food components (Gutierrez et al. 2008). Considering these reasons, EOs have been extensively tested, *in vitro*, for their antimicrobial activity against spoilage and pathogenic microorganisms incorporated in edible films and coatings. Specifically, chitosan or alginate films with garlic EO showed strong antimicrobial activity against *Escherichia coli*, *Staphylococcus aureus*, *Salmonella* Typhimurium, *Listeria monocytogenes*, and *Bacillus cereus* according to agar diffusion method (Pranoto et al. 2005a,b). Seydim and Sarikus (2006) managed to control similar groups of pathogenic bacteria and spoilage microorganisms such as *Lactobacillus plantarum* by using whey protein-based films with garlic, rosemary, and oregano EOs (1.0–4.0% w/v). A more recent study reported the strong antimicrobial efficacy of chitosan, carboxymethyl-cellulose or casein-based films bearing oleoresins of onion, capsicum, cranberry, garlic, oregano EOs and oregano EO + carvacrol against *L. monocytogenes* growth (estimation by measuring inhibitory zones) (Ponce et al. 2008). Sánchez-González et al. (2011) reported 100% inhibition of *E. coli, L. monocytogenes* and *S. aureus* growth by applying chitosan-based films containing tea tree EO, hydroxypropyl-methylcellulose (HPMC)-based films with tea tree EO and HPMC-based films with bergamot EO, respectively. Chitosan films carrying bergamot EO also showed 2.5 log CFU/g reduction on *Penicillium italicum* compared to controls (Sánchez-González et al. 2010). Films produced by blend of starch and alginate containing lemongrass EO inhibited *E. coli* O157:H7 growth (Maizura et al. 2007).

Nisin

Nisin, a bacteriocin produced by certain strains of *Lactococcus lactis* subsp. *lactis*, is classified as GRAS antimicrobial agent. Nisin shows great thermal stability and increased solubility and activity at low pH values (Kristo et al. 2008). The mode of action of nisin includes pore formation in cell membranes through interaction with lipid II, a precursor of peptidoglucan biosynthesis (Bauer and Dicks 2005). Nisin has been extensively studied for its antimicrobial effects incorporated in edible films and coatings. Specifically, HPMC films with nisin (10^4 IU) showed efficacy against microorganisms in the order: *Listeria* spp.> *Enterococcus* spp.> *Staphylococcus* spp.> *Bacillus* spp. according to agar diffusion method (Imran et al. 2010). Moreover, Basch et al. (2012) reported great inhibitory activity against *Listeria innocua* (5 log CFU/g reduction) compared to controls, by applying 3000 IU/mL nisin in tapioca-starch films, alone or in mixture with HPMC. According to literature, the tested nisin concentrations incorporated on edible films and coatings range from 500 to 18000 IU/mL or g (Millette et al. 2007, Gadang et al. 2008, Martins et al. 2010).

Lactoperoxidase system

Lactoperoxidase system (LPOS) is an enzyme that has shown bactericidal effect on gram-negative bacteria, bacteriostatic effect on gram-positive bacteria and antifungal activity (Min and Krochta 2005, Seifu et al. 2004). LPOS is a hemoprotein present in milk, tears, and saliva and consists of lactoperoxidase, thiocyanate, and hydrogen peroxide. LPOS action is more effective at low pH values and temperatures (Naidu 2003).

Even though LPOS has been tested against *L. monocytogenes, S. aureus, E. coli,* and *Salmonella* Enteritidis, its antimicrobial activity incorporated in edible films and coatings has only recently been investigated (Seifu et al. 2004, Touch et al. 2004). In fact, the inhibitory effect of LPOS (0.1–4% w/v) and its components have been recorded into whey protein films against *L. monocytogenes* (4.2 log CFU/cm^2 in agar and ≤ 3 log CFU/cm^2 in smoked salmon), *Salmonella enterica*, and *E. coli* O157:H7 (4.2 log CFU/cm^2) (Min and Krochta 2005, Min et al. 2005a,b). Yener et al. (2009) produced alginate films 2 per cent (w/v) with 0.2 to 0.6 mg/g LPOS, H_2O_2 (0.2, 0.4, and 0.8 mM) and KSCN (1, 2, and 4 mM) and showed that the decreasing order of studied bacteria resistance to LPOS was: *E. coli*, *L. innocua*, and *Pseudomonas fluorescens*. Antimicrobial effect of LPOS (5% v/v) incorporated into chitosan-based films (0.5, 1 and 1.5% w/w) has also been tested on phytopathogenic bacterial strains and fungi by applying disc diameter and puncture techniques, respectively, indicating in many assays 100% inhibition compared to chitosan film alone (Cissé et al. 2012).

Chitosan

Several studies have reported chitosan's ability for inactivation or inhibition of gram-positive and gram-negative bacteria as well as fungi and yeasts in concentrations ranging from 1–2 per cent (w/v) (Vásconez et al. 2009, Jiang et al. 2011a,b, Avila-Sosa et al. 2012). The mode of antimicrobial action of chitosan has not been fully elucidated. However, Alishahi and Aïder (2012) reviewed the possible mechanisms that have been proposed as follows: (i) interactions between the positively charged chitosan amine groups and the negatively charged microbial cell membranes, leading to the leakage of intracellular constituents; (ii) chitosan may act as a water-binding agent and may inhibit various enzymes by blocking their active centers; (iii) chitosan complexes metal ions, e.g., copper, zinc, ferric may inhibit microbial growth; (iv) formation of an impervious polymeric layer on the surface of the cell, thereby blocking the entry of nutrients into the cell; (v) penetration of mainly low-molecular weight chitosan into the cytosol of the microorganism to bind DNA, resulting in interference with the synthesis of mRNA and proteins (vi) and the adsorption and flocculation of electronegative substances in the cell by chitosan, disturbing the physiological activities of the microorganisms and causing death (Kong et al. 2010).

Chemical antimicrobials. Chemical preservatives include alcohols, ozone, hydrogen peroxide, antibiotics, and weak organic acids, i.e., acetic, benzoic, lactic, citric, malic, tartaric, propionic, fumaric, and sorbic acids and their salts, i.e., sorbates, benzoates, propionates, phosphates. Organic acids may inhibit growth of bacterial and fungal cells, while some, i.e., sorbic acid also reduce the rate of spore germination (Cha and Chinnan 2004). The antimicrobial mechanism of organic acids is mainly based on the penetration of membranes in their undissociate form, followed by the dissociation that is taking place inside the cells and finally the acidification of cytoplasm (Beales 2004). Eswaranandam et al. (2004) studied the antimicrobial activity of citric, lactic, malic, and tartaric acids (0.9, 1.8, and 2.6% w/w) in combination with nisin (205 IU/g of protein) incorporated in soy protein film against *L. monocyteges*, *Salmonella* Gaminara, and *E. coli* O157:H7 and reported that *L. monocytogenes* was inhibited by all acids, *S.* Gaminara by malic and tartaric, while *E. coli* by citric and tartaric. Pyla et al. (2010) reported the antimicrobial activity of corn-starch films with tannic acid against *E. coli* O157:H7 and *L. monocytogenes*.

Potassium sorbate and sodium benzoate are the two organic acids salts that are more widely used as preservative agents. Kristo et al. (2008) showed that sodium caseinate films containing potassium sorbate (10 and 25%-weight basis of film dry matter) were strong inhibitors of *L. monocytogenes* growth, whereas films containing sodium lactate were slightly effective and only at the highest concentrations (40%). Several studies have dealt with films or coatings with sodium lactate, alone or in combination with other antimicrobials in concentrations ranging from 0–2.4 per cent (w/w) and showed inhibition against *L. monocytogenes* by 1.3–4.9 log CFU/g on meat and fish products, at 4 and 22°C (Juck et al. 2010, Jiang et al. 2011a, Ye et al. 2011). Sayanjali et al. (2011) showed that carboxymethyl cellulose films with potassium sorbate inhibited *Aspergillus flavus* and *A. parasiticus* using agar diffusion method. Ehivet et al. (2011) characterized the antimicrobial properties of sweet potato starch-based films containing oregano EO exhibiting greater inhibitory effect against *S.* Enteritidis and *E. coli* O157:H7 than *L. monocytogenes*.

In situ studies

Meat products and poultry. Temperature, oxygen, endogenous enzymes, moisture, light and microorganisms, may influence shelf-life, quality or safety of fresh and processed (Ready-To-Eat, RTE) meat and poultry. The aforementioned factors, alone or in combination, may result in undesirable changes of sensory properties, e.g., colour, odour, texture, and flavor. However, growth of microorganisms is possibly the most important reason for meat spoilage or safety.

The application of edible films and coatings in meat is already known since 16th century, when similar preservation methods were used, i.e., fats were used as coatings for meat pieces (Kester and Fennema 1986). The application of coatings or films on fresh and RTE meat and poultry may prevent moisture loss from tissues during storage, retain juices of fresh meat cuts, inhibit lipid oxidation and discolouration, restrict volatile flavor loss and ensure product safety by inhibiting growth of pathogens (Table 1) (Quintavalla and Vicini 2002).

The technology of using edible films and coatings as carriers of antimicrobials to extend the shelf-life of fresh meat and poultry products

Table 1. Application of antimicrobial edible films and coatings to extend shelf-life and enhance food safety of fresh and processed meat and poultry.

Biopolymer	Antimicrobial	Product	Microorganisms	Indicative effect	Reference
POLYSACCHARIDES					
Coatings of: Starch (15% w/v) Chitosan (2% w/v) Alginate (1% w/v) Pectin (1% w/v)	Sodium lactate (2% w/v) Sodium diacetate (0.25% w/v)	Roasted turkey	*L. monocytogenes*	*L. monocytogenes* populations were *ca.* 2.1–4.0 log CFU/cm^2 lower than controls	Jiang et al. 2011a
Coatings of: Chitosan (0.5% or 1% w/v)	Acetic and lactic acid (0.5% or 1% w/v)	Roast beef	*L. monocytogenes*	Acetic acid/CH coatings reduced *L. monocytogenes* counts by 2.0–3.0 log CFU/g, while lactic acid/CH coatings by 1.0 log CFU/g, compared to controls	Beverly et al. 2008
Films of: Chitosan (2% w/v)	Zataria multiflora Boiss EO (0.5% and 1% w/w) Grape seed extract (1% w/w)	Mortadella-type sausages	TVC LAB *L. monocytogenes*	• TVC and LAB were the most sensitive (0.1–1.1 log CFU/g) and resistant groups (0.1–0.7 log CFU/g) • CH/EOs films reduced *L. monocytogenes* by 1.4 log CFU/g compared to controls	Moradi et al. 2011
Films of: Alginate (3% w/v)	Oregano, winter savory, and cinnamon EOs (1% w/v)	Bologna ham slices	*S.* Typhimurium *L. monocytogenes*	Bologna: • Oregano-based films were the most effective • *L. monocytogenes* was the most sensitive on EOs films (<10 CFU/ cm^2) Ham: • *L. monocytogenes* was highly resistant in wrapped ham • Cinnamon-based films reduced *S.* Typhimurium by 1.9 log CFU/cm^2 after 5-days storage	Oussalah et al. 2007
Films and beads of: Alginate (3% w/v)	Nisin (500 and 1000 IU/mL)	Beef	*S. aureus*	Films with nisin reduced *S. aureus* 0.9–1.9 log CFU/cm^2, while its encapsulation on beads caused a reduction of *ca.* 1.0 log CFU/cm^2 compared to controls after 7 days	Millette et al. 2007

Table 1. contd....

Table 1. contd.

Biopolymer	Antimicrobial	Product	Microorganisms	Indicative effect	Reference
Films of: κ-carrageenan (2% w/w)	Ovotransferrin (25 mg/g) EDTA (5 mM) Potassium sorbate (10 mg/g)	Chicken breast	*E. coli* TVC	Highest reductions were reported by κ-carrageenan + EDTA + Ovotransferrin (2.7 log CFU/g) and carrageenan +EDTA+PS (2.1 log CFU/g) compared to controls at day 7	Seol et al. 2009
PROTEINS					
Films of: Whey protein isolate (5 and 8 % w/w)	Oregano EO (0.5, 1.0, and 1.5 % w/w)	Fresh beef	TVC LAB *Pseudomonas* spp.	Significant inhibition of *Pseudomonas* spp. and TVC by reducing the specific growth rate, while the growth of LAB was completely inhibited at 1.5% EO	Zinoviadou et al. 2009
BLENDS					
Coatings or films of: Chitosan (2% w/v)/ Sodium caseinate (2.5% w/v)	Chitosan (2% w/v)	Salami	TVC Psychrotrophs Yeasts and moulds	Coatings or films with CH and SC/CH showed reductions of 3.3–4.8 log CFU/g, with respect to controls	Moreira et al. 2011a
Coatings of: Tapioca starch (1.35% w/v)/ hsian-tsao leaf gum (0.15% w/v)	Green tea extracts (1% w/v)	Pork slices	*Bacillus cereus*	After 48 hr storage: Coatings with extracts reduced 4.3 and 4.2 log CFU/g of *B. cereus* compared to controls	Chiu and Lai 2010
Films of: Chitosan (2% w/v)/ HPMC (3% w/v)	Nisin (500 IU/cm2) Sodium lactate (0.01 g/cm^2) Sodium diacetate (0.0025 g/cm^2) Potassium sorbate (0.003 g/cm^2) Sodium benzoate (0.001 g/cm^2)	Ham steak	*L. monocytogenes*	On day 10: • At room temperature: CH (7.0 log CFU/cm^2)> >CH+SB>CH+Nisin> >CH+SD>CH+PS>CH+SL • At 4°C: CH (6.5 log CFU/cm^2)> CH+SL (2.0 log CFU/cm^2)	Ye et al. 2008a

SL: Sodium lactate, SD: Sodium diacetate, CH: Chitosan, PS: Potassium sorbate, LAB: Lactic acid bacteria, SC: Sodium caseinate, SB: Sodium benzoate has been widely explored. Oussalah et al. (2004) evaluated the preservation ability of milk-protein films containing 0.1 per cent (w/v) oregano and pimento EOs, alone or in mixture (1:1) on beef muscle slices at 4°C. The oregano-containing films provided the greatest antimicrobial activity against *E. coli* O157:H7 and *Pseudomonas* spp. The results from the study of Ravishankar et al. (2009) showed that the application of apple-pectin films containing carvacrol and cinnamaldehyde (0.5%, 1.5% or 3.0% w/w) may significantly reduce *E. coli* O157:H7 (1.8–6.8 CFU/g) and *S. enterica* (1.6–4.6 log CFU/g) on chicken breasts stored at 23°C and under anaerobic conditions after 72 hr, compared to non-coated samples. The reduction of *E. coli* O157:H7 (0.2–3.0 log CFU/g) and *S. enterica* (0.1–3.3 log CFU/g) at 4°C was lower than at 23°C. The same research group reported that the aforementioned antimicrobial compounds may also reduce *Campylobacter jejuni* populations on chicken breasts during storage at 4 (0.2–6.0 log CFU/g) and 23°C (0.4-1.9 log CFU/g) compared to controls (Mild et al. 2011). Moreover, Emiroğlu et al. (2010) reported that soy protein (5% w/v) films enriched with thyme and oregano EOs (5% v/v), alone or in mixture (1:1) did not significantly affect TVC, lactic acid bacteria and *Staphylococcus* spp. when applied on ground beef patties, whereas reductions in coliform (*ca.* 2.0 log CFU/g) and *Pseudomonas* spp. (*ca.* 1.0 log CFU/g) counts were observed, compared to unwrapped samples.

Extensive research has also been carried out about the antimicrobial effect of edible films and coatings on RTE meat and poultry products. Different concentrations of carvacrol and cinnamaldehyde (0.5%, 1.5% or 3.0% w/w) incorporated in apple-pectin films reduced *L. monocytogenes* on ham slices at 4 and 23°C (up to 2.3 log CFU/g) (Ravishankar et al. 2009). More recently, the antimicrobial effects of carvacrol and cinnamaldehyde (1.5 and 3% w/v), incorporated into apple, carrot, and hibiscus puree-pectin (3% w/v) films against *L. monocytogenes* on contaminated ham and bologna were tested and the results showed that growth reduction by apple-films was greater than by carrot or hibiscus-films (Ravishankar et al. 2012). Given the relatively high prevalence of *L. monocytogenes* in RTE is of great concern, the group of Juck et al. (2010) reported that alginate coatings supplemented with sodium lactate (2.4%)/potassium sorbate (0.3%) delayed the growth of *L. monocytogenes* reaching populations of 4.3 log CFU/g on poached turkey and 6.5 log CFU/g on roasted deli turkey, while populations of controls were 9.9 and 7.9 log CFU/g, respectively. Cagri et al. (2002) found that the whey protein films with 0.5–1.0 per cent (w/v) p-aminobenzoic acid and/or sorbic acid decreased *L. monocytogenes*, *E. coli*, and *S.* Typhimurium populations by 3.4–4.1, 3.1–3.6, and 3.1–4.1 log CFU/cm^2, respectively, on bologna and sausage slices, after 21 d at 4°C. Whey protein coatings incorporated with natural extracts, nisin, and organic acids, alone or mixed proved promising inhibitors, reducing the populations of *S.* Typhimurium by 1.0 log CFU/g,

L. monocytogenes by 2.3 logs CFU/g, and *E. coli* O157:H7 by 4.6 logs CFU/g on turkey frankfurter system compared to controls (Gadang et al. 2008).

Fish products. Intensive research has been performed in the last decade regarding the shelf-life and safety of fish and fish products. Due to their high content of polyunsaturated fatty acids, high water activity, abundant free amino acids, almost neutral pH, and the presence of autolytic enzymes, marine products are considered as highly perishable.

Given the above, there has been an increased interest in the application of edible films and coatings as a means to extend shelf-life and ensure safety of these products (Table 2). Alishahi and Aïder (2012) report that chitosan may be an excellent antimicrobial applied alone or combined with others in sea food products, i.e., herring, cod, cold-smoked salmon and trout, resulting in reductions 1–3 log CFU/g compared to controls (Ojagh et al. 2009, Jiang et al. 2011b). Moreover, Gómez-Estaca et al. (2007) coated cold-smoked sardines with gelatin-based films (4% w/v) enriched with oregano (1.25%) and rosemary extracts (20%) and showed that growth of TVC, on day 16, was retarded 2 log and 1.5 log, respectively, compared to uncoated samples. Lu et al. (2010) tested, in snakehead fish fillets, alginate coatings (20 mg mL^{-1}) with cinnamon EO (10 µL mL^{-1}), EDTA (150 µg mL^{-1}) and nisin (2000 IU mL^{-1}) alone and with their mixes against *Pseudomonas* spp., TVC and psychrotrophic counts during storage at 4°C. The inhibitory effect of those antimicrobials on TVC followed the order: cinnamon, EDTA and nisin or cinnamon (5.5 log CFU/g) > nisin and EDTA (1.5 log CFU/g) compared to controls.

Regarding safety, Song et al. (2012) reported that wrapped salmon with films of barley bran protein/gelatin composite containing grapefruit seed extract caused a reduction of 0.5 logs CFU/g on *E. coli* O157:H7 and *L. monocytogenes* after 15 days of storage at 4°C. Moreover, Ye et al. (2011) showed that prior frozen storage enhances the effect of alginate-based coatings and chitosan-based films incorporated with sodium lactate (1.2 or 2.4% w/w) or sodium diacetate (0.125 or 0.25% w/w) against *L. monocytogenes* on cold-smoked during subsequent refrigerated storage, showing reductions ranging from 0.5–4.5 log CFU/cm^2 compared to uncoated samples.

Dairy products. The complex microenvironments (fat, proteins, salts, etc.) and the variety of dairy products' composition, i.e., cheese, yogurt along with hygienic requirements during handling and storage often reduces their quality and shelf-life; therefore edible films and coatings may be applied to different types of cheeses (soft, semi-hard and cream). Specifically, Cerqueira et al. (2010) coated a commercial semi-hard cheese with a galactomannan coating (0.5% w/v) and observed that water loss and colour changes were decreased compared to controls, while a slight reduction of TVC and yeasts-

Table 2. Application of antimicrobial edible films and coatings to extend shelf-life and enhance food safety of fish products.

Biopolymer	**Antimicrobial**	**Product**	**Microorganisms**	**Indicative effect**	**Reference**
POLYSACCHARIDES					
Coatings and films of: Chitosan (2% w/v)	Sodium lactate Sodium diacetate Potassium sorbate	Cold-smoked salmon	*L. monocytogenes*	1) CH-films with 1.2% SL/0.25% SD or 2.4% SL ≥ 1.3 log CFU/ cm^2 2) CH-coatings with 1.2% SL/0.25% SD or 0.15% PS/ O.125% SD ≥ 2.8 log CFU/cm^2	Jiang et al. 2011b
Coatings of: Chitosan (2% w/v)	Chitosan (2% w/v)	Silver carp	TVC	2.0 log CFU/g reduction compared to controls	Fan et al. 2009
Coatings of: Calcium alginate (1% w/v)	Oyster lysozyme Hen lysozyme Nisin	Smoked salmon	*L. monocytogenes* *Salmonella anatum*	Reduction of 2.2–2.8 log CFU/g compared to controls	Datta et al. 2008
PROTEINS					
Coatings and films of: Whey protein	Lactoperoxidase (11, 17, 23, 29 mg/g film)	Cold-smoked salmon	*L. monocytogenes* TVC	Coatings of 0.7% (w/w) LPOS reduced *L. monocytogenes* and TVC, ≥3.0 and 1.0 log CFU/g, respectively	Min et al. 2005b
BLENDS					
Films of: Gelatine (6 % w/v)/ chitosan (2 % w/v)	Clove (0.75 mL/g biopolymer)	Cod fillets	TVC H_2S-producers organisms lactic acid bacteria *Pseudomonas* spp. Enterobacteriaceae	Growth of Enterobacteriaceae and H_2S-producers were reduced < limit detection, while LAB remained constant (4 log CFU/g) during storage compared to controls (8 log CFU/g)	Gómez-Estaca et al. 2010

SL: Sodium lactate, SD: Sodium diacetate, PS: Potassium sorbate, CH: Chitosan, LPOS: Lactoperoxidase, LAB: Lactic acid bacteria moulds (approx. 0.4 CFU/g) was recorded at 4 and 20°C. Galactomannan film incorporated with nisin (50 IU g^{-1}) has also been studied on a soft Italian cheese, named Ricotta, and the results showed that treatments increased CO_2 and decreased O_2 permeability; decreased weight loss; improved elongation at break point; and delayed growth of *L. monocytogenes* by 2.0 log CFU/g (Martins et al. 2010). Chitosan alone or with other antimicrobials has been tested on different types of cheese aiming at prolonging their shelf-life, such as Mozzarella (Altieri et al. 2005), Emmental (Coma et al. 2002), Portuguese Saloio (Fajardo et al. 2010), Apulia spreadable cheese (Gammariello et al. 2008) and Cheddar (Moreira et al. 2011a). Moreover, Ramos et al. (2012) reported that whey protein (10% w/w) coating with 20 g L^{-1} chitooligosaccharide and 6 g L^{-1} lactic acid in laboratory manufactured cheese proved successful in inhibiting *Staphylococcus* spp., *Pseudomonas* spp., Enterobacteriaceae, and yeasts and moulds (< 2.0 log CFU/g) as compared to controls (3.7–6.6 log CFU/g) after 60 days of storage at 10°C.

Fresh produce. Due to the multiple processing stages, fresh or minimally processed fruits and vegetables may become highly perishable; therefore they are required combined antimicrobial treatments (in the context of multiple hurdles) to extend shelf-life and ensure safety (Robson et al. 2008). Considering also that consumers usually judge food quality based on appearance and freshness at the time of purchase, mild preservative methods such as edible films and coatings can be promising (Fig. 6). Between coatings and films, edible coatings are the most commonly studied and applied on fresh produce.

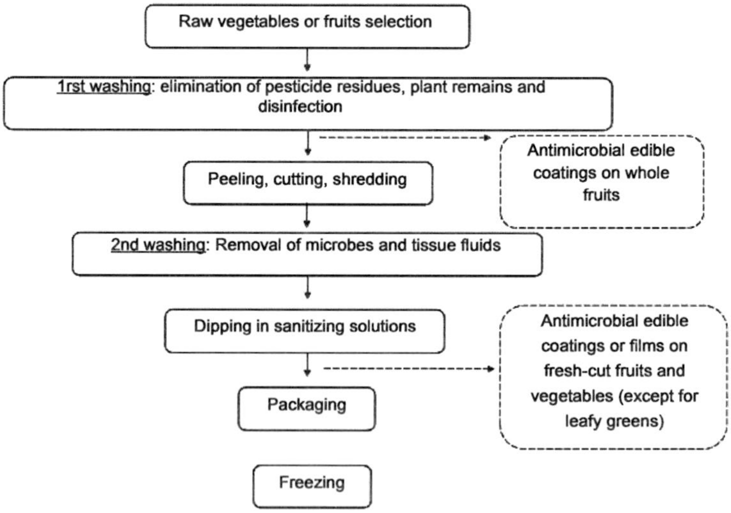

Figure 6. Flow diagram of whole or fresh-cut fruits and vegetables (except for leafy greens) with suggestions of intermediate stages in which interventions with antimicrobial edible coatings may be applied (based on Corbo et al. 2010).

Fruits

Edible coatings have been successfully used for their potential for fruit preservation (Table 3). According to literature, edible coatings have been used on the whole fruit (e.g., waxes to enhance color and mechanical properties) and on fruit pieces (cubes, slices) in fruit based-salads. Even though several studies have reported the beneficial effect of applying edible coatings (i.e., reduction of weight loss and respiration rate, maintenance of colour and firmness), their functionality can be expanded by incorporating antimicrobial agents (Navarro-Tarazaga et al. 2011).

Polysaccharides-based coatings with or without antimicrobial, e.g., *trans*-cinnamaldehyde or antioxidant, e.g., ascorbic acid, compounds have been applied to a variety of fruits, i.e., papaya, apples, mangos, to reduce moisture loss, oxidation or tissue disintegration and improve color and total appearance (Tapia et al. 2007, Chiumarelli et al. 2011, Brasil et al. 2012, Pan et al. 2013, Hamzah et al. 2013). Chitosan coatings are usually applied alone or combined with other polymers on fruit products such as strawberries, table grapes, fresh-cut papaya and cantaloupe as antimicrobial coatings and may delay ripening and decrease transpiration rates in fruits and vegetables (Romanazzi et al. 2002, 2007, Tapia et al. 2007, Vu et al. 2011). Chien et al. (2007) reported that chitosan-based coatings (0.5–2.0% w/v) on mango slices may retard water loss, extend sensory quality by increasing the soluble solid content, titratable acidity and ascorbic acid content and by reducing TVC *ca.* 1.0 log CFU/g, compared to controls. More recently, study of Rojas-Graü et al. (2007) successfully extended the shelf-life of fresh-cut apples by 2 weeks compared to controls by applying alginate and gellan-based edible coatings. Moreover, the results of Maftoonazad and Ramaswamy (2008) showed that pectin/beeswax coating on avocados prolonged their shelf-life to over one month at 10°C compared to controls, while the research team of Moalemiyan (2011) reported that pectin-based coatings reduced the rate of color development, texture softening, weight loss, CO_2 evolution and pH of mangos compared to uncoated samples. Carrageenan coating (0.5% w/w) has been also tested in combination with subsequent immersion of fresh-cut bananas into chemical solution (0.5% ascorbic acid, 2% calcium chloride, and 0.75% cysteine) showing that coated fruits were maintained fresh (<2.0 log CFU/g) for 5 days at 5°C compared to controls (4.0 log CFU/g) (Bico et al. 2009). Alginate coatings (2% w/v) with malic acid (2% w/v) extended shelf-life of fresh-cut melon (> 9.6 days) compared to controls where shelf-life was > 3.6 days, while the application of cinnamon, palmarosa, and lemongrass EOs (0.3 and 0.7% v/v) coatings prolonged the shelf-life of melon > 21 days at 5°C (Raybaudi-Massilia et al. 2008). Regarding lipid-based coatings, Saucedo-Pompa et al. (2009) showed that damage caused by *Colletotrichum gloeosporioides*, the main

Table 3. Application of antimicrobial edible coatings to extend shelf-life and enhance food safety of fruits.

Biopolymer	**Antimicrobial**	**Product**	**Microorganisms**	**Indicative effect**	**Reference**
POLYSACCHARIDES					
Chitosan (0.01 and 0.02 g mL^{-1})	Chitosan (0.01 and 0.02 g mL^{-1})	Fresh-cut papaya	TVC Yeasts and moulds	Coatings suppressed TVC (<b.l.d-3.3 .log CFU/g), and yeast and moulds (1.6–4.3 log CFU/g) growth, compared to controls (4.5–5.5 log CFU/g) after 14-day storage at 5°C	González-Aguilar et al. 2009
Chitosan (2% w/v)	0.02% (w/v) of: Red thyme Oregano Limonene Peppermint EOs	Strawberries	TVC Yeasts and moulds	After 14 days, coated fruits with chitosan and EOs reached 60% decay, while controls were at 100%	Vu et al. 2011
Chitosan (0.1 and 0.5 w/v)	Chitosan + ethanol (10 or 20% v/v)	Table grapes	*Botrytis cinerea*	0.5% chitosan + 10 or 20% ethanol was the most effective treatment for controlling the decay with respect to controls	Romanazzi et at. 2007
Chitosan (2.0 g L^{-1})	Chitosan (2.0 g L^{-1})	Fresh-cut apples	*Penicillium* sp. *Alternaria* sp.	Coated apples showed lower proportion of infected area (40%) compared to controls (70%) after 10-day storage	Garrido Assis and de Britto 2011
Methylcellulose (3% w/v)	Chitosan (0, 1, 1.5, and 2% w/v)	Fresh-cut cantaloupe	TVC Psychrotrophs Yeast and moulds Coliforms *E. coli* *S. aureus* *Salmonella* sp.	Chitosan reduced growth of native flora by 3.9 log CFU/g, while counts of *S. aureus*, *Salmonella* sp. and *E. coli* were <10 CFU g^{-1}	Krasaekoopt and Mabumrung 2008

Tapioca starch (1.35% w/v) + tsao leaf gum (0.15% w/v)	Green tea extracts (1% w/v)	Fruit-based salads	TVC Yeasts and moulds Psychrotrophs	Reduction of native flora on coated samples by 2 log CFU/g compared to controls	Chiu and Lai 2010
BLENDS					
Chitosan/MC (1.5% w/v)	Vanillin	Fresh-cut pineapple and cantaloupe	*E. coli* *S. cerevisiae*	Vanillin films decreased populations by 4 logs CFU/g compared to controls.	Sangsuwan et al. 2008
Chitosan (2% w/w)/ pectin (2% w/w)	Mix of 2% w/v β-cyclodextrin/ *trans*-cinnamaldehyde	Fresh-cut papaya	TVC Total psychrotrophic Yeasts and moulds	Coating extended shelf-life of papaya from 5 to 15 days.	Brasil et al. 2012

phytopathogenic fungus for avocados, was reduced significantly, while changes in the appearance and weight loss were lower in waxed fruits (candelilla wax with 0.01% ellagic acid) compared to controls.

Edible coatings have been also tested for their ability to constitute the applied fruits safe. Kenney and Beuchat (2002) investigated the survival of *E. coli* O157:H7 and *Salmonella* Muenchen on apples waxed by commercial products and showed that waxing reduced populations of *E. coli* O157:H7 and *S.* Muenchen by > 1.5 log CFU/apple. Total control of *E. coli* DH5a on fresh tomatoes have been also reported by Ramos-García et al. (2012) applying a formulation of chitosan (1%) + beeswax (0.1%) + lime EO (0.1%). According to Raybaudi-Massilia et al. (2008), alginate coating with 0.7 % v/v palmarosa EO reduced the population of *S.* Enteritidis by 1.7 log CFU/g after 21-day storage at 5°C.

Vegetables

Edible coatings have been studied on several fresh or minimally processed vegetables, e.g., carrots, potatoes, broccoli; however their study on leafy greens (e.g., cabbage, lettuce) is not an objective due to the expected difficulty of their application on such type of vegetables (Table 4). Similar as in case of fruits, research for application of edible coatings on vegetables is focused on polysaccharides. The application of starch (4% w/w)/chitosan (0.5% and 1.5%) coatings on minimally processed carrot slices may significantly inhibit TVC by 1.3 log CFU/g, yeasts and moulds by 2.5 log CFU/g, and psychrotrophic counts by 1.3 log CFU/g compared to controls at 10°C (Durango et al. 2006). Potassium sorbate (1% w/v) incorporated in guar gum (1% w/v), pea and potato starches (4% w/v) coatings, which were subsequently applied to cucumbers, stored at 4°C, showed 1.0–1.5 log CFU/g reduction on yeast and mould populations compared to controls, after 25-days storage (Mehyar et al. 2014). Blend of tapioca starch/decolorized Hsian-tsao leaf gum containing green tea extracts (1% w/v) on romaine hearts showed not only strong antimicrobial activity (*ca.* 3 log CFU/g reduction) against *S. aureus* compared to untreated samples (*ca.* 6 log CFU/g) but in case of *L. monocytogenes* no viable populations were observed on coated samples after 48 hr (Chiu and Lai 2010). *S. aureus* growth was also inhibited by using banana flour/chitosan films on fresh-cut asparagus, baby corn, and Chinese cabbage by 0.9, 1.5, 1.0 log CFU/g, respectively, after 3 days storage at 10°C (Pitak and Rakshit 2011).

Table 4. Application of antimicrobial edible coatings to extend the shelf-life and enhance food safety of vegetables.

Biopolymer	Antimicrobial	Product	Microorganisms	Indicative effect	Reference
POLYSACCHARIDES					
Alginate (0.5% w/v)	Potassium sorbate (15% w/v)	Potato	TVC	Initial microbial load was decreased by 1.0 log CFU/cm² during storage at 5°C	Mitrakas et al. 2008
Chitosan (2% w/v)	Chitosan	Fresh-cut broccoli	TVC Psychrotrophs Yeasts and moulds LAB Coliforms *E. coli* O157:H7	Chitosan caused reductions of 1.5–2.5 log CFU/g on native flora and *E. coli* O157:H7 compared to controls.	Moreira et al. 2011b
Chitosan (0.5% v/v)	Chitosan	Carrot sticks	TVC LAB Yeasts and moulds *E. coli*	Native flora populations were maintained very low	Simões et al. 2009
Cassava starch	Potassium sorbate	Pumpkin cylinders	TVC LAB Yeasts and moulds	TVC, LAB, yeasts and moulds growth was prevented	Garcia et al. 2008
BLENDS					
Chitosan (2% w/v)/ Casein (2.5% w/v)	Chitosan	Carrots	TVC Psychrotrophs Yeast and moulds	Reduction of native flora by 1.0–4.5 log CFU/g with respect to controls.	Moreira et al. 2011a

PS: Potassium sorbate, LAB: Lactic acid bacteria

Future Trends

Given the consumer demand for high quality, nutritious, and fresh products, the need for alternative and mild methods of extending shelf-life and improving safety is evident. Edible films and coatings containing antimicrobials are among these types of methods. Even though there are numerous studies regarding the antimicrobial activity of edible films and coatings; the related factors are not fully elucidated, since it is a dynamic technology that needs improvement. Specifically, concerning the functionality of antimicrobial compound, stability, bioavailability, and gradual release of antimicrobial are the most important factors. Moreover, from the aspect of biopolymer, the improvement of mechanical properties, water vapor transport and gas permeability are issues that still need investigation. Moreover, in the last few years, the following challenging objectives which have been raised for studying may result in a new generation of edible films and coatings:

1. The investigation of newer methods for applying biopolymers on foods, i.e., nanodispersion (spray drying-based emulsion-evaporation method) (Shah et al. 2012).
2. The compatibility with other techniques, i.e., high pressures, electric fields, ultrasound, microwave radiation, and gamma radiation to induce the mechanical and gas barrier properties.
3. The encapsulation of probiotic cultures in biopolymers to improve their survival in low acid foods (Gerez et al. 2012, Sohail et al. 2012).

In conclusion, the study of antimicrobial packaging should also be focused on updating the food industry for such technologies (financial benefits) and ensuring consumer acceptance. However, further research is needed to evaluate the feasibility of antimicrobial edible films and coatings application on an industrial scale and their commercialization.

Keywords: edible films and coatings, polysaccharides, proteins, lipids, antimicrobial packaging

References

Aider, M. 2010. Chitosan application for active bio-based films production and potential in the food industry: Review. LWT- Food Sc. Technol. 43: 837–842.

Alishahi, A. and M. Aïder. 2011. Applications of chitosan in the seafood industry and aquaculture: A Review. Food Bioprocess. Tech. 5: 817–830.

Altieri, C., C. Scrocco, M. Sinigaglia and M.A. Del Nobile. 2005. Use of chitosan to prolong mozzarella cheese shelf-life. J. Dairy Sci. 88 (8): 2683–2688.

Anker, M., J. Berntsen, A.M. Hermansson and M. Stading. 2001. Improved water vapor barrier of whey protein films by addition of an acetylated monoglyceride. Innov. Food Sci. Emerg. 3: 81–92.

Avila-Sosa, R., E. Palou, M.T. Jiménez-Munguía, G.V. Nevárez-Moorillón, A.R. Navarro Cruz and A. Lopez-Málo. 2012. Antifungal activity by vapor contact of essential oils added to amaranth, chitosan, or starch edible films. Int. J. Food Microbiol. 153(1–2): 660–672.
Aydinli, M., M. Tutas and Ö.A. Bozdemir. 2004. Mechanical and light transmittance properties of locust bean gum-based edible films. Turk. J. Chem. 28: 163–171.
Baldwin, E.A. 1994. Edible coatings for fresh fruits and vegetables: Past, present, and future. *In*: J.M. Krochta, E.A. Baldwin and M. Nisperos-Carriedo (eds.). Edible films and coatings to improve food quality. Technomic Publishing Company Inc., Lancaster, PA, pp. 25–64.
Baldwin, E. 1999. Surface treatments and edible coating in food preservation. *In*: M.S. Rahman (ed.). Handbook of Food Preservation. Marcel Dekker Inc., New York, pp. 577–609.
Barreto, P.L.M., J. Roeder, J.S. Crespo, G.R. Maciel, H. Terenzi, A.T.N. Piresa and V. Soldia. 2003. Effect of concentration, temperature and plasticizer content on rheological properties of sodium caseinate and sodium caseinate/sorbitol solutions and glass transition of their films. Food Chem. 82: 425–431.
Basch, C.Y., R.J. Jagus and S.K. Flores. 2012. Physical and antimicrobial properties of tapioca starch-HPMC edible films incorporated with nisin and/or potassium sorbate. Food Bioprocess. Tech. 42(4): 511–515.
Bauer, R. and L.M.T. Dicks. 2005. Mode of action of lipid II-targeting antibiotics. Int. J. Food Microbiol. 101: 201–216.
Beales, N. 2004. Adaptation of microorganisms to cold temperatures, weak acid preservatives, low pH, and osmotic stress: A Review. Compr. Rev. Food Sci. F. 3: 1–20.
Beverly, R.L., M.E. Janes, W. Prinyawiwatkula and H.K. Nob. 2008. Edible chitosan films on ready-to-eat roast beef for the control of *Listeria monocytogenes*. Food Microbiol. 25: 534–537.
Bico, S.L.S., M.F.J. Raposo, R.M.S.C. Morais and A.M.M.B. Morais. 2009. Combined effects of chemical dip and/or carrageenan coating and/or controlled atmosphere on quality of fresh-cut banana. Food Control 20: 508–514.
Bourtoom, T. and M.S. Chinnan. 2008. Preparation and properties of rice starch-chitosan blend biodegradable film. LWT-Food Sci. Technol. 41: 1633–641.
Bourtoom, T. 2009. Review article. Edible protein films: Properties enhancement. Int. Food Res. J. 16: 1–9.
Brasil, I.M., C. Gomes, A. Puerta-Gomez, M.E. Castell-Perez and R.G. Moreira. 2012. Polysaccharide-based multilayered antimicrobial edible coating enhances quality of fresh-cut papaya. LWT- Food Sci. Technol. 47(1): 39–45.
Burt, S. 2004. Essential oils: their antibacterial properties and potential applications in foods—A review. Int. J. Food Microbiol. 94(3): 223–253.
Cagri, A., Z. Ustunol and E.T. Ryser. 2002. Inhibition of three pathogens on bologna and summer sausage slices using antimicrobial edible films. J. Food Sci. 67: 2317–2324.
Campos, C.A., L.N. Gerschenson and S.K. Flores. 2011. Development of edible films and coatings with antimicrobial activity. Food Bioprocess. Technol. 4: 849–875.
Caprioli, I., M. O'Sullivan and F.J. Monahan. 2008. Use of sodium caseinate/glycerol edible films to reduce lipid oxidation in sliced turkey meat. Eur. Food Res. Technol. 228(3): 433–440.
Cerqueira, M.A., M.J. Souza-Gallagher, I. Macedo, R. Rodriguez-Aguilera, B.W.S. Souza, J.A. Teixeira and A.A. Vicente. 2010. Use of galactomannan edible coating application and storage temperature for prolonging shelf-life of regional cheese. J. Food Eng. 97: 87–94.
Cha, D.S. and M.S. Chinnan. 2004. Biopolymer-based antimicrobial packaging: a review. Crit. Rev. Food Sci. 44(4): 223–237.
Chan, E.-S. 2011. Preparation of Ca-alginate beads containing high oil content: Influence of process variables on encapsulation efficiency and bead properties. Carbohyd. Polym. 84(4): 1267–1275.
Chen, H. 2002. Formation and properties of casein films and coatings. *In*: A. Gennadios (eds.). Protein-based films and coatings. Boca Raton, FL: CRC Press LLC. pp. 181–211.

Chen, C.-H., W.-S. Kuo and L.-S. Lai. 2009. Rheological and physical characterization of film-forming solutions and edible films from tapioca starch/decolorized hsian-tsao leaf gum. Food Hydrocolloid. 23(8): 2132–2140.

Chien, P.-J., S. Fuu and F.-I. Yang. 2007. Effects of edible chitosan coating on quality and shelf-life of sliced mango fruit. J. Food Eng. 78(1): 225–229.

Chiu, P.-E. and L.-S. Lai. 2010. Antimicrobial activities of tapioca starch/decolorized hsian-tsao leaf gum coatings containing green tea extracts in fruit-based salads, romaine hearts and pork slices. Int. J. Food Microbiol. 139: 23–30.

Chiumarelli, M., C.C. Ferrari, C.I.G.L. Sarantópoulos and M.D. Hubinger. 2011. Fresh cut "Tommy Atkins" mango pre-treated with citric acid and coated with cassava (*Manihot esculenta* Crantz) starch or sodium alginate. Innov. Food Sci. Emerg. 12(3): 381–387.

Chiumarelli, M. and M.D. Hubinger. 2012. Stability, solubility, mechanical and barrier properties of cassava starch-Carnauba wax edible coatings to preserve fresh-cut apples. Food Hydrocolloid. 28(1): 59–67.

Cissé, M., A.C. Kouakou, D. Montet, G. Loiseau and M.N. Ducamp-Collin. 2012. Antimicrobial and physical properties of edible chitosan films enhanced by lactoperoxidase system. Food Hydrocolloid. 30(2): 576–580.

Coma, V., A. Martial-Gros, S. Garreau, A. Copinet, F. Salin and A. Deschamps. 2002. Edible antimicrobial films based on chitosan matrix. J. Food Sci. 67(3): 1162–1169.

Coma, V. 2008. Bioactive packaging technologies for extended shelf-life of meat based products. Meat Sci. 78: 90–103.

Corbo, M.R., B. Speranza, D. Campaniello, D.D. Amato and M. Sinigaglia. 2010. Fresh-cut fruits preservation : current status and emerging technologies. Appl. Microbiol. 1: 143–1154.

Datta, S., M.E. Janes, J. Xue, Q.-G. Losso and J.F. La Peyre. 2008. Control of *Listeria monocytogenes* and *Salmonella anatum* on the surface of smoked salmon coated with calcium alginate coating containing oyster lysozyme and nisin. J. Food Sci. 73(2): M67–71.

Dragich, A.M. and J.M. Krochta. 2010. Whey protein solution coating for fat-update reduction in deep fried chicken breast strips. J. Food Sci. 75: S43–S47.

Durango, M., N.F. Soares and N.J. Andrade. 2006. Microbiological evaluation of an edible antimicrobial coating on minimally processed carrots. Food Control 17(5): 336–341.

EC Regulation No 1935/2004. Materials and articles intended to come into contact with food and repealing Directives 80/590/EEC and 89/109/EEC. 27 October 2004.

EC Regulation No 450/2009. Active and intelligent materials and articles intended to come into contact with food. 29 May 2009.

EC Regulation No 749/2012. Classification of certain goods in the Combined Nomenclature. 14 August 2012.

Ehivet, F.E., B. Min, M.-K. Park and J.-H. Oh. 2011. Characterization and antimicrobial activity of sweet potato starch-based edible film containing origanum (*Thymus capitatus*) oil. J. Food Sci. 76(1): C178–C184.

Emiroğlu, Z.K., P.Y. Gökçe, K.C. Betül and C. Kezban. 2010. Antimicrobial activity of soy edible films incorporated with thyme and oregano essential oils on fresh ground beef patties. Meat Sci. 86: 283–288.

Eswaranandam, S., N.S. Hettiarachchy and M.G. Johnson. 2004. Antimicrobial activity of citric, lactic, malic, or tartaric acids and nisin-incorporated soy protein film against *Listeria monocytogenes*, *Escherichia coli* O157:H7 and *Salmonella* gaminara. J. Food Sci. 69: 79–84.

Fajardo, P., J.T. Martins, C. Fuciños, L. Pastrana, J.A. Teixeira and A.A. Vicente. 2010. Evaluation of a chitosan-based edible film as carrier of natamycin to improve the storability of Saloio cheese. J. Food Eng. 101(4): 349–356.

Fan, W., J. Sun, Y. Chen, J. Qiu, Y. Zhangc and Y. Chi. 2009. Effects of chitosan coating on quality and shelf-life of silver carp during frozen storage. Food Chem. 115(1): 66–70.

FDA. 2006. Food additives permitted for direct addition to food for human consumption 21CFR172, subpart C. Coatings, films and related substances.

Flores, S., L. Famá, A.M. Rojas, S. Goyanes and L. Gerschenson. 2007. Physical properties of tapioca-starch edible films: influence of filmmaking and potassium sorbate. Food Res. Int. 40: 257–265.

Flores, S.K., D. Costa, F. Yamashita, L.N. Gerschenson and M.V. Grossmann. 2010. Mixture design for evaluation of potassium sorbate and xanthan gum effect on properties of tapioca starch films obtained by extrusion. Mater Sci. Eng. C 30: 196–202.

Gadang, V.P., N.S. Hettiarachchy, M.G. Johnson and C. Owens. 2008. Evaluation of antibacterial activity of whey protein isolate coating incorporated with nisin, grape seed extract, malic acid, and EDTA on a Turkey frankfurter system. J. Food Sci. 73(8): M389–394.

Gammariello, D., S. Chillo, M. Mastromatteo, S. Di Giulio, M. Attanasio and M.A. Del Nobile. 2008. Effect of chitosan on the rheological and sensorial characteristics of Apulia spreadable cheese. J. Dairy Sci. 91(11): 4155–4163.

García, M., M. Martino and N. Zaritizky. 2000. Lipid addition to improve barrier properties of edible starch-based films and coatings. J. Food Sci. 65(6): 941–947.

Garcia, M., V. Bifani, C.A. Campos, M.N. Martino, P. Sobral, S.K. Flores et al. 2008. Edible coating as an oil barrier or active system. *In*: Gutiérrez Lopez, Barbosa-Cánovas, Welti-Chanes and Parada Arias (eds.). Food Engineering: Integrated Approaches. New York: Springer, pp. 225–241.

Garcia, L.C., L. Mendes Pereira, C.I.G. de Luca Sarantópoulos and M. Dupas Hubinger. 2012. Effect of antimicrobial starch edible coating on shelf-life of fresh strawberries. Packag. Technol. Sci. 45(7): 413–425.

Garrido-Assis, O.B. and D. de Britto. 2011. Evaluation of the antifungal properties of chitosan coating on cut apples using a non-invasive image analysis technique. Polym. Int. 60(6): 932–936.

Gennadios, A., T.H. McHugh, C.L. Weller and J.M. Krochta. 1994. Edible films and coatings based on proteins. *In*: J.M. Krochta, E.A. Baldwin and M. Nisperos-Carriedo (eds.). Edible films and coatings to improve food quality. Technomic Publishing Company, Inc. Lancaster, PA, pp. 201–277.

Gerez, C.L., G. Font de Valdez, M.L. Gigante and C.R.F. Grosso. 2012. Whey protein coating bead improves the survival of the probiotic *Lactobacillus rhamnosus* CRL 1505 to low pH. Lett. Appl. Microbiol. 54(6): 552–556.

Gialamas, H., K.G. Zinoviadou, C.G. Biliaderis and K.P. Koutsoumanis. 2010. Development of a novel bioactive packaging based on the incorporation of *Lactobacillus sakei* into sodium-caseinate films for controlling *Listeria monocytogenes* in foods. Food Res. Int. 43(10): 2402–2408.

Gomez-Estaca, J., P. Montero, B. Gimenez and M. Gomezguillen. 2007. Effect of functional edible films and high pressure processing on microbial and oxidative spoilage in cold-smoked sardine (*Sardina pilchardus*). Food Chem. 105(2): 511–520.

Gómez-Estaca, J., A. López de Lacey, M.E. López-Caballero, M.C. Gómez-Guillén and P. Montero. 2010. Biodegradable gelatin-chitosan films incorporated with essential oils as antimicrobial agents for fish preservation. Food Microbiol. 27: 889–896.

Gonzalez-Aguilar, G.A., E. Valenzuela-Soto, J. Lizardi-Mendoza, F. Goycoolea, M.A. Martínez-Téllez, M.A. Villegas-Ochoa, I.N. Monroy-García and J.F. Ayala-Zavala. 2009. Effect of chitosan coating in preventing deterioration and preserving the quality of fresh-cut papaya "Maradol". J. Food Sci. Agric. 89: 15–23.

Gounga, M.E., S.-Y. Xu, Z. Wang and W.G. Yang. 2008. Effect of whey protein isolate-pullulan edible coatings on the quality and shelf-life of freshly roasted and freeze-dried Chinese chestnut. J. Food Sci. 73(4): E155–E161.

Guilbert, S. and N. Gontard. 1995. Edible and biodegradable food packaging. *In*: P. Ackermann, M. Jägerstad and T. Ohlsson (eds.). Foods and packaging materials. Chemical interactions. The Royal Society of Chemistry. England, pp. 159–168.

Guilbert, S., N. Gontard and L.G.M. Gorris. 1996. Prolongation of the shelf-life of perishable food products using biodegradable films and coatings. LWT-Food Sci. Technol. 29: 10–17.

Gutierrez, J., C. Barry-Ryan and P. Bourke. 2008. The antimicrobial efficacy of plant essential oil combinations and interactions with food ingredients. Int. J. Food Microbiol. 124: 91–97.

Hagenmaier, R. 2000. Evaluation of a polyethylene-candelilla coating for 'Valencia' oranges. Postharvest Biol. Technol. 19(2): 147–154.

Hagenmaier, R. 2002. The flavor of mandarin hybrids with different coatings. Postharvest Biol. Technol. 24(1): 79–87.

Hamzah, H.M., A. Osman, C.P. Tan and F. Mohamad Ghazali. 2013. Carrageenan as an alternative coating for papaya (*Carica papaya* L. cv. Eksotika). Postharvest Biol. Technol. 75: 142–146.

Han, J.H. 2002. Protein-based edible films and coatings carrying antimicrobial agents. *In*: A. Gennadios (ed.). Protein-based films and coatings. CRC Press, Boca Raton, FL, USA, pp. 485–499.

Han, J.H. and A. Gennadios. 2005. Edible films and coatings: A review. *In*: J.H. Han (ed.). Innovations in Food Packaging. Elsevier Academic Press, Amsterdam, pp. 239–262.

Imran, M., S. El-Fahmy, A.-M. Revol-Junelles and S. Desobry. 2010. Cellulose derivative based active coatings: Effects of nisin and plasticizer on physico-chemical and antimicrobial properties of hydroxypropyl methylcellulose films. Carbohyd. Polym. 81(2): 219–225.

Jeon, Y.J., J.Y.V.A. Kamil and F. Shahidi. 2002. Chitosan as an edible invisible film for quality preservation of herring and Atlantic cod. J. Agr. Food Chem. 50: 5167–5178.

Jiang, Z., H. Neetoo and H. Chen. 2011a. Efficacy of freezing, frozen storage and edible antimicrobial coatings used in combination for control of *Listeria monocytogenes* on roasted turkey stored at chiller temperatures. Food Microbiol. 28: 1394–1401.

Jiang, Z., H. Neetoo and H. Chen. 2011b. Control of *Listeria monocytogenes* on cold-smoked salmon using chitosan-based antimicrobial coatings and films. J. Food Sci. 76(1): M22–M26.

Jiménez, A., M.J. Fabra, P. Talens and A. Chiralt. 2012. Effect of re-crystallization on tensile, optical and water vapor barrier properties of corn starch films containing fatty acids. Food Hydrocolloid. 26(1): 302–310.

Juck, G., H. Neetoo and H. Chen. 2010. Application of an active alginate coating to control the growth of *Listeria monocytogenes* on poached and deli turkey products. Int. J. Food Microbiol. 142(3): 302–308.

Kaplan, D.L., J.M. Mayer, D. Ball, J. McCassie, A.L. Allen and P. Stenhouse. 1993. Fundamental of biodegradable polymer. *In*: C. Ching, D. Kaplan and E. Thomas (eds.). Biodegradable Polymers and Packaging. Technomic Publishing, Lancaster, pp. 1–42.

Karbowiak, T., F. Debeaufort and A. Voilley. 2007. Influence of thermal process on structure and functional properties of emulsion-based edible films. Food Hydrocolloid. 21(5–6): 879–888.

Kenney, S.J. and L.R. Beuchat. 2002. Survival of *Escherichia coli* O157:H7 and *Salmonella* Muenchen on apples as affected by application of commercial fruit waxes. Int. J. Food Microbiol. 77(3): 223–231.

Kester, J.J. and O.R. Fennema. 1986. Edible films and coatings: A review. Food Technol. 40(12): 47–59.

Kong, M., X.G. Chen, K. Xing and H.J. Park. 2010. Antimicrobial activity of chitosan and mode of action: A state of the art review. Int. J. Food Microbiol. 144: 51–63.

Krasaekoopt, W., J. Mabumrung and J .Kasetsart. 2008. Microbiological evaluation of edible coated fresh-cut cantaloupe. Nat. Sci. 42(3): 552–557.

Kristo, E., K.P. Koutsoumanis and C.G. Biliaderis. 2008. Thermal, mechanical and water vapor barrier properties of sodium caseinate films containing antimicrobials and their inhibitory action on *Listeria monocytogenes*. Food Hydrocolloid. 22(3): 373–386.

Krochta, J.M. 2002. Proteins as raw materials for films and coatings: Definitions, current status and opportunity. *In*: A. Gennadios (ed.). Protein-based films and coatings. CRC, Boca Raton, FL, pp. 1–42.

Longares, A., F.J. Monahan, E.D. O'Riordan and M. O'Sullivan. 2004. Physical properties and sensory evaluation of WPI films of varying thickness. LWT-Food Sci. Technol. 37: 545–550.

Lu, F., Y. Ding, X. Ye and D. Liu. 2010. Cinnamon and nisin in alginate-calcium coating maintain quality of fresh northern snakehead fish fillets. LWT-Food Sci. Technol. 43: 1331–1335.

Maftoonazad, N. and H.S. Ramaswamy. 2008. Effect of pectin-based coating on the kinetics of quality change associated with stored avocados. J. Food Proc. Preserv. 32(4): 621–643.

Maftoonazad, N., H.S. Ramaswamy and M. Marcotte. 2007. Moisture sorption, behavior, and effect of moisture content and sorbitol on thermo-mechanical and barrier properties of pectin based edible films. Int. J. Food Eng. 3(4): Article No. 10.

Maizura, M., A. Fazilah, M.H. Norziah and A.A. Karim. 2007. Antibacterial activity and mechanical properties of partially hydrolyzed sago starch-alginate edible film containing lemongrass oil. J. Food Sci. 72(6): C324–330.

Mali, S., M.V.E. Grossman, M.A. Garcia, M.N. Martino and N.E. Zaritzky. 2006. Effects of controlled storage on thermal, mechanical and barrier properties of plasticized films from different starch sources. J. Food Eng. 75(4): 453–460.

Martins, J.T., M. Cerqueira, B.W.S. Souza, M.D. Carmo Avides and A. Vicente. 2010. Shelf-life extension of ricotta cheese using coatings of galactomannans from nonconventional sources incorporating nisin against *Listeria monocytogenes*. J. Agr. Food Chem. 58(3): 1884–1891.

McHugh, T.H. and J.M. Krochta. 1994. Milk-protein based edible films and coatings. Food Technol. 48(1): 97–103.

Mehyar, G.F., H.M. Al-Qadiri and B.G. Swanson. 2014. Edible coatings and retention of potassium sorbate on apples, tomatoes and cucumbers to improve antifungal activity during refrigerated storage. J. Food Proc. Preserv. 38(1): 175–182.

Mild, R.M., L. Joens, M. Friedman et al. 2011. Antimicrobial edible apple films inactivate antibiotic resistant and susceptible *Campylobacter jejuni* strains on chicken breast. J. Food Sci. 76(3): M163–168.

Millette, M., C. Le Tien, W. Smoragiewicz and M. Lacroix. 2007. Inhibition of *Staphylococcus aureus* on beef by nisin-containing modified alginate films and beads. Food Control 18(7): 878–884.

Min, S. and J. Krochta. 2005. Inhibition of *Penicillium commune* by edible whey protein films incorporating lactoferrin, lactoferrin hydrolysate, and lactoperoxidase systems. J. Food Sci. 70(2): M87–M94.

Min, S., L.J. Harris and J.M. Krochta. 2005a. Antimicrobial effects of lactoferrin, lysozyme, and the lactoperoxidase system and edible whey protein films incorporating the lactoperoxidase system against *Salmonella enterica and Escherichia coli* O157:H7. J Food Sci. 70(7): M332–338.

Min, S., L.J. Harris and J.M. Krochta. 2005b. *Listeria monocytogenes* inhibition by whey protein films and coatings incorporating the lactoperoxidase system. J. Food Sci. 70: M317–24.

Mitrakas, G.E., K.P. Koutsoumanis and H.N. Lazarides. 2008. Impact of edible coating with or without anti-microbial agent on microbial growth during osmotic dehydration and refrigerated storage of a model plant material. Innov. Food Sci. Emerg. 9(4): 550–555.

Moalemiyan, M., H.S. Ramaswamy and N. Maftoonazad. 2011. Pectin-based edible coating for shelf-life extension of Ataulfo Mango. J. Food Process. Eng. 35(4): 572–600.

Moradi, M., H. Tajik, S.M. Razavi Rohani and A.R. Oromiehie. 2011. Effectiveness of *Zataria multiflora* Boiss essential oil and grape seed extract impregnated chitosan film on ready-to-eat mortadella-type sausages during refrigerated storage. J. Sci. Food Agr. 91(15): 2850–2857.

Moreira, M.D.R., M. Pereda, N.E. Marcovich and S.I. Roura. 2011a. Antimicrobial effectiveness of bioactive packaging materials from edible chitosan and casein polymers: assessment on carrot, cheese, and salami. J. Food Sci. 76(1): M54–63.

Moreira, M.D.R., S.I. Roura and A. Ponce. 2011b. Effectiveness of chitosan edible coatings to improve microbiological and sensory quality of fresh cut broccoli. LWT-Food Sci. Technol. 44(10): 2335–2341.

Naidu, A.S. 2003. Antimicrobials from animals. *In*: S. Roller (ed.). Natural Antimicrobials for the Minimal Processing of Foods. CRC Press, Boca Raton, FL, pp. 133–149.

Navarro-Tarazaga, M.L., A. Massa and M.B. Pérez-Gago. 2011. Effect of beeswax content on hydroxypropyl methylcellulose-based edible film properties and postharvest quality of coated plums (Cv. Angeleno). LWT-Food Sci. Technol. 44(10): 2328–2334.

Nualkaekul, S., D. Lenton, M.T. Cook, V.V. Khutoryanskiy and D. Charalampopoulos. 2012. Chitosan coated alginate beads for the survival of microencapsulated *Lactobacillus plantarum* in pomegranate juice. Carbohyd. Polym. 90(3): 1281–1287.

Ojagh, S.M., M. Rezaei, S.H. Razavi and S.M.H. Hosseini. 2009. Effect of chitosan coatings enriched with cinnamon oil on the quality of refrigerated rainbow trout. Food. Chem. 120: 193–198.

Oussalah, M., S. Caillet, S. Salmieri, L. Saucier and M. Lacroix. 2004. Antimicrobial and antioxidant effects of milk protein-based film containing essential oils for the preservation of whole beef muscle. J. Agr. Food Chem. 52: 5598–5605.

Oussalah, M., S. Caillet, S. Salmiéri, L. Saucier and M. Lacroix. 2006. Antimicrobial effects of alginate-based film containing essential oils for the preservation of whole beef muscle. J. Food Protect. 69(10): 2364–2369.

Oussalah, M., S. Caillet, S. Salmiéri, L. Saucier and M. Lacroix. 2007. Antimicrobial effects of alginate-based films containing essential oils on *Listeria monocytogenes* and *Salmonella* Typhimurium present in bologna and ham. J. Food Protect. 70(4): 901–908.

Pagella, C., G. Spigno and D.M. De Faveri. 2002. Characterization of starch-based edible coatings. Food Bioprod. Process. 80: 193–198.

Pan, S.-Y., C.-H. Chen and L.-S. Lai. 2013. Effect of tapioca starch/ decolorized Hsian-tsao leaf gum-based active coatings on the qualities of fresh-cut apples. Food Bioprocess. Technol. 6(8): 2059–2069.

Pérez-Gago, M.B. and J.M. Krochta. 2005. Emulsion and bi-layer edible films. *In*: J.H. Han (ed.). Innovations in food packaging. Elsevier Science & Technology Book, Amsterdam, pp. 304–361.

Pitak, N. and S.K. Rakshit. 2011. Physical and antimicrobial properties of banana flour/chitosan biodegradable and self sealing films used for preserving Fresh-cut vegetables. LWT-Food Sci. Technol. 44(10): 2310–2315.

Ponce, A.G., S.I. Roura, C.E. del Valle and M.R. Moreira. 2008. Antimicrobial and antioxidant activities of edible coatings enriched with natural plant extracts: *In vitro* and *in vivo* studies. Postharvest Biol. Technol. 49(2): 294–300.

Pranoto, Y., V.M. Salokhe and S.K. Rakshit. 2005a. Physical and antibacterial properties of alginate-based edible film incorporated with garlic oil. Food Res. Int. 38(3): 267–272.

Pranoto, Y., S.K. Rakshit and V.M. Salokhe. 2005b. Enhancing antimicrobial activity of chitosan films by incorporating garlic oil, potassium sorbate, and nisin. Lebensm. Wiss. Technol. 8: 859–65.

Pyla, R., T.J. Kim, J.L. Silva and Y.S. Jung. 2010. Enhanced antimicrobial activity of starch-based film impregnated with thermally processed tannic acid, a strong antioxidant. Int. J. Food Microbiol. 137(2–3): 154–160.

Quintavalla, S. and L. Vicini. 2002. Antimicrobial food packaging in meat industry. Meat Sci. 62: 373–380.

Ramos, O.L., J.C. Fernandes, S.I. Silva, M.E. Pintado and F.X. Malcata. 2012. Edible films and coatings from whey proteins: a review on formulation, and on mechanical and bioactive properties. Crit. Rev. Food Sci. 52(6): 533–552.

Ramos-García, M., E. Bosquez-Molina, J. Hernández-Romano, G. Zavala-Padilla, E. Terrés-Rojas, I. Alia-Tejacal, L. Barrera-Necha, M. Hernández-López and S. Bautista-Banos. 2012. Use of chitosan-based edible coatings in combination with other natural compounds, to control *Rhizopus stolonifer* and *Escherichia coli* DH5α in fresh tomatoes. Crop Prot. 38: 1–6.

Ravishankar, S., L. Zhu, C.W. Olsen, T.H. McHugh and M. Friedman. 2009. Edible apple film wraps containing plant antimicrobials inactivate food-borne pathogens on meat and poultry products. J. Food Sci. 74(8): M440–445.

Ravishankar, S., D. Jaroni, L. Zhu, C. Olsen, T. McHugh and M. Friedman. 2012. Inactivation of *Listeria monocytogenes* on ham and bologna using pectin-based apple, carrot, and hibiscus edible films containing carvacrol and cinnamaldehyde. J. Food Sci. 77(7): M377–382.

Raybaudi-Massilia, R.M., J. Mosqueda-Melgar and O. Martín-Belloso. 2008. Edible alginate-based coating as carrier of antimicrobials to improve shelf-life and safety of fresh-cut melon. Int. J. Food Microbiol. 121: 313–327.

Reinoso, E., G.S. Mittal and L.-T. Lim. 2008. Influence of whey protein composite coatings on plum (*Prunus Domestica* L.) fruit quality. Food Bioprocess. Technol. 1: 314–325.

Rhim, J.W. 2004. Physical and mechanical properties of water resistant sodium alginate films. Lebensm. Wiss. Technol. 37: 323–330.

Robson, M.G., N.F. Ferreira Soares, D. Alvarenga Botrel and G.L. de Almeida. 2008. Characterization and effect of edible coatings on minimally processed garlic quality. Carbohyd. Polym. 72(3): 403–409.

Rodriguez, M., J. Oses, K. Ziani and J.I. Mate. 2006. Combined effect of plasticizers and surfactants on the physical properties of starch based edible films. Food Res. Int. 39: 840–884.

Rodriguez-Turienzo, L., A. Cobos, V. Moreno Amado, J.M. Caride Vieites and O. Diaz. 2011. Whey protein-based coatings on frozen Atlantic salmon (*Salmo salar*): Influence of the plasticiser and the moment of coating on quality preservation. Food Chem. 128: 187–194.

Rojas-Graü, M.A., R.M. Raybaudi-Massilia, R.C. Soliva-Fortuny, R.J. Avena-Bustillos, T.H. McHugh and O. Martín-Belloso. 2007. Apple puree alginate edible coating as carrier of antimicrobial agents to prolong shelf-life of fresh-cut apples. Postharvest Biol. Technol. 45: 254–264.

Rojas-Graü, M.A., R. Soliva-Fortuny and O. Martín-Belloso. 2009. Edible coatings to incorporate active ingredients to fresh-cut fruits: a review. Trends Food Sci. Technol. 20(10): 438–447.

Romanazzi, G., O.A. Karabulut and J.L. Smilanick. 2007. Combination of chitosan and ethanol to control postharvest gray mold of table grapes. Postharvest Biol. Technol. 45: 134–140.

Sánchez-González, L., C. Pastor, M. Vargas, A. Chiralt and C. González-Martínez. 2011. Effect of hydroxypropylmethylcellulose and chitosan coatings with and without bergamot essential oil on quality and safety of cold-stored grapes. Postharvest Biol. Technol. 60(1): 57–63.

Sánchez-González, L., M. Cháfer, A. Chiralt and C. González-Martínez. 2010. Physical properties of edible chitosan films containing bergamot essential oil and their inhibitory action on *Penicillium italicum*. Carbohyd. Polym. 82(2): 277–283.

Sanderson, G.R. 1981. Polysaccharides in foods. Food Technol. 35(7): 50–52, 54–57, 83.

Sangsuwan, J., N. Rattanapanone and P. Rachtanapun. 2008. Effect of chitosan/methyl cellulose films on microbial and quality characteristics of fresh-cut cantaloupe and pineapple. Postharvest Biol. Technol. 49: 403–410.

Saucedo-Pompa, S., R. Rojas-Molina, A.F. Aguilera-Carbó, A. Saenz-Galindo, H. de La Garza, D. Jasso-Cantú and C.N. Aguilar. 2009. Edible film based on candelilla wax to improve the shelf-life and quality of avocado. Food Res. Int. 42(4): 511–515.

Sayanjali, S., B. Ghanbarzadeh and S. Ghiassifar. 2011. Evaluation of antimicrobial and physical properties of edible film based on carboxymethyl cellulose containing potassium sorbate on some mycotoxigenic *Aspergillus* species in fresh pistachios. LWT-Food Sci. Technol. 44(4): 1133–1138.

Schou, M., A. Longares, C. Montesinos-Herrero, F.J. Monahan, D. O'Riordan and M. O'Sullivan. 2005. Properties of edible sodium caseinate films and their applications as food wrapping. Lebensm. Wiss. Technol. 38: 605–610.

Seifu, E., E.M. Buys, E.F. Donkin and I.-M. Petzer. 2004. Antibacterial activity of the lactoperoxidase system against food-borne pathogens in Saanen and South African Indegenous goat milk. Food Control 15: 447–452.

Seol, K.-H., D.-G. Lim, A. Jang, C. Jo and M. Lee. 2009. Antimicrobial effect of κ-carrageenan-based edible film containing ovotransferrin in fresh chicken breast stored at 5°C. Meat Sci. 83: 479–483.

Seydim, A.C. and G. Sarikus. 2006. Antimicrobial activity of whey protein based edible films incorporated with oregano, rosemary and garlic essential oils. Food Res. Int. 39: 639–644.

Shah, B., S. Ikeda, P. Michael Davidson and Q. Zhong. 2012. Nanodispersing thymol in whey protein isolate-maltodextrin conjugate capsules produced using the emulsion-evaporation technique. J. Food Eng. 113(1): 79–86.

Simões, A.D.N., J.A. Tudela, A. Allende, R. Puschmann and M.I. Gil. 2009. Edible coatings containing chitosan and moderate modified atmospheres maintain quality and enhance phytochemicals of carrot sticks. Postharvest Biol. Technol. 51: 364–370.

Soares, R.M.D., A.M.F. Lima, R.V.B. Oliveira, A.T.N. Pires and V. Soldi. 2005. Thermal degradation of biodegradable edible films based on xanthan and starches from different sources. Polym. Degrad. Stabil. 90(3): 449–454.

Sohail, A., M.S. Turner, E.K. Prabawati, A.G. Coombes and B. Bhandari. 2012. Evaluation of *Lactobacillus rhamnosus* GG and *Lactobacillus acidophilus* NCFM encapsulated using a novel impinging aerosol method in fruit food products. Int. J. Food Microbiol. 157(2): 162–166.

Song, H.Y., Y.J. Shin and K.B. Song. 2012. Preparation of a barley bran protein-gelatin composite film containing grapefruit seed extract and its application in salmon packaging. J. Food Eng. 113(4): 541–547.

Sothornvit, R. and N. Pitak. 2007. Oxygen permeability and mechanical properties of banana films. Food Res. Int. 40(3): 365–370.

Suppakul, P., J. Miltz, K. Sonneveld and S. W. Bigger. 2003. Active packaging technologies with emphasis on antimicrobial packaging and its applications. J. Food Sci. 68: 408–420.

Tapia, M.S., M. Rojas-Graü, F.J. Rodríguez, J. Ramírez, A. Carmona and O. Martin-Belosso. 2007. Alginate- and gellan-based edible films for probiotic coatings on fresh-cut fruits. J. Food Sci. 72(4): E190–196.

Touch, V., S. Hayakawa, S. Yamada and S. Kaneko. 2004. Effects of a lactopeoxidase-thiocyanate-hydrogen peroxide system on *Salmonella enteritidis* in animal or vegetable foods. Int. J. Food Microbiol. 93: 175–183.

Valencia-Chamorro, S., L. Palou, M. Del Río and M.B. Pérez-Gago. 2011. Antimicrobial edible films and coatings for fresh and minimally processed fruits and vegetables: a review. Crit. Rev. Food Sci. 51(9): 872–900.

Vargas, M., C. Pastor, A. Chiralt, D.J. McClements and C. González-Martínez. 2008. Recent advances in edible coatings for fresh and minimally processed fruits. Crit. Rev. Food Sci. 48(6): 496–511.

Vásconez, M.B., S.K. Flores, C.A. Campos, J. Alvarado and L.N. Gerschenson. 2009. Antimicrobial activity and physical properties of chitosan-tapioca starch based edible films and coatings. Food Res. Int. 42: 762–769.

Villalobos, R., J. Chanona, P. Hernandez, G. Gutiaerrez and A. Chiralt. 2005. Gloss and transparency of hydroxypropyl methylcellulose films containing surfactants as affected by their microstructure. Food Hydrocolloid. 19: 53–61.

Vu, K.D., R.G. Hollingsworth, E. Leroux, S. Salmieri and M. Lacroix. 2011. Development of edible bioactive coating based on modified chitosan for increasing the shelf-life of strawberries. Food Res. Int. 44: 198–203.

Whistler, R.L. and J.R. Daniel. 1985. Carbohydrates. *In*: O.R. Fennema (ed.). Food Chemistry. Marcel Dekker Inc., New York, pp. 69–137.
Williams, P.A. and G.O. Phillips. 2000. Handbook of hydrocolloids. *In*: G. Phillips and P. Willians (eds.). CRC Press, Cambridge, England, pp. 1–19.
Ye, M., H. Neetoo and H. Chen. 2008a. Control of *Listeria monocytogenes* on ham steaks by antimicrobials incorporated into chitosan-coated plastic films. Food Microbiol. 25: 260–268.
Ye, M., H. Neetoo and H. Chen. 2008b. Effectiveness of chitosan-coated plastic films incorporating antimicrobials in inhibition of *Listeria monocytogenes* on cold-smoked salmon. Int. J. Food Microbiol. 127: 235–240.
Ye, M., H. Neetoo and H. Chen. 2011. Prior frozen storage enhances the effect of edible coatings against *Listeria monocytogenes* on cold-smoked salmon during subsequent refrigerated storage. J. Appl. Microbiol. 111: 865–876.
Yener, F.Y.G., F. Korel and A. Yemenicioğlu. 2009. Antimicrobial activity of lactoperoxidase system incorporated into cross-linked alginate films. J. Food Sci. 74(2): M73–79.
Ziani, K., I. Fernandez Pan, M. Royo and J. Maté. 2009. Antifungal activity of films and solutions based on chitosan. Food Hydrocolloid. 23: 2309–2314.
Zinoviadou, K.G., K.P. Koutsoumanis and C.G. Biliaderis. 2009. Physico-chemical properties of whey protein isolate films containing oregano oil and their antimicrobial action against spoilage flora of fresh beef. Meat Sci. 82(3): 338–345.
Zivanovic, S., S. Chi and A.F. Draughon. 2005. Antimicrobial activity of chitosan films enriched with essential oils. J. Food. Sci. 70: 45–50.

Part B

Modern Techniques in Food Microbiology

CHAPTER 10

Old Targets, New Weapons: Food Microbial Communities Revealed With Molecular Tools

Aspasia Nisiotou,[1] *Foteini F. Parlapani,*[2] *Konstantinos A. Kormas*[2] and *Ioannis S. Boziaris*[2,*]

Introduction

The exploration of food-associated microbiota has been historically based on the comparison of morphological and physiological characteristics of isolated microorganisms with those of type or typical strains. Food microbiologists have applied several generic and specialized culture media for the isolation of cultivable populations and routinely performed a series of tests to record the morphological and phenotypic responses of isolates. The main reference texts applied for the identification and characterization of microorganisms are those of Krieg and Holt (1984) or Dworkin et al. (2006) for bacteria, and Kurtzman et al. (2011) or Barnett et al. (2000) for yeasts. Phenotypical identification of isolates served food microbiologists for years and formed the established knowledge for food ecosystems till the late 1980s. However, this methodology encounters two main drawbacks. First, phenotypic characteristics depend largely on culture conditions that often lead to inaccurate identification, while it is also laborious since multiple tests are typically required to identify microorganisms at the species level.

[1] Hellenic Agricultural Organisation "Demeter" (NAGREF) wine Institute of Athens S. Venizelou 1, 14123, Lykovrisi, Athens, Greece.
[2] School of Agricultural Sciences, University of Thessaly, Fitoko street, 38446, N. Ionia, Volos, Greece.
* Corresponding author: boziaris@uth.gr

Second, not all the microorganisms present in an environmental sample can be recovered on culture media (Hugenholtz 2002). To date, there are more than 30 bacterial and 4 archaeal phyla, several of which host exclusively yet-uncultivated representatives (GenBank, accessed on 01 May 2013, Garrity et al. 2007). This indicates that when we investigate cultivated microorganisms in biodiversity studies, it is very likely we miss a great deal of the existing structural and functional diversity. Hence, phenotypic criteria cannot be solely applied for unambiguous characterization of the food-associated microbiota.

During the last decades, the introduction of molecular biology techniques has provided new insights into the exploration of food-associated microbial populations and significantly ameliorated our understanding on the complexity of these microbial communities. Evolving DNA-based technologies provide distinct advantages over traditional identification schemes in the accuracy of identification of microbial assemblages and may overcome the obstacles associated with cultivation. With respect to microbial identification and also quantification, molecular techniques are relatively rapid and generate a great amount of genetic information, thus providing a high degree of resolution in reduced amount of time. Moreover, the relevant molecular methodology has provided unprecedented insight into functional diversity, i.e., the discovery of novel biochemical pathways and the assignment of specific metabolic processes to cultivated species or yet-uncultivated cells.

Assessment of Food Microbiota by Culture-dependent Molecular Methods

The so-called "culture-dependent" molecular techniques are currently widely applied in food ecosystems for elucidating the magnitude of microbial community composition and community structure (sensu Prosser 2012). However, their efficiency in describing the entire microbial community, is somewhat restricted by the fact that a proportion of the total microorganisms in an environmental sample is often non-cultivable and thus undetectable. Nevertheless, this inherent limitation of culture-based methods may not diminish their merit for certain applications, particularly when the isolation of microorganisms is desired. Culture-based molecular methods are indispensable to precisely identify and characterise key players in food samples, such as strains responsible for fermentation, spoilage or flavour-associated microorganisms. In this chapter, some of the most important culture-dependent molecular approaches for assessing microbial diversity in food microbial communities are presented, emphasizing on their suitability, robustness and strengths as well as on possible weaknesses. Table 1 presents an overview of studies employing culture-dependent methods for the analysis of microbial communities in foods.

Whole-genome molecular techniques

Unlike the PCR-based methods, which target a single or a few loci typically corresponding to an extremely small portion of the genome, the whole-genome molecular approaches evaluate the total genomic DNA of a microorganism. Thus, they provide a higher resolution in identifying different species/strains. However, whole genome analysis techniques are often laborious, time-consuming and costly. The current development of high-throughput technologies such as next generation sequencing (NGS) has dramatically reduced the time and cost required for a more comprehensive analysis of microbial communities.

DNA-DNA reassociation. The DNA-DNA reassociation technique is applied to determine nuclear DNA (nDNA) relatedness between microbial genomes. As such, it has been used to resolve phylogenetic relatedness among microbes in environmental samples or in pure cultures. For reassociation comparisons, genomic DNA of two strains is first denatured by applying heating or a denaturing reagent. Annealing of complementary DNA strands is then induced by decreasing temperature or removing the denaturant. The rate of DNA re-association depends on the genome diversity and is favoured at low genomic complexity. A value of 70% nDNA complementarity has been accepted as the baseline for defining the boundaries of a bacterial species (Goris et al. 2007). Kurtzman (1987) after reviewing data from nDNA reassociation studies of various ascomycetous yeasts, concluded that the above 70% nDNA complementarity baseline may also apply as the criterion for discriminating yeast species. This technique may not be quite suitable for routine discrimination studies since it requests several pairwise comparisons of different isolates, while the resolution provided is limited to the genetic distance of only closely related species. However, it has succeeded in the identification of food isolates when ribotyping approaches failed (Koort et al. 2005, 2006, Gavini et al. 2006). Thus, it should be combined with other more rapid species identification techniques at least to confirm species delimitation. More recently, the DNA reassociation technique has been adapted for whole microbial community analysis. However, it has not yet found wide application in food samples, probably due to the high cost of equipment needed and the time-consuming, laborious sample processing.

Genomic GC content. Since the earliest microbial diversity studies it has been noted that microorganisms, in particular prokaryotes, exhibit dramatic differences in the percentage of guanine and cytosine nucleotides in their genomes (GC content), which varies from less than 20% to about 75%. The determination of the GC content is accomplished after the enzymatic degradation of DNA into nucleotides (Mesbah et al. 1989), followed by HPLC analysis of the obtained nucleoside mixture. The differences in GC content among taxa provide an opportunity to resolve phylogenetic and taxonomic relationships, particularly when our knowledge on sequence data is limited. Identical GC content between two isolates suggests high similarity, while dissimilar values point to different species. However, unrelated isolates may exhibit similar GC content values, thus, there is often uncertainty as per the threshold value for species boundaries. In general, a difference of less than 5% in GC content is adequate to assign two microorganisms to the same species and those differing by more than 10% GC should not be classified within the same genus (Goodfellow and O'Donnell 1993). The ambiguity of data interpretation and the development of more informative molecular methods based on sequence analysis have diminished the use of GC % in phylogenetic studies. Nevertheless, the method is one of the necessary taxonomic characteristics required for the description of novel species isolated from food samples (Kim et al. 2011, Shin et al. 2011, Whon et al. 2010) or any other habitat. Its usage should be perceived with caution as it is known that species or even strains might have variable highly—or not—GC contents. For example, *Vibrio* spp. GC contents varies between 37 and 47% while for *Salmonella* spp. this value is rather constant (ca. 52%) (Microbial Genome Resources, accessed on 01 May 2013, http://www.ncbi.nlm.nih.gov/genomes/MICROBES/microbial_taxtree.html).

Restriction analysis-PFGE. Electrophoretic karyotyping of genomic DNA using pulse field gel electrophoresis (PFGE) allows the separation of large DNA molecules. It overcomes the limitations of conventional electrophoresis by using a different to standard approach so that to resolve large DNA molecules by maintaining their mobility. While in standard electrophoresis the direction of the electric field is constant, in PFGE there is a periodical alteration of the field orientation. This shift forces the DNA molecules within the gel to redirect towards the newly formed positively charged pole. The time required to complete the orientation increases with the size of the molecule, while the maximum electrophoretic mobility is achieved when the molecules are fully oriented in the direction of the field. At the end, the resulting electrophoretic mobility depends on the frequency of applied electric fields. The pulses are selected to optimize the intervals in which the molecules are re-oriented causing transverse movements and achieving separations up 12 Mb (Orbach et al. 1988).

In microbial typing, PFGE is typically used to separate intact yeast chromosomes or large DNA fragments produced after restriction digestion of bacterial or yeast DNA by rare cutting restriction endonucleases. The method has been applied for the detection of foodborne pathogens, e.g., *Listeria monocytogenes* in fermented sausages (Ferreira et al. 2011) and diced-celery salad (Gaul et al. 2012), *Salmonella* Enteritidis isolated from food (Campioni et al. 2012), non-O157 shiga toxin-producing *Escherichia coli* in retail ground beef and pork (Ju et al. 2012), and also to describe different bacterial or yeast communities in various foods, such as lactic acid bacteria (LAB) in meat and meat products (Tran et al. 2011) and in wine (Ruiz et al. 2008), *Brochothrix thermosphacta* in pork (Papadopoulou et al. 2012), Enterobacteriaceae in beef (Doulgeraki et al. 2011), yeasts in fruit pulp (Duarte et al. 2009) and in rice wine (Jeyaram et al. 2008). The higher discriminatory power of PFGE compared to other typing methods is often reported for food-associated isolates. PFGE showed greater resolution power over rep-PCR (Tran et al. 2011) and differential display-PCR (Ruiz et al. 2008) in typing LAB isolates. When compared to RAPD-, microsatellite- or Sau-PCR analysis, PFGE of NotI digested yeast DNA was superior for the discrimination of *Dekkera bruxellensis* wine and beer strains (Miot-Sertier and Lonvaud-Funel 2006). PFGE is ideal to resolve the complexity of food-associated populations, which are usually formed by a relatively high number of individuals (Ruiz et al. 2008, Tran et al. 2011, Papadopoulou et al. 2012). However, it should be considered that protocols for DNA extraction and electrophoresis are laborious and time consuming and the banding profiles are highly complex to interpret and to communicate between different laboratories.

Whole genome sequencing. The whole-genome sequencing analysis is considered the most informative method for molecular surveys of microorganisms from food samples as it provides the near-complete genetic repertoire of a cell. This approach includes (1) extraction of genomic DNA from pure cultures, (2) DNA fragmentation into random short pieces, (3) construction of a sequencing library (e.g., fosmid or BAC libraries), (4) sequencing of fragments by Sanger dideoxy terminator or next generation sequencing technology, and (5) sequence assembly, which is aligning and merging fragments.

Early complete genome information from diverse food-relevant lactobacilli (Boekhorst et al. 2004, Klaenhammer et al. 2003), streptococci (Bolotin et al. 2004, Hols et al. 2005) and bifidobacteria (Klijn et al. 2005) opened a new era on large-scale comparative genomics studies to analyse common or variable coding sequences, to identify single nucleotide polymorphisms (SNPs), viral insertions, transposable elements, and to elucidate evolutionary history and phylogenetic relationships among species or strains. Whole genome sequence comparisons is quite informative for comparison among yeast species of genera like *Saccharomyces*, since there are several heterothallic species that interbreed, and thus it would shed light on the origin of interspecific hybrids and the speciation process (Kurtzman 2011). Recently, the whole genomes of six industrial wine and beer brewing strains of *S. cerevisiae* were sequenced (Borneman et al. 2011).

The genome-wide analysis and comparison with laboratory, pathogenic, bioethanol, and "natural" isolates revealed several genetic differences, novel open reading frames (ORFs) and genomic rearrangements, that could explain the phenotypic differences observed among strains. The wealth of information acquired from the accumulating whole genome sequencing studies is invaluable to reveal new microbial processes and design novel food preservation strategies for the food industry.

Apart from genotyping means, whole-genome sequencing can unravel important aspects of molecular mechanisms and processes of food-relevant microorganisms with potential applications in food industry. Functional genomics integrated with data obtained from transcriptomics, proteomics and metabolomics will ameliorate our understanding on gene regulation and the discovery of key metabolic components. Such knowledge is crucial towards the optimization of pathways leading to the development of product privileged functionalities, such as flavor, taste, texture, health-promoting components and other functional ingredients (Smid et al. 2005, Sieuwerts et al. 2008).

PCR-based techniques

The polymerase chain reaction (PCR) has greatly facilitated the identification of microbial species or strains—and in cases their abundance, as well—in food samples and the characterization of the structure of the respective microbial communities. The power of PCR technology relies on its rapidity, sensitivity, specificity and robustness. During PCR, a single or multiple nucleotide sequences are selectively amplified out of a complex mixture of DNA molecules. The targeted sequences may be specific genes or other loci, like repetitive DNA regions of known sequence or arbitrary sequences. PCR is used in several molecular typing techniques, such as restriction fragment length polymorphism (RFLP) analysis, random amplified polymorphic DNA (RAPD), amplified fragment length polymorphism (AFLP), simple sequence repeats (SSRs), automated approach for ribosomal intergenic spacer analysis (ARISA) or direct sequencing of informative genes and further phylogenetic analysis. Most PCR-based fingerprinting methods are rapid, easy to use, relatively inexpensive, and moderately reproducible.

Gene sequence analysis. The identification and phylogenetic analysis of food related taxa based on sequence analysis has been greatly ameliorated by recent advances in DNA sequencing technology, like automated and next generation sequencers. Gene sequence analysis can be applied to resolve closely related species or even strains within a species (Weisburg et al. 1991, Kurtzman 2011). The respective protocols start with total DNA extraction from pure cultures, followed by PCR amplification by using universal primers. After sequencing the amplicons, sequences are compared with other reference sequences through the Basic Local Alignment Search Tool (BLAST) in the National Center for Biotechnology Information Library (http://www.ncbi.nlm.nih.gov/blast/Blast.cgi), The Ribosomal Data Base Project (http://rdp.cme.msu.edu/) (Cole et al. 2007) or other public databases.

Among different genes or genetic loci, sequencing of the ribosomal DNA (rDNA) region has been established as the "gold standard" for indisputable identification of food-associated isolates. The interest in rRNA genes for species identification relies on two important properties: (1) ribosomes are present in all cellular organisms and appear to share a common origin, thus providing an evolutionary history shared by all organisms, (2) some regions within the rDNA are highly conserved within a species but contain informative variation amongst species and in some cases among strains. In prokaryotes, the rRNA genes are organized in a single operon, composed of the small (16S) and the large (23S) ribosomal subunit genes separated by the internal transcribed spacer (ITS). Sequence analysis of the 16S rRNA gene is the method of choice for bacterial species identification, since it is universally distributed and contains sufficient variation to discriminate between species and in a few cases among strains (Weisburg et al. 1991). By comparing the degree of DNA-DNA hybridization relatedness with 16S rDNA similarities within and among various bacterial species it has been suggested that a 16S rDNA sequence similarity of ≥97% may serve as a criterion for conspecificity (Wayne et al. 1987, Stackebrandt and Göbel 1994). However, it should be underlined that there is no universal agreement for this "threshold value", since the within-species sequence variation may be lower than 97%, while the resolution of 16S rDNA analysis between closely related species is generally low (Coenye and Vandamme 2004, and references therein). Recently, and especially after the expansion of NGS, the ribosomal multilocus sequence typing (rMLST) has been proposed as a more effective typing method for bacteria (Jolley et al. 2012). It involves the genetic variation of 53 genes encoding the subunits of the bacterial ribosomal proteins, a clearly increased number of loci compared to the more usual six to eight loci usually analysed before (Maiden 2006). Besides rDNA region, additional genes of taxonomic value are also applied for the identification of food-associated microorganisms. The most common include the *recA* gene (Felis et al. 2001, Torriani et al. 2001) and the *rpoB* gene (Rantsiou et al. 2004, Renouf et al. 2006).

In eukaryotes, the rRNA genes are organized as tandem repeated transcription units, about 100 to 200 copies, each composed by the small ribosomal subunit gene (18S) and two of the large ribosomal subunit genes (25S–28S and 5.8S). Between the small and the large subunits there are the two internal transcribed spacers (ITS1 and ITS2), flanking the 5.8S rRNA gene, while the transcription units are separated by the non-transcribed intergenic spacers (IGS), also referred to as non-transcribed spacers (NTS). The nucleotide sequences of both the rRNA genes and ITS spacers have been extensively used to analyse phylogenetic relationships in eukaryotes, because they are highly conserved within a species but differ between species (White et al. 1990, James et al. 1994, Eickbush and Eickbush 2007). The D1/D2 domain of the 26S rRNA gene, in particular, is variant enough to identify and differentiate among most ascomycetous and basidiomycetous yeasts (Kurtzman and Robnett 1998, Fell et al. 2000, Scorzetti et al. 2002, Kurtzman and Robnett 2003) and practically is well-applied in all known yeast species. The internal transcribed spacers (ITS1 and ITS2) and the non-transcribed spacer (NTS) regions are highly variable between or within species. The complex non-coding ITS spacers are more variable and when combined with the coding and more conserved 5.8S rRNA gene, collectively often referred to as ITS region, are quite useful in measuring close fungus genealogical relationships, since they exhibit greater interspecific differences than either the 18S or the 26S rRNA gene (White et al. 1990, Cai et al. 1996). These regions may be particularly useful in the identification of species that are poorly resolved by the D1/D2 domain. Other genes that have been used in yeast species identification include the translation elongation factor 1α, actin-1, RNA polymerase II, pyruvate decarboxylase, β-tubulin, and cytochrome oxidase II (Belloch et al. 2000, Daniel et al. 2001, Daniel and Meyer 2003, Kurtzman and Robnett 2003). Some of them are superior to rRNA genes for resolving relationships at various taxonomic levels, for instance in Ascomycota (Schoch et al. 2009). However, for the most of protein-coding markers there is little standardization, while the available primers usually amplify a narrow taxonomic range (Schoch et al. 2012).

In conclusion, gene sequencing appears as a robust and reliable approach for species identification and phylogenetic studies. Although DNA Sanger sequencing is considered expensive and hence not applicable for routine identification or for large sets of isolates, emerging NGS technologies have lowered the sequencing cost, rendering it widely accessible for this kind of applications.

PCR-based restriction fragment length polymorphism. PCR-based restriction fragment length polymorphism (PCR-RFLP) analysis, or simply restriction enzyme analysis (REA), applies restriction endonucleases to detect differences in homologous DNA sequences between microbial genomes. Restriction analysis produces a collection of DNA fragments of different lengths that may be separated and visualized by agarose gel electrophoresis. Sizes of restriction fragments are deduced by comparison with standard molecular weight markers. Fragment size variations used to distinguish microorganisms are due to insertions or deletions (indels) and nucleotide substitutions or sequence rearrangements in restriction sites. With respect to the DNA region subjected to RFLP analysis, differences at the inter- and intra-specific level can be resolved. The method is straightforward and fast and allows comparisons between multiple samples.

As in gene sequencing analysis, rDNA is the preferred region for detecting RFLPs in both yeasts and bacteria. The amplified ribosomal DNA restriction enzyme analysis (ARDRA) uses specific primers to amplify the complete or partial 16S, 23S or intergenic spacer in bacterial rDNA (Navarro et al. 1992, Heyndrickx et al. 1996). In ARDRA, one or two frequently cutting restriction endonucleases are typically applied. The method has found wide application in the identification of many food-associated bacteria, e.g., homo-fermentative lactobacilli from cheese (Bouton et al. 2002), gram-positive cocci from dairy food products (Devirgiliis et al. 2010), *Oenococcus oeni* from wine (Solieri et al. 2010), and *Staphylococcus* species from cheese (Hoppe-Seyler et al. 2004).

Several regions have been applied in PCR-RFLP analysis for yeast species identification in food systems including the non-transcribed spacer (NTS) (Baleiras Couto et al. 1996, Pulvirenti et al. 2001, Caruso et al. 2002, Capece et al. 2003), the 18S (Smole Mozina et al. 1997, Redzepovic et al. 2002, Capece et al. 2003) or the 26S rDNA genes (van Keulen et al. 2003, Balerias Couto 2005). The most frequently applied region is the 5.8S-ITS rDNA, usually amplified with the ITS1/ITS4 primer pair (White et al. 1990). Amplicons are subjected to restriction reaction by three or four endonucleases, such as CfoI, HaeIII, HinfI, HpaII, ScrFI, TaqI, NdeII, DdeI, DraI and MboII. The restriction patterns of numerous yeast species have been published and may be used as reference for species identification (Esteve-Zarzoso et al. 1999). This approach has been applied for the identification of various yeasts from grapes (Nisiotou and Nychas 2007), wine (Nisiotou et al. 2007), fermenting olives (Nisiotou et al. 2010, Arroyo-Lopez et al. 2012) or sourdough fermentations (Valmorri et al. 2010).

Random Amplified Polymorphic DNA. The random amplified polymorphic DNA (RAPD) method (Welsh and McClelland 1990) is one of the most widely applied approaches in genetic fingerprinting of food isolates. RAPD uses PCR with single short oligonucleotides of random sequence (about 10–15 bp in length). Amplification produces a mixture of PCR products resulting from the annealing of primers in multiple complementary sites within the genome. Amplicons are resolved by agarose gel electrophoresis and the banding profiles provide a picture of the genome's organisation. Hence, different banding profiles of individuals reflect differences between their genomes, whether members of the same or different species. Although the method is quite simple and rapid, it is highly sensitive to even small changes in PCR conditions and DNA template. Therefore, consideration should be taken in the optimization of the reaction conditions and in the interpretation of the results. Both the primers and the PCR conditions should be rigorously evaluated and the use of several sets of primers should be considered in setting up a fingerprinting experiment.

This method has been extensively applied in food samples and several studies have described the characterization of different yeast or bacterial populations associated with foods. RAPD analysis has been applied for typing lactic acid bacteria (LAB) from olives (Arroyo-Lopez et al. 2012), *O. oeni* from cider (Sanchez et al. 2012), *Carnobacterium maltaromaticum* strains from meat (Casaburi et al. 2011), *Bacillus* spp. and *Anoxybacillus flavithermus* from powdered milk (Reginensi et al. 2011), *Staphylococcus aureus* from dairy products (Morandi et al. 2010) and pseudomonads from retail-displayed beef (Aslam and Service 2008). In a recent study, the resolution of *Bacillus cereus* foodborne isolates was improved by combining three RAPD fingerprints produced with three different primers in a single pattern (Martínez-Blanch et al. 2011). The typing efficiency for *Lactobacillus sanfranciscensis* isolates from sour dough was also increased by combining the conventional RAPD (single primer) and multiplex RAPD banding profiles into a unique RAPD pattern (Venturi et al. 2012). RAPD analysis has also been applied for typing different yeast populations in foods, such as *Candida zemplinina* strains from grapes and wines (Tofalo et al. 2012), *Saccharomyces cerevisiae* from wine (Capece et al. 2010), *Debaryomyces hansenii* cheese isolates (Del Bove et al. 2009) and *Brettanomyces bruxellensis* from wines (Agnolucci et al. 2009). Deschuyffeleer et al. (2011) tested eight primers from the OPA series (Operon Technologies; http://www.operon.com) and concluded that OPA 05, OPA 09 and OPA 13 provided polymorphic banding patterns for typing of *Pichia anomala* and *Saccharomycopsis fibuligera* isolates from bread. However, when mitochondrial DNA restriction and RAPD patterns of *D. hansenii* and *Candida zeylanoides* from dry cured meat were compared, it was shown that isolates characterized by mtDNA restriction analysis as different species had the same RAPD profiles (Andrade et al. 2010).

Amplified Fragment Length Polymorphism. Amplified Fragment Length Polymorphism (AFLP) is another widely used molecular technique (Vos et al. 1995) for DNA fingerprinting of microorganisms. AFLP uses a selective PCR of restriction fragments ligated to oligonucleotide adaptors. Initially, genomic DNA is digested with two restriction enzymes, typically a frequently cutting (e.g., MseI) and a less frequently cutting endonuclease (e.g., EcoRI). The digested fragments are ligated to oligonucleotide adaptors, which hybridize to the restriction site sequences. Selective amplification is then accomplished by using primer sets complementary to the adaptor and restriction site sequences. The resolution of amplified fragments is typically performed on denaturing polyacrylamide gels. More recently, Sau-PCR, a modification of the AFLP technique, was introduced in microbial typing (Corich et al. 2005). In this technique, microbial DNA is digested with Sau3AI, a frequently cutting endonuclease, which leaves a 4-base long sticky end on digested fragments. Amplification is then applied by using primers complementary to the restriction site sequence. Sau-PCR was proposed as a more direct and rapid enzyme-based approach than AFLP considering the use of a single enzymatic reaction, the absence the ligase treatment step and the use of agarose gels to resolve amplicons.

Compared to RAPD, AFLP exhibits high reproducibility and sensitivity. Therefore, it provides a robust means for detecting DNA polymorphisms in microbial genomes of food-associated microorganisms. It has been successfully used in typing of bovine and food isolates of *Staphylococcus aureus* (Salasia et al. 2011), coagulase-negative *Staphylococcus* species from milk (Piessens et al. 2011), *Oenococcus oeni* strains from wine (Cappello et al. 2010), *Lactobacillus plantarum* isolates from vegetables and fruits (Di Cagno et al. 2010), *Listeria monocytogenes* from different foods (Parisi et al. 2010). The discrimination capacity of AFLP for *Listeria monocytogenes* was higher than serotyping (Parisi et al. 2010). When compared to other methods such as the Biolog System and RAPD, AFLP exhibited greater resolution for *L. plantarum* strains (Di Cagno et al. 2010). It was also applied for discriminating closely related strains of different yeast species associated with grape juice (de Barros Lopes et al. 1999), of *Saccharomyces sensu stricto* yeasts (Azumi and Goto-Yamamoto 2001), yeast isolates from Mexican agave-distilled beverages and grape musts (Flores Berrios et al. 2005), and *Dekkera bruxellensis* strains from red wine (Curtin et al. 2007).

Rep-PCR fingerprinting. Repetitive element sequence (rep)-PCR is a PCR fingerprinting technique with primers that anneal to repetitive DNA sequences located throughout the bacterial genome. Primers for rep-PCR were invented to target different sequences (elements) and thus there are different variations of the method: the repetitive extragenic palindromic (REP-PCR) sequences, the enterobacterial repetitive intergenic consensus (ERIC-PCR) sequence, the BOX element (BOX-PCR) and the polytrinucleotide $(GTG)_5$ sequence of diverse bacterial species (Versalovic et al. 1991, 1994, Koeuth et al.1995). During the PCR amplification, primers hybridize DNA between consecutive repetitive elements and produce a mixture of amplicons of different sizes that can be resolved in agarose gel. The resulting banding patterns provide species- and strain-specific genomic fingerprints.

Rep-PCR is a rapid fingerprinting technique of high discriminatory capacity. Genomic fingerprints are stable and can be combined with rapid cell lysis methods (Versalovic et al. 1991). $(GTG)_5$-PCR has found broad application in typing lactic acid bacteria from cheese (Solieri et al. 2012), *Lactobacillus rhamnosus* cheese strains (Bove et al. 2011), *Brochothrix* spp. from meat, poultry and fish (Xu et al. 2010), *Brochothrix thermosphacta* from pork (Papadopoulou et al. 2012), and *Bacillus cereus* strains from market food products (Samapundo et al. 2011). Typing of *Vibrio parahaemolyticus* from cockles (Zulkifli et al. 2009) and *Enterobacter sakazakii* strains from infant formula milk (Ye et al. 2010) have been achieved by ERIC-PCR. The efficiency of different repetitive sequences for typing of diverse food-associated populations has been also evaluated. ERIC-PCR was more discriminative than rep-PCR for typing of *V. cholerae* from edible ice (Waturangi et al. 2012). ERIC-PCR and $(GTG)_5$-PCR gave similar resolution in typing acetic acid bacteria from grapes (Valera et al. 2011). To achieve a higher level of resolution, more than one Rep-PCR fingerprints for a set of isolates may be attempted. By combining data from REP-PCR, ERIC-PCR and BOX-PCR in a single composite fingerprint, it was shown that the rep-PCR technique had greater discriminatory power than PFGE or MLST in typing of *Campylobacter jejuni* isolates from poultry (Wilson et al. 2009).

With the advent of new generation sequencing technologies and as complete genome sequences of bacterial species are becoming available, rep elements can be searched for in individual species genomes to design specific primers. In a recent study, Chaves et al. (2011) analysed genetic diversity of *Bacillus cereus* food isolates with specific rep primers for the *B. cereus* group (Reyes-Ramirez and Ibarra 2005).

Microsatellites. Microsatellites, also referred to as Simple Sequence Repeats (SSRs), are short DNA sequences (usually of 1–6 bp in length) of tandem repeated motifs observed in most known organisms. The SSRs are typically distributed randomly throughout the genome, spanning both coding and non-coding regions (van Belkum et al. 1998, Tóth et al. 2000). These sequence repeats are highly prone to misalignment, known as slipped-strand mispairing, during DNA replication, resulting into length polymorphism of the repeated units by either insertion or deletion mutations (Levinson and Gutman 1987, Strand et al. 1993). When an SSR is associated with a single genetic locus and shows intra-species variation it is often called "variable number of tandem repeat" (VNTR). The observed DNA length polymorphism of SSRs may serve as an invaluable molecular marker for genotyping and assessing genetic diversity within either bacterial or eukaryotic species. Briefly, PCR primers flanking the variable SSR region are designed to target well-conserved regions among strains of a species or, more rarely, among different species. The size or the sequence of amplicons, i.e., polymorphism in the repeat unit number, is documented through gel or capillary electrophoresis and/or sequencing. In order to differentiate among closely related strains of bacterial species, which are not easily resolved by other molecular marker techniques combined VNTRs from several loci may be multiplexed and analyzed simultaneously, a methodology referred to as Multi-Locus VNTR Analysis (MLVA). The development of VNTR markers for a certain species is dependent on previous knowledge of genome sequence. This limitation has restricted the wide use of VNTR, and thus MLVA, in bacterial discrimination studies. However, as more large-scale nucleotide sequence data are beginning to be available, MLVA is becoming a common sub-typing method. Hitherto, MLVA has been successfully used for typing food-associated bacterial isolates, such as non-O157 verocytotoxigenic *Escherichia coli* isolated from raw beef products (Franci et al. 2011) or *Staphylococcus aureus* in dairy products (Morandi et al. 2010).

Microsatellite fingerprinting has also greatly facilitated the resolution of genomic variation within and among yeast populations. SSRs vary extensively among or even within yeast species, ranging from 10 to more than 2000 bp in length. PCR amplification with primers complementary to sequences flanking the repeating motifs may generate simple and informative banding patterns. Several microsatellites have been described for detecting polymorphic loci that can be used for strain discrimination (Gallego et al. 1998, Hennequin et al. 2001, Perez et al. 2001, Legras et al. 2005, Masneuf-Pomarede et al. 2007). SSRs were used to explore genetic variation of *S. boulardii* from probiotic products (Vanhee et al. 2010), of *S. cerevisiae* from grapes (Schuller and Casal 2007, Schuller et al. 2012) and to discriminate commercial wine strains (Vaudano and Garcia-Moruno 2008). SSR-based fingerprinting exhibits equal discriminatory capacity to the mtDNA RFLP, interdelta sequence or karyotype analysis (Perez et al. 2001, Masneuf-Pomarede et al. 2007, Schuller and Casal 2007) and higher to that of the RFLP-ITS or -NTS (Perez et al. 2001). An important advantage of SSR is its portability, since all results can be expressed as a number of repeats (Masneuf-Pomarede et al. 2007).

In conclusion, the current wealth of sequence data originating from whole-genome sequence initiatives will help to develop informative SSRs from more species. Specific algorithms and bioinformatics tools publicly available, such as Sputnik (http://services.cbib.u-bordeaux2.fr/pise/sputnik.html), Repearmasker (http://www.repeatmasker.org/), Tandem repeats finder (http://tandem.bu.edu/trf/trf.html), WebSat (http://wsmartins.net/websat/), MISA (http://pgrc.ipk-gatersleben.de/misa/), and MICAS (http://180.149.48.108/micas/index.php), may ameliorate the development of SSR markers and may strongly support their wide-spread use in diverse genera.

Table 1. Overview of studies employing culture-dependent methods for the analysis of microbial communities in foods.

Method	Microbial target	Food	Reference
Restriction Enzyme Analysis-PFGE	*Brochothrix thermosphacta*	Pork	Papadopoulou et al. (2012)
	Enterobacteriaceae	Beef	Doulgeraki et al. (2011)
	Escherichia coli	Beef & pork	Ju et al. (2012)
	Lactic acid bacteria	Meat & meat products	Tran et al. (2011)
	Lactic acid bacteria	Wine	Ruiz et al. (2008)
	Listeria monocytogenes	Fermented meat sausages	Ferreira et al. (2011)
	Listeria monocytogenes	Diced-celery salad	Gaul et al. (2012)
	Salmonella Enteritidis	Foods & human feces	Campioni et al. (2012)
	Yeasts	Fruit pulp	Duarte et al. (2009)
	Yeasts	Rice wine	Jeyaram et al. (2008)
Amplified Ribosomal DNA Restriction Enzyme Analysis (ARDRA)	Gram-positive cocci	Dairy products	Devirgiliis et al. (2010)
	Homofermentative lactobacilli	Cheese	Bouton et al. (2002)
	Oenococcus oeni	Wine	Solieri et al. (2010)
	Staphylococcus spp.	Cheese	Hoppe-Seyler et al. (2004)
PCR-based Restriction Fragment Length Polymorphism (PCR-RFLP)	Yeasts	Grapes	Nisiotou and Nychas (2007)
	Yeasts	Wine	Nisiotou et al. (2007)
	Yeasts	Fermenting olives	Nisiotou et al. (2010)
	Yeasts	Fermenting olives	Arroyo-Lopez et al. (2012)
	Yeasts	Sour dough fermentations	Valmorri et al. (2010)
Random Amplified Polymorphic DNA (RAPD)	*Bacillus cereus*		Martínez-Blanch et al. (2011)
	Bacillus spp. & *Anoxybacillus flavithermus*	Powdered milk	Reginensi et al. (2011)
	Brettanomyces bruxellensis	Wines	Agnolucci et al. (2009)
	Candida zemplinina	Grapes & wines	Tofalo et al. (2012)
	Carnobacterium maltaromaticum	Meat	Casaburi et al. (2011)
	Debaryomyces hansenii	Cheese	Del Bove et al. (2009)

Table 1. contd....

Table 1. contd....

Method	Microbial target	Food	Reference
	Lactic acid bacteria	Olives	Arroyo-Lopez et al. (2012)
	Oenococcus oeni	Cider	Sanchez et al. (2012)
	Pseudomonads	retail-displayed beef	Aslam and Service (2008)
	Saccharomyces cerevisiae	Wine	Capece et al. (2010)
	Staphylococcus aureus	Dairy products	Morandi et al. (2010)
Amplified Fragment Length Polymorphism (AFLP)	*Dekkera bruxellensis* strains	Red wine	Curtin et al. (2007)
	Lactobacillus plantarum	Vegetables & fruits	Di Cagno et al. (2010)
	Listeria monocytogenes	Foods	Parisi et al. (2010)
	Oenococcus oeni	Wine	Cappello et al. (2010)
	Saccharomyces sensu stricto		Azumi and Goto-Yamamoto (2001)
	Staphylococcus aureus	Bovine & food	Salasia et al. (2011)
	Yeasts	Grape juice	de Barros Lopes et al. (1999)
	Yeasts	Mexican agave-distilled beverages & grape musts	Flores Berrios et al. (2005)
	Staphylococcus spp.	Milk	Piessens et al. (2011)
$(GTG)_5$-PCR	*Bacillus cereus*	Market food products	Samapundo et al. (2011)
	Brochothrix spp.	Meat, poultry & fish	Xu et al. (2010)
	Brochothrix thermosphacta	Pork	Papadopoulou et al. (2012)
	Lactic Acid Bacteria	Cheese	Solieri et al. (2012)
	Lactobacillus rhamnosus	Cheese	Bove et al. (2011)
ERIC-PCR	Acetic acid bacteria	Grapes	Valera et al. (2011)
	Enterobacter sakazakii	Infant formula milk	Ye et al. (2010)
	Vibrio cholerae	Edible ice	Waturangi et al. (2012)
	Vibrio parahaemolyticus	Cockles	Zulkifli et al. (2009)
Multi-Locus VNTR Analysis (MLVA)	Non-O157 verocytotoxigenic *Escherichia coli*	Raw beef products	Franci et al. (2011)
	Staphylococcus aureus	Dairy products	Morandi et al. (2010)
Microsatellites (SSRs)	*S. boulardii*	Probiotic products	Vanhee et al. (2010)
	S. cerevisiae	Grapes	Schuller and Casal (2007), Schuller et al. (2012)
	S. cerevisiae	Wine	Vaudano and Garcia-Moruno (2008)
PCR-DGGE	Wine bacteria	Fermented grape juice	Nisiotou et al. (2011)

Assessment of Food Microbiota by Culture-independent Molecular Methods

The culture-dependent strategy may underestimate the richness in food samples, which is more realistically estimated by culture-independent methods. Culture-independent methods provide a more actual picture of the richness of bacterial communities than culture-dependent methods because of its ability to detect species that are missed from plates.

Targeting the diversity of the 16S rRNA gene sequence is currently the most common approach for community analysis due to the phylogenetic properties and the large amount of both variable and conserved regions within this gene. The 16S rRNA gene (approx. 1,550 bp long) is universal in bacteria and hence phylogenetic relationships can be assessed among all taxa (Woese et al. 1985). Separation of 16S ribosomal DNA (rDNA) by fingerprinting (pattern analysis) or sequencing analysis are applied regarding accurate separation of taxa.

In culture-dependent methods, microorganisms grown on selective or general growth media plates can easily be isolated and identified by conventional and/or molecular biology techniques. In culture-independent methods, the exploration of microorganisms occurs by direct extraction of DNA from the substrate, e.g., food sample. In that case, both cultivable and uncultivable microorganisms that escape the standard plating approaches can be revealed. Extraction of DNA directly from samples is the first challenge, involving cell lysis, followed by separation of the nucleic acids from other sample components (e.g., polysaccharides, fat, carbohydrates, proteins or salts). The extraction of DNA is followed by PCR amplification of 16S rRNA gene. The number of PCR cycles needs to be optimized in order to reduce some of the PCR limitations, such as the production of chimeric sequences (Wang and Wang 1997) and the preferential amplification of abundant phylotypes (Chandler et al. 1997) that decrease the reliability of the determined microbial diversity in samples. Finally, the detection of 16S rRNA genes from the different bacteria present in the mixed amplification product is accomplished. The microbial composition can be mainly evaluated through fingerprinting or cloning of the PCR products (PCR-amplified 16S rRNA fragments).

In fingerprinting, there are few electrophoresis approaches used to distinguish among PCR products of different populations present in a sample. The most common techniques applied in food microbiology are the terminal restriction fragment length polymorphism (T-RFLP) which separates products by size, the denaturing gradient gel electrophoresis (DGGE) or the thermal gradient gel electrophoresis (TGGE) which separate products by sequence composition (GC-content) and the single-strand conformation polymorphism (SSCP) which separates single-stranded nucleic acids based on subtle differences in sequence (often in a single base pair).

Terminal restriction fragment length polymorphism (T-RFLP)

T-RFLP analyses variation among amplified 16S rRNA genes from different bacteria and offers information on microbial community (Liu et al. 1997, Clement et al. 1998, Osborn et al. 2000). The amplified products are cut with restriction enzymes (endonucleases) and then, the digested products are mainly separated by gel or capillary electrophoresis. Upon analysis, only the individual terminal restriction fragments (T-RFs), end-labelled 5′ and 3′ restriction fragments, are detected. In gel electrophoresis, products can be cloned and sequenced (Rudi et al. 2004). In capillary electrophoresis, electrophoretograms (peak profiles) are collected and analyzed by appropriate fragment analysis softwares, such as GeneScan (Applied Biosystems) (Rademaker et al. 2005, 2006, Nieminen et al. 2011). Forward and reverse terminal fragment lengths of the dominant bands are predicted based on available DNA sequences or web-based T-RFLP analysis programs.

Hitherto, the application of T-RFLP for the analysis of food microbial communities is rather limited (Table 2). The T-RFLP method has been shown to be a useful tool enabling rapid and high-throughput characterization of psychrotrophic bacterial communities in MAP minced meat (Nieminen et al. 2011). It has also been used for typing of lactic acid bacteria and bacilli in wine and beer (Bokulich and Mills 2012) and for the analysis of yeast communities in wine fermentation (Bokulich et al. 2012). It has been also applied to follow microbial population dynamics during yoghurt and hard cheese fermentation and ripening (Rademaker et al. 2006), as well as to describe the microbial composition of salmon and coalfish under MAP (Rudi et al. 2004).

Denaturing gradient gel electrophoresis (DGGE) and thermal gradient gel electrophoresis (TGGE)

DGGE (Myers et al. 1987) and TTGE (Yoshino et al. 1991) use electrophoresis in a denaturing gradient gel to distinguish PCR amplicons (usually PCR-amplified 16S rRNA products) of the same size but with different sequence composition (Jany and Barbier 2008). In a polyacrylamide gel, the denaturing conditions are provided by chemical denaturants (e.g., urea and formamide) or temperature in DGGE or TGGE, respectively (Justé et al. 2008). Separation

Table 2. Overview of studies employing culture-independent methods for the analysis of microbial communities in foods.

Methodologies	Food	Target locus	Reference
	Pozol balls	16S rDNA, V3	Ampe et al. (1999)
	Sausages	16S rDNA, V1-V3	Cocolin et al. (2004)
	Fermented sausages	16S rDNA, V1	Rantsiou et al. (2005)
	Beef under air & MAP	16S rDNA V6-V8	Ercolini et al. (2006)
	Farmed halibut & Atlantic cod in MAP	16S rDNA, V3	Hovda et al. (2007a,b)
	Fermenting grape must	26S rDNA, D1/ D2	Nisiotou et al. (2007)
Denaturing gradient gel electrophoresis (DGGE) & Sequencing	'Alheira' traditional Portuguese fermented sausage	16S rDNA, V3	Albano et al. (2008)
	Milk & cheese	16S rDNA,V3	Giannino et al. (2009)
	Beef under air & VP	16S rDNA, V6-V8	Pennacchia et al. (2011)
	Casein cheese (starter-free cheese made from cow's milk)	16S rDNA, V3	Alegría et al. (2009)
	Beef under VP	16S rDNA	Brightwell et al. (2009)
	'doenjang' Korean fermented soybean paste	16S rDNA, V3	Kim et al. (2009)
	'meju' fermented cooked soya beans	16S rDNA, V3	Lee et al. (2010)
	Pork under VP	16S rDNA, V3	Jiang et al. (2010)
	Atlantic mackerel	16s rDNA, V3	Svanevik and Lunestad (2011)
DGGE, Cloning & Sequencing	Pecorino Crotonese cheese	16S rDNA, V3	Randazzo et al. (2010)
	Various types of meat & cheeses	16S rDNA, various regions	Cocolin et al. (2007)
	'kunu-zaki' & 'ogi', cereal foods	16S rDNA, V3	Oguntoyinbo et al. (2011)
	Raw cow milk	16S rDNA, V3	Raats et al. (2011)

Table 2. contd....

Table 2. contd.

Methodologies	Food	Target locus	Reference
	Fermented fish products	16S rDNA, V3	An et al. (2010)
Thermal gradient gel electrophoresis (TGGE), Cloning & Sequencing	Cooked and peeled shrimps	16S rDNA, V3	Jaffrès et al. (2009)
	Darfiyeh cheese	16S rDNA, V3	Serhan et al. (2009)
	Raw salmon steaks under VP & MAP	16s rDNA, V3	Macé et al (2012)
Cloning & Sequencing	Pork under VP	16S rDNA	Schirmer et al. (2009)
	Milk	16S rDNA	Rasolofo et al. (2010)
	Seabream	16s rDNA	Parlapani et al (2013)
Terminal restriction fragment length polymorphism (T-RFLP), Cloning & Sequencing	Salmon and Coalfish under MAP	16S rDNA	Rudi et al. (2004)
T-RFLP Sequencing	Wine & beer	16S rDNA	Bokulich and Mills (2012)
T-RFLP Peak profile	MAP minced meat	16S rDNA	Nieminen et al. (2011)
	Cheese	16S rDNA	Rademaker et al. (2005)
	Yoghurt & hard cheese	16S rDNA	Rademaker et al. (2006)
Single-strand conformation polymorphism (SSCP) Peak profile	Cheese	16S rDNA, V3	Duthoit et al. (2003)
	Cheese	16S rDNA, V3	Duthoit et al. (2005a)
	Cheese	16S rDNA, V2,V3	Duthoit et al. (2005b)
Cloning & Sequencing, SSCP Peak profile	Cheese	16S rDNA, V3	Feurer et al. (2004)
Cloning, SSCP Peak profile	Cheese	18S rDNA,V4	Callon et al. (2006)
SSCP Peak profile Cloning & Sequencing	Milk & cheese	16S rDNA,V3	Delbés et al. (2007)

Table 2. contd....

Table 2. contd.

Methodologies	Food	Target locus	Reference
Pyrosequencing	Fermented seafood	16S rDNA, V3	Roh et al (2010)
	Beef under various packaging conditions	16S rDNA, V1-V3	Ercolini et al. (2011)
	'Cheonggukjang' traditional fermented soybean food product	16S rDNA, V1-V2	Nam et al. (2012a)
	'Kochujang' fermentation, traditional Korean fermented food	16S rDNA, V1-V2	Nam et al. (2012b)
	Korean fermented soybean pastes	16S rDNA, V1-V2	Nam et al. (2012c)
	'Narezushi' fermented food, Japanese sushi	16S rDNA, V1-V2	Koyanagi et al. (2011)
	'Kimchi', made of fermented vegetables	16S rDNA, V1-V3	Park et al. (2012)
	'saeu-jeot' fermentation, traditional Korean salted seafood	16S rDNA, V1-V3	Jung et al. (2013)
	Spinach leaves	16S rDNA, V4	Lopez-Velasco et al. (2011)
	Banana	16S rDNA	Rossmann et al. (2012)

of fragments is based on the decreased electrophoretic mobility of melted double stranded DNA molecules in polyacrylamide gel with the gradient of the denaturing factor (e.g., urea). DNA fragments of the same length but with different base pair compositions may show different melting behavior and hence, the different distances that the PCR products migrate can produce a distinct fingerprint profile (Muyzer 1999). To identify main fragments obtained by typing, correct positioned bands are excised from gels, purified and sequenced. After sequencing, NCBI BLAST searches are conducted to find the closest known relatives to the partial 16S rRNA gene sequences (approx. 200–700 bp). PCR-DGGE/TGGE are the most extensively molecular techniques used to directly determine microbial communities in dairy products,cheese, meat, seafood and fermenting grape must, among others (see references in Table 2). In addition, these techniques have been used for the differentiation of cultivated pure-culture microorganisms in different foods (Ampe et al. 1999, Lopez et al. 2003, Nisiotou et al. 2011, Martins et al. 2012).

Single-strand conformation polymorphism (SSCP)

SSCP analysis is used to detect sequence variations among different DNA fragments (usually PCR-amplified 16S rRNA gene fragments) through electrophoretic mobility differences. SSCP separates different DNA fragments of similar length on the basis of conformational differences in folded single-stranded products, which are visualized on gels or as peaks using an automated sequencer (Justé et al. 2008, Randazzo et al. 2009, Quigley et al. 2011). After gel electrophoresis, dominant bands may be excised from SSCP gels, purified and sequenced (Rossmann et al. 2012). In capillary electrophoresis, peak profiles are collected and analyzed using appropriate softwares, such as GeneScan (Applied Biosystems). To analyze the different profiles, the ratio of the area of one peak to the sum of the areas of all peaks is calculated (Duthoit et al. 2003, 2005a,b, Feurer et al. 2004, Callon et al. 2006, Delbés et al. 2007). The discriminatory ability of SSCP pattern analysis is generally most secure for fragments up to 700 bp in size. SSCP has been applied in determining microbiota in cheese (Duthoit et al. 2003, 2005a,b, Feurer et al. 2004, Callon et al. 2006, Delbés et al. 2007), raw milk (Delbés et al. 2007), goat milk (Callon et al. 2007) and banana samples (Rossmann et al. 2012).

16S rRNA gene analysis, cloning and sequencing

An alternative culture-independent method for microbial community analysis in foods is the 16S rRNA gene analysis followed by cloning and sequencing. Sequencing analysis of the whole 16S rRNA gene (approx. 1,550 bp) provides far more precise phylogenetic information than the above described molecular techniques. The 16S rRNA gene analysis combined with cloning and sequencing has been successfully used to determine the microbial communities in different food samples (Table 2). Undoubtedly, it may provide information for food microbial communities that often escaped attention by using other approaches.

PCR amplification of 16S rRNA genes using universal primers, followed by sequencing and phylogenetic analysis, is the commonly used culture-independent method. While PCR amplicons of 16S rRNA gene generated from pure cultures may be sequenced directly, in the case of bulk genomic DNA extracts from microbial communities, the cloning step has to be included. Upon amplification by using universal primers, a mixture of 16S rRNA gene copies corresponding to the microorganisms present in the sample is obtained. The cloned genes are sequenced and phylogenetic affiliation is determined by using bioinformatics tools. New microbial species have been revealed that may contribute to food spoilage (Schirmer et al. 2009, Rasolofo et al. 2010, Parlapani et al. 2013). Additionally, Rasolofo et al. (2010) reported clones associated with contamination from the farm environment, that had never been revealed before by other molecular approaches in milk.

Next generation sequencing (NGS)

Recently the introduction of molecular techniques using genetic information directly obtained from samples, provided a new way to analyze the 'genetic' diversity (phylotypes) of organisms from practically any natural or engineered sample. Regarding prokaryotes, most studies target on the 16S rRNA gene diversity as a proxy for diversity analysis through the newly emerged sequencing technology platforms like 454 pyrosequencing and Illumina (Ronaghi et al. 1998). The investigation of the 16S rRNA gene diversity by Sanger sequencing is not cost-and-time-effective when aiming to reveal the in-depth diversity of a sample since a large number of environmental clones needs to be analyzed. The solution came with the pyrosequencing technology (Edwards et al. 2006, Quince et al. 2009, Novais and Thorstenson 2011). The novelty of this fast and reliable methodology lies in its high discriminatory power of resulting sequences, even for very short fragments of the target gene (Jonasson et al. 2002). Although the resulting sequences from pyrosequencing are short enough (ca. 400 bp) to prevent full phylogenetic analysis, as in the case of the complete 16S or 18S rRNA genes, discrimination for bacteria to species level has proven efficient using very short (630 bp) sequences (Monstein et al. 2001).

Recent advances in sequencing technology changed the way in food microbiota exploration. The barcoded pyrosequencing analysis using 454 pyrosequencing with barcoded primers targeting different regions (e.g., V1–V2, V1–V3, V6–V8) of the bacterial 16S rRNA gene has been used to better understanding microbial diversity in various foods (see references in Table 2). Pyrosequencing analysis revealed hundreds of sequences compared to other molecular approaches such as DGGE (Roh et al. 2010, Ercolini et al. 2011, Nam et al. 2012a). Various species, never targeted by culture media, were revealed by using pyrosequencing in foods (Ercolini et al. 2011). In fermented foods, pyrosequencing analysis also allows better understanding of changes in microbial succession during fermentation processes (Koyanagi et al. 2011, Nam et al. 2012a,b,c, Park et al. 2012, Jung et al. 2013).

Conclusions and Perspectives

We now have ample and powerful tools to reveal the structural and functional diversity of practically any microbial habitat. Although this technological advantage has been recognized and is already being applied in several aspects of microbiology and microbial ecology, with subsurface microbiology and gut microbiology being the two most benefited fields, food microbiology has lagged in this direction, although this has just started to change (O'Flaherty and Klaenhammer 2011, Ercolini 2013). Of all the major categories of food, seafood seems to have attracted less attention for being investigated with the latest molecular tools (Roh et al. 2010, Koyanagi et al. 2011, Ercolini 2013, Jung et al. 2013). The ongoing technological progress in sequencing along with the bioinformatics boom, allow us not only to see which microorganisms are there but also what they are doing. As RNA-based, i.e., metatranscriptomics, and proteomics come closer providing the methodological framework for analysis of any sample at the metabolic level, it is realistic to think microbial communities not as assemblages of species (structural diversity) but rather as pools of metabolic processes (functional diversity). The effectiveness of these tools becomes more promising if we integrate the ecosystems perception to food samples. In this case, food microbiologists should lean on the accumulated knowledge gained from the application of these novel techniques from macroecology but also from other microbial habitats. Food samples can be considered as ecosystems where biological relations exist and as such should be investigated. Replicated sampling, microbial community assembly, function and succession are now within reach. These ecological features can provide us with realistic information on competition and synergism and whole community metabolic repertoires. Ultimately, such information can set secure foundations for effective monitoring, control or promotion of food microorganisms setting, thus, the potential for unprecedented progress on food bioengineering.

Keywords: food microbiota, PCR, 16S rRNA, T-RFLP, DGGE, TGGE, SSCP, DNA-DNA reassociation, restriction analysis-PFGE, RAPD, AFLP, fingerprinting, microsatellites, next generartion sequencing, pyrosequencing, molecular markers, metagenomics

References

Agnolucci, M., I. Vigentini, G. Capurso, A. Merico, A. Tirelli, C. Compagno, R. Foschino and M. Nuti. 2009. Genetic diversity and physiological traits of *Brettanomyces bruxellensis* strains isolated from Tuscan Sangiovese wines. Int. J. Food. Microbiol. 130: 238–244.

Albano, H., I. Henriques, A. Correia, T. Hogg and P. Teixeira. 2008. Characterization of microbial population of 'Alheira' (a traditional Portuguese fermented sausage) by PCR-DGGE and traditional cultural microbiological methods. J. Appl. Microbiol. 105: 2187–2194.

Alegría, Á., P. Álvarez-Martín, N. Sacristán, E. Fernández, S. Delgado and B. Mayo. 2009. Diversity and evolution of the microbial populations during manufacture and ripening of Casín, a traditional Spanish, starter-free cheese made from cow's milk. Int. J. Food Microbiol. 136: 44–51.

Altschul, S.F., T.L. Madden, A.A. Schäffer, J. Zhang, Z. Zhang, W. Miller and D.J. Lipman. 1997. Gapped BLAST and PSI-BLAST: a new generation of protein data-base search programs. Nucleic Acids Res. 25: 3389–3402.

Ampe, F., N.B. Omar, C. Moizan, C. Wacher and J.-P. Guyot. 1999. Polyphasic study of the spatial distribution of microorganisms in mexican pozol, a fermented maize dough, demonstrates the need for cultivation-independent methods to investigate traditional fermentations. Appl. Environ. Microbiol. 65: 5464–5473.

An, C., H. Takahashi, B. Kimura and T. Kuda. 2010. Comparison of PCR-DGGE and PCR-SSCP analysis for bacterial flora of Japanese traditional fermented fish products, *aji-narezushi* and *iwashi-nukazuke*. J. Sci. Food Agric. 90: 1796–1801.

Andrade, M.J., M. Rodríguez, E. Casado and J.J. Córdoba. 2010. Efficiency of mitochondrial DNA restriction analysis and RAPD-PCR to characterize yeasts growing on dry-cured Iberian ham at the different geographic areas of ripening. Meat Sci. 84: 377–383.

Aquilanti, L., S. Santarelli, G. Silvestri, A. Osimani, A. Petruzzelli and F. Clementi. 2007. The microbial ecology of a typical Italian salami during its natural fermentation. Int. J. Food Microbiol. 120: 136–145.

Arroyo-López, F.N., J. Bautista-Gallego, J. Domínguez-Manzano, V. Romero-Gil, F. Rodriguez-Gómez, P. García-García, A. Garrido-Fernández and R. Jiménez-Díaz. 2012. Formation of lactic acid bacteria and yeasts communities on the olive surface during Spanish-style Manzanilla fermentations. Food Microbiol. 32: 295–301.

Aslam, M. and C. Service. 2008. Genetic characterization of spoilage pseudomonads isolated from retail-displayed beef. Lett. Appl. Microbiol. 47: 153–157.

Azumi, M. and N. Goto-Yamamoto. 2001. AFLP analysis of type strains and laboratory and industrial strains of *Saccharomyces sensu stricto* and its application to phenetic clustering. Yeast 18: 1145–1154.

Baleiras Couto, M.M., B. Eijsma, H. Hofstra, J.H.J. Huis in't Veld and J.M.B.M. van der Vossen. 1996. Evaluation of molecular typing techniques to assign genetic diversity among *Saccharomces cerevisiae* strains. Appl. Environ. Microbiol. 62: 41–46.

Baleiras Couto, M.M., R.G. Reizinho and F.L. Duarte. 2005. Partial 26S rDNA restriction analysis as a tool to characterise non-*Saccharomyces* yeasts present during red wine fermentations. Int. J. Food Microbiol. 102: 49–56.

Barnett, J.A., R.W. Payne and D. Yarrow. 2000. Yeasts: Characteristics and identification. Third edition, Cambridge University Press.

Boekhorst, J., R.J. Siezen, M.C. Zwahlen, D. Vilanova, R.D. Pridmore, A. Mercenier, M. Kleerebezem, W.M. de Vos, H. Brüssow and F. Desiere. 2004. The complete genomes of *Lactobacillus plantarum* and *Lactobacillus johnsonii* reveal extensive differences in chromosome organization and gene content. Microbiology 150: 3601–3611.

Belloch, C., A. Querol, M.D. Garcia and E. Barrio. 2000. Phylogeny of the genus *Kluyveromyces* inferred from the mitochondrial cytochrome-c oxidase II gene. Int. J. Syst. Evol. Microbiol. 50: 405–416.

Bokulich, N.A. and D.A. Mills. 2012. Differentiation of mixed lactic acid bacteria communities in beverage fermentations using targeted terminal restriction fragment length polymorphism. Food Microbiol. 31: 126–132.

Bokulich, N., C.F. Hwang, S. Liu, K. Boundy-Mills and D.A. Mills. 2012. Profiling the Yeast communities of wine using terminal restriction fragment length polymorphism analysis. Am. J. Enol. Vitic. 63: 2185–194.

Bolotin, A., B. Quinquis, P. Renault, A. Sorokin, S.D. Ehrlich, S. Kulakauskas, A. Lapidus, E. Goltsman, M. Mazur, G.D. Pusch, M. Fonstein, R. Overbeek, N. Kyprides, B. Purnelle, D. Prozzi, K. Ngui, D. Masuy, F. Hancy, S. Burteau, M. Boutry, J. Delcour, A. Goffeau and P. Hols. 2004. Complete sequence and comparative genome analysis of the dairy bacterium *Streptococcus thermophilus*. Nat. Biotechnol. 22: 1554–1558.

Borneman, A.R., B.A. Desany, D. Riches, J.P. Affourtit, A.H. Forgan, I.S. Pretorius, M. Egholm and P.J. Chambers. 2011. Whole-genome comparison reveals novel genetic elements that characterize the genome of industrial strains of *Saccharomyces cerevisiae*. PLoS Genet. 7: e1001287.

Bouton, Y., P. Guyot, E. Beuvier, P. Tailliez and R. Grappin. 2002. Use of PCR-based methods and PFGE for typing and monitoring homofermentative lactobacilli during Comté cheese ripening. Int. J. Food Microbiol. 76: 27–38.

Bove, C.G., J. De Dea Lindner, C. Lazzi, M. Gatti and E. Neviani. 2011. Evaluation of genetic polymorphism among *Lactobacillus rhamnosus* non-starter Parmigiano Reggiano cheese strains. Int. J. Food Microbiol. 144: 569–72.

Brightwell, G., R. Clemens, K. Adam, S. Urlich and J. Boerema. 2009. Comparison of culture-dependent and independent techniques for characterisation of the microflora of peroxyacetic acid treated, vacuum-packaged beef. Food Microbiol. 26: 283–288.

Cai, J., I.N. Roberts and M.D. Collins. 1996. Phylogenetic relationships among members of the ascomycetous yeast genera *Brettanomyces, Dekkera* and *Kluyveromyces* deduced by small subunit rRNA gene sequences. Int. J. Syst. Bacteriol. 46: 542–549.

Callon, C., C. Delbés, F. Duthoit and M. Montel. 2006. Application of SSCP-PCR fingerprinting to profile the yeast community in raw milk Salers cheeses. Syst. Appl. Microbiol. 29: 172–180.

Callon, C., F. Duthoit, C. Delbés, M. Ferrand, Y. Le Frileuxc, R. De Crémoux and M. Montel. 2007. Stability of microbial communities in goat milk during a lactation year: Molecular approaches. Syst. Appl. Microbiol. 30: 547–560.

Campioni, F., A.M. Moratto Bergamini and J.P. Falcão. 2012. Genetic diversity, virulence genes and antimicrobial resistance of *Salmonella* Enteritidis isolated from food and humans over a 24-year period in Brazil. Food Microbiol. 32: 254–264.

Capece, A., G. Salzano and P. Romano. 2003. Molecular typing techniques as a tool to differentiate non-*Saccharomyces* wine species. Int. J. Food Microbiol. 84: 33–39.

Capece, A., R. Romaniello, G. Siesto, R. Pietrafesa, C. Massari, C. Poeta and P. Romano. 2010. Selection of indigenous *Saccharomyces cerevisiae* strains for Nero d'Avola wine and evaluation of selected starter implantation in pilot fermentation. Int. J. Food Microbiol. 144: 187–92.

Cappello, M.S., G. Zapparoli, D. Stefani and A. Logrieco. 2010. Molecular and biochemical diversity of *Oenococcus oeni* strains isolated during spontaneous malolactic fermentation of Malvasia Nera wine. Syst. Appl. Microbiol. 33: 461–467.

Caruso, M., A. Capece, G. Salzano and P. Romano. 2002. Typing of *Saccharomyces cerevisiae* and *Kloeckera apiculata* strains from Aglianico wine. Lett. Appl. Microbiol. 34: 323–328.

Casaburi, A., A. Nasi, I. Ferrocino, R. Di Monaco, G. Mauriello, F. Villani and D. Ercolini. 2011. Spoilage-related activity of *Carnobacterium maltaromaticum* strains in air-stored and vacuum-packed meat. Appl. Environ. Microbiol. 77: 7382–7393.

Chandler, D.P., J.K. Fredrickson and F.J. Brockman. 1997. Effect of PCR template concentration on the composition and distribution of total community 16S rDNA clone libraries. Mol. Ecol. 6: 475–482.

Chaves, J.Q., E.S. Pires and A.M. Vivoni. 2011. Genetic diversity, antimicrobial resistance and toxigenic profiles of *Bacillus cereus* isolated from food in Brazil over three decades. Int. J. Food Microbiol. 147: 12–16.

Chenu, J.W., J.M. Cox and A. Pavic. 2012. Classification of *Salmonella enterica* serotypes from Australian poultry using repetitive sequence-based PCR. J. Appl. Microbiol. 112: 185–96.

Clement, B.G., L.E. Kehl, K.L. DeBord and C.L. Kitts. 1998. Terminal restriction fragment patterns (TRFPs), a rapid PCR based method for the comparison of complex bacterial communities. J. Microbiol. Methods 31: 135–142.

Cocolin, L., K. Rantsiou, L. Iacumin, R. Urso, C. Cantoni and G. Comi. 2004. Study of the ecology of fresh sausages and characterization of populations of lactic acid bacteria by molecular methods. Appl. Environ. Microbiol. 70: 1883–1894.

Cocolin, L., A. Diez, R. Urso, K. Rantsiou, G. Comi, I. Bergmaier and C. Beimfohr. 2007. Optimization of conditions for profiling bacterial populations in food by culture-independent methods. Int. J. Food Microbiol. 120: 100–109.

Coenye, T. and P. Vandamme. 2004. Use of the genomic signature in bacterial classification and identification. System. Appl. Microbiol. 27: 175–185.

Cole, J.R., B. Chai, R.J. Farris, Q. Wang, A.S. Kulam-Syed-Mohideen, D.M. McGarrell, A.M. Bandela, E. Cardenas, G.M. Garrity and J.M. Tiedje. 2007. The ribosomal database project (RDP-II): introducing myRDP space and quality controlled public data. Nucleic Acids Res. 35: D169–D172.

Corich, V., A. Mattiazzi, E. Soldati, A. Carraro and A. Giacomini. 2005. Sau-PCR, a novel amplification technique for genetic fingerprinting of microorganisms. Appl. Environ. Microbiol. 71: 6401–6406.

Curtin, C.D., J.R. Bellon, P.A. Henschke, P.W. Godden and M.A. de Barros Lopes. 2007. Genetic diversity of *Dekkera bruxellensis* yeasts isolated from Australian wineries. FEMS Yeast Res. 7: 471–481.

Daniel, H.M. and W. Meyer. 2003. Evaluation of ribosomal RNA and actin gene sequences for the identification of ascomycetous yeasts. Int. J. Food Microbiol. 86: 61–78.

Daniel, H.M., T.C. Sorrell and W. Meyer. 2001. Partial sequence analysis of the actin gene and its potential for studying the phylogeny of *Candida* species and their teleomorphs. Int. J. Syst. Evol. Microbiol. 51: 1593–1606.

de Barros Lopes, M., S. Rainieri, P.A. Henschke and P. Langridge. 1999. AFLP fingerprinting for analysis of yeast genetic variation. Int. J. Syst. Bacteriol. 49: 915–924.

Delbés, C., L. Ali-Mandjee and M.C. Montel. 2007. Monitoring bacterial communities in raw milk and cheese by culture-dependent and -independent 16S rRNA gene-based analyses. Appl. Environ. Microbiol. 73: 1882–1891.

Delbés, C. and M.C. Montel. 2005. Design and application of Staphylococcus-specific single strand conformation polymorphism-PCR analysis to monitor Staphylococcus populations diversity and dynamics during production of raw milk cheese. Lett. Appl. Microbiol. 41: 169–174.

Del Bove, M., M. Lattanzi, P. Rellini, C. Pelliccia, F. Fatichenti and G. Cardinali. 2009. Comparison of molecular and metabolomic methods as characterization tools of Debaryomyces hansenii cheese isolates. Food Microbiol. 26: 453–459.

Deschuyffeleer, N., K. Audenaert, S. Samapundo, S. Ameye, M. Eeckhout and F. Devlieghere. 2011. Identification and characterization of yeasts causing chalk mould defects on par-baked bread. Food Microbiol. 28: 1019–1027.

Devirgiliis, C., S. Barile, A. Caravelli, D. Coppola and G. Perozzi. 2010. Identification of tetracycline- and erythromycin-resistant Gram-positive cocci within the fermenting microflora of an Italian dairy food product. J. Appl. Microbiol. 109: 313–23.

Di Cagno, R., G. Minervini, E. Sgarbi, C. Lazzi, V. Bernini, E. Neviani and M. Gobbetti. 2010. Comparison of phenotypic (Biolog System) and genotypic (random amplified polymorphic DNA-polymerase chain reaction, RAPD-PCR, and amplified fragment length polymorphism, AFLP) methods for typing *Lactobacillus plantarum* isolates from raw vegetables and fruits. Int. J. Food Microbiol. 143: 246–253.

Doulgeraki, A.I., S. Paramithiotis and G.J. Nychas. 2011. Characterization of the Enterobacteriaceae community that developed during storage of minced beef under aerobic or modified atmosphere packaging conditions. Int. J. Food Microbiol. 145: 77–83.

Duarte, W.F., D.R. Dias, G.V. de Melo Pereira, I.M. Gervasio and R.F. Schwan. 2009. Indigenous and inoculated yeast fermentation of gabiroba (*Campomanesia pubescens*) pulp for fruit wine production. J. Ind. Microbiol. Biotechnol. 36: 557–569.

Duthoit, F., C. Callon, L. Tessier and M.C. Montel. 2005b. Relationships between sensorial characteristics and microbial dynamics in 'Registered Designation of Origin' Salers cheese. Int. J. Food Microbiol. 103: 259–270.

Duthoit, F., J. Godon and M. Montel. 2003. Bacterial community dynamics during production of registered designation of origin Salers cheese as evaluated by 16S rRNA single-strand conformation polymorphism analysis. Appl. Environ. Microbiol. 69: 3840–3848.

Duthoit, F., L. Tessier and M.C. Montel. 2005a. Diversity, dynamics and activity of bacterial populations in 'Registered Designation of Origin' Salers cheese by single-strand conformation polymorphism analysis of 16S rRNA genes. J. Appl. Microbiol. 98: 1198–1208.

Dworkin, M., S. Falkow, E. Rosenberg, K.H. Schleifer and E. Stackenbrandt. 2006. The prokaryotes. Third edition. A handbook on the biology of Bacteria. Springer, Berlin.

Edwards, R., B. Rodriguez-Brito, L. Wegley, M. Haynes, M. Breitbart, D. Peterson, M. Saar, S. Alexander, E.J. Alexander and F. R. 2006. Using pyrosequencing to shed light on deep mine microbial ecology. BMC Genomics 7: 57.

Eickbush, T.H. and D.G. Eickbush. 2007. Finely orchestrated movements: evolution of the ribosomal RNA genes. Genetics 175: 477–485.

Ercolini, D., F. Russo, E. Torrieri, P. Masi and F. Villani. 2006. Changes in the spoilage related microbiota of beef during refrigerated storage under different packaging conditions. Appl. Environ. Microbiol. 72: 4663–4671.

Ercolini, D., I. Ferrocino, A. Nasi, M. Ndagijimana, P. Vernocchi, A. La Storia, L. Laghi, G. Mauriello, M.E. Guerzoni and F. Villani. 2011. Monitoring of Microbial Metabolites and bacterial diversity in beef stored under different Packaging conditions. Appl. Environ. Microb. 77(20): 7372–7381.

Ercolini, D. 2013. High-throughput sequencing and metagenomics: moving forward in the culture-independent analysis of food microbial ecology. Appl. Environ. Microb. 79: 3148–3155.

Esteve-Zarzoso, B., C. Belloch, F. Uruburu and A. Querol. 1999. Identification of yeasts by RFLP analysis of the 5.8S rRNA gene and the two ribosomal internal transcribed spacers. Int. J. Syst. Bacteriol. 49: 329–337.

Felis, G.E., F. Dellaglio, L. Mizzi and S. Torriani. 2001. Comparative sequence analysis of a *recA* gene fragment brings new evidence for a change in the taxonomy of the *Lactobacillus casei* group. Int. J. Syst. Evol. Microbiol. 51: 2113–2117.

Fell, J.W., T. Boekhout, A. Fonseca, G. Scorzetti and A. Statzell-Tallman. 2000. Biodiversity and systematics of basidiomycetous yeasts as determined by large subunit rDNA D1/D2 domain sequence analysis. Int. J. Syst. Evol. Microbiol. 50: 1351–1371.

Ferreira, V., J. Barbosa, M. Stasiewicz, K. Vongkamjan, A. Moreno Switt, T. Hogg, P. Gibbs, P. Teixeira and M. Wiedmann. 2011. Diverse geno- and phenotypes of persistent *Listeria monocytogenes* isolates from fermented meat sausage production facilities in Portugal. Appl. Environ. Microbiol. 77: 2701–2715.

Feurer, C., F. Irlinger, H.E. Spinnler, P. Glaser and T. Vallaeys. 2004. Assessment of the rind microbial diversity in a farmhouse-produced vs a pasteurized industrially produced soft red-smear cheese using both cultivation and rDNA-based methods. J. Appl. Microbiol. 97: 546–556.

Flores Berrios, E.P., J.F. Alba González, J.P. Arrizon Gavino, P. Romano, A. Capece and A. Gschaedler Mathis. 2005. The uses of AFLP for detecting DNA polymorphism, genotype identification and genetic diversity between yeasts isolated from Mexican agave-distilled beverages and from grape musts. Lett. Appl. Microbiol. 41: 147–52.

Foschino, R., D. Nucera, G. Volponi, C. Picozzi, M. Ortoffi and M.T. Bottero. 2008. Comparison of *Lactococcus garvieae* strains isolated in northern Italy from dairy products and fishes through molecular typing. J. Appl. Microbiol. 105: 652–662.

Franci, T., A.M. Sanso, A.V. Bustamante, P.M. Lucchesi and A.E. Parma. 2011. Genetic characterization of non-O157 verocytotoxigenic *Escherichia coli* isolated from raw beef products using multiple-locus variable-number tandem repeat analysis. Foodborne Pathog. Dis. 8: 1019–1023.

Gallego, F.J., M.A. Perez, I. Martinez and P. Hidalgo. 1998. Microsatellites obtained from database sequences are useful to characterize *Saccharomyces cerevisiae* strains. Am. J. Enol. Vitic. 49: 350–351.

Garrity, M.G., T.G. Lilburn, J.R. Cole, S.H. Harrison, J. Euzéby and B.J. Tindall. 2007. Taxonomic Outline of the Bacteria and Archaea, Release 7.7 March 6, 2007.

Gaul, L.K., N.H. Farag, T. Shim, M.A. Kingsley, B.J. Silk and E. Hyytia-Trees. 2012. Hospital-acquired listeriosis outbreak caused by contaminated diced celery-texas, 2010. Clin. Infect. Dis. doi: 10.1093/cid/cis817.

Gavini, F., V. Delcenserie, K. Kopeinig, S. Pollinger, H. Beerens, C. Bonaparte and M. Upmann. 2006. *Bifidobacterium* species isolated from animal feces and from beef and pork meat. J. Food Prot. 69: 871–877.

Giammanco, G.M., A. Aleo, I. Guida and C. Mammina. 2011. Molecular epidemiological survey of *Citrobacter freundii* misidentified as *Cronobacter* spp. (*Enterobacter sakazakii*) and *Enterobacter hormaechei* isolated from powdered infant milk formula. Foodborne Pathog. Dis. 8: 517–25.

Giannino, M.L., M. Marzotto, F. Dellaglio and M. Feligini. 2009. Study of microbial diversity in raw milk and fresh curd used for Fontina cheese production by culture-independent methods. Int. J. Food Microbiol. 130: 188–195.

Goodfellow, M. and A.G. O'Donnell. 1993. Roots of bacterial systematics. *In*: M. Goodfellow and A.G. O'Donnell (eds.). Handbook of new bacterial systematics. Academic, London, pp. 3–54.

Goris, J., K.T. Konstantinidis, J.A. Klappenbach, T. Coenye, P. Vandamme and J.M. Tiedje. 2007. DNA–DNA hybridization values and their relationship to whole-genome sequence similarities. Int. J. Syst. Evol. Microbiol. 57: 81–91.

Hennequin, C., A. Thierry, G.F. Richard, G. Lecointre, H.V. Nguyen, C. Gaillardin and B. Dujon. 2001. Microsatellite typing as a new tool for identification of *Saccharomyces cerevisiae* strains. J. Clin. Microbiol. 39: 551–559.

Heyndrickx, M., L. Vauterin, P. Vandamme, K. Kersters and P. De Vos. 1996. Applicability of combined amplified ribosomal DNA restriction analysis (ARDRA) patterns in bacterial phylogeny and taxonomy. J. Microbiol. Methods 26: 247–259.

Hols, P., F. Hancy, L. Fontaine, B. Grossiord, D. Prozzi, N. Leblond-Bourget, B. Decaris, A. Bolotin, C. Delorme, S. Dusko Ehrlich, E. Guédon, V. Monnet, P. Renault and M. Kleerebezem. 2005. New insights in the molecular biology and physiology of *Streptococcus thermophilus* revealed by comparative genomics. FEMS Microbiol. Rev. 29: 435–463.

Hovda, M.B., B.T. Lunestad, M. Sivertsvik and J.T. Rosnes. 2007a. Characterisation of the bacterial flora of modified atmosphere packaged farmed Atlantic cod (*Gadus morhua*) by PCR-DGGE of conserved 16S rRNA gene regions. Int. J. Food Microbiol. 117: 68–75.

Hovda, M.B., M. Sivertsvik, B.T. Lunestad, G. Lorentzen and J.T. Rosnes. 2007b. Characterisation of the dominant bacterial population in modified atmosphere packaged farmed halibut (*Hippoglossus hippoglossus*) based on 16S rDNA-DGGE. Food Microbiol. 24: 362–37.

Hoppe-Seyler, T.S., B. Jaeger, W. Bockelmann, W.H. Noordman, A. Geis and K.J. Heller. 2004. Molecular identification and differentiation of *Staphylococcus* species and strains of cheese origin. Syst. Appl. Microbiol. 27: 211–218.

Hugenholtz, P. 2002. Exploring prokaryotic diversity in the genomic era. Genome Biol. 3: Reviews 0003.

Jaffrès, E., D. Sohier, F. Leroi, M.F. Pilet, H. Prévost, J.-J. Joffraud and X. Dousset. 2009. Study of the bacterial ecosystem in tropical cooked and peeled shrimps using a polyphasic approach. Int. J. Food Microbiol. 131: 20–29.

James, S.A., M.D. Collins and N. Roberts. 1994. Genetic relationships among species of the yeast genus *Zygosaccharomyces* as revealed by small-subunit rRNA gene sequences. Yeast 20: 871–881.

Jany, J.-L. and G. Barbier. 2008. Culture-independent methods for identifying microbial communities in cheese. Food Microbiol. 25: 839–848.

Jeyaram, K., W.M. Singh, A. Capece and P. Romano. 2008. Molecular identification of yeast species associated with 'Hamei'-a traditional starter used for rice wine production in Manipur, India. Int. J. Food Microbiol. 124: 115–125.

Jiang, Y., F. Gao, X.L. Xu, Y. Su, K.P. Ye and G.H. Zhou. 2010. Changes in the bacterial communities of vacuum-packaged pork during chilled storage analyzed by PCR–DGGE. Meat Sci. 86: 889–895.

Jung, J.Y., S.H. Lee, H.J. Lee and C.O. Jeon. 2013. Microbial succession and metabolite changes during fermentation of saeu-jeot: Traditional Korean salted seafood. Food Microbiol. 34: 360–368.

Justé, A., B.P.H.J. Thomma and B. Lievens. 2008. Recent advances in molecular techniques to study microbial communities in food-associated matrices and processes. Food Microbiol. 25: 745–761.

Jolley, K.A., C.M. Bliss, J.S. Bennett, H.B. Bratcher, C. Brehony, F.M. Colles, H. Wimalarathna, O.B. Harrison, S.K. Sheppard, A.J. Cody and M.C.J. Maiden. 2012. Ribosomal multilocus sequence typing: universal characterization of bacteria from domain to strain. Microbiology 158: 1005–1015.

Jonasson, J., M. Olofsson and H.J. Monstein. 2002. Classification, identification and subtyping of bacteria based on pyrosequencing and signature matching of 16S rDNA fragments. Acta Pathol. Microbiol. Immunol. Scand. 110: 263–272.

Ju, W., J. Shen, Y. Li, M.A. Toro, S. Zhao, S. Ayers, M.B. Najjar and J. Meng. 2012. Non-O157 Shiga toxin-producing *Escherichia coli* in retail ground beef and pork in the Washington D.C. area. Food. Microbiol. 32: 371–277.

Kim, J., M.J. Jung, S.W. Roh, Y.D. Nam, K.S. Shin and J.W. Bae. 2011. *Virgibacillus alimentarius* sp. nov., isolated from a traditional Korean food. Int. J. Syst. Evol. Microbiol. 61: 2851–2855.

Kim, T.-W., J.-H. Lee, S.-E. Kim, M.-H. Park, H.C. Chang and H.-Y. Kim. 2009. Analysis of microbial communities in doenjang, a Korean fermented soybean paste, using nested PCR-denaturing gradient gel electrophoresis. Int. J. Food Microbiol. 131: 265–271.

Klijn, A., A. Mercenier and F. Arigoni. 2005. Lessons from the genomes of bifidobacteria. FEMS Microbiol. Rev. 29: 491–509.

Kleerebezem, M., J. Boekhorst, R. van Kranenburg, D. Molenaar, O.P. Kuipers, R. Leer, R. Tarchini, S.A. Peters, H.M. Sandbrink, M.W. Fiers, W. Stiekema, R.M. Lankhorst, P.A. Bron, S.M. Hoffer, M.N. Groot, R. Kerkhoven, M. de Vries, B. Ursing, W.M. de Vos and R.J. Siezen. 2003. Complete genome sequence of *Lactobacillus plantarum* WCFS1. Proc. Natl. Acad. Sci. USA 100: 1990–1995.

Koeuth, T., J. Versalovic and J.R. Lupski. 1995. Differential subsequence conservation of interspersed repetitive *Streptococcus pneumoniae* BOX elements in diverse bacteria. Genome Res. 5: 408–418.

Koort, J., T. Coenye, E.M. Santos, C. Molinero, I. Jaime, J. Rovira, P. Vandamme and J. Björkroth. 2006. Diversity of *Weissella viridescens* strains associated with "Morcilla de Burgos". Int. J. Food Microbiol. 109: 164–168.

Koort, J., T. Coenye, P. Vandamme and J. Björkroth. 2005. *Streptococcus parauberis* associated with modified atmosphere packaged broiler meat products and air samples from a poultry meat processing plant. Int. J. Food Microbiol. 106: 318–323.

Koyanagi, T., M. Kiyohara, H. Matsui, K. Yamamoto, T. Kondo, T. Katayama and H. Kumagai. 2011. Pyrosequencing survey of the microbial diversity of 'narezushi', an archetype of modern Japanese sushi. Lett. Appl. Microbiol. 53: 635–640.

Krieg, N.R. and J.G. Holt (ed.). 1984. Bergey's manual of systematic bacteriology. The Williams & Wilkins Co., Baltimore, Md.

Kurtzman, C.P. 1987. Prediction of biological relatedness among yeasts from comparisons of nuclear DNA complementarity. Stud. Mycol. 30: 459–468.

Kurtzman, C.P. 2011. Recognition of yeast species from gene sequence comparisons. Open Appl. Informat. J. 5: 20–29.

Kurtzman, C.P. and C.J. Robnett. 1998. Identification and phylogeny of ascomycetous yeasts from analysis of nuclear large subunit (26S) ribosomal DNA partial sequences. Antonie van Leeuwenhoek 73: 331–371.

Kurtzman, C.P. and C.J. Robnett. 2003. Phylogenetic relationships among yeasts of the '*Saccharomyces* complex' determined from multigene sequence analyses. FEMS Yeast Res. 3: 417–432.

Kurtzman, C.P., J.W. Fell and T. Boekhout. 2011. The Yeasts, a Taxonomic Study Elsevier 5th Edition.

Larpin-Laborde, S., M. Imran, C. Bonaïti, N. Bora, R. Gelsomino, S. Goerges, F. Irlinger, M. Goodfellow, A.C. Ward, M. Vancanneyt, J. Swings, S. Scherer, M. Guéguen and N. Desmasures. 2011. Surface microbial consortia from Livarot, a French smear-ripened cheese. Can. J. Microbiol. 57: 651–660.

Lee, J.-H., T.-W. Kim, H. Lee, H.C. Chang and H.-Y. Kim. 2010. Determination of microbial diversity in meju, fermented cooked soya beans, using nested PCR-denaturing gradient gel electrophoresis. Lett. Appl. Microbiol. 51: 388–394.

Legras, J.L., O. Ruh, D. Merdinoglu and F. Karst. 2005. Selection of hypervariable microsatellite loci for the characterization of *Saccharomyces cerevisiae* strains. Int. J. Food Microbiol. 102: 73–83.

Levinson, G. and G.A. Gutman. 1987. Slipped-strand mispairing: a major mechanism for DNA sequence evolution. Mol. Biol. Evol. 4: 203–221.

Liu, W.T., T.L. Marsh, H. Cheng and L.J. Forney. 1997. Characterization of microbial diversity by determining terminal restriction fragment length polymorphism of genes encoding 16S rRNA. Appl. Environ. Microbiol. 63: 4516–4522.

Lopez, I., F. Ruiz-Larrea, L. Cocolin, E. Orr, T. Phister, M. Marshall, J. Vander Gheynst and D.A. Mills. 2003. Design and evaluation of PCR primers for analysis of bacterial populations in wine by denaturing gradient gel electrophoresis. Appl. Environ. Microbiol. 69: 6801–6807.

Lopez-Velasco, G., G.E. Welbaum, R.R. Boyer, S.P. Mane and M.A. Ponder. 2011. Changes in spinach phylloepiphytic bacteria communities following minimal processing and refrigerated storage described using pyrosequencing of 16S rRNA amplicons. J. Appl. Microbiol. 110: 1203–1214.

Macé, S., J. Cornet, F. Chevalier, M. Cardinal, M.-F. Pilet, X. Dousset and J.-J. Joffraud. 2012. Characterisation of the spoilage microbiota in raw salmon (*Salmo salar*) steaks stored under vacuum or modified atmosphere packaging combining conventional methods and PCR-TTGE. Food Microbiol. 30: 164–172.

Maiden, M.C. 2006. Multilocus sequence typing of bacteria. Annu. Rev. Microbiol. 60: 561–588.

Martínez-Blanch, J.F., G. Sánchez, E. Garay and R. Aznar. 2011. Evaluation of phenotypic and PCR-based approaches for routine analysis of *Bacillus cereus* group foodborne isolates. Antonie Van Leeuwenhoek 99: 697–709.

Martins, G., C. Miot-Sertier, B. Lauga, O. Claisse, A. Lonvaud-Funel, G. Soulas and I. Masneuf-Pomarède. 2012. Grape berry bacterial microbiota: Impact of the ripening process and the farming system. Int. J. Food Microbiol. 158: 93–100.

Masneuf-Pomarede, I., C. Le Jeune, P. Durrens, M. Lollier, M. Aigle and D. Dubourdieu. 2007. Molecular typing of wine yeast strains *Saccharomyces bayanus* var. *uvarum* using microsatellite markers. Syst. Appl. Microbiol. 30: 75–82.

Mesbah, M., U. Premachandran and W.B. Whitman. 1989. Precise measurement of the G+C content of deoxyribonucleic acid by high-performance liquid chromatography. Int. J. Syst. Bacteriol. 39: 159–167.

Miot-Sertier, C. and A. Lonvaud-Funel. 2007. Development of a molecular method for the typing of *Brettanomyces bruxellensis* (*Dekkera bruxellensis*) at the strain level. Journal of Applied Microbiology 102: 555–562.

Monstein, H.J., S. Nikpour-Badr and J. Jonasson. 2001. Rapid molecular identification and subtyping of Helicobacter pylori by pyrosequencing of the 16S rDNA variable V1 and V3 regions. FEMS Microbiol. Lett. 199: 103–107.

Morandi, S., M. Brasca, R. Lodi, L. Brusetti, C. Andrighetto and A. Lombardi. 2010. Biochemical profiles, restriction fragment length polymorphism (RFLP), random amplified polymorphic DNA (RAPD) and multilocus variable number tandem repeat analysis (MLVA) for typing *Staphylococcus aureus* isolated from dairy products. Res. Vet. Sci. 88: 427–35.

Muyzer, G. 1999. DGGE/TGGE a method for identifying genes from natural ecosystems. Curr. Op. Microbiol. 2: 317–322.

Myers, R.M., T. Maniatis and L.S. Lerman. 1987. Detection and localization of single base changes by denaturing gradient gel electrophoresis. Method. Enzymol. 155: 501–527.

Nam, Y.-D., S.-H. Yi and S.-I. Lim. 2012a. Bacterial diversity of cheonggukjang, a traditional Korean fermented food, analyzed by barcoded pyrosequencing. Food Control 28: 135–142.

Nam, Y.-D., S.-L. Park and S.-I. Lim. 2012b. Microbial Composition of the Korean traditional food "*kochujang*" analyzed by a massive sequencing technique. J. Food Sci. 77(4): M250–M256.

Nam, Y.-D., S.-Y. Lee and S.-I. Lim. 2012c. Microbial community analysis of Korean soybean pastes by next-generation sequencing. Int. J. Food Microbiol. 155: 36–42.

Navarro, E., P. Simonet, P. Normand and R. Bardin. 1992. Characterization of natural populations of *Nitrobacter* spp. using PCR/RFLP analysis of the ribosomal intergenic spacer. Arch. Microbiol 157: 107–115.

Nieminen, T.T., E. Vihavainen, A. Paloranta, J. Lehto, L. Paulin, P. Auvinen, M. Solismaa and K.J. Björkroth. 2011. Characterization of psychrotrophic bacterial communities in modified atmosphere-packed meat with terminal restriction fragment length polymorphism. Int. J. Food Microbiol. 144: 360–366.

Nisiotou, A.A. and G.-J.E. Nychas. 2007. Yeast populations residing on healthy or *Botrytis*-infected grapes from a vineyard in Attica, Greece. Appl. Environ. Microbiol. 73: 2765–2768.

Nisiotou, A.A., A.E. Spiropoulos and G.-J.E. Nychas. 2007. Yeast community structures and dynamics in healthy and *Botrytis*-affected grape must fermentations. Appl. Environ. Microbiol. 73: 6705–6713.

Nisiotou, A.A., N. Chorianopoulos, G.J. Nychas and E.Z. Panagou. 2010. Yeast heterogeneity during spontaneous fermentation of black Conservolea olives in different brine solutions. J. Appl. Microbiol. 108: 396–405.

Nisiotou, A.A., K. Rantsiou, V. Iliopoulos, L. Cocolin and G.J. Nychas. 2011. Bacterial species associated with sound and Botrytis-infected grapes from a Greek vineyard. Int. J. Food Microbiol. 145: 432–36.

Novais, R.C. and Y.R. Thorstenson. 2011. The evolution of Pyrosequencing® for microbiology: From genes to genomes. J. Microbiol.l Meth. 86: 1–7.

Oguntoyinbo, F.A., P. Tourlomousis, M.J. Gasson and A. Narbad. 2011. Analysis of bacterial communities of traditional fermented West African cereal foods using culture independent methods. Int. J. Food Microbiol. 145: 205–210.

O'Flaherty, S. and T.R. Klaenhammer. 2011. The impact of omic technologies on the study of food microbes. Annu. Rev. Food Sci. Technol. 2: 353–371.

Orbach, M.J., D. Vollrath, R.W. Davis and C. Yanofsky. 1988. An electrophoretic karyotype of Neurospora crassa. Molecular and Cellular Biology 8: 1469–1473.

Osborn, A.M., E.R.B. Moore and K.N. Timmis. 2000. An evaluation of terminal-restriction fragment length polymorphism (T-RFLP) analysis for the study of microbial community structure and dynamics. Environ. Microbiol. 2: 39–50.

Papadopoulou, O.S., A.I. Doulgeraki, C. Botta, L. Cocolin and G.-J.E. Nychas. 2012. Genotypic characterization of *Brochothrix thermosphacta* isolated during storage of minced pork under aerobic or modified atmosphere packaging conditions. Meat Science 92: 735–738.

Parlapani, F.F., A. Meziti, K. Ar Kormas and I.S. Boziaris. 2013. Indigenous and spoilage microbiota of farmed sea bream stored in ice identified by phenotypic and 16S rRNA gene analysis. Food Microbiol. 33: 85–89.

Parisi, A., L. Latorre, G. Normanno, A. Miccolupo, R. Fraccalvieri, V. Lorusso and G. Santagada. 2010. Amplified fragment length polymorphism and multi-locus sequence typing for high-resolution genotyping of *Listeria monocytogenes* from foods and the environment. Food Microbiol. 27: 101–108.

Park, E.-J., J. Chun, C.-J. Cha, W.-S. Park, C.O. Jeon and J.-W. Bae. 2012. Bacterial community analysis during fermentation of ten representative kinds of kimchi with barcoded pyrosequencing. Food Microbiol. 30: 197–204.

Pennacchia, C., D. Ercolini and F. Villani. 2011. Spoilage-related microbiota associated with chilled beef stored in air or vacuum pack. Food Microbiol. 28: 84–93.

Perez, M.A., F.J. Gallego, I. Martinez and P. Hidalgo. 2001. Detection, distribution and selection of microsatellites (SSRs) in the genome of the yeast *Saccharomyces cerevisiae* as molecular markers. Lett. Appl. Microbiol. 33: 461–466.

Piessens, V., E. Van Coillie, B. Verbist, K. Supré, G. Braem, A. Van Nuffel, L. De Vuyst, M. Heyndrickx and S. De Vliegher. 2011. Distribution of coagulase-negative *Staphylococcus* species from milk and environment of dairy cows differs between herds. J Dairy Sci. 94: 2933–2944.

Prosser, J.I. 2012. Ecosystem processes and interactions in a morass of diversity. FEMS Microbiol. Ecol. 81: 507–519.

Pulvirenti, A., C. Caggia, C. Restuccia, M. Gullo and P. Giudici. 2001. DNA fingerprinting methods used for identification of yeasts isolated from Sicilian sourdoughs. Annals of Microbiol. 51: 107–120.

Quigley, L., O. O'Sullivan, T.P. Beresford, R.P. Ross, G.F. Fitzgerald and P.D. Cotter. 2011. Molecular approaches to analysing the microbial composition of raw milk and raw milk cheese. Int. J. Food Microbiol. 150: 81–94.

Quince, C., A. Lanzen, T.P. Curtis, R.J. Davenport, N. Hall, I.M. Head, L.F. Read and W.T. Sloan. 2009. Accurate determination of microbial diversity from 454 pyrosequencing data. Nat. Meth. 6: 639–641.

Raats, D., M. Offek, D. Minz and M. Halpern. 2011. Molecular analysis of bacterial communities in raw cow milk and the impact of refrigeration on its structure and dynamics. Food Microbiol. 28: 465–471.

Rademaker, J.L.W., J.D. Hoolwerf, A.A. Wagendorp and M.C. te Giffel. 2006. Assessment of microbial population dynamics during yoghurt and hard cheese fermentation and ripening by DNA population fingerprinting. Int. Dairy J. 16: 457–466.

Rademaker, J.L.W., M. Peinhopf, L. Rijnen, W. Bockelmann and W.H. Noordman. 2005. The surface microflora dynamics of bacterial smear-ripened Tilsit cheese determined by T-RFLP DNA population fingerprint analysis. Int. Dairy J. 15: 785–794.

Randazzo, C.L., C. Caggia and E. Neviani. 2009. Application of molecular approaches to study lactic acid bacteria in artisanal cheeses. J. Microbiol. Methods 78: 1–9.

Randazzo, C.L., I. Pitino, A. Ribbera and C. Caggia. 2010. Pecorino Crotonese cheese: Study of bacterial population and flavour compounds. Food Microbiol. 27: 363–374.

Rantsiou, K., G. Comi and L. Cocolin. 2004. The *rpoB* gene as a target for PCR-DGGE analysis to follow lactic acid bacterial population dynamics during food fermentations. Food Microbiol. 21: 481–487.

Rantsiou, K., R. Urso, L. Iacumin, C. Cantoni, P. Cattaneo, G. Comi and L. Cocolin. 2005. Culture-dependent and -independent methods to investigate the microbial ecology of Italian fermented sausages. Appl. Environ. Microbiol. 71: 1977–1986.

Rasolofo, E.A., D. St-Gelais, G. LaPointe and D. Roy. 2010. Molecular analysis of bacterial population structure and dynamics during cold storage of untreated and treated milk. Int. J. Food Microbiol. 138: 108–118.

Redzepovic, S., S. Orlic, S. Sikora, A. Majdak and I.S. Pretorius. 2002. Identification and characterisation of *Saccharomyces cerevisiae* and *Saccharomyces paradoxus* strains isolated from Croatian vineyards. Lett. Appl. Microbiol. 35: 305–310.

Reginensi, S.M., M.J. González, J.A. Olivera, M. Sosa, P. Juliano and J. Bermúdez. 2011. RAPD-based screening for spore-forming bacterial populations in Uruguayan commercial powdered milk. Int. J. Food Microbiol. 148: 36–41.

Renouf, V., O. Claisse and A. Lonvaud-Funel. 2006. *rpoB* gene: A target for identification of LAB cocci by PCR-DGGE and melting curves analysis in real-time PCR. J. Microbiol. Meth. 67: 162–170.

Reyes-Ramirez, A. and J.E. Ibarra. 2005. Fingerprinting of *Bacillus thuringiensis* type strains and isolates by using *Bacillus cereus* group-specific repetitive extragenic palindromic sequence-based PCR analysis. Appl. Environ. Microbiol. 71: 1346–1355.

Ronaghi, M., M. Uhlén and P. Nyrén. 1998. A sequencing method based on real-time pyrophosphate. Science 281: 363–365.

Roh, S.W., K.H. Kim, Y.D. Nam, H.W. Chang, E.J. Park and J.W. Bae. 2010. Investigation of archaeal and bacterial diversity in fermented seafood using barcoded pyrosequencing. The ISME Journal 4: 1–16.

Rossmann, B., H. Móller, K. Smalla, S. Mpiira, J. Baptist Tumuhairwe, C. Staver and G. Berga. 2012. Banana-associated microbial communities in uganda are highly diverse but dominated by *Enterobacteriaceae*. Appl. Environ. Microbiol. 78(14): 4933–4941.

Rudi, K., T. Maugesten, S.E. Hannevik and H. Nissen. 2004. Explorative Multivariate Analyses of 16S rRNA Gene Data from Microbial Communities in Modified-Atmosphere-Packed Salmon and Coalfish. Appl. Environ. Microb. 70(8): 5010–5018.

Ruiz, P., P.M. Izquierdo, S. Sesena and M.L. Palop. 2008. Intraspecific genetic diversity of lactic acid bacteria from malolactic fermentation of Cencibel wines as derived from combined analysis of RAPD-PCR and PFGE patterns. Food Microbiol. 25: 942–948.

Salasia, S.I., S. Tato, N. Sugiyono, D. Ariyanti and F. Prabawati. 2011. Genotypic characterization of *Staphylococcus aureus* isolated from bovines, humans, and food in Indonesia. J. Vet. Sci. 12: 353–61.

Samapundo, S., M. Heyndrickx, R. Xhaferi and F. Devlieghere. 2011. Incidence, diversity and toxin gene characteristics of *Bacillus cereus* group strains isolated from food products marketed in Belgium. Int. J. Food Microbiol. 150: 34–41.

Sanchez, A., M. Coton, E. Coton, M. Herrero, L.A. Garcia and M. Diaz. 2012. Prevalent lactic acid bacteria in cider cellars and efficiency of *Oenococcus oeni* strains. Food Microbiol. 32: 32–37.

Schirmer, B.C., E. Heir and S. Langsrud. 2009. Characterization of the bacterial spoilage flora in marinated pork products. J. Appl. Microbiol. 106: 2106–2116.

Schoch, C.L., G.H. Sung, F. López-Giráldez, J.P. Townsend, J. Miadlikowska, V. Hofstetter, B. Robbertse, P.B. Matheny, F. Kauff, Z. Wang, C. Gueidan, R.M. Andrie, K. Trippe, L.M. Ciufetti, A. Wynns, E. Fraker, B.P. Hodkinson, G. Bonito, J.Z. Groenewald, M. Arzanlou, G.S. de Hoog, P.W. Crous, D. Hewitt, D.H. Pfister, K. Peterson, M. Gryzenhout, M.J. Wingfield, A. Aptroot, S.O. Suh, M. Blackwell, D.M. Hillis, G.W. Griffith, L.A. Castlebury, A.Y. Rossman, H.T. Lumbsch, R. Lücking, B. Büdel, A. Rauhut, P. Diederich, D. Ertz, D.M. Geiser, K. Hosaka, P. Inderbitzin, J. Kohlmeyer, B. Volkmann-Kohlmeyer, L. Mostert, K. O'Donnell, H. Sipman, J.D. Rogers, R.A. Shoemaker, J. Sugiyama, R.C. Summerbell, W. Untereiner, P.R. Johnston, S. Stenroos, A. Zuccaro, P.S. Dyer, P.D. Crittenden, M.S. Cole, K. Hansen, J.M.Trappe, R. Yahr, F. Lutzoni and J.W. Spatafora. 2009. The Ascomycota tree of life: A phylum-wide phylogeny clarifies the origin and evolution of fundamental reproductive and ecological traits. Syst. Biol. 58: 224–239.

Schoch, C.L., K.A. Seifert, S. Huhndorf, V. Robert, J.L. Spouge, C.A. Levesque and W. Chen. 2012. Nuclear ribosomal internal transcribed spacer (ITS) region as a universal DNA barcode marker for Fungi. Proc. Natl. Acad. Sci. USA 109: 6241–6246.

Schuller, D. and M. Casal. 2007. The genetic structure of fermentative vineyard-associated *Saccharomyces cerevisiae* populations revealed by microsatellite analysis. Antonie Van Leeuwenhoek 91: 137–50.

Schuller, D., F. Cardoso, S. Sousa, P. Gomes, A.C. Gomes, M.A. Santos and M. Casal. 2012. Genetic diversity and population structure of *Saccharomyces cerevisiae* strains isolated from different grape varieties and winemaking regions. PLoS One 7: e32507.

Scorzetti, G., J.W. Fell, A. Fonseca and A. Statzell-Tallman. 2002. Systematics of basidiomycetous yeasts: a comparison of large subunit D1/D2 and internal transcribed spacer rDNA regions. FEMS Yeast Res. 2: 495–517.

Serhan, M., C. Cailliez-Grimal, F. Borges, A.-M. Revol-Junelles, C. Hosri and J. Fanni. 2009. Bacterial diversity of Darfiyeh, a Lebanese artisanal raw goat's milk cheese. Food Microbiol. 26: 645–652.

Shin, N.R., M.S. Kim, M.J. Jung, S.W. Roh, Y.D. Nam, E.J. Park and J.W. Bae. 2011. *Leucobacter celer* sp. nov., isolated from Korean fermented seafood. Int. J. Syst. Evol. Microbiol. 61: 2353–2357.

Sieuwerts, S., de F.A. Bok, J. Hugenholtz and J.E. van Hylckama Vlieg. 2008. Unraveling microbial interactions in food fermentations: from classical to genomics approaches. Appl. Environ. Microbiol. 74: 4997–5007.

Smid, E.J., D. Molenaar, J. Hugenholtz, W.M. de Vos and B. Teusink. 2005. Functional ingredient production: application of global metabolic models. Curr. Opin. Biotechnol. 16: 190–197.

Smole Mozina, S., D. Dlauchy, T. Deak and P. Raspor. 1997. Identification of *Saccharomyces sensu stricto* and *Torulaspora* yeasts by PCR ribotyping. Lett. Appl. Microbiol. 24: 311–315.

Solieri, L., A. Bianchi and P. Giudici. 2012. Inventory of non starter lactic acid bacteria from ripened Parmigiano Reggiano cheese as assessed by a culture dependent multiphasic approach. Syst. Appl. Microbiol. 35: 270–277.

Solieri, L., F. Genova, M. De Paola and P. Giudici. 2010. Characterization and technological properties of *Oenococcus oeni* strains from wine spontaneous malolactic fermentations: a framework for selection of new starter cultures. J. Appl. Microbiol. 108: 285–98.

Stackebrandt, E. and B.M. Göbel. 1994. Taxonomic note: a place for DNA-DNA reassociation and t65 rRNA sequence analysis in the present species definition in bacteriology. Int. J. Syst. Bacteriol. 44: 846–849.

Strand, M., T.A. Prolla, R.M. Liskay and T.D. Petes. 1993. Destabilisation of tracts of simple repetitive DNA in yeast by mutations affecting DNA mismatch repair. Nature 365: 274–276.

Svanevik, C.M. and B.T. Lunestad. 2011. Characterisation of the microbiota of Atlantic mackerel (*Scomber scombrus*). Int. J. Food Microbiol. 151: 164–170.

Tofalo, R., M. Schirone, S. Torriani, K. Rantsiou, L. Cocolin, G. Perpetuini and G. Suzzi. 2012. Diversity of *Candida zemplinina* strains from grapes and Italian wines. Food Microbiol. 29: 18–26.

Torriani, S., G.E. Felis and F. Dellaglio. 2001. Differentiation of *Lactobacillus plantarum, L. pentosus*, and *L. paraplantarum* by *recA* gene sequence analysis and multiplex PCR assay with *recA*gene-derived primers. Appl. Environ. Microbiol. 67: 3450–3454.

Tóth, G., Z. Gáspári and J. Jurka. 2000. Microsatellites in different eukaryotic genomes: survey and analysis. Genome Res. 10: 967–81.

Tran, K.T.M., B.K. May, P.M. Smooker, T.T.H. Van and P.J. Coloe. 2011. Distribution and genetic diversity of lactic acid bacteria from traditional fermented sausage. Food Res. Int. 44: 338–344.

Valera, M.J., F. Laich, S.S. González, M.J. Torija, E. Mateo and A. Mas. 2011. Diversity of acetic acid bacteria present in healthy grapes from the Canary Islands. Int. J. Food Microbiol. 151: 105–112.

Valmorri, S., R. Tofalo, L. Settanni, A. Corsetti and G. Suzzi. 2010. Yeast microbiota associated with spontaneous sourdough fermentations in the production of traditional wheat sourdough breads of the Abruzzo region (Italy). Antonie Van Leeuwenhoek 97: 119–129.

van Belkum, A., S. Scherer, L. van Alphen and H. Verbrugh. 1998. Short-sequence DNA repeats in prokaryotic genomes. Microbiol. Mol. Biol. Rev. 62: 275–293.

Van Hoorde, K., T. Verstraete, P. Vandamme and G. Huys. 2008. Diversity of lactic acid bacteria in two Flemish artisan raw milk Gouda-type cheeses. Food Microbiol. 25: 929–935.

van Keulen, H., D.G. Lindmark, E.K. Zeman and W. Gerlosky. 2003. Yeasts present during spontaneous fermentation of Lake Erie Chardonnay, Pinot Gris and Riesling. Antonie van Leeuwenhoek 83: 149–154.

Vanhee, L.M., F. Goemé, H.J. Nelis and T. Coenye. 2010. Quality control of fifteen probiotic products containing *Saccharomyces boulardii*. J. Appl. Microbiol. 109: 1745–752.

Vaudano, E. and E. Garcia-Moruno. 2008. Discrimination of *Saccharomyces cerevisiae* wine strains using microsatellite multiplex PCR and band pattern analysis. Food Microbiol. 25: 56–64.

Vegas, C., E. Mateo, A. González, C. Jara, J.M. Guillamón, M. Poblet, M.J. Torija and A. Mas. 2010. Population dynamics of acetic acid bacteria during traditional wine vinegar production. Int. J. Food Microbiol. 138(1–2): 130–36.

Venturi, M., S. Guerrini, L. Granchi and M. Vincenzini. 2012. Typing of *Lactobacillus sanfranciscensis* isolates from traditional sourdoughs by combining conventional and multiplex RAPD-PCR profiles. Int. J. Food Microbiol. 156: 122–126.

Vera, A., M.H. Ly-Chatain, V. Rigobello and Y. Demarigny. 2012. Description of a French natural wheat sourdough over 10 consecutive days focussing on the lactobacilli present in the microbiota. Antonie Van Leeuwenhoek. 101: 369–377.

Versalovic, J., M. Schneider, F.J. De Bruij and J.R. Lupski. 1994. Genomic fingerprinting of bacteria using repetitive sequence based PCR (rep-PCR). Meth. Mol. Cell. Biol. 5: 25–40.

Versalovic, J., T. Koeuth and J.R. Lupski. 1991. Distribution of repetitive DNA sequences in eubacteria and application to fingerprinting of bacterial genomes. Nucleic Acids Res. 19: 6823–6831.

Vos, P., R. Hogers, M. Bleeker, M. Reijans, T. van de Lee, M. Hornes, A. Frijters, J. Pot, J. Peleman, M. Kuiper and M. Zabeau. 1995. AFLP: a new technique for DNA fingerprinting. Nucleic Acids Res. 23: 4407–4414.

Wang, G.C.-Y. and Y. Wang. 1997. Frequency of formation of chimeric molecules is a consequence of PCR coamplification of 16S rRNA genes from mixed bacterial genomes. Appl. Environ. Microb. 63: 4645–4650.

Waturangi, D.E., I. Joanito, Y. Yogi and S. Thomas. 2012. Use of REP- and ERIC-PCR to reveal genetic heterogeneity of *Vibrio cholerae* from edible ice in Jakarta, Indonesia. Gut Pathog. 4: 2–9.

Wayne, L.G., D.J. Brenner, R.R. Colwell, P.A.D. Grimont, O. Kandler, M.I. Krichevsky, L.H. Moore, W.E.C. Moore, R.G.E. Murray, E. Stackebrandt, M.P. Starr and H.G. Trüper. 1987. Report of the ad hoc committee on reconciliation of approaches to Bacterial Systematics. Int. J. Syst. Bacteriol. 37: 463–464.

Weisburg, W.G., S.M. Barns, D.A. Pellettier and D.J. Lane. 1991. 16S ribosomal DNA amplification for phylogenetic studies. J. Bacteriol. 173: 697–703.

Welsh, J. and M. McClelland. 1990. Fingerprinting genomes using PCR with arbitrary primers. Nucleic Acids Res. 18: 7213–7218.

White, T.J., T. Bruns, S. Lee and J. Taylor. 1990. Amplification and direct sequencing of fungal ribosomal RNA genes for phylogenetics. *In*: M.A. Innis, D.H. Gelfand, J.J. Sninsky and T.J. White (eds.). PCR protocols, a guide to methods and applications. Academic Press, San Diego, Calif, pp. 315–322.

Whon, T.W., M.J. Jung, S.W. Roh, Y.D. Nam, E.J. Park, K.S. Shin and J.W. Bae. 2010. *Oceanobacillus kimchii* sp. nov. isolated from a traditional Korean fermented food. J. Microbiol. 48: 862–866.

Wilson, M.K., A.B. Lane, B.F. Law, W.G. Miller, L.A. Joens, M.E. Konkel and B.A. White. 2009. Analysis of the pan genome of *Campylobacter jejuni* isolates recovered from poultry by pulsed-field gel electrophoresis, multilocus sequence typing (MLST), and repetitive sequence polymerase chain reaction (rep-PCR) reveals different discriminatory capabilities. Microb. Ecol. 58: 843–55.

Woese, C.R., E. Stackebrandt, T.J. Macke and G.E. Fox. 1985. A phylogenetic definition of the major eubacterial taxa. Syst. Appl. Microbiol. 6: 143–151.

Xu, Y.Zh., A. Anyogu, L.I. Ouoba and J.P. Sutherland. 2010. Genotypic characterization of *Brochothrix* spp. isolated from meat, poultry and fish. Lett. Appl. Microbiol. 51: 245–251.

Ye, Y., Q. Wu, X. Xu, X. Yang, X. Dong and J. Zhang. 2010. The phenotypic and genotypic characterization of *Enterobacter sakazakii* strains from infant formula milk. J. Dairy Sci. 93: 2315–2320.

Yoshino, K., K. Nishigaki and Y. Husimi. 1991. Temperature sweep gel electrophoresis: a simple method to detect point mutations. Nucleic Acid Res. 19: 3153.

Zulkifli, Y., N.B. Alitheen, R. Son, A.R. Raha, L. Samuel, S.K. Yeap and M. Nishibuchi. 2009. Random amplified polymorphic DNA-PCR and ERIC PCR analysis on *Vibrio parahaemolyticus* isolated from cockles in Padang, Indonesia. Int. Food Res. J. 16: 141–150.

CHAPTER 11

The Use of qPCR-based Methods to Identify and Quantify Food Spoilage Microorganisms

Daniel M. Linares, Beatriz del Rio, Begoña Redruello, María Fernández, María Cruz Martín, Victor Ladero and *Miguel A. Alvarez**

Introduction

Whereas the contamination of foods by pathogenic microorganisms represents a serious threat to public health, the microbiological spoilage of food, although sometimes associated with compromised food safety, is seen as more of an inconvenience. The negative impact of non-pathogenic spoilage microorganisms (NPSMs) on food quality can, however, cause significant economic losses to the food industry. In today's competitive global market, food needs to reach consumers in optimum condition—any change in their smell, taste or appearance can lead to rejection by consumers. The USDA Economic Research Service estimates that, in 1995, over 96 billion pounds of food were lost to US retailers, food services and consumers as a consequence of food spoilage (Doyle 2007). Current consumer demand for milder processing procedures and fresher products with lower salt

Instituto de Productos Lácteos de Asturias, IPLA-CSIC, Paseo Rio Linares s/n, 33300 Villaviciosa, Asturias, Spain.
Email: maag@ipla.csic.es
* Corresponding author

concentrations, etc. (which makes preservation more difficult) could increase such losses significantly.

Microbes begin breaking down food molecules for their own metabolic needs long before any organoleptic alterations become obvious. Their rapid detection, identification and enumeration is, therefore, of the utmost importance to the food industry and related agencies; detection allows action to be taken in time to avoid negative impacts (Martínez et al. 2011). Classical microbiological methods require multiple culturing steps for microbes to be isolated and identified, and they may not provide results quickly enough for appropriate interventions to be made. Further, these methods are not always appropriate since some spoilage microorganisms fail to grow in culture media (Martínez et al. 2011). Improving the food industry's ability to react to contamination by NPSMs therefore requires new methods able to rapidly detect, identify and quantify them. In recent years, real time quantitative PCR (qPCR) has gained much acceptance in this respect, and would appear to have a promising future in the improvement of food quality surveillance (Postollec et al. 2011). The interest in this technique is reflected in the increasing number of methods available for detecting spoilage microorganisms (Fernández et al. 2006b, Ladero et al. 2008, 2010a,b), but also for i) detecting and quantifying pathogens, including viruses and parasites (Levin 2004, Querol and Fleet 2006, Makino et al. 2010), ii) monitoring the presence of antibiotic resistance genes in food-related scenarios (Manuzon et al. 2007), iii) detecting toxin-encoding genes (Fischer et al. 2007), iv) preventing ingredient fraud and monitoring the unintended contamination of special foods (e.g., gluten-free foods) (Sandberg et al. 2003, Mafra et al. 2008, Ballin et al. 2009), v) bacteriophage-detection in the dairy industry (del Rio et al. 2008), vi) detecting genetically modified organisms (Rodríguez-Lázaro et al. 2007), and vii) detecting allergens (Köppel et al. 2010).

Not all NPSMs spoil all food products; some microbial groups preferentially affect certain types of food. Potentially important food spoilage microorganisms include yeasts (*Candida, Saccharomyces, Zygosaccharomyces, Dekkera/Brettanomyces*) and moulds (*Aspergillus, Mucor, Penicillium, Geotrichum, Botrytis*), Gram-negative rod shaped bacteria (such as *Pseudomonas, Aeromonas, Vibrio, Acinetobacter, Citrobacter, Serratia, Proteus, Escherichia, Enterobacter, Erwinia, Klebsiella*), Gram-positive spore forming bacteria (*Bacillus* and *Clostridium*), lactic acid bacteria (e.g., *Streptococcus, Leuconostoc, Enterococcus* and *Pediococcus*), other Gram-positive bacteria (e.g., *Brochothrix thermosphacta, Flavobacterium, Micrococcus* spp.), Indeed, some microorganisms, e.g., fecal bacteria, such as coliforms or enterococci, and coliphages/enteroviruses, are used as bioindicators in the assessment of microbiological quality during food manufacture (Jay 2001). Their presence above certain thresholds indicates inadequate hygiene and the possible presence of pathogens (Mossel et al. 1995). Bacteriophages infecting lactic acid bacteria can also be considered as NPSMs, since they can affect the microbial balance of food fermentations, which, in turn, often leads to food spoilage.

Microbial populations appear in food because they are present in its raw materials or contaminate it at a later stage. In some cases, they are actually desirable, taking part in fermentation and thus in securing the organoleptic properties of the final product. However, in some environments, certain strains can produce undesirable compounds or promote changes leading to food decay. The total absence of NPSMs is generally regarded as unattainable.

qPCR is an appropriate tool for identifying and quantifying the NPSMs present in food. The results can be used to predict the risk of spoilage by a given microorganism, allowing preventive or corrective actions to be taken at different steps during processing, e.g., the introduction of stricter hygiene, the use of disinfectants, or even the use of heat treatments. Such early detection of NPSMs would help optimize the use of resources and reduce food losses.

This chapter examines the potential of qPCR in the detection and quantification of some of the major groups of NPSMs, and investigates its possible contribution to food stability and shelf life assessment against the background of other techniques currently employed.

Conventional and Molecular Microbiology Methods for NPSMs Detection

Current safety regulations make it essential that foods be examined for the presence of microorganisms (Frazier and Westhoff 2001). The different types of method used to monitor food microbiota, including NPSMs, can be broadly grouped into two classes: conventional and molecular.

Conventional methods

These methods are all based on the incorporation of food samples into nutrient media followed by incubation to allow any microorganisms to grow. The simplest form of growth assessment is visual. If enumeration is required, this can be performed by standard plate counting (SPC), the most probable number (MPN) technique, or direct microscopic counting (DMC) (Black 1996). Plate counting is used to calculate the number of viable cells or colony forming units (cfu) per unit volume or mass of a sample. When microbial loads are not sufficiently high to provide reliable estimates of population size by SPC (a situation commonly encountered in food quality and water sanitation studies), or when microorganisms will not grow on agar, other methods have to be used. The MPN technique estimates the number of viable organisms in a sample based on probability. Ten-fold dilutions of the sample are transferred to tubes containing a broth medium. After incubation, those tubes that show growth are recorded and compared to a standard table of results that indicates the contamination level of the product. In direct microscopic counting, diluted samples are transferred to a counting device (a haemocytometer or Neubauer chamber) consisting of a volumetric grid divided into cubes of known volume. One can then accurately count the particles in a known volume and use the figure obtained to determine the concentration of the entire sample. This method enumerates both viable and non viable cells.

Over the last 30 years, faster methods for counting microbes have been developed, based on physical processes (impedance, microcalorimetry or flow cytometry), biochemical reactions, immunological recognition (enzyme-linked immunoabsorbent assay (ELISA), and on radioimmunoassay (RIA) techniques (reviewed in detail in Fung 2000). Some have been very successful and have gained a significant share of the market. However, all these methods need a relatively large number of living microbial cells to be present, meaning culturing is required prior to any examination. Indeed, all the methods so far described require this step. At least 24 h are necessary for this, but up to five days may be required, a delay that limits the role such microbiological testing methods can play in food safety and quality assurance. In some cases, optimizing variables such as the incubation temperature, the incubation time or the oxygen concentration can further elongate the process (Gammon et al. 2007). Due to the required enrichment step, the quantification of the initial population level is not possible but only to record absence or presence. Moreover, since non-cultivatable cells cannot be detected, only a proportion of the microbiota present may be identified. Even in food matrices, in which cultivable microorganisms predominate, a large proportion of the active microbial community may remain undetected (Justé et al. 2008).

Molecular microbiological methods

Unfortunately, traditional identification methods based on laborious morphological and physiological tests usually lack discriminatory power. This can lead to misidentifications (van der Vossen et al. 1996). Molecular techniques that focus on the genes carried by microorganisms rather than on physical or metabolic characteristics potentially offer great advantages in the identification and specific detection of NPSMs (Mayo et al. 2008). Certainly, such DNA-based technologies are not influenced by the environmental conditions under which the cells are grown (Ness et al. 1993).

Although now largely obsolete due to the rise of inexpensive DNA sequencing technologies, restriction fragment length polymorphism (RFLP) analysis was the first DNA profiling technique that was cheap enough to see widespread use. In such analyses, typically used to distinguish between strains of a species, DNA is cleaved by restriction enzymes and the resulting fragments separated by electrophoresis. The different banding patterns (reflecting polymorphisms) may be observed after transferring the DNA from the gel to a blotting filter and hybridizing with labeled DNA-probes (generally based on highly conserved genes coding for rRNA) (Kelly et al. 1995, Caggia et al. 2001). Pulsed field gel electrophoresis (PFGE) allows the separation of large DNA molecules via the continuous reorientation of the electric field during gel electrophoresis. This also affords the possibility of typing microorganisms (Swartz and Cantor 1984, Zhang and Holley 1999).

The polymerase chain reaction (PCR). The aforementioned techniques are, however, rather laborious and time-consuming, since considerable amounts of relatively pure DNA or RNA are required, which again means a prior culturing step is necessary. In addition, these methods are generally suitable for the typing of NPSMs but cannot actually be used for their detection in an investigative sense. They are not, therefore, appropriate for making routine identifications in the food industry. The polymerase chain reaction (PCR), however, is generally rapid and easy to perform. With this technique, a specific DNA-fragment is amplified by a thermocyclic process (Mullis et al. 1986). Only a small amount of DNA is required, and it does not necessarily need to be highly purified. Different PCR techniques are available. One of the most popular is Random Amplified Polymorphic DNA (RAPD) analysis. In RAPD assays, patterns are generated by the amplification of random DNA segments with arbitrary small primers, and their subsequent resolution in a gel. This allows NPSMs to be typed (Williams et al. 1990).

Another useful technique is denaturing gradient gel electrophoresis (DGGE), a fingerprinting method that generates a genetic diversity pattern via the efficient separation of PCR amplicons that may vary by as little as the modification of a single base pair (Muyzer et al. 1993, 1998, Sheffield et al. 1989).

Both RAPD and DGGE discriminate at the species level, but unfortunately patterns are not always reproducible. PCR methods for the accurate detection of specific microorganisms are therefore currently under development. DNA amplicons of the expected size confirm the presence of a given NPSM or gene in a sample. Various PCR methods have been developed for detecting certain spoilage bacteria, such as tyramine-producing lactic acid bacteria in cheese (Fernández et al. 2006a), *Brochothrix* (Grant et al. 1993) and *Carnobacterium* in meat (Brooks et al. 1992), alcohol-tolerant lactic acid bacteria in wine and beer (Tsuchiya et al. 1992, Nakagawa et al. 1994), and *Clostridium tyrobutyricum* spores in cheese (Herman et al. 1995). In fact, several types of microorganism can be simultaneously detected by combining primer pairs in the same reaction (multiplex PCR). This has been used to detect different bacteriophages in milk (del Rio et al. 2007) and food-borne bacteria that can potentially produce histamine, tyramine or putrescine (Marcobal et al. 2005, Coton et al. 2010).

Quantitative Real-Time PCR: A Powerful Tool for Inspecting Food Quality

Unlike food pathogens, NPSMs are permissible in some foods up to a certain number. It may, therefore, be essential to quantify the initial size of a particular NPSM population in order to assess the spoilage risk it poses and to predict a food's shelf-life. This is not easy with classical PCR methods. Quantitative real-time PCR (qPCR), however, is particularly suited to such quantification tasks. Indeed, it is able to detect small numbers of specific microorganisms.

The basics and chemistry of qPCR

qPCR is a powerful tool that can be used to detect and quantify a DNA target of a particular NPSM. Data are collected over the entire reaction, the quantification of the target DNA being based on the cycle in which it becomes detectable rather than the amount accumulated at the end of the reaction. Such monitoring of the reaction is made possible via the use of fluorescent chemicals that label the synthesized DNA, and a fluorescence detector (incorporated into the thermocycler) that measures the emitted fluorescence. As a result, qPCR results are expressed in terms of "threshold cycle" (Ct), i.e., the cycle in which the fluorescence becomes detectable by the equipment. The Ct is inversely proportional to the initial number of copies of the specific target DNA: the larger the initial number of target DNA copies, the earlier the fluorescence threshold will be reached, and hence the lower the Ct value (Fig. 1).

Two different types of chemistry can be used in qPCR fluorescence: (1) non-specific dyes that fluoresce when they intercalate with any double-stranded DNA (dsDNA), and (2) sequence-specific DNA probes consisting of oligonucleotides labeled with a reporter that fluoresce only when a probe hybridizes with its target. Non-specific fluorescent dyes include ethidium bromide, YO-PRO-I, SYBR green I, SYBR Gold, BEBO, BOXTO, LCGrenn and SYT09. These molecules are DNA minor-groove binders that emit a strong fluorescent signal only when associated with dsDNA and

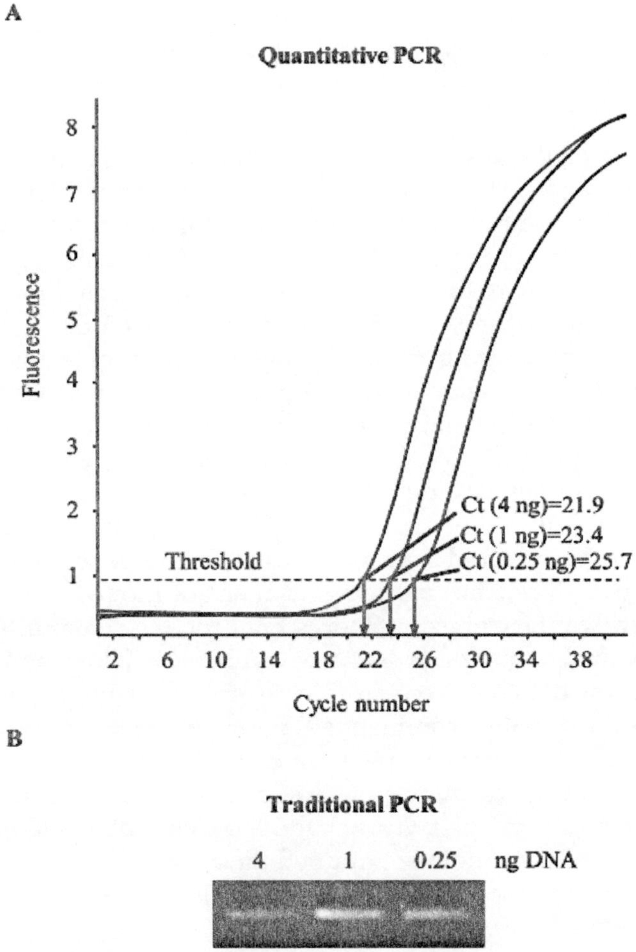

Figure 1. Analysis of different quantities (0.25, 1 and 4 ng) of bacterial genomic DNA by qPCR and traditional PCR using similar thermal profiles. (A) Amplification plot of genomic DNA analyzed by qPCR. The corresponding Ct value for each quantity of DNA is indicated. (B) Agarose gel electrophoresis of the amplicons generated by traditional PCR amplification.

are exposed to the appropriate wavelength. The flexibility and low cost of non-specific fluorescent dyes, and the accurate results they return, has made them by far the most commonly used (especially SYBR Green). However, the formation of primer dimers is common. Such dimers, or indeed any other non-specific amplified DNA, result in quantification errors. These problems can be solved, at least in part, by melting curve analysis using appropriate software; such analysis is strongly recommended when this type of fluorophore is used (Martínez et al. 2011). Further, since the melting curve of a given amplicon may be characteristic, its analysis may allow the identification of a microorganism (Ladero et al. 2012).

The use of fluorescent probes adds an additional level of specificity to qPCR reactions. These are oligonucleotides complementary to a nucleotide sequence between the targets of the two primers; they fluoresce only when bound to their targets. This greater specificity solves the problems associated with the use of non-specific fluorescent dyes. Different types of probe have been developed, including TaqMan, Molecular Beacon, Scorpion, Eclipse and DANP probes, all of them based on the presence of a fluorophore and a quencher. In the absence of target DNA, the probe does not hybridize and fluorescence remains quenched through fluorescence resonance energy transfer (FRET). When the target DNA is present, however, the primers and the probe hybridize to their target sequences and the DNA polymerase starts the amplification process. When the probe is reached by the polymerase, 5′–3′ exonuclease activity degrades it. As a result, the fluorophore is released from the quencher and starts to fluoresce (Logan et al. 2009). For both, non-specific dyes and DNA probes, the Ct is dependent on the initial concentration of DNA, thus allowing its quantification.

Another advantage of using fluorescent probes is that multiple probes labeled with different fluorophores can be combined to allow the quantitative detection of more than one target in a single reaction (multiplex qPCR). This feature offers the (highly recommended) possibility of including an internal positive amplification control in the same reaction. Since many inhibitory compounds may be present, the use of such controls is particularly useful in food analysis (Levin 2004). Further, the TaqMan probe method has been optimized for FAST technology (Applied Biosystems, Warrington, United Kingdom), allowing the detection and quantification of NPSMs in just 30 min (del Rio et al. 2008).

Requirements for optimum qPCR performance

qPCR methods can be designed to allow the simultaneous detection and quantification of DNA. Quantification is undertaken by comparing the sample Ct value against a standard curve usually prepared from serial dilutions of a target DNA solution of known concentration. Following the amplification of the standard dilution series, the Ct values obtained are represented against the concentration of the dilutions used. If the efficiency of the amplification does not change over the range of the template concentrations employed, the plotting of these points should generate a straight line. The comparison of the Ct values of the unknown samples to this standard curve allows the quantification of the sample.

Accuracy is of great importance in microbiological analysis. Quantification requires the optimal design or selection of the primers, probes, the target DNA, and heat-resistant DNA polymerase taking part in the PCR process. Special attention should also be paid to the possible presence of inhibitors and to the efficiency of DNA extraction (Martínez et al. 2011, Schneider et al. 2009).

Primer and probe design is the most challenging step in preparing for any qPCR analysis. The sequence region to be used as the target must be considered carefully. For optimal performance, the region spanned by the primers (measured from the 5′end of each) should be 60–120 bp in length for probe-based methods, and 100–400 bp in length if a fluorescent dye is used. The general points to be taken into account when designing primers for conventional PCR should be also kept in mind here. The selection of the target gene is very important. It should be highly conserved across species for broad-based detection strategies, and one should ensure that the primers bind equally well to the target DNA of all the species in mind. However, a sequence for specifically detecting each species or even strain is required; this requires the target sequence show no homology across related species or strains (Hanna et al. 2005). In any case it is advisable to invest all the time required to design the oligonucleotides well. When designing a fluorescent probe, it is necessary to ensure that the fluorophore and quencher pair are compatible. In addition, when designing multiplexed reactions, the fluorophores and quenchers chosen for the different targets should minimize the spectral overlap to avoid possible crosstalk.

The extraction of the DNA from food samples is one of the most critical steps in ensuring reproducibility. Nowadays, most procedures employ commercial extraction kits, each adapted for use with a specific food matrix. Internal controls can be used to determine the effects of the food matrix on qPCR efficiency. In this regard, detection based on fluorescence probes is recommendable since it allows the incorporation of an internal control in the reaction (Levin 2004). In some cases, methods can be optimized so that no DNA purification is necessary; rather, a small sample of the food can be used directly as a template (Martín et al. 2008).

The thermocycle profile is quite similar to that used in conventional PCR, although the shorter size of the amplification products usually makes the elongation step unnecessary. The concentration of DNA, probes and primers, as well as the PCR conditions, should be optimal in order to guarantee high sensitivity (a low detection limit) but also to ensure a broad dynamic range (efficiency). If multiplex PCR is performed, all the primers and probes used in the reaction should be of similar length, Tm and GC content. Moreover, the relative primer concentrations need to be optimized; it is important that the lowest possible concentrations be used.

qPCR Assays for the Detection and Quantification of Food Spoilage Microorganisms

In recent years, great advances have been made in qPCR applications for food microbiology. This is reflected in the increasing number of research articles on the subject (Reynisson et al. 2008, Renard et al. 2008, Broda et al. 2009, Rawsthorne and Phister 2009, Prevost et al. 2010, Makino et al. 2010, Gray et al. 2011). Nowadays, qPCR is being increasingly used to monitor food quality, which may be affected by microbial spoilage (Venkitanarayanan et al. 1997, Rawsthorne et al. 2006, Pennachia et al. 2009). Table 1 lists a number of studies that demonstrate the ability of qPCR for the detection of fastidious organisms, their toxin-encoding genes etc., in food.

The introduction of qPCR to the food industry requires technical staff be trained in its use, the adaptation of protocols described in the scientific literature for use in analytical laboratories, and the availability of reagents in kit form at reasonable prices. Several commercial kits for the detection of the main foodborne pathogens exist and have already been validated for use in the laboratory. However, although many oligonucleotides have been designed for the identification of NPSMs, most have not yet reached the market. As far as we know, only a few commercial kits for the detection of spoilage microorganisms are available at present (Table 2).

In the BAX® System Yeast and Mold Assay the target is the ribosomal RNA genes (rRNA). A Foodproof® Beer Screening Kit uses a mixture of primers and hybridisation probes based on conserved and variable 16S–23S spacer regions. The Foodproof® Beer Screening Kit is based on real-time PCR and melting curve analysis that allows the detection and identification of different bacteria in the same PCR-run. The kit contains all reagents needed for the detection of beer spoilage bacteria including UNG (Uracil N-Glycosylase) to prevent carryover contamination. The Foodproof® *S. cerevisiae* var. *diastaticus* Detection Kit allows the detection of this yeast in beer. Foodproof® *Dekkera* Quantification kit and Foodproof® *Alicyclobacillus* Detection Kit uses a fluorogenic probe to enable the detection of a specific PCR product as it accumulates during PCR (TaqMan®-based products in a 5′nuclease assay). In the Hygiene Screening System the method is based on fluorescence-coupled PCR technology. The kits from Gen-ial mainly use SYBR Green I for the product detection. A new First Beer P1 HybProbe Kit is based on hybridization probes and detects both beer-spoilage bacteria and yeasts in a single test.

Table 1. qPCR-based methods described in the literature for monitoring spoilage in different food products. In each case the target gene and spoilage microorganism are indicated, as well as the qPCR fluorescent marker employed. (* ITS: Internal Transcribed Spacer region including the 5.8 rRNA gene; FRNA: F-specific RNA; IR: intergenic region).

Microorganism	Food	Gene target	qPCR fluorescent marker	Reference
Yeasts and Molds				
Yeasts and Molds	Yogurts and pasteurized food products	*act*	SYBR Green	Bleve et al. 2003
Yeasts and Molds	Fruit juice	*CS1*	SYBR Green	Casey et al. 2004
Yeasts	Orange juice	5.8S-ITS*	SYBR Green	Renard et al. 2008
Yeasts	Wine	26S rRNA	SYBR Green	Hierro et al. 2006
Molds	Orange juice	18S rRNA	Taqman probe	Wan et al. 2006
Saccharomyces cerevisiae	Wine	RAPD-fragment	SYBR Green	Martorell et al. 2005
Saccharomyces cerevisiae	Wine	*mrp2*	Taqman probe	Salinas et al. 2009
Brettanomyces	Wine	*rad4*	SYBR Green	Tessonnière et al. 2009
Hanseniaspora sp.	Wine/Juice	D1/D2 26S rRNA	SYBR Green	Phister et al. 2007
Zygosaccharomyces bailli	Wine and fruit juices	26S rRNA	SYBR Green	Rawsthorne and Phister 2006
Dekkera bruxellensis	Wine	26S rRNA	SYBR Green	Phister and Mills 2003
Dekkera anomala	Beer/cola/cider	26S rRNA	SYBR Green	Gray et al. 2011
Bacteria				
Firmicutes Phylum	Beer	16S rRNA	Taqman probe	Haakensen et al. 2008
Total viable bacteria	Refrigerated fish	16S rRNA	SYBR Green	Lee and Levin 2007
Strictly anaerobic bacteria of the class *Clostridia*	Beer	16S rRNA	SYBR Green	Juvonen et al. 2008
Lactic acid bacteria cocci	Fermented beverages	*rpoB*	SYBR Green	Renouf et al. 2006
Lactic acid bacteria	Fermented milk	16S rRNA	SYBR Green	Furet et al. 2004
Lactic acid bacteria	Beer	*horA*	Taqman probe	Haakensen et al. 2007
Lactic acid bacteria	Wine	16S rRNA	SYBR Green	Neeley et al. 2005

Table 1. contd....

Table 1. contd.

Microorganism	**Food**	**Gene target**	**qPCR fluorescent marker**	**Reference**
Lactococcus lactis subsp. *cremoris*	Milk fermented	16S rRNA	SYBR Green	Grattepanche et al. 2005
Thermophilic bacilli	Milk powder	16S rRNA	Taqman probe	Rueckert et al. 2005
Thermophilic bacilli	Milk powder	*spo0A*	SYBR Green	Rueckert et al. 2006
Alicyclobacillus spp.	Apple juice	*shc*	Taqman probe	Luo et al. 2004
Alicyclobacillus spp.	Juice products	16S rRNA	Taqman probe	Connor et al. 2005
Bacillus spp.	Milk	*hblC*	Molecular Beacon NASBA	Gore et al. 2003
Brettanomyces bruxellensis and ropy *Pediococcus damnosus* strains	Wine	*Rad4/dps*	SYBR Green	Delaherche et al. 2004
Clostridium tyrobutyricum	Milk	*fla*	Taqman probe	López-Enríquez et al. 2007
Enterococcus gilvus	Cheese	*pheS*	Taqman probe	Zago et al. 2009
Lactobacillus sakei	Meat/Fermented sausages	16-23S rRNA ITS	Taqman probe	Martín et al. 2006
Brochothrix thermosphacta	Raw meat	16S rRNA	SYBR Green	Pennachia et al. 2009
Pseudomonas	Fish	*carA*	SYBR Green	Reynisson et al. 2008
Obesumbacterium proteus	Beer	16S rRNA	SYBR Green	Koivula et al. 2006
Gluconobacter and *Gluconacetobacter*	Electrolyte replacement drink	16S rRNA	Taqman probe	Gammon et al. 2007
Acetic acid bacteria	Wine	16S rRNA	SYBR Green	González et al. 2006
Xylella fastidiosa	Wine	16-23S rRNA ITS	Taqman probe	Schaad et al. 2002
Leuconostoc mesenteroides	Meat products	23S rRNA	Taqman probe	Elizaquivel et al. 2008
Bacillus cereus, Bacillus subtilis, Bacillus licheniformis	Sterile food	16S rRNA	Taqman probe	Fernández-No et al. 2011
Brochothrix thermosphacta	Cooked shrimp and salmon	16S rRNA	SYBR Green	Mamlouk et al. 2012
Escherichia coli	Water	*clpB*	NASBA Molecular beacon	Heijnen and Medema 2009

Escherichia coli	Tuna/mince beef/oyster	*uidA*	Taqman probe	Takahashi et al. 2009
Escherichia coli	Fresh produce/water	*uidA*	Molecular beacons	Sandhya et al. 2008
Lactose-fermenting *Enterobacteriaceae*	Dairy products	*lacZ*	SYBR Green	Martín et al. 2010
Gram + biogenic amine producers				
Histamine	Cheese	*hdcA*	SYBR Green	Fernández et al. 2006b
Histamine, tyramine and putrescine	Wine	*hdcA/tdc/odc/agdi*	SYBR Green	Nannelli et al. 2008
Tyramine	Dairy products	*tdc*	SYBR Green	Ladero et al. 2010b
Tyramine	Fermented food	*tdc*	SYBR Green	Torriani et al. 2008
Putrescine	Dairy products	*aguD-aguA* IR	SYBR Green	Ladero et al. 2012
Gram – biogenic amine producers				
Histamine	Fish	*hdc*	Taqman probe	Bjornsdottir-Butler et al. 2011
Viruses				
FRNA bacteriophage (enteric viruses)	Oysters	NoV	Taqman probe	Flannery et al. 2009
Enteric viruses	Shellfish	NV/AdV/TTV/TLMV	SYBR Green	Myrmel et al. 2004
Enteric viruses	Berries and vegetables	NV	Taqman probe	Butot et al. 2007
Bacteriophages of *Streptococcus thermophilus*	Milk	*orf1510/orf18*	Taqman probe	del Rio et al. 2008
Bacteriophages of *Lactobacillus delbrueckii*	Milk	*mur/lysA*	Taqman probe	Martín et al. 2008
Bacteriophages of *Lactococcus lactis*	Dairy facilities	*orf6/mcp*	SYBR Green	Verreault et al. 2011

Table 2. Commercially available qPCR kits for the detection and quantification of spoilage microorganisms in food products and beverages.

Kit name	**Target species**	**Sample**	**Manufacturer**
BAX® System Yeast and Mold	Yeasts and moulds	Prepared food	DuPont Qualicon
Foodproof® Detection System	*Enterobacteriaceae*	Prepared food	Biotecon Diagnostics GmbH
Foodproof® Beer Screening Kit	*Lactobacillus* *Megasphaera* *Pectinatus* *Pediococcus*	Beer	Biotecon Diagnostics GmbH
Foodproof® *Saccharomyces cerevisiae* var. *diastaticus* Detection Kit	*Saccharomyces cerevisiae* var. *diastaticus*	Beverages	Biotecon Diagnostics GmbH
Foodproof® *Dekkera* Quantification Kit	*Dekkera bruxelensis* *Dekkera anomala*	Alcoholic and non-alcoholic beverages	Biotecon Diagnostics GmbH
Foodproof® *Alicyclobacillus* Detection Kit	*Alicyclobacillus*	All relevant food	Biotecon Diagnostics GmbH
Hygiene Screening System	*Corynebacterium* *Staphylococcus* *Macrococcus* *Micrococcus* *Kocuria* *Kytococcus*	Environmental samples	Biotecon Diagnostics GmbH
Primermix P1 Screening	*Acetobacter* *Lactobacillus* *Megasphaera* *Pectinatus* *Pediococcus* *Selenomonas*	Beer	Gen-ial
GEN-IAL® P1 Hyb Probe Screening	*Lactobacillus* *Pediococcus* *Pectinatus* *Megasphaera* Yeasts	Beer or beverage samples	Gen-ial

Final Remarks and Future Perspectives

The use of qPCR for the quantification of NPSMs in food offers many advantages over other microbiological, biochemical and molecular techniques. The versatility, rapidity, and sensitivity of this technique, together with its capacity to quantify target organisms within complex matrices, suggest it to have a future in improving the safety and quality of food products. However, its use for NPSM quantification in food samples must overcome a number of technical, economic and legal hurdles before it becomes routine.

Although most of the technical difficulties affecting the use of qPCR in food matrices have been solved for pathogen detection (Levin 2004, Hanna et al. 2005, Cocolin et al. 2011), there are still some problems to address. The main challenge is developing a DNA extraction method that can be used across the range of food textures, compositions (concentrations of proteins, sugars and fat) and physical states (solid, liquid, etc.). In addition, when working with food samples, special attention must be paid to the possible presence of inhibitors. The best means of avoiding this inconvenience is the incorporation of an internal control (Levin 2004). This would also allow the identification of false negatives. The selection of the DNA extraction method must be made taking into account the type of microorganisms to be detected since their quantification can vary by up to two log units depending of the lysis method used (Cheng and Griffiths 2003). In addition, the nucleic acid extraction method employed must be adapted to the analyzed food matrix in order to ensure its efficiency and that it can render DNA preparations of repeatable quantity and quality (Demeke and Jenkins 2010).

For legal reasons, a method adapted for use with different foodstuffs must usually be accredited. Unfortunately, most of the qPCR methods available for use with NPSMs have not been standardized as ISO or APHA norms, a major inconvenience for the food industry. A growing number of such methods have been accredited, however, for the detection of foodborne pathogens.

Although the investment necessary to start up a qPCR laboratory remains high, reaction costs have been reduced in recent years. Further, it must be taken into account that qPCR greatly reduces the time needed to detect spoilage microorganisms. The identification and quantification of NPSMs by conventional microbiological or biochemical methods takes days, whereas with qPCR this can be done in a few hours. This may be crucial for taking appropriate, timely decisions regarding the final use of a foodstuff (Martínez et al. 2011). Moreover, in some qPCR systems up to 384 samples can be tested at any time, with no need for post-amplification processing. The speed and efficiency of analysis may also improve as qPCR technology evolves. For instance, FAST technology can reduce assay times enormously (del Rio et al. 2008, Martínez et al. 2011), and its automation could reduce the number of personnel required to perform analyses and increase the number of samples that can be processed per unit time (high throughput). The potential of qPCR should also expand as the technology is developed. For instance, current commercial technologies can discriminate up to five different fluorescent dyes, allowing for the simultaneous detection of four different spoilage organisms plus an internal control. In the future, many more types of organism may be detected at the same time.

Finally, qPCR quantification could serve to help construct microbial growth prediction models for NPSMs. These could be very useful in the design of quantitative detection protocols, hazard analysis and critical control point systems, and help in the making of accurate predictions of shelf life (Gram and Dalgaard 2002).

Conclusions

qPCR has proven its usefulness in basic microbiological research. Its capacity to amplify DNA from a wide range of sample types makes it an ideal system for use in different microbiological disciplines (Mackay 2004). Consequently, it has aroused great interest in food microbiology and food safety circles (Hanna et al. 2005, Martínez et al. 2011), and a number of commercial kits have been developed and validated as methods for detecting food-borne pathogens. With respect to the quantification of foodborne spoilage microorganisms, qPCR could eventually be routinely used to improve the safety of foodstuffs and help establish better quality control in food processing facilities.

Acknowledgements

This work was funded by the Ministry of Economy and Competivity, Spain (AGL2010-18430). DML and BDR are the recipients of a JAE-Doc contract (CSIC). The authors thank Adrian Burton for linguistic assistance.

Keywords: spoilage microorganisms, PCR, qPCR, amplification, amplicons, threshold cycle, fluorescent dyes, TaqMan

References

Ballin, N.Z., F.K. Vogensen and A.H. Karlsson. 2009. Species determination—Can we detect and quantify meat adulteration? Meat Sci. 83: 165–174.

Bjornsdottir-Butler, K., J.L. Jones, R. Benner and W. Burkhardt. 2011. Development of a real-time PCR assay with an internal amplification control for detection of gram-negative histamine producing bacteria in fish. Food Microbiol. 28: 356–363.

Black, J.G. 1996. Microbiology. Principles and Applications. Third Edition. Prentice Hall. Upper Saddle River, New Jersey, pp. 140–144.

Bleve, G., L. Rizzotti, F. Dellaglio and S. Torriani. 2003. Development of reverse transcription (RT)-PCR and real-time RT-PCR assays for rapid detection and quantification of viable yeasts and molds contaminating yogurts and pasteurized food products. Appl. Environ. Microbiol. 69: 4116–4122.

Broda, D.M., J.A. Boerema and G. Brightwell. 2009. Sources of psychrophilic and psychrotolerant clostridia causing spoilage of vacuum-packed chilled meats, as determined by PCR amplification procedure. J. Appl. Microbiol. 107: 178–186.

Brooks, J.L., A.S. Moore, R.A. Patchett, M.D. Collins and R.G. Kroll. 1992. Use of the polymerase chain reaction and oligonucleotide probes for the rapid detection and identification of *Carnobacterium* species from meat. J. Appl. Bacteriol. 72: 294–301.

Butot, S., T. Putallaz and G. Sánchez. 2007. Procedure for rapid concentration and detection of enteric viruses from berries and vegetables. Appl. Environ. Microbiol. 73: 186–192.

Caggia, C., C. Restuccia, A. Pulvirenti and P. Giudici. 2001. Identification of *Pichia anomala* isolated from yoghurt by RFLP of the ITS region. Int. J. Food Microbiol. 71: 71–73.

Casey, G.D. and A.D. Dobson. 2004. Potential of using real-time PCR-based detection of spoilage yeast in fruit juice—a preliminary study. Int. J. Food Microbiol. 91: 327–335.

Cheng, Z. and M.W. Griffiths. 2003. Rapid detection of *Campylobacter jejuni* in chicken rinse water by melting-peak analysis of amplicons in real-time polymerase chain reaction. J. Food Prot. 66: 1343–1352.

Cocolin, L., A. Rajkovic, K. Rantsiou and M. Uyttendaele. 2011. The challenge of merging food safety diagnostic needs with quantitative PCR platforms. Trends Food Sci. Technol. 22: S30–S38.

Connor, C.J., H. Luo, B.B. Gardener and H.H. Wang. 2005. Development of a real-time PCR-based system targeting the 16S rRNA gene sequence for rapid detection of *Alicyclobacillus* spp. in juice products. Int. J. Food Microbiol. 99: 229–235.

Coton, M., A. Romano, G. Spano, K. Ziegler, C. Vetrana, C. Desmarais, A. Lonvaud-Funel, P. Lucas and E. Coton. 2010. Occurrence of biogenic amine-forming lactic acid bacteria in wine and cider. Food Microbiol. 27: 1078–1085.

del Rio, B., A.G. Binetti, M.C. Martín, M. Fernández, A.H. Magadán and M.A. Alvarez. 2007. Multiplex PCR for the detection and identification of dairy bacteriophages in milk. Food Microbiol. 24: 75–81.

del Rio, B., M.C. Martín, N. Martínez, A.H. Magadán and M.A. Álvarez. 2008. Multiplex fast real-time PCR for quantitative detection and identification of cos- and pac-type *Streptococcus thermophilus* bacteriophages. Appl. Environ. Microbiol. 74: 4779–4781.

Delaherche, A., O. Claisse and A. Lonvaud-Funel. 2004. Detection and quantification of *Brettanomyces bruxellensis* and 'ropy' *Pediococcus damnosus* strains in wine by real-time polymerase chain reaction. J. Appl. Microbiol. 97: 910–915.

Demeke, T. and G.R. Jenkins. 2010. Influence of DNA extraction methods, PCR inhibitors and quantification methods on real-time PCR assay of biotechnology-derived traits. Anal. Bioanal. Chem. 396: 1977–1990.

Doyle, M.E. 2007. Microbial food spoilage: Losses and control strategies. Food Research Institute, University of Wisconsin-Madison, WI 53706.

Elizaquível, P., E. Chenoll and R. Aznar. 2008. A TaqMan-based real-time PCR assay for the specific detection and quantification of *Leuconostoc mesenteroides* in meat products. FEMS Microbiol. Lett. 278: 62–71.

Fernández, M., A.B. Flórez, D.M. Linares, B. Mayo and M.A. Álvarez. 2006a. Early PCR detection of tyramine-producing bacteria during cheese production. J. Dairy Res. 73: 318–321.

Fernández, M., B. del Río, D.M. Linares, M.C. Martín and M.A. Álvarez. 2006b. Real-Time polymerase chain reaction for quantitative detection of histamine-producing bacteria: use in cheese production. J. Dairy Sci. 89: 3763–3769.

Fernández-No, I.C., M. Guarddon, K. Böhme, A. Cepeda, P. Calo-Mata and J. Barros-Velázquez. 2011. Detection and quantification of spoilage and pathogenic *Bacillus cereus*, *Bacillus subtilis* and *Bacillus licheniformis* by real-time PCR. Food Microbiol. 28: 605–610.

Fischer, A., C. von Eiff, T. Kuczius, K. Omoe, G. Peters and K. Becker. 2007. A quantitative real-time immuno-PCR approach for detection of staphylococcal enterotoxins. J. Mol. Med. 85: 461–469.

Flannery, J., S. Keaveney and W. Doré. 2009. Use of FRNA bacteriophages to indicate the risk of Norovirus contamination in Irish oysters. J. Food Protec. 72: 2358–2362.

Frazier, W.C. and D.C. Westhoff. 2001. Food microbiology. Tata Mcgraw-Hill Eds. New Delhi.

Fung, D. 2000. Rapid methods and automation in microbiology. Compreh. Rev. Food Sci. Food Safety 1: 3–22.

Furet, J.P., P. Quénée and P. Tailliez. 2004. Molecular quantification of lactic acid bacteria in fermented milk products using real-time quantitative PCR. Int. J. Food Microbiol. 97: 197–207.

Gammon, K.S., S. Livens, K. Pawlowsky, S.J. Rawling, S. Chandra and A.M. Middleton. 2007. Development of real-time PCR methods for the rapid detection of low concentrations of *Gluconobacter* and *Gluconacetobacter* species in an electrolyte replacement drink. Lett. Appl. Microbiol. 44: 262–267.

González, A., N. Hierro, M. Poblet, A. Mas and J.M. Guillamón. 2006. Enumeration and detection of acetic acid bacteria by real-time PCR and nested PCR. FEMS Microbiol. Lett. 254: 123–128.

Gore, H.M., C.A. Wakeman, R.M. Hull and L.L. McKillip. 2003. Real-time molecular beacon NASBA reveals *hblC* expression from *Bacillus* spp. in milk. Biochem. Biophys. Res. Commun. 311: 386–390.

Gram, L. and P. Dalgaard. 2002. Fish spoilage bacteria-problems and solutions. Curr. Opin. Biotech. 13: 262–266.

Grant, K.A., J.H. Dickinson, M.J. Payne, S. Campbell, M.D. Collins and R.G. Kroll. 1993. Use of the polymerase chain reaction and 16S rRNA sequences for the rapid detection of *Brochothrix* spp. in foods. J. Appl. Bacteriol. 74: 260–267.

Grattepanche, F., C. Lacroix, P. Audet and G. Lapointe. 2005. Quantification by real-time PCR of *Lactococcus lactis* subsp. *cremoris* in milk fermented by a mixed culture. Appl. Microbiol. Biotechnol. 66: 414–21.

Gray, S.R., H. Rawsthorne, B. Dirks and T.G. Phister. 2011. Detection and enumeration of *Dekkera anomala* in beer, cola, and cider using real-time PCR. Lett. Appl. Microbiol. 52: 352–359.

Haakensen, M., C.M. Dobson, H. Deneer and B. Ziola. 2008. Real-time PCR detection of bacteria belonging to the Firmicutes Phylum. Int. J. Food Microbiol. 125: 236–241.

Haakensen, M.C., L. Butt, B. Chaban, H. Deneer, B. Ziola and T. Dowgiert. 2007. *horA*-specific real-time PCR for detection of beer-spoilage lactic acid bacteria. J. Am. Soc. Brew. Chem. 65: 157–165.

Hanna, S.E., C.J. Connor and H.H. Wang. 2005. Real-time polymerase chain reaction for the food microbiologist: technologies, applications, and limitations. J. Food Sci. 70: R49–R53.

Heijnen, L. and G. Medema. 2009. Method for rapid detection of viable *Escherichia coli* in water using real-time NASBA. Water Res. 43: 3124–32.

Herman, L.M., J.H. De Block and G.M. Waes. 1995. A direct PCR detection method for *Clostridium tyrobutyricum* spores in up to 100 milliliters of raw milk. Appl. Environ. Microbiol. 61: 4141–4146.

Hierro, N., B. Esteve-Zarzoso, A. González, A. Mas and J.M. Guillamón. 2006. Real-time quantitative PCR (QPCR) and reverse transcription-QPCR for detection and enumeration of total yeasts in wine. Appl. Environ. Microbiol. 72: 7148–55.

Jay, J.M. 2001. Indicator organisms in foods. *In*: Y.H. Hui, M.D. Pierson and J.R. Gorham (eds.). Foodborne disease handbook (2nd ed.). New York: Marcel Dekker.

Justé, A., B.P. Thomma and B.A. Lievens. 2008. Recent advances sin molecular techniques to study microbial communities in food-associated matrices and processes. Food Microbiol. 25: 745–761.

Juvonen, R., T. Koivula and A. Haikara. 2008. Group-specific PCR-RFLP and real-time PCR methods for detection and tentative discrimination of strictly anaerobic beer-spoilage bacteria of the class *Clostridia*. Int. J. Food Microbiol. 125: 162–169.

Kelly, W.J., R.V. Asmundson, G.L. Harrison and C.M. Huang. 1995. Differentiation of dextran-producing *Leuconostoc* strains from fermented rice cake using PFGE. Int. J. Food Microbiol. 26: 345–352.

Koivula, T.T., R. Juvonen, A. Haikara and M.L. Suihko. 2006. Characterization of the brewery spoilage bacterium *Obesumbacterium proteus* by automated ribotyping development of PCR methods for its biotype 1. J. Appl. Microbiol. 100: 398–406.

Köppel, R., V. Dvorak, F. Zimmerli, A. Breitenmoser, A. Eugster and H.U. Waiblinger. 2010. Two tetraplex real-time PCR for the detection and quantification of DNA from eight allergens in food. Eur. Food Res. Technol. 230: 367–374.

Ladero, V., D.M. Linares, M. Fernández and M.A. Álvarez. 2008. Real time quantitative PCR detection of histamine-producing lactic acid bacteria in cheese: relation with histamine content. Food Res. Int. 41: 1015–1019.

Ladero, V., M. Fernández, I. Cuesta and M.A. Álvarez. 2010a. Quantitative detection and identification of tyramine-producing enterococci and lactobacilli in cheese by multiplex qPCR. Food Microbiol. 27: 933–939.

Ladero, V., N. Martínez, M.C. Martín, M. Fernández and M.A. Álvarez. 2010b. qPCR for quantitative detection of tyramine-producing bacteria in dairy products. Food Res. Int. 43: 289–295.

Ladero, V., C. Cañedo, M. Perez, M.C. Martín, M. Fernández and M.A. Álvarez. 2012. Multiplex qPCR for the detection and quantification of putrescine-producing lactic acid bacteria in dairy products. Food Control 27: 307–313.

Lee, J.L. and R.E. Levin. 2007. Quantification of total viable bacteria on fish fillets by using ethidium bromide monoazide real-time polymerase chain reaction. Int. J. Food Microbiol. 118: 312–317.

Levin, R.E. 2004. The application of real-time PCR to food and agricultural systems. A review. Food Biotechnol. 18: 97–133.

Logan, J., K. Edwards and N. Saunders. 2009. Real-Time PCR: Current technology and applications (1st ed.). Caister Academic Press. Norfolk, UK.

López-Enríquez, L., D. Rodríguez-Lázaro and M. Hernández. 2007. Quantitative detection of *Clostridium tyrobutyricum* in milk by Real-Time PCR. Appl. Environ. Microbiol. 73: 3747–3751.

Luo, H., A.E. Yousef and H.H. Wang. 2004. A real-time polymerase chain reaction-based method for rapid and specific detection of spoilage *Alicyclobacillus* spp. in apple juice. Lett. Appl. Microbiol. 39: 376–382.

Mackay, I.M. 2004. Real-time PCR in the microbiology laboratory. Clin. Microbiol. Infect. 10: 190–212.

Mafra, I., I. Ferreira, M. Beatriz and P.P. Oliveira. 2008. Food authentication by PCR-based methods. Eur. Food Res. Technol. 227: 649–665.

Makino, H., J. Fujimoto and K. Watanabe. 2010. Development and evaluation of a real-time quantitative PCR assay for detection and enumeration of yeasts of public health interest in dairy products. Int. J. Food Microbiol. 140: 76–83.

Mamlouk, K., S. Macé, M. Guilbaud, E. Jaffrès, M. Ferchichi, H. Prévost, M.F. Pilet and X. Dousset. 2012. Quantification of viable *Brochothrix thermosphacta* in cooked shrimp and salmon by real-time PCR. Food Microbiol. 30: 173–179.

Manuzon, M.Y., S.E. Hanna, H. Luo, Z. Yu, W.J. Harper and H.H. Wang. 2007. Quantitative assessment of the tetracycline resistance gene pool in cheese samples by real-time TaqMan PCR. Appl. Environ. Microbiol. 73: 1676–1677.

Marcobal, A., B. de las Rivas, M.V. Moreno-Arribas and R. Muñoz. 2005. Multiplex PCR method for the simultaneous detection of histamine-, tyramine-, and putrescine-producing lactic acid bacteria in foods. J. Food Prot. 68: 874–878.

Martín, B., A. Jofré, M. Garriga, M. Pla and T. Aymerich. 2006. Rapid quantitative detection of *Lactobacillus sakei* in meat and fermented sausages by real-time PCR. Appl. Environ. Microbiol. 72: 6040–6048.

Martín, M.C., B. del Río, N. Martínez, A.H. Magadán and M.A. Álvarez. 2008. Fast real-time polymerase chain reaction for quantitative detection of *Lactobacillus delbrueckii* bacteriophages in milk. Food Microbiol. 25: 978–982.

Martín, M.C., N. Martínez, B. del Río, V. Ladero, M. Fernández and M.A. Álvarez. 2010. A novel real-time polymerase chain reaction-based method for the detection and quantification of lactose-fermenting *Enterobacteriaceae* in the dairy and other food industries. J. Dairy Sci. 93: 860–867.

Martínez, N., M.C. Martín, A. Herrero, M. Fernández, M.A. Álvarez and V. Ladero. 2011. qPCR as a powerful tool for microbiol food spoilage quantification: Significance for food quality. Trends Food Sci. Technol. 22: 367–376.

Martorell, P., A. Querol and M.T. Fernández-Espinar. 2005. Rapid identification and enumeration of *Saccharomyces cerevisiae* cells in wine by real-time PCR. Appl. Environ. Microbiol. 71: 6823–30.

Mayo, B., M. Marzotto, A.B. Flórez and S. Torriani. 2008. Culture independent microbial techniques in dairy microbiology: The state of the art. *In*: B. Mayo, P. López and G. Pérez-Martínez. Molecular aspects of lactic acid bacteria for traditional and new applications. Research Signpost, Kerala, India, pp. 1–27.

Mossel, D.A., J.E. Corry, C.B. Struijk and R.M. Baird. 1995. Essentials of the microbiology of foods: A textbook for advanced studies. England: Wiley.

Mullis, K., F. Faloona, S. Scharf, R. Saiki, G. Horn and H. Erlich. 1986. Specific enzymatic amplification of DNA *in vitro*: the polymerase chain reaction. Cold Spring Harb Symp. Quant. Biol. 51: 263–273.

Muyzer, G., E.C. de Waal and A.G. Uitterlinden. 1993. Profiling of complex microbial populations by denaturing gradient gel electrophoresis analysis of polymerase chain reaction-amplified genes coding for 16S rRNA. Appl. Environ. Microbiol. 59: 695–700.

Muyzer, G. and S. Kornelia. 1998. Application of denaturing gradient gel electrophoresis (DGGE) and temperature gradient gel electrophoresis (TGGE) in microbial ecology. Antonie van Leeuwenhoek 73: 127–141.

Myrmel, M., E.M. Berg, E. Rimstad and B. Grinde. 2004. Detection of enteric viruses in shellfish from the Norwegian coast. Appl. Environ. Microbiol. 70: 2678–2684.

Nakagawa, T., M. Shimada, H. Mukai, K. Asada, I. Kato, K. Fujino and T. Sato. 1994. Detection of alcohol-tolerant Hiochi bacteria by PCR. Appl. Environ. Microbiol. 60: 637–640.

Nannelli, F., O. Claisse, E. Gindreau, G. de Revel, A. Lonvaud-Funel and P.M. Lucas. 2008. Determination of lactic acid bacteria producing biogenic amines in wine by quantitative PCR methods. Lett. Appl. Microbiol. 47: 594–599.

Neeley, E.T., T.G. Phister and D.A. Mills. 2005. Differential real-time PCR assay for enumeration of lactic acid bacteria in wine. Appl. Environ. Microbiol. 71: 8954–8957.

Ness, F., F. Lavallee, D. Dubourdieu, M. Aigle and L. Dulau. 1993. Identification of yeast strains using the polymerase chain reaction. J. Sci. Food Agric. 62: 89–94.

Pennacchia, C., D. Ercolini and F. Villani. 2009. Development of a Real-Time PCR assay for the specific detection of *Brochothrix thermosphacta* in fresh and spoiled raw meat. Int. J. Food Microbiol. 134: 230–236.

Phister, T.G. and D.A. Mills. 2003. Real-Time PCR assay for detection and enumeration of *Dekkera bruxellensis* in wine. Appl. Environ. Microbiol. 69: 7430–7434.

Phister, T.G., H. Rawsthorne, C.M. Lucy Joseph and D.A. Mills. 2007. Real-Time PCR assay for detection and enumeration of *Hanseniaspora* species from wine and juice. Am. J. Enol. Vitic. 58: 229–233.

Postollec, F., H. Falentin, S. Pavan, J. Combrisson and D. Sohier. 2011. Recent advances in quantitative PCR (qPCR) applications in food microbiology. Food Microbiol. 28: 848–861.

Prevost, S., S. Andre and F. Remize. 2010. PCR detection of thermophilic spore-forming bacteria involved in canned food spoilage. Curr. Microbiol. 61: 525–533.

Querol, A. and G.H. Fleet. 2006. Yeast in foods and beverages. Berlin: Springer-Verlag. (Chapter 12).

Rawsthorne, H. and T.G. Phister. 2006. A real-time PCR assay for the enumeration and detection of *Zygosaccharomyces bailii* from wine and fruit juices. Int. J. Food Microbiol. 112: 1–7.

Rawsthorne, H. and T.G. Phister. 2009. Detection of viable *Zygosaccharomyces bailii* in fruit juices using ethidium monoazide bromide and real-time PCR. Int. J. Food Microbiol. 131: 246–50.

Renard, A., P. Gómez di Marco, M. Egea-Cortines and J. Weiss. 2008. Application of whole genome amplification and quantitative PCR for detection and quantification of spoilage yeasts in orange juice. Int. J. Food Microbiol. 126: 195–201.

Renouf, V., O. Claisse and A. Lonvaud-Funel. 2006. *rpoB* gene: A target for identification of LAB cocci by PCR-DGGE and melting curves analyses in real time PCR. J. Microbiol. Methods 67: 162–170.

Reynisson, E., H.L. Lauzon, H. Magnusson, G.O. Hreggvidsson and V.T. Marteinsson. 2008. Rapid quantitative monitoring method for the fish spoilage bacteria *Pseudomonas*. J. Environ. Monit. 10: 1357–1362.

Rodríguez-Lázaro, D., B. Lombard, H. Smith, A. Rzezutka, M. D'Agostino, R. Helmuth, A. Schroeter, B. Malorny, A. Miko, B. Guerra, J. Davison, A. Kobilinsky, M. Hernández, Y. Bertheau and N. Cook. 2007. Trends in analytical methodology in food safety and quality: monitoring microorganisms and genetically modified organisms. Trends Food Sci. Technol. 18: 306–319.

Rueckert, A., R.S. Ronimus and H.W. Morgan. 2005. Development of a rapid detection and enumeration method for thermophilic bacilli in milk powders. J. Microbiol. Methods 60: 155–167.

Rueckert, A., R.S. Ronimus and H.W. Morgan. 2006. Development of a real-time PCR assay targeting the sporulation gene, *spo0A*, for the enumeration of thermophilic bacilli in milk powder. Food Microbiol. 23: 220–230.

Salinas, F., D. Garrido, A. Ganga, G. Veliz and C. Martínez. 2009. Taqman real-time PCR for the detection and enumeration of *Saccharomyces cerevisiae* in wine. Food Microbiol. 26: 328–332.

Sandberg, M., L. Lundberg, M. Ferm and I.M. Yman. 2003. Real time PCR for the detection and discrimination of cereal contamination in gluten free foods. Eur. Food Res. Technol. 217: 344–349.

Sandhya, S., W. Chen and A. Mulchandani. 2008. Molecular beacons: a real-time polymerase chain reaction assay for detecting *Escherichia coli* from fresh produce and water. Anal. Chim. Acta 614: 208–212.

Schaad, N.W., D. Opgenorth and P. Gaush. 2002. Real-Time polymerase chain reaction for one-hour on-site diagnosis of Pierce's disease of grape in early season asymptomatic vines. Phytopathol. 92: 721–728.

Schneider, S., J. Enkerli and F. Widmer. 2009. A generally applicable assay for the quantification of inhibitory effects on PCR. J. Microbiol. Methods 78: 351–353.

Schwartz, D.C. and C.R. Cantor. 1984. Separation of yeast chromosome-sized DNAs by pulsed field gradient gel electrophoresis. Cell 37: 67–75.

Sheffield, V.C., D.R. Cox, L.S. Lerman and R.M. Myers. 1989. Attachment of a 40-base-pair G + C-rich sequence (GC-clamp) to genomic DNA fragments by the polymerase chain reaction results in improved detection of single-base changes. Proc. Natl. Acad. Sci. USA 86: 232–236.

Takahashi, H., B. Kimura, Y. Tanaka, J. Shinozaki, T. Suda and T. Fujii. 2009. Real-time PCR and enrichment culture for sensitive detection and enumeration of *Escherichia coli*. J. Microbiol. Methods 79: 124–127.

Tessonnière, H., S. Vidal, L. Barnavon, H. Alexandre and F. Remize. 2009. Design and performance testing of a real-time PCR assay for sensitive and reliable direct quantification of *Brettanomyces* in wine. Int. J. Food Microbiol. 129: 237–243.

Torriani, S., V. Gatto, S. Sembeni, R. Tofalo, G. Suzzi, N. Belletti, F. Gardini and S. Bover-Cid. 2008. Rapid detection and quantification of tyrosine decarboxylase gene (*tdc*) and its expression in gram-positive bacteria associated with fermented foods using PCR-based methods. J. Food Protec. 71: 93–101.

Tsuchiya, Y., H. Kaneda, Y. Kano and S. Koshino. 1992. Detection of beer spoilage organisms by polymerase chain reaction technology. Am. Soc. Brew. Chem. J. 50: 64–67.

van der Vossen, J.M. and H. Hofstra. 1996. DNA based typing, identification and detection systems for food spoilage microorganisms: development and implementation. Int. J. Food Microbiol. 33: 35–49.

Venkitanarayanan, K.S., C. Faustman, J.F. Crivello, M.I. Khan, T.A. Hoagland and B.W. Berry. 1997. Rapid estimation of spoilage bacterial load in aerobically stored meat by a quantitative polymerase chain reaction. J. Appl. Microbiol. 82: 359–364.

Verreault, D., L. Gendron, G.M. Rousseau, M. Veillette, D. Massé, W.G. Lindsley, S. Moineau and C. Duchaine. 2011. Detection of airborne lactococcal bacteriophages in cheese manufacturing plants. Appl. Environ. Microbiol. 77: 491–497.

Wan, K., A.E. Yousef, S.J. Schwartz and H.H. Wang. 2006. Rapid, specific, and sensitive detection of spoilage molds in orange juice using a real-time Taqman PCR assay. J. Food Prot. 69: 385–390.

Williams, J.G., A.E. Kubelik, K.J. Levak, J.A. Rafalski and S.C. Tingey. 1990. DNA polymorphisms amplified by arbitrary primers are useful as genetic markers. Nucleic Acids Res. 18: 6531–6535.

Zago, M., B. Bonvini, D. Carminati and G. Jirafa. 2009. Detection and quantification of *Enterococcus gilvus* in cheese by real-time PCR. Sys. Appl. Microbiol. 32: 514–521.

Zhang, G. and R.A. Holley. 1999. Development and PFGE monitoring of dominance among spoilage lactic acid bacteria from cured meats. Food Microbiol. 16: 633–644.

CHAPTER 12

Advances in the Detection of Food-borne Pathogens Using Molecular Methods

Luca Cocolin and Kalliopi Rantsiou*

Introduction

The presence of pathogens is a serious problem that foodstuff producing industries have to face on a daily basis. Foodborne pathogens can survive during processing or they can come into contact with the product, causing recontamination or cross-contamination. Food products that contain pathogens represent a risk for human health since the consumption of a contaminated product may lead to health problems for the consumer. For these reasons, there is a great need for food industries to have rapid, sensitive, specific and reliable detection methods.

Routine approaches for testing food for microbiological parameters, based on culturing bacteria, have largely remained unchanged over the last 20 years and are still the gold standard for the detection of bacterial pathogens and indicator of organisms. In the current EU legislation (EU Regulation 2073/2005, last amended in 1441/2007 [Anon, 2005, 2007]), culture-based approaches are identified as reference methods, although alternative techniques can be used, provided that they are appropriately validated and produce equivalent results. In conventional microbiological methods, the presence of food-borne pathogens is revealed through their

Dipartimento di Scienze Agrarie, Forestali e Alimentari, Università di Torino, Via Leonardo da Vinci 44, Grugliasco, 10095 Torino, Italy.
* Corresponding author: lucasimone.cocolin@unito.it

capability to grow in enriched and selective media, and form colonies on selective/differential agar plates. These methods include the identification of pure culture isolates using a series of morphological, biochemical, immunological and other tests. Since the 80s, significant efforts have been made by researchers to develop improved culture media for the detection and isolation of food-borne pathogens and, especially in the 90s, a number of chromogenic media were developed that were able to presumptively identify the pathogenic microorganisms directly on the plate, on the basis of the specific aspect and color of the colony (Manafi 2000). These evolutions are still under way. Traditional microbiological methods do not fulfill the requirements requested by operators in the food business, primarily because of the time needed to obtain the result on the presence or absence of the specific pathogen. For *Salmonella* spp., at least 4 to 5 days are necessary before the presence of this bacterium can be confirmed in food products. Moreover, it is widely accepted and recognized that microbiological methods, based on cultivation, can lead to biases related to the use of the laboratory media. Culture-dependent methods are not able to recover populations that have entered a viable non-culturable (VNC) state (Rowan 2004, Alessandria et al. 2010) and cells that are numerically less abundant may be inhibited by the other major populations present in the system. The microbiological picture that is obtained through the use of methods based on cultivation may be completely different from the real ecological situation (Hugenholtz et al. 1998).

Conventional PCR

In the last few decades, new approaches to the study microorganisms in food systems have taken advantage of the technological advancements in the field of molecular biology. The polymerase chain reaction (PCR) was invented in the late 80s (Mullis et al. 1986), and a new area was thereby opened in microbiological analysis, especially when pathogens have to be detected. The philosophy behind the molecular methods no longer concerns cultivation, but the direct analysis of the nucleic acids. In this way, the total DNA and/or RNA are extracted from the sample and are subsequently subjected to analysis in which a specific target, which is often specific for the microorganism under investigation, is searched for and amplified. PCR allows target micro-organisms to be detected, regardless of their physiological state. Consequently, stressed or injured cells can be detected by these methods, and thus the previously described false-negative results obtained with traditional microbiological methods are avoided. In addition, the time required for the analysis is much shorter than for cultural methods. The application of PCR based methods in food microbiology has allowed protocols to be developed that are faster, more sensitive and more specific than the traditional microbiological methods that are used for the detection and identification of foodborne pathogens. The presence of pathogens in food can be determined in under 3 hours (this includes the extraction of the nucleic acids and the amplification process), a period that can reach a maximum of 24 hours if an enrichment step is included when it is deemed necessary to detect low numbers of pathogenic microorganisms (1–100 colony forming units [cfu]/10g). If, prior to DNA or RNA extraction, an enrichment step is performed, the method can no longer be considered culture-independent. However, in the case of food systems, enrichment steps help dilute the compounds that are present in the samples which are inhibitory towards the activity of the DNA polymerase, the enzyme responsible for the synthesis of DNA molecules during PCR. In dairy products and meat, for example, lipids, proteases, divalent cations, carbohydrates, or a host of undefined organic materials can drastically interfere with PCR assay detection sensitivity (Bickley et al. 1996, Lantz et al. 1998).

Traditional PCR protocols, in which the amplification products are detected by means of gel electrophoresis, suffer from the great disadvantage of being exclusively qualitative. A semi-quantification is possible if a densitometric analysis is performed using a calibration curve obtained from known amounts of DNA. However, this kind of approach cannot be applied, for example, for the quantification of the number of a specific pathogen present in a food sample. This is because the detection of the signal in conventional PCR takes place at the end of the amplification cycle, when the reaction reaches a plateau (Fig. 1). For this reason, a quantification based on densitometry would not be able to discriminate between 10^3 and 10^5 cells per ml or g.

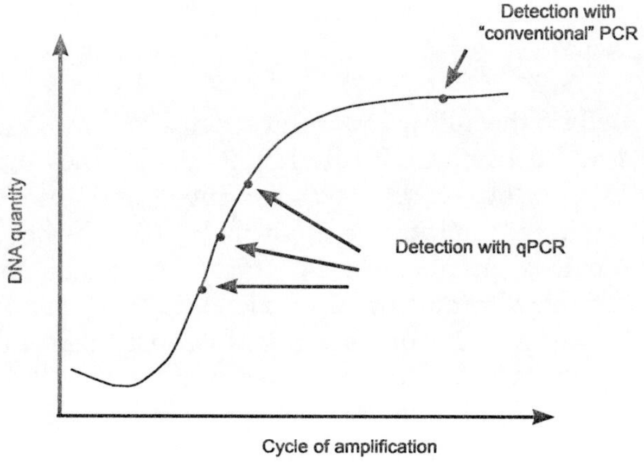

Figure 1. Monitoring of the amplification product by means of conventional PCR and quantitative PCR.

In addition, PCR methods can lead to some disadvantages that prevent their extensive use in the food sector. Since the target for the amplification is mainly represented by DNA, molecules characterized by high stability, even after cell death, a positive result by PCR does not necessarily indicate the presence of live pathogenic microorganisms in food. This evidence is extremely important for processed foods, i.e., thermally processed, where the production includes a step in which food-borne pathogens are eliminated. This pitfall of the PCR has been addressed in different ways by researchers and it is generally accepted that if positive results are obtained after an enrichment step, then live populations that were able to multiply are present, and it is their DNA that was subsequently amplified by PCR (Rossen et al. 1992). Moreover, a number of approaches have been designed for PCR itself to overcome the detection of dead cells. One approach is the detection of mRNA in reverse-transcriptase PCR format and the exclusion of dead cells on the basis of their permeable membrane to intercalating dyes, such as ethidium bromide monoazide (EMA), in, e.g., EMA-PCR format. None of the approaches has completely and always been successful (Nogva et al. 2003). There is a need for a significant investment, by the food business operator, in terms of equipment and trained personnel, even though, in the last few years, the market has seen a significant decrease in the prices of thermal cyclers, dedicated to the amplification of DNA. Lastly, PCR protocols lack legal value, due to the fact that the competent authorities do not allow the official testing of pathogenic microorganisms in foodstuffs by means of PCR. At present, there are a few PCR protocols that have been validated by international validation organizations (i.e., ISO and AFNOR), however none of them has received legal recognition, and they cannot therefore be used in case of official analyses.

Quantitative PCR

PCR became a method that allows quantification, after certain technological advancements had been made. In the late 90s, machines that were able to detect the PCR product while it was produced and amplified became available, thus making quantitative analysis by PCR protocols possible. This technique, called quantitative PCR (qPCR) or Real time PCR (Rt PCR), has revolutionized molecular approaches in food microbiology. Not only is it possible to quantify one specific microorganism in food, but it is also possible to study its behavior as a consequence of the influence of the environment (i.e., food composition, temperature, pH, oxygen, etc.) (Cocolin et al. 2011).

PCR assays that measure reaction progress during each amplification, rather than after reaching a plateau, represent an attractive means of obtaining real-time quantitative data for the rapid and sensitive detection of DNA (or RNA). Higuchi et al. (1992, 1993) pioneered the analysis of PCR kinetics by constructing a system that detects PCR products as they accumulate. This "real-time" system includes intercalator ethidium bromide in each amplification reaction, an adapted thermal cycler to irradiate the samples with ultraviolet light, and the detection of the resulting fluorescence with a computer-controlled cooled CCD camera. Amplification produces increasing amounts of double-stranded DNA, which binds ethidium bromide, resulting in an increase in fluorescence. By plotting the increase in fluorescence versus cycle number, the system produces amplification plots that provide a more complete picture of the PCR process than assaying product accumulation after a fixed number of cycles.

PCR reaction progress is measured by means of data acquisition at regular intervals (Fig. 1), thanks to the presence of molecules that produce fluorescence which correlates to the amount of produced PCR product. Relative fluorescence units are plotted as function of cycle number on the x-axis of a typical plot. Currently, a number of qPCR chemistries are commercially available for use in PCR. These can be divided into those that are not sequence specific, such as DNA minor groove-binding dyes, and those that are sequence specific and might even afford simultaneous detection and confirmation of the target amplicon during the PCR reaction (McKillip and Drake 2004).

Fluorescence data generated by qPCR assays are generally collected from PCR cycles that occur within the exponential amplification portion of the reaction, where conditions are optimal and fluorescence accumulates in proportion to the amplicon. This is in contrast with signal detection from the endpoint of the reaction, where the final amount of amplicon may have been affected by inhibitors, poorly optimized reaction conditions or saturation effects due to the presence of an excess of double-stranded amplicons. In fact, there may be no relationship between the initial template and the final amplicon concentrations. Because the emissions from fluorescent chemistries are temperature-dependent, data are generally acquired only once/cycle, at the same temperature.

The fractional cycle number at which the real-time fluorescence signal mirrors progression of the reaction above the background noise is used as an indicator of successful target amplification. This is usually called the threshold cycle (C_t). C_t is defined as the PCR cycle in which the gain in fluorescence generated by the accumulating amplicon exceeds ten standard deviations of the mean baseline fluorescence, when data taken from cycles 3–15 are used. C_t is proportional to the number of target copies present in the sample and is assumed to represent equal amounts of amplicon in each tube, since the C_t value represents the fractional cycle number of each sample at a single fluorescence intensity value. In practice, the C_t value is calculated after the definition of a noise band that excludes data from early

PCR cycles which cannot be distinguished from background noise. The final C_t value is the fractional cycle at which a single fluorescence value (usually at or close to the noise band) intersects the plotted PCR curve of each sample. The accuracy of C_t depends on the concentration and nature of the fluorescence-generation component, the amount of template initially present, the sensitivity of the platform, and the ability of the platform to discriminate specific fluorescence from background noise (Mackay 2004).

One application of qPCR in food microbiology is the detection and quantification of pathogens. While "conventional" PCR can only be used to determine the presence or the absence of a specific pathogen in food, it is also possible to perform a quantification (in terms of cells or colony forming units [cfu] per g or ml) with qPCR. Moreover, if the target nucleic acid is RNA, it is also possible to understand how many cells of the pathogen are alive in the analyzed sample.

The construction of the calibration curve is essential for the quantification. In the case of food samples, it should be pointed out that, because of the extreme heterogeneity of the food matrices, it is not possible to construct a unique calibration curve to use in the case of quantification, but specific curves, one for each food matrix to be considered in the analysis, should be prepared. Alternatively, DNA extracted from *in vitro* cultures can be used to construct the calibration curves. However, the efficiency of the amplification should be corrected by a factor that takes into consideration the influence of the food matrix.

The first step in the construction of calibration curves is the inoculation of the food samples with a known number of cells or cfu per g or ml. Subsequently, an aliquot (usually 1 ml) is subjected to DNA extraction and then amplified by qPCR. The obtained signals will depend on the amount of DNA loaded in the PCR reaction, and this is directly correlated to the number of cells inoculated in the sample. The C_t values obtained from the amplification can be correlated to the number of cells, or cfu, inoculated in the samples, and in this way a calibration curve is constructed.

It is possible to determine important parameters that will give an indication of the quality of the qPCR protocol from this kind of experiment. The first one is the quantification limit, defined as the smallest number of cells that the protocol can quantify. This is different from the sensitivity limit, which represents the smallest number of cells that the method can detect, but not quantify. Considering the calibration curves, the quantification limit represents the lowest value of cells that stand into the linear range of the curve, while the sensitivity limit falls outside this range. The slope of the curve will also give an important parameter, defined as efficiency, which is calculated mathematically as being equal to $10^{-1/\mathrm{slope}}$.

As reported above, both DNA and RNA can be targeted in quantification assays. If the DNA is extracted and analyzed, the calibration curves can be used to quantify the total number of pathogen cells without any differentiation between dead and live cells. On the other hand, if the RNA is sampled, the calibration curve offers the possibility of quantifying only the cells that are alive. In the latter case, the extracted RNA will obviously have to be reverse transcribed and then amplified. This aspect is extremely important in the case of foodborne pathogens. The approach of food industries should be completely different towards foodstuffs when they only contain dead pathogens or live cells that are able to cause health problems for the final consumer.

One major problem related to the construction of calibration curves in food samples is the availability of nucleic acids extraction methods that result in PCR-grade DNA and RNA. The necessity of removing all inhibitory compounds from the prep becomes even more important when considering qPCR. Because of its great sensitivity, the presence of small quantities of inhibitors may result in lower efficiencies and, therefore, an underestimation of the real quantity of nucleic acids present in the sample.

When designing standard curves, one needs to bear in mind the insensitivity of PCRs to the physiological status of cells, while the destiny of cells (dying or surviving and growing) in food samples depends on their physiological status. Therefore, cultures used for the determination of standard curves still need to be as realistically stressed as possible.

Generally, the limit of quantification is situated at approximately 10^3–10^4 cells per gram, whereas the maximum tolerable level is set at 100/g. The explanation for this relatively low performance of qPCR is not related to the PCR technique itself (which is intrinsically sensitive), but rather to the problems of sample preparation and DNA extraction from food samples. As previously mentioned, this is related to the PCR inhibitors present in food, but also to the use of small input volumes to PCR compared to the large test portions that have to be analyzed. The quantification limit of qPCR, as reported above, is thus still too high for practical application purposes, since the contamination is low (usually less than 100 cells/g) for most samples taken along the food chain. For this reason, the main application of qPCR is for qualitative presence/absence testing, which is performed after an enrichment step. From this point of view, it is possible to question the use of qPCR for qualitative purposes. Only in a few cases (compared with the total number of papers published over the last 10 years) have authors put any effort into developing qPCR protocols that could be used for quantification purposes without enrichment. Some examples of these studies are reported in Table 1, in which the target organism, the food matrices in which the microorganism was quantified and the detection limit are reported.

Apart from being used for the quantification of microorganisms, qPCR has recently shown good applicability in the environmental control of foodborne pathogens. Cross-contamination during or post-production

Table 1. Examples of studies in which the target food-borne pathogen(s) was quantified by qPCR (modified from Cocolin and Rantsiou 2012).

Microorganism	Food matrix	Limit of quantification	Reference
Foodborne pathogens:			
Bacillus cereus	Liquid egg Infant formula	60 cfu/ml	Martinez-Blach et al. 2009
Campylobacter jejuni and *Campylobacter* spp.	Chicken meat	10 cfu/g	Rantsiou et al. 2010 Melero et al. 2011
	Chicken carcass rinse	100 cfu/ml	Josefsen et al. 2004
	Chicken carcass rinse	2x10^3 cfu/g	Botteldoorn et al. 2008
Clostridium botulinum	Rice	100 cells/g	Takahashi et al. 2010
	Milk Dairy foods	10^3 cfu/g	Singh et al. 2009
Escherichia coli O157:H7	Vegetable leaves	2.3 x 10^3 cfu/g	Ibekwe et al. 2006
	Ground beef	10^4 cfu/g	Wang et al. 2009
Escherichia coli STEC	Meat and meat products Milk and milk products	100 cfu/ml or g	Rantsiou et al. 2012a Oses et al. 2010
Hepatitis A	Raspberries	50 pfu/25 g fresh raspberry 100 pfu/25 g frozen raspberries	Blaise-Boisseau et al. 2010
Klebsiella pneumoniae	Infant formula	2x10^3 cfu/25 g	Sun et al. 2010
Listeria monocytogenes	Meat and meat products Milk and milk products Ready to eat salads	10^3 to 10^4 cfu/ml or g	Rantsiou et al. 2008 Alessandria et al. 2010
	Salmon	10^3 cfu/g	Rodriguez-Lazaro et al. 2005
	Meat	10^3 cfu/g	Rodriguez-Lazaro et al. 2004
	Salads	10 cfu/g	Berrada et al. 2006

Mycobacterium avium subsp. *paratuberculosis*	Milk	5 to 6 cells/ml	Slana et al. 2008
Salmonella enterica *Listeria monocytogenes*	Yogurt	10 cells/10 g	D'Urso et al. 2009
Salmonella enterica	Pork meat (pig carcass swab)	4.4 x 10^2 cfu/ml	Lofstrom et al. 2011
	Seafood	20 cfu/g	Kumar et al. 2010
Salmonella typhimurium	Meat products	60 cfu/g	Cheng et al. 2009
Staphylococcus aureus	Milk Meat	10^3 cfu/ml or g	Chiang et al. 2007
Vibrio alginolyticus	Seafood	100 cells/1 ml tissue homogenate	Zhou et al. 2007
Vibrio vulnificus	Shellfish Drinkwater	10^2 to 10^3 cells/g or ml	Gordon et al. 2008
Yersinia pestis	Milk Ground beef	10 to 10^3 cfu/ml 10^2 to 10^5 cfu/g	Amoako et al. 2010

represents one of the most important factors of contamination by foodborne pathogens. Contamination after the production step, especially for specific microorganisms, such as *Listeria monocytogenes*, *Salmonella*, *Staphylococcus aureus* and *Escherichia coli*, often becomes the most important aspect as far as food contamination is concerned. An important role is played by the operator. However, the environment, and more specifically the equipment, represent key factors that need to be controlled. In this context, qPCR can be considered a relevant tool to exploit because of its capability to quantify foodborne pathogens in a culture-independent way. Several papers have been published on this subject over the last couple of years. Studies on *L. monocytogenes* in dairy processing plants (Alessandria et al. 2010), Shiga-like producing *E. coli* (STEC) in the lamb production food chain (Oses et al. 2010) and *Campylobacter jejuni* poultry processing lines (Melero et al. 2011), have underlined how qPCR can be used to quantify loads on pathogens in plants. The results of these studies have highlighted that it is vital for food producing companies to control the contamination routes of foodborne pathogens in their plants. This can be achieved through the implementation of correct sampling schemes and adequate cleaning and disinfection procedures. In addition, the choice of the analytical procedure should also be considered as a relevant factor. When the environmental conditions influence the fitness and behavior of the pathogen, such as in brines (with 15–20% w/v salt) or in equipment swabs (where cells have to respond to several stresses such as disinfectants, starvation and dried conditions), molecular methods should be used to quantify pathogens correctly. The application of molecular methods can result in a better comprehension of the spread of a specific pathogen in a processing plant, thereby allowing the implementation of corrective actions to eliminate or decrease the risk associated with its presence in the final product.

Multiplex PCR Assays

Although one set of primers is used in conventional PCR to amplify one target molecule, when several couples of primers are used various targets can be amplified. These multiplex PCR (mPCR) assays offer the advantage of allowing a simultaneous amplification of different DNA molecules, which are then detected in gel electrophoresis because they usually possess different molecular weights. One important aspect that should be considered when optimizing mPCR is the compatibility of the primers that are used for the amplification. Oligonicleotides should not hybridize between each other to form dimers and should not be able to anneal to non-specific regions. Moreover, they should possess similar annealing temperatures in order to allow a simultaneous hybridization of the target sequence. mPCR has been used extensively in the food field and the studies that are available in the literature can be divided into two main groups. In the first group, mPCR protocols have been developed to detect several food-borne pathogens in a single tube reaction, while the second one is represented by amplifications of several genes in one pathogenic organism. The literature is rich in papers that describe the concurrent amplification of food-borne pathogens, however *L. monocytogenes*, *Salmonella* spp. and *E. coli* represent the microorganisms that have most frequently been the targets of mPCR assays (Maurer 2011). mPCR, by amplifying multiple target genes in one specific pathogen, is becoming a typing method. This possibility is particularly interesting for those foodborne pathogens that are characterized by a pathogenic mechanism in which multiple genes are involved. In this context, pathogenic *E. coli* have represented the perfect model for mPCR development and application. Recently, Rantsiou et al. (2012a) have applied mPCR to determine the genotype of Shiga toxin producing *E. coli* isolated from food products of animal origin. In this case, an mPCR assay that was able to test for the presence of *stx1*, *stx2*, *eaeA* and *hly* genes was exploited, and a wide variety of combinations was found within the investigated strains. Interestingly, the *stx2* gene was detected more often than *stx1*, and this was of particular importance since it had previously been proposed that Stx2 was the most common virulence factor associated with severe human health problems.

mPCR has even been adapted to qPCR. In this case, instead of detecting specific-sized amplicons, the test is developed to detect multiple-colored fluorescent dyes associated with distinct probes for pathogens. While qualitative multiplex qPCR protocols have been published (Elizaquível and Aznar 2008), the challenge of simultaneously quantifying different foodborne pathogens still remains.

Other DNA Amplification Techniques

Currently, PCR is the most common technique used for nucleic acid amplification. Nevertheless, other techniques that are able to amplify specific nucleic acid fragments have been developed. Nucleic acid sequence-based amplification (NASBA), loop-mediated isothermal amplification (LAMP) and the ligase chain reaction (LCR) are examples of such techniques. While the first two methods offer the advantage of being carried out at a constant temperature, thereby avoiding the necessity of purchasing expensive equipment for the thermal cycling that is necessary in PCR amplification, LCR is cyclic. NASBA uses three enzymes: a reverse transcriptase, RNaseH and a T7 RNA polymerase, which act together to amplify sequences from an original single stranded RNA molecule. This method specifically allows viable cells to be detected, since the amplification target is the RNA. The method has been used to detect a range of pathogens in foods including

Campylobacter spp., *L. monocytogenes*, *S. enterica*, *Mycobacterium avium* subsp. *paratuberculosis* and others (Cook 2003, Rodriguez-Lazaro et al. 2006). In the LAMP method, four primers that recognize six regions of the target DNA are used, so the specificity is high. The method uses a polymerase to amplify target DNA (or RNA, after a reverse transcription step) following an autocycling strand displacement mechanism that produces detectable products in approximately 1 hour (Tomita et al. 2008). LAMP has the potential to detect foodborne pathogens (Sibley et al. 2012). LCR uses a thermostable enzyme to bind two oligonucleotides that are immediately adjacent to each other. In the first cycle of LCR, two sets of oligonucleotide pairs are annealed to the target sequence and if complementary base pairing exists at the ligation junction, the ligase joins the pair, producing a longer product that is hybridized to the template. After ligation, a new denaturation takes place, which causes the denaturation of the ligated product from the target, and this is followed by a cooling step to allow a new annealing and ligation of other oligonucleotides. Since a single mismatch in the pair junction prevents ligation, the entire target sequence that has to be amplified must be known (Vanrompay 2000).

Studying the Behavior of Food-borne Pathogens: Expression of the Virulence Gene

Nowadays, the field of biotechnology, with the recent advancements in nucleic acid analyses, offers a number of choices. As far as food safety and foodborne pathogens are concerned, one aspect that can be explored is the behavior of the microorganisms in the food matrices. This approach takes into consideration the *in situ* expression of specific traits, namely virulence and stress responses. By virulence factors we mean both the pathogenicity characteristics that are directly related to disease symptoms (e.g., the Shiga-like toxin produced by pathogenic *E. coli*) and the foodborne pathogen attributes that allow them to survive and respond to environmental stresses encountered during food processing, distribution and more importantly, after ingestion and within the human body. These factors will determine whether a bacterium will eventually be able to cause disease. Several genes encoding potential virulence factors can be identified from DNA sequences and genome mining. The goal now is to determine whether these genes are expressed in any of the stages 'from farm to fork', and if so, whether the synthesized proteins act in the predicted way and contribute to the disease.

Global approaches that permit scientists to look into the total RNA synthesized in the cells (transcriptomics, Chatterjee et al. 2006) and the total proteins produced by the cells (proteomics), at any given moment or condition, are now available. These are powerful techniques that allow one to pinpoint target genes that may require further investigation. On the basis of the studies conducted so far, it has been shown that strain heterogeneity in virulence potential exists within the same species of pathogenic bacteria. Furthermore, the environmental conditions that pathogenic microorganisms encounter in food influence their stress response capabilities, and enhance survival and possibly their overall virulence potential (Alessandria et al. 2013, Rantsiou et al. 2012b).

The technologies that are used in transcriptomics are: microarrays, in order to evaluate gene expression events, which provide information on the differentially expressed genes (global) and Reverse transcription qPCR (RT-qPCR), to quantify (and confirm) the differential expression of the most important genes. Generally, it is recognized that qPCR is the appropriate method when one needs to study a moderate number of genes in a number of samples that ranges from small to hundreds. On the contrary, microarrays offer the possibility of conducting whole genome discovery experiments in a small number of samples (VanGuilder et al. 2008). Furthermore, a trend has recently been developed towards more function-focused sub-arrays that target specific cellular functions, for example virulence regulons for pathogens or metabolic regulons of interest for technologically important microorganisms. This trend can be applied to a larger number of samples and facilitates the interpretation of the obtained data (Rantsiou et al. 2012c). It is important to always bear in mind that data obtained by means of microarrays, or subarrays (which can be considered of a qualitative nature) need to be validated by RT-qPCR.

The most common approach employed in the study of the expression of a potential virulence target gene is the definition of the comparative (or relative) gene expression (Sue et al. 2004, Werbrouck et al. 2006). This approach is schematically presented in Fig. 2. The expression of the target gene(s) is measured, by means of qPCR, under controlled conditions and under conditions that resemble the food environment. The expression of the target gene may be increased, decreased or kept stable. Up-regulation of a target gene under conditions that resemble the food environment can be a clear indication that this gene is important for the adaptation of the microorganism to this particular environment. On the contrary, a gene that is down-regulated is not considered essential in these conditions.

The application of *omics* disciplines, such as genomics, transcriptomics, proteomics and/or metabolomics, offers significant opportunities for advancement by improving the understanding of the virulence determinants of pathogenic bacteria (Yoshida et al. 2001). The results of studies targeting gene expression to clarify foodborne pathogen behavior will be useful for risk assessment in order to improve food safety (Brul 2007).

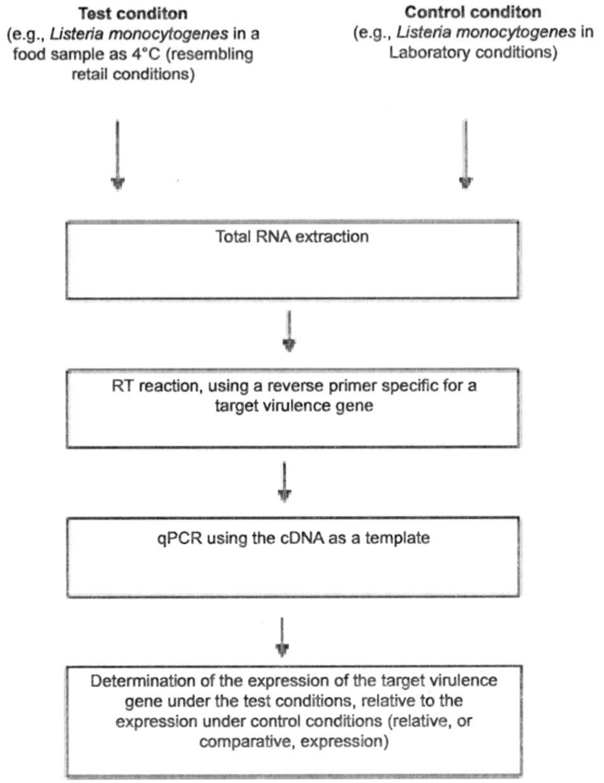

Figure 2. Schematic representation of the experimental approach for the study of virulence gene expression, employing qPCR.

Conclusions and Future Perspective

Important improvements are currently being made in food microbiology analysis, thanks to the availability of techniques that, never as before, can make significant information available regarding the safety of foodstuffs. PCR, which was initially characterized by extreme rapidity, compared to traditional microbiological methods, but which was not adequate for the enumeration of pathogen cells, has today become a quantitative technique, due to the possibility of monitoring the amplification of the PCR product in real time.

Such DNA and RNA based assays offer versatility in reducing the incidence of false negatives in quality assurance screening measures, sensitivity in processing heterogeneous food matrices, specificity in differentiating among closely related bacterial strains, and speed, as in the case of many of the real-time fluorescent chemistries currently on the market. Although some of these nucleic acid based technologies are unlikely to completely replace conventional culture, biochemical, or antibody-based testing methods, most offer a significant opportunity for high throughput and reliability, advantages that offset the initial equipment costs and ongoing maintenance or reagent expenses. Molecular diagnostics also offers a reliable initial screening tool so that follow-up, using conventional methods, is only necessary when a sample tests positive for the pathogen of interest.

Keywords: pathogens detection, PCR, qPCR, multiplex PCR, Nucleic acid sequence-based amplification (NASBA), loop-mediated isothermal amplification (LAMP), ligase chain reaction (LCR), micro-arrays, virulence genes

References

Alessandria, V., K. Rantsiou, P. Dolci, G. Zeppa and L. Cocolin. 2013. A comparison of gene expression of Listeria monocytogenes *in vitro* and the soft cheese Crescenza. Int. J. Dairy Technol. 66: 83–89.

Alessandria, V., K. Rantsiou, P. Dolci and L. Cocolin. 2010. Molecular methods to assess *Listeria monocytogenes* route of contamination in a dairy processing plant. Int. J. Food Microbiol. 141S: S156–162.

Amoako, K.K., N. Goji, T. Macmillan, K.B. Said, S. Druhan, E. Tanaka and E.G. Thomas. 2010. Development of multitarget real-time PCR for the rapid, specific, and sensitive detection of *Yersinia pestis* in milk and ground beef. J. Food Prot. 73: 18–25.

Anonymous. 2007. Commission Regulation (EC) No 1441/2007 of 5 December 2007 amending Regulation (EC) No 2073/2005 on microbiological criteria for foodstuffs.

Anonymous. 2005. Commission Regulation (EC) No 2073/2005 of 15 November 2005 on microbiological criteria for foodstuffs.

Berrada, H., J.M. Soriano, Y. Pico and J. Manes. 2006. Quantification of *Listeria monocytogenes* in salads by real time quantitative PCR. Int. J. Food Microbiol. 107: 202–206.

Bickley, J., J. Short, D.G. McDowell and H.C. Parkes. 1996. Polymerase chain reaction (PCR) detection of *Listeria monocytogenes* in diluted milk and reversal of PCR inhibition caused by calcium ions. Lett. Appl. Microbiol. 22: 153–158.

Blaise-Boisseau, S., C. Hennechart-Collette, L. Guillier and S. Perelle. 2010. Duplex real-time qRT-PCR for the detection of hepatitis A virus in water and raspberries using the MS2 bacteriophage as a process control. J. Virol. Methods 166: 48–53.

Botteldoorn, N., E. Van Coillie, V. Piessens, G. Rasschaert, L. Debruyne, M. Heyndrickx, L. Herman and W. Messens. 2008. Quantification of *Campylobacter* spp. in chicken carcass rinse by real-time PCR. J. Appl. Microbiol. 105: 1909–1918.

Brul, S. 2007. Systems and food microbiology. *In:* S. Brul, S. van Gerwen and M. Zwietering (eds.). Modelling microorganisms in food, 1st edition. Woodhead Publishing Limited, Cambridge, pp. 250–288.

Chatterjee, S.S., H. Hossain, S. Otten, C. Kuenne, K. Kuchmina, S. Machata, E. Domann, T. Chakraborty and T. Hain. 2006. Intracellular gene expression profile of *Listeria monocytogenes*. Infect. Immun. 74: 1323–1338.

Cheng, C.Y., J.R. Chi, S.R. Lin, C.C. Chou and C.C. Huang. 2009. Rapid quantification of *Salmonella* Typhimurium inoculated to meat products by real-time PCR. Acta Vet. Hung. 57: 25–38.

Chiang, Y.C., C.M. Fan, W.W. Liao, C.K. Lin and H.Y. Tsen. 2007. Real-time PCR detection of *Staphylococcus aureus* in milk and meat using new primers designed from the heat shock protein gene *htrA* sequence. J. Food Prot. 70: 2855–2859.

Cocolin, L. and K. Rantsiou. 2012. Quantitative polymerase chain reaction in food microbiology. *In:* F. Martin (ed.). Quantitative Real Time PCR in Applied Microbiology. Caister Academic Press, Norwich, pp. 149–160.

Cocolin, L., A. Rajkovic, K. Rantsiou and M. Uyttendaele. 2011. The challenge of merging food safety diagnostic needs with quantitative PCR platforms. Trends Food Sci. Technol. 22: S30–38.

Cook, N. 2003. The use of NASBA for the detection of microbial pathogens in food and environmental samples. J. Microbiol. Methods 53: 165–174.

D'Urso, O.F., P. Poltronieri, S. Marsigliante, C. Storelli, M. Hernandez and D. Rodriguez-Lazaro. 2009. A filtration-based real-time PCR method for the quantitative detection of viable *Salmonella enterica* and *Listeria monocytogenes* in food samples. Food Microbiol. 26: 311–316.

Elizaquível, P. and R. Aznar. 2008. A multiplex RTi-PCR reaction for simultaneous detection of *Escherichia coli* O157:H7, *Salmonella* spp. and *Staphylococcus aureus* on fresh, minimally processed vegetables. Food Microbiol. 25: 705–713.

Gordon, K.V., M.C. Vickery, A. DePaola, C. Staley and V.J. Harwood. 2008. Real-time PCR assays for quantification and differentiation of *Vibrio vulnificus* strains in oysters and water. Appl. Environ. Microbiol. 74: 1704–1709.

Higuchi, R., C. Fockler, G. Dollinger and R. Watson. 1993. Kinetic PCR: Real time monitoring of DNA amplification reactions. Biotechnology 11: 1026–1030.

Higuchi, R., G. Dollinger, P.S. Walsh and R. Griffith. 1992. Simultaneous amplification and detection of specific DNA sequences. Biotechnology 10: 413–417.

Hugenholtz, P., B.M. Goebel and N.R. Pace. 1998. Impact of culture-independent studies on the emerging phylogenetic view of bacterial diversity. J. Bacteriol. 180: 4765–4774.

Ibekwe, A.M., P.J. Shouse and C.M. Grieve. 2006. Quantification of survival of *Escherichia coli* O157:H7 on plants affected by contaminated irrigation water. Eng. Life Sci. 6: 566–572.

Josefsen, M.H., N.R. Jacobsen and J. Hoorfar. 2004. Enrichment followed by quantitative PCR both for rapid detection and as a tool for quantitative risk assessment of food-borne thermotolerant campylobacters. Appl. Environ. Microbiol. 70: 3588–3592.

Kumar, R., P.K. Surendran and N. Thampuran. 2010. Rapid quantification of salmonella in seafood using real-time PCR assay. J. Microbiol. Biotechnol. 20: 569–573.

Lantz, P.G., R. Knutsson, Y. Blixt, A. Al-Soud, E. Borch and P. Radstrom. 1998. Detection of pathogenic *Yersinia enterocolitica* in enrichment media and pork by multiplex PCR: a study of sample preparation and PCR-inhibitory components. Int. J. Food Microbiol. 45: 93–105.

Lofstrom, C., J. Schelin, B. Norling, H. Vigre, J. Hoorfar and P. Radstrom. 2011. Culture-independent quantification of *Salmonella enterica* in carcass gauze swabs by flotation prior to real-time PCR. Int. J. Food Microbiol. 145: S103–109.

Mackay, I.M. 2004. Real-time PCR in microbiology laboratory. Clin. Microbiol. Infec. 10: 190–212.

Manafi, M. 2000. New developments in chromogenic and fluorogenic culture media. Int. J. Food Microbiol. 60: 205–218.

Martínez-Blanch, J.F., G. Sánchez, E. Garay and R. Aznar. 2009. Development of a real-time PCR assay for detection and quantification of enterotoxigenic members of *Bacillus cereus* group in food samples. Int. J. Food Microbiol. 135: 15–21.

Maurer, J.J. 2011. Rapid detection and limitations of molecular techniques. Annu. Rev. Food Sci. Technol. 2: 12.1–12.21.

McKillip, J.L. and M. Drake. 2004. Real-time nucleic acid based detection methods for pathogenic bacteria in food. J. Food Prot. 67: 823–832.

Melero, B., L. Cocolin, K. Rantsiou, I. Jaime and J. Rovira. 2011. Comparison between conventional and qPCR methods for enumerating *Campylobacter jejuni* in a poultry processing plant. Food Microbiol. 28: 1353–1358.

Mullis, K., F. Faloona, S. Scharf, R. Saiki, G. Horn and H. Erlich. 1986. Specific enzymatic amplification of DNA *in vitro*: the polymerase chain reaction. Cold Spring Harb. Symp. Quant. Biol. 51: 263–373.

Nogva, H., S.M. Dromtorp, H. Nissen and K. Rudi. 2003. Ethidium monoazide for DNA-based differentiation of viable and dead bacteria by 5′-nuclease PCR. Biotechiques 34: 804–813.

Oses, S.M., K. Rantsiou, L. Cocolin, I. Jaime and J. Rovira. 2010. Prevalence and quantification of shiga-toxin producing *Escherichia coli* along the lamb food chain by quantitative PCR. Int. J. Food Microbiol. 141S: S163–169.

Rantsiou, K., V. Alessandria and L. Cocolin. 2012a. Prevalence of Shiga toxin—producing *Escherichia coli* in food products of animal origin as determined by molecular methods. Int. J. Food Microbiol. 154: 37–43.

Rantsiou, K., M. Mataragas, V. Alessandria and L. Cocolin. 2012b. Expression of virulence genes of *Listeria monocytogenes* in food. J. Food Saf. 32: 161–168.

Rantsiou, K., A. Greppi, M. Garosi, A. Acquadro, M. Mataragas and L. Cocolin. 2012c. Strain dependent expression of stress response and virulence genes of *Listeria monocytogenes* in food as determined by microarray. Int. J. Food Microbiol. 152: 116–122.

Rantsiou, K., C. Lamberti and L. Cocolin. 2010. Survey of *Campylobacter jejuni* in retail chicken meat products by application of a quantitative PCR protocol. Int. J. Food Microbiol. 141S: S75–79.

Rantsiou, K., V. Alessandria, R. Urso, P. Dolci and L. Cocolin. 2008. Detection, quantification and vitality of *Listeria monocytogenes* in food as determined by quantitative PCR. Int. J. Food Microbiol. 121: 99–105.

Rodriguez-Lazaro, D., M. Hernández, M. D'Agostino and N. Cook. 2006. Application of nucleic acid sequence based amplification (NASBA) for the detection of viable foodborne pathogens: progress and challenges. J. Rapid Methods Autom. Microbiol. 14: 218–236.

Rodriguez-Lazaro, D., A. Jofré, T. Aymerich, M. Garriga and M. Pla. 2005. Rapid quantitative detection of *Listeria monocytogenes* in salmon products: evaluation of pre-real-time PCR strategies. J. Food Prot. 68: 1467–1471.

Rodriguez-Lazaro, D., A. Jofré, T. Aymerich, M. Hugas and M. Pla. 2004. Rapid quantitative detection of *Listeria monocytogenes* in meat products by realtime PCR. Appl. Environ. Microbiol. 70: 6299–6301.

Rossen, L., P. Norskov, K. Holmstrom and O.F. Rasmussen. 1992. Inhibition of PCR by components of food samples, microbial diagnostic assays and DNA-extraction solutions. Int. J. Food Microbiol. 17: 37–45.

Rowan, N.J. 2004. Viable but non culturable forms of food and waterborne bacteria: Quo Vadis? Trends Food Sci. Tech. 15: 462–467.

Sibley, C.D., G. Peirano, L. Deirdre and S. Church. 2012. Molecular methods for pathogen and microbial community detection and characterization: current and potential application in diagnostic microbiology. Infect. Genet. Evol. 12: 505–521.

Singh, J., V.K. Batish and S. Grover. 2009. A scorpion probe-based real-time PCR assay for detection of *E. coli* O157:H7 in dairy products. Foodborne Pathog. Dis. 6: 395–400.

Slana, I., P. Kralik, A. Kralova and I. Pavlik. 2008. On-farm spread of *Mycobacterium avium* subsp. *paratuberculosis* in raw milk studied by IS900 and F57 competitive real time quantitative PCR and culture examination. Int. J. Food Microbiol. 128: 250–257.

Sue, D., D. Fink, M. Wiedmann and K.J. Boor. 2004. σ^B-dependent gene induction and expression in *Listeria monocytogenes* during osmotic and acid stress conditions simulating the intestinal environment. Microbiol. 150: 3843–3855.

Sun, F., D. Wu, Z. Qiu, M. Jin, X. Wang and J. Li. 2010. Development of real-time PCR systems based on SYBR Green for the specific detection and quantification of *Klebsiella pneumoniae* in infant formula. Food Control 21: 487–91.

Takahashi, H., C. Takakura and B.O.N. Kimura. 2010. A quantitative real-time PCR method for monitoring *Clostridium botulinum* type A in rice samples. J. Food Prot. 73: 688–694.

Tomita, N., Y. Mori, H.T. Kanda and T. Notomi. 2008. Loop-mediated isothermal amplification (LAMP) of gene sequences and simple visual detection of products. Nat. Protoc. 3: 877–882.

VanGuilder, H.J., K.E. Vrana and W.M. Freeman. 2008. Twenty-five years of quantitative PCR for gene expression analysis. Biotechniques 44: 619–626.

Vanrompay, D. 2000. Advances in nucleic acid-based diagnosis. Semin. Avian and Exot. Pet 9: 2–13.

Wang, L., Y. Li and A. Mustapha. 2009. Detection of viable *Escherichia coli* O157:H7 by ethidium monoazide real-time PCR. J. Appl. Microbiol. 107: 1719–1728.

Werbrouck, H., K. Grijspeerdt, N. Botteldoorn, E. Van Pamel, N. Rijpens, J. Van Damme, M. Uyttendaele, L. Herman and E. Van Coillie. 2006. Differential *inlA* and *inlB* expression and interaction with human intestinal and liver cells by *Listeria monocytogenes* strains of different origins. Appl. Environ. Microbiol. 72: 3862–3871.

Yoshida, K., K. Kobayashi, Y. Miwa, C.M. Kang, M. Matsunaga, H. Yamaguchi, S. Tojo, M. Yamamoto, R. Nishi, N. Ogasawara, T. Nakayama and Y. Fujita. 2001. Combined transcriptome and proteome analysis as a powerful approach to study genes under glucose repression in *Bacillus subtilis*. Nucl. Acids Res. 29: 683–692.

Zhou, S., Z. Hou, N. Li and Q. Qin. 2007. Development of a SYBR Green I real-time PCR for quantitative detection of *Vibrio alginolyticus* in seawater and seafood. J. Appl. Microbiol. 103: 1897–1906.

CHAPTER 13

MALDI-TOF Mass Spectrometry, a Rapid and Reliable Method for the Identification of Bacterial Species in Food-Microbiology Laboratories

M. Quintela-Baluja, K. Böhme, I.C. Fernández-No, M.E. Alnakip, S. Caamaño, J. Barros-Velázquez and *Pilar Calo-Mata**

Introduction

Food safety is one of the highest public health priorities at the national, community and international levels. The globalization of the food supply and the increased complexity of the food chain have recently defined the closely related areas of "food quality" and "food safety" as important priorities in the food sector. Food scandals that have occurred over the last 20 years, an increase in international food trading and the identification of emerging risks have led to an increasing demand for safer products that do not present health hazards for the consumer. One of the major aims of "food safety" is to avoid health risks that are caused by the consumption of food products, including microbiological hazards, pesticide residues,

Department of Analytical Chemistry, Nutrition and Food Science, School of Veterinary Sciences, University of Santiago de Compostela, 27002 Lugo, Spain.
* Corresponding author: p.calo.mata@usc.es

misuse of food additives and the presence of chemical or biological toxin contaminants, all of which render the food adulterated and unfit for human consumption. In contrast, "food quality" encompasses all attributes that influence the value of a food product. Such attributes can be negative (i.e., spoilage, contamination with filth, discoloration and off-odors) or positive (i.e., origin, color, flavor, texture and the processing method of the food) (FAO 2003).

To ensure the safety of the global food supply and minimize socio-economic losses due to food safety threats, the early identification of vulnerabilities in the food chain and the implementation of effective prevention systems are essential. The World Health Organization (WHO) has launched an initiative to estimate the burden of foodborne illness on a global scale (Stein et al. 2007). This initiative involves experts of the Foodborne Epidemiology Reference Group (FERG), which assembles and assesses global evidence regarding foodborne disease epidemiology.

Microbial Food Safety

Food products not only represent an important nutrient source for those who consume them, they are also an excellent environment for microbial growth. The development of microbial growth in food is mainly due to intrinsic factors, such as pH, moisture content, available nutrients and the structure of the food. Extrinsic factors such as temperature, relative humidity and gases (CO_2, O_2) also play an important role. While chemical residues and additives typically enter the food chain at less predictable intervals, microbial contamination of food products can occur at any time throughout the production process and the most effective ways to control those hazards can vary.

Microorganisms are frequently used in food production technology and in the manufacturing of gastronomic products, including wine, beer, cheese, pickles, sausages, soy sauce and other products. However, the contamination of food products with undesired microorganisms represents a problem of global concern because the growth and metabolism of microorganisms can cause serious foodborne illnesses and rapid spoilage of the food products. The quality and safety of a food product, and thus the acceptance of it by consumers, greatly depends on the presence and nature of the microorganisms. In addition to molds and yeasts, bacteria are the principle agents responsible for food spoilage and foodborne illnesses (Blackburn 2006). Different microorganisms can proliferate and adulterate the product depending on the preservation method. Microbial risk assessment therefore requires a thorough understanding of the entire food production process. Food products naturally contain endogenous microbiota that can induce spoilage. In some cases, food products may also contain pathogenic bacterial species that can be dormant for an initial time period, but ultimately cause disease in a matter of days or weeks.

Food spoilage bacteria

Spoilage microbes are often common inhabitants of soil, water, or the intestinal tracts of animals, and they may be dispersed through the air, water or by the activity of small animals, particularly insects. Food spoilage can be defined as a disagreeable change in the normal state of a food product. It can occur at any time during the course of food production, transport, storage, or manipulation. Food is considered an optimal medium for bacterial growth. If the conditions are appropriate, bacteria will grow rapidly and produce changes in appearance and flavor and may develop unpleasant odors, tastes and/or textures. The changes vary according to the type of food degraded, but they can be summarized by examining the content of the major nutrients found in all food: proteins, carbohydrates, and fats. A spoiled food has lost its original nutritional value, texture and/or flavor, and it can become harmful to people when it is unsuitable for consumption. A variety of factors determine whether microbial growth will preserve or spoil foods, as shown in Table 1. The microbial spoilage of food products represents an important problem for the food industry, as it results in high economic losses, especially when food is stored under incorrect refrigeration conditions. Spoilage bacteria normally do not cause illness; however, when consumed in large amounts they can cause gastrointestinal disturbances (Blackburn 2006).

Table 1. Factors and parameters that affect microbial spoilage.

<table>
<tr><th colspan="4">FACTORS AFFECTING MICROBIAL SPOILAGE</th></tr>
<tr><th>Intrinsic factors (foodstuffs)</th><th>Extrinsic factors (foodstuffs)</th><th>Implicit parameters</th><th>Processing factors</th></tr>
<tr><td>Physical-chemical properties:
-Water activity
-Redox potential
-pH</td><td>Storage Temperature</td><td>Interaction microorganisms
-Synergism
-Antagonism
-Commensalism</td><td rowspan="3">-Pasteurization
-Packaging
-Irradiation
-Washing
-Slicing</td></tr>
<tr><td>Chemical composition
-Nutrient materials
-Limiting factor (vitamins)
-Inhibitory substances</td><td>Relative humidity</td><td rowspan="2">Growth rate</td></tr>
<tr><td>Biological structure</td><td>Gaseous atmosphere</td></tr>
</table>

There are different bacterial species that can cause spoilage of food (Fig. 1). Thus, the storage of fresh foodstuffs, such as fish and meat, at refrigeration temperatures can result in the growth of *Pseudomonas* species (spp.), including spoilage species such as *Pseudomonas fragi* and *Pseudomonas putida*. A light preservation and change in the atmosphere (e.g., by vacuum-packaging), may inhibit the growth of these bacterial species and favor the growth of lactic acid bacteria (LAB), Enterobacteriaceae, *Bacillus* spp. and *Clostridium* spp. The last two genera are able to produce spores that can survive heat treatments and germinate after a pasteurization process; thus, these species are an important issue in the canning industry and for the safety of food. Spoilage species may be food-specific; for example, the presence of *Erwinia* spp. has been reported in products of vegetal origin. In addition, seafood products are commonly spoiled by species such as *Pseudomonas* spp., *Shewanella* spp. and/or *Photobacterium* spp. In general, bacteria can spoil different foods depending on the physical-chemical preservation profile of the food (Gram et al. 2002).

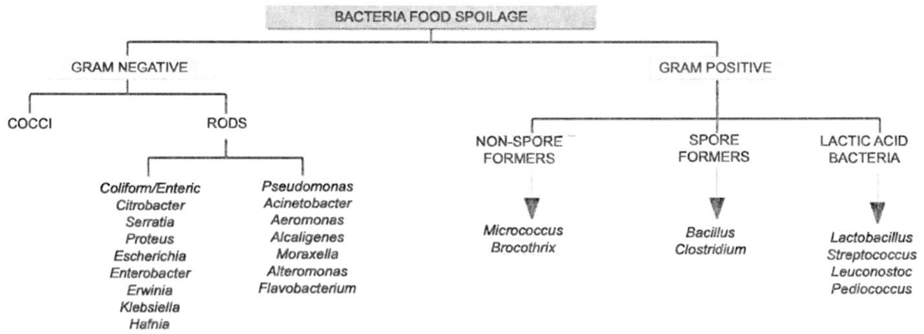

Figure 1. Different bacteria species that can cause food spoilage.

Foodborne diseases

There are two primary types of food-related diseases: foodborne infections and food intoxications (Fig. 2). Since 1942, the number of recognized foodborne pathogens has increased more than five-fold. In most cases, these pathogens are simply agents that we can now describe based on an improved understanding of microbial diversity. Foodborne diseases are caused by agents that enter the body through the ingestion of food followed by growth inside the host organism, which can subsequently lead to complications such as tissue invasion and/or the release of toxins. Food can transmit disease from person to person and serve as a growth medium for bacteria that cause food poisoning. The global incidence of foodborne diseases is difficult to estimate; however, in 2005 alone it was reported that 1.8 million people died from diarrheal diseases. All of these

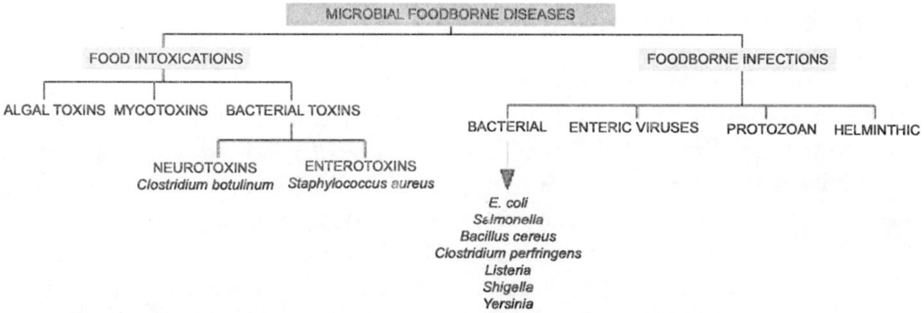

Figure 2. Food related diseases: foodborne infections and food intoxications.

foodborne diseases are associated with poor hygienic practices. Whether by water or food transmission, the fecal-oral route of contamination is often utilized by bacterial species, and a great proportion of these cases can be attributed to its consumption. In industrialized countries, the percentage of the population suffering from food-borne diseases each year is reportedly as high as 30% (World Health Organization 2007). Pathogenic bacteria often do not change the color, odor, taste or texture of a food product, rendering it difficult to recognize if the product has been contaminated. Microbial growth in food products can also result in food intoxication through the consumption of microbial toxins. Intoxication produces symptoms shortly after the food is consumed because no growth of the disease-causing microorganism is required. Toxins produced in food can be associated with the microorganism itself or can be released from the microorganisms. In general, toxins do not alter the appearance, odor or flavor of the food product. Common bacteria that produce toxins include *Staphylococcus aureus* and *Clostridium botulinum*. In some cases, bacteria such as *Clostridium perfringens*, cause illness by releasing toxins into the gut upon ingestion of a high concentration of vegetative cells. Bacterial food poisoning is commonly caused by pathogenic bacterial species such as *Escherichia coli*, *Salmonella* spp., *Listeria monocytogenes*, *Staphylococcus aureus*, *Bacillus cereus*, *Clostridium perfringens*, *Campylobacter* spp., *Shigella* spp., *Streptococcus* spp., *Vibrio cholerae*, *Vibrio parahaemolyticus*, *Vibrio vulnificus*, *Yersinia enterocolitica* and *Yersinia pseudotuberculosis*. The term emerging foodborne pathogens may refer to new pathogens that emerge due to changing ecology or technology that connects a potential pathogen with the food chain or those that emerge *de novo* by the transfer of mobile virulence factors (Tauxe 2002). Emerging foodborne pathogens include *E. coli* O157:H7, *Aeromonas hydrophila*, *Aeromonas caviae*, *Aeromonas sobria*, *Mycobacterium* spp., vancomycin-resistant enterococci, non-gastric *Helicobacter* spp., *Enterobacter sakazakii*, non-jejuni/coli species of *Campylobacter*, and non-O157 Shiga toxin- producing *E. coli*.

Detection of Food-related Bacteria

Some of the major problems with regards to ensuring food safety include the need to identify microorganisms in a rapid and unequivocal way and the need to predict and enhance the shelf-life of food products, which is especially important for the wide-scale distribution of perishable foods. Standard culture techniques may require days or weeks for the identification of pathogens. The varied chemical and physical composition of food and the low ratio of food spoilage and foodborne microorganisms to the background microflora make the isolation and identification of foodborne pathogens complicated. In order to control and minimize the microbial hazard, several methods have been designed to achieve bacterial identification. These methods are grouped broadly into two classes: conventional and rapid methods (Fig. 3). The conventional methodologies include culturing, microscopic, physical, chemical and immunological techniques, while the rapid methods include the use of DNA based techniques. After years of development, new technologies based on mass spectrometry have become a powerful tool for the rapid and accurate identification of microorganisms.

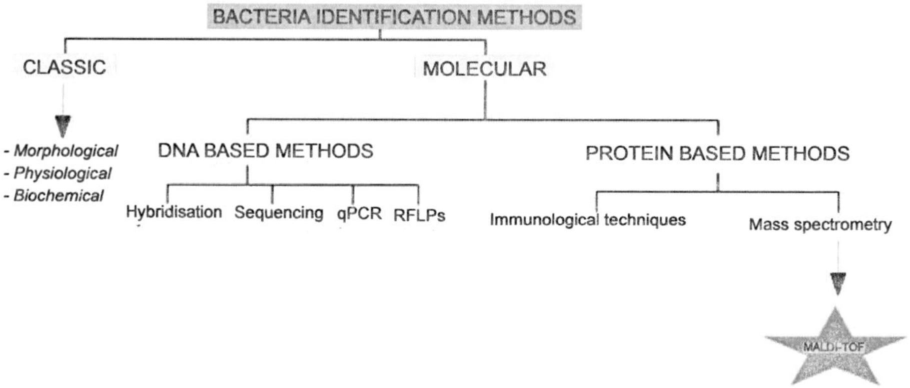

Figure 3. Designed methods for bacterial identification.

Bacterial identification by conventional methods

Traditionally, bacterial species have been identified by classic tools relying on culturing processes coupled with morphological, physiological, and biochemical characterization. Phenotypic identification is based on the direct comparison of phenotypic characteristics of unknown bacteria with those of a known type. The reliability of this type of identification is directly proportional to the number of similar phenotypic characteristics.

When classifying microorganisms, all known characteristics are taken into account. However, certain characteristics are selected and used for the purpose of identification. Primary identification usually involves a few simple assays such as the determination of colony morphology, gram staining, growth conditions, and catalase and oxidase testing (Duerden et al. 1998). Testing the requirements for growth includes the ability to grow in the presence or absence of oxygen and the ability to grow in different culture media. For a better approximation of bacterial identification, other laborious techniques are employed, such as microscopic observation, hemolysis assays and biochemical analyses, such as tests for the presence or activity of aminopeptidase, urease, indol, oxido-fermentation, coagulase, or resistance to different substances, etc. To make these analyses faster and less laborious, miniaturized biochemical assays with multi-test galleries are utilized. With these arrays, inoculation, incubation and culture can be carried out within a minimal space and the whole process can be automated. By using these tests, it is usually possible to characterize the bacterial genus and even the species to which an unknown strain likely belongs.

Conventional identification methods are widely used despite some disadvantages, including slow turnaround time, laborious lab techniques, and the requirement that the organisms be cultivated *in vitro*. Furthermore, some strains exhibit unique biochemical characteristics that do not fit previously described characteristic patterns of any known genus or species.

In the last few decades, the progress of microbial identification has turned to more rapid and sensitive methods, including antibody-based assays and DNA-based methods which have been combined with important advances in bioinformatics tool development. Thus, some methodologies such as Enzyme-Linked Immuno-Sorbent Assay (ELISA) or Polymerase Chain Reaction (PCR) have already become classic identification methods. The development of PCR protocols coupled with sequencing tools has provided a large amount of information that has already been deposited into public databases and is freely available. Recently, the development of rapid and high sensitivity techniques, such as real-time PCR (qPCR), DNA microarrays and biosensors, have led to the replacement of traditional culturing methods in the field of bacterial identification for clinical diagnostics and the food sector (Feng et al. 2007, Mohania et al. 2008). Furthermore, Fourier transform infrared spectrometry (FT-IR) has been described as a new method for rapid and reliable bacterial identification (Sandt et al. 2006). At the same time, proteomic tools, such as mass spectrometry, were introduced for the identification of microorganisms (Klaenhammer et al. 2007).

Bacterial identification by DNA-based methods

Between the 1960s and 1990, taxa from nearly all areas of the microbial kingdom were intensely studied using a variety of biochemical analytical methods. Currently, in the field of bacterial diagnostics, traditional culturing methods have been replaced by molecular techniques that detect specific DNA or RNA sequences with oligonucleotide probes specific for the pathogen of interest. These methods offer an advantage to traditional methods in that they are faster, more sensitive and more accurate. These methods were initially explored using rRNA oligonucleotide-cataloguing; however, because DNA sequencing has become more accessible, the comparative analysis of the small subunit of the ribosomal RNA molecule has replaced it (Ludwig and Klenk 2001). By the 1990s, the introduction of PCR had enabled the global study of an abundant and diverse range of microbial taxa. However, these methods are more costly when compared to conventional techniques due to the requirement for specific industrial equipment and more qualified personnel.

Despite the plasticity of microbial genomes and the frequency of horizontal gene transfer, microbial species retain a large number of stable traits that enable the assignment of isolates to a given taxon. Among the most fluid genomes, full genome analysis indicates that changes are often superimposed upon a backbone of core genes that are indispensable to a particular species. The sequencing of the 16S rRNA gene is a common tool used for bacterial identification. Universal primers are designed that can bind to conserved regions in order to amplify the variable regions, which is achieved via PCR. Sequences of the 16S rRNA gene have been determined for an extremely large number of species and are accessible through freely available DNA databases, such as the GenBank database of the National Center for Biotechnology Information (NCBI). For bacterial identification, the 16S rRNA sequence of an unknown strain is compared to the database of published sequences, and the most similar bacterial sequences are determined using the bioinformatics tool BLAST. In addition to the 16S rRNA gene, other genes, such as the 23S rRNA, the intergenic spacer region of 16S–23S (ITS), rpoB (subunit BETA of RNA polymerase) and gyrB (subunit β of the DNA gyrase) can be used as indicators of evolutionary history and species diversity. However, these genes have not yet been applied to bacterial identification, due to the lack of ample databases as a reference.

More recently, real-time PCR (qPCR) has been introduced for the detection and quantification of several of the main foodborne pathogenic bacteria. As a consequence, a large number of qPCR procedures are currently available for the specific detection and quantification of many foodborne bacteria, such as *Salmonella* spp., *L. monocytogenes* and *S. aureus* (Hein et al. 2001, Malorny et al. 2004, Elizaquível and Aznar 2008).

The DNA microarray is the current state-of-the-art DNA-based technique, consisting of gene arrays that can hybridize multiple DNA targets simultaneously; thus, these arrays have enormous potential for the detection and identification of pathogens. Due to an increase in the number of complete microbial genome sequences available, DNA microarrays are becoming a common tool in many areas of microbial research, particularly in the field of bacterial identification for clinical diagnostics and in the food sector (Severgnini et al. 2011). The microarray is a powerful, sensitive and specific technology that allows an accurate identification based on single target detection and can distinguish between subtle differences in the genome.

Mass spectrometry for bacterial identification

Mass spectrometry has been used traditionally for the chemical analysis of low molecular weight organic compounds. In 1968 Yamada reported the classification of acetic acid bacteria based on mass spectrometry measurements (Yamada et al. 1968). However, Anhalt and Fenselau were the first to apply electron ionization to identify small molecules released by heating dehydrated bacteria in 1975 (Anhalt and Fenselau 1975). Excepting GC/MS analysis of fatty acid methyl esters and a recent resurgence in interest in pyrolysis MS, most of the early mass spectrometry methods for bacterial characterization are no longer used today. One reason for the decline in the use of these methods was the development of alternative approaches aimed at the analysis of polar biomolecules by mass spectrometry. In the past few years, the emergence of gentler ionization processes permitted the determination of higher molecular masses, which was well suited for the analysis of polar biomolecules, including peptides and proteins. Depending on the ionization source, mass spectrometry for bacterial analysis can be divided into two sets of techniques: electrospray ionization (ESI) and matrix-assisted laser desorption/ionization (MALDI).

Since its introduction by Karas et al. (1987), matrix-assisted laser desorption ionization time-of-flight mass spectrometry (MALDI-TOF MS) has proved to be a technique with high potential for microorganism identification due to its rapidity, reduced costs and minimal sample preparation when compared to conventional biochemical and molecular techniques (Seng et al. 2009, Giebel et al. 2010). Matrix material is mixed with the biological sample and upon irradiation with a laser. Molecules in the sample are ionized and desorbed to form a plume of gaseous ions. Currently, the most common method used to detect this plume of ions is a TOF (time-of-flight) analyzer, in which accelerated ions obtain a mass/charge-dependent velocity when moving from the ion source through an electric field. After acceleration, the ions drift through a "field-free" region where their different velocities, depending on their mass and charge, separate them as they "race" towards the detector. In the detector region, the separated ions are counted and converted into a signal for each *m/z* value proportional to the number of ions present. The resulting output of such analyses is a mass spectral profile, representing molecular masses of ions in the original plume (Fig. 4). The difference between separation in time and space is one of the major differences between TOF and many others mass spectrometers. With TOF mass spectrometers, the entire process can be repeated very rapidly. Many spectra can be obtained in a few seconds and typically several spectra are averaged to produce a final recorded mass spectrum with an enhanced signal-to-noise ratio.

Initially, several studies showed that MALDI-TOF MS may be used for the rapid analysis of biological components using intact bacterial cells, significantly reducing the preparation time and the amount of biomass required for analysis (Claydon et al. 1996, Holland et al. 1996, Krishnamurthy and Ross 1996, Arnold and Reilly 1998, Haag et al. 1998, Wang et al. 1998, Welham et al. 2000, Domin et al. 1999, Lynn et al. 1999). The spectral patterns analyzed are generally surface components with interesting properties related to the bacterial physiology (electron transport, signal transduction,

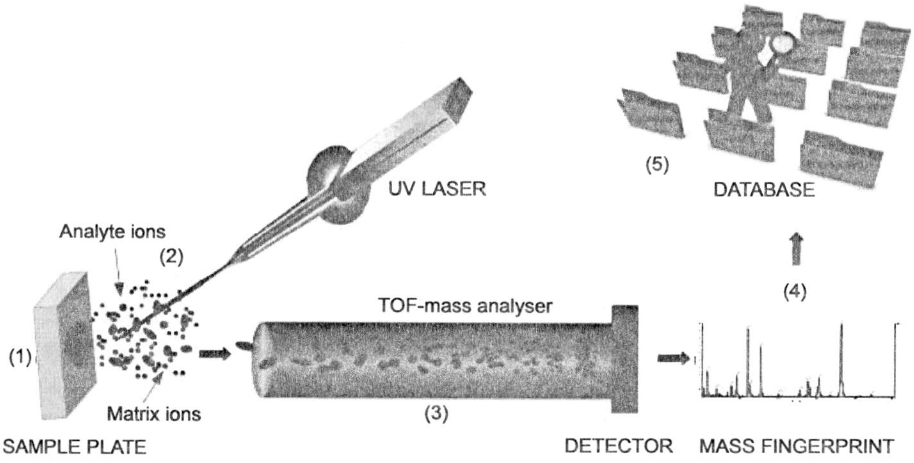

Figure 4. Technical description of MALDI-TOF MS. (1) The sample is mixed with a matrix and placed in a metal plate. (2) After crystallization of the matrix and microbial material, the metal plate is introduced into the mass spectrometer and is bombarded with brief laser pulses. (3) The desorbed and ionized molecules are accelerated through an electrostatic field and ejected through a metal flight tube subjected to vacuum until they reach a detector, with smaller ions traveling faster than larger ions. (4) Thus, bioanalytes separated according to their TOF create a mass spectrum that is composed of mass to charge ratio (m/z) peaks with varying intensities. (5) A spectrum is a microbial signature that is compared to a database for identification at the species or genus level.

Color image of this figure appears in the color plate section at the end of the book.

etc.), virulence and pathogenicity (toxin assembly, hemagglutinins, ligands, binding receptors, etc.) of the sample. The variability in the surface properties of each bacterial species caused by changes in environmental parameters required a readjustment of the protocol for use in microbial identification. In this sense, a pretreatment of intact bacterial cells with trifluoroacetic acid (TFA) disrupts the outer polymeric layers of the cell wall without disintegrating the cell. Studies in which bacterial cells were treated with trypsin and lysozyme before analysis by MALDI-TOF MS showed that the spectral patterns detected after pretreatment with TFA were generally attributed to intracellular proteins (Conway et al. 2001). In further studies, the proteins detected by MALDI-TOF MS were identified, and most peaks in the spectral profiles were found to be derived from ribosomal proteins (Ryzhov and Fenselau 2001). As a result, reproducible mass spectral profiles can be obtained, irrespective of the culture medium used to grow the cells or the duration of the growth, in contrast to the initial methods which required the procedure to be rigorously standardized.

Bacterial identification is carried out by either identifying ion biomarker masses that are correlated with theoretically determined protein masses obtained from databases or by comparing the whole spectral profile to a spectral reference database (Baar 2000). In the first approach, bacterial strains are identified by determining the masses of biomarker proteins against sequence-derived masses (Demirev et al. 2004). A number of studies have been performed to determine which protein biomarkers are the most useful for bacterial species identification by MALDI-TOF MS (Demirev et al. 2004, Fagerquist et al. 2005). A critical challenge of protein database searching is the necessity of high mass accuracy because some proteins have very similar masses. Furthermore, identification is limited to well-characterized microorganisms with known protein sequences available in proteomic databases (Dare 2006). It should also be noted that peaks with similar *m/z* values in mass spectra do not necessarily represent the same protein, and therefore, a single peak is not sufficient for characterization in most cases.

The second approach, also referred to as the "fingerprint approach", is the most commonly applied bacterial identification technique. It relies on the spectral differences of bacterial species to identify an unknown species by comparison of its spectral profile to a reference database of known spectral profiles (Giebel et al. 2010, Mazzeo et al. 2006). This approach allows the differentiation of bacterial strains, due to the highly specific nature of the obtained spectral profiles, or "fingerprints". For this purpose, it is not necessary to identify the proteins but just to determine a number of characteristic peaks that are representative of the corresponding species and/or genus. The mass range of 2000 to 20,000 Da generated high inter-specific variability and simultaneously promoted high intra-specific similarity for most bacterial species. In lower mass ranges, e.g., 500–2500 Da, the variability between mass fingerprints of different species can be either too low for species differentiation, or mass fingerprints of closely related isolates can be very dissimilar, although this mass range had been proposed for identification in early studies (Claydon et al. 1996). This complication depends mostly on the group of microorganisms subjected to analysis, as some bacterial species may produce a high diversity of secondary metabolites, and thus, strain-specific mass patterns are observed (Welker and Moore 2011, Neuhof et al. 2007). For those species that generally lack respective gene clusters, mass patterns appear similar for different species. In the mass range above 20,000 Da, generally only a limited number of mass signals are recorded by MALDI-TOF MS of intact bacterial cells, despite the large diversity of proteins present in living cells (Fig. 5).

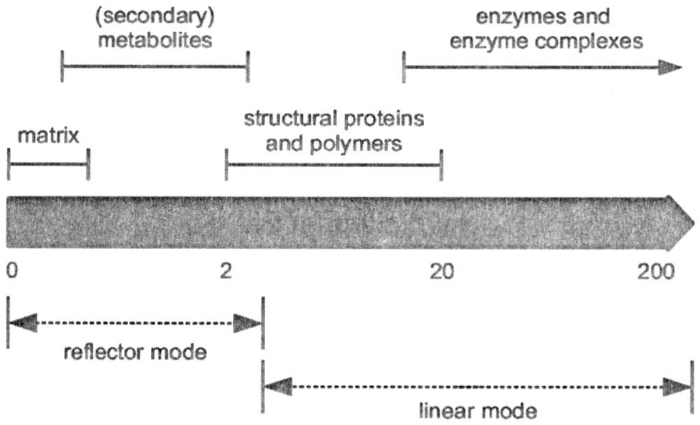

Figure 5. Types of cellular compounds detected in different mass ranges and TOF-modes by intact-cell MS.

MALDI-TOF MS Fingerprinting, a Rapid and Reliable Method for Bacterial Identification in Food

MALDI-TOF MS fingerprinting has proven to be an applicable method for bacterial identification at the genus, species and even strain level. It has been used routinely for bacterial identification in the clinical sector, and it is demonstrated to be a rapid, cost effective and accurate technique that achieved correct species identification in over 92% of cases. This is a significantly higher accuracy than is achieved with conventional biochemical systems or even 16S rRNA sequencing for bacterial identification (Seng et al. 2009, Bizzini and Greub 2010). Several authors agree that the cost of bacterial identification by MALDI-TOF MS fingerprinting is around two-thirds less than conventional methods, when considering the cost of materials and staff (Hsieh et al. 2008, Nassif 2009). Furthermore, it has several advantages over other genomic methods, such as DNA-microarrays, because fewer steps are necessary to achieve bacterial identification, and thus, fewer errors are introduced throughout the analysis process. Another advantage of MALDI-TOF mass fingerprinting is the effortless analysis of results because extensive data processing and statistical analysis are not required, as is the case with other rapid methods for bacterial identification such as FT-IR and DNA-microarrays.

Until recently, MALDI-TOF MS techniques for the identification and typing of microorganisms had remained confined to research laboratories and specific species identification was limited by the accessibility of spectra. In the last few years, the availability of MALDI-TOF MS devices and the reduction of costs enabled the use of these instruments in clinical, food and environmental microbiology laboratories. Some authors conclude that the identification of microbial isolates by whole-cell mass spectrometry is one of the latest tools contributing to a revolution in microbial diagnostics, with the potential to end many of the time-consuming and labor-intensive identification procedures that have been used for decades. However, to increase the reliability of the method, correct identification at the species or strain level should be routinely achieved (Bizzini and Greub 2010). In this regard, bacterial differentiation at the species level is not always possible using the commonly applied methods for bacterial identification. For example, the analysis of the 16S rRNA gene is complicated in some cases due to the high similarity of sequences between species of the same genus, such as *Bacillus* spp. and *Pseudomonas* spp. However, the correct identification of the corresponding species is of great importance for food safety and quality because the pathogenic and spoilage characteristics can vary significantly. A higher discrimination potential has been described for MALDI-TOF MS fingerprinting, allowing for the differentiation and correct identification of close bacterial species and even strains of the same species (Keys et al. 2004, Donohue et al. 2006, Vargha et al. 2006, Quintela-Baluja et al. 2013, Alnakip et al. 2013). Recent studies of MALDI-TOF MS have correlated the presence of some shared biomarker peaks of strains of the same species with their isolation source (Giebel et al. 2008, Quintela-Baluja et al. 2013). Quintela-Baluja et al. (2013), and it was observed that the spectral profiles of *Enterococcus* strains isolated from different food samples exhibited variation in some peaks when compared to the reference strains obtained from culture collections. In addition to the importance of identifying protein biomarkers that could facilitate the rapid identification of an unknown strain at the genus and species level, the existence of common and distinct peaks for a limited number of strains within the same species represents crucial information for bacterial classification at the strain level. Bacterial typing of closely related strains can aid in the understanding of microbial population dynamics and the description of different groups of strains or ecotypes. As MALDI-TOF MS has a higher discrimination potential, more information could be extracted using this technique than from other phenotypic and genotypic methods such as 16S rRNA sequencing. These results suggested that MALDI-TOF MS fingerprinting of environmental isolates has the potential to be a rapid and accurate bacterial source tracking tool.

In recent years, several reports have shown the feasibility of using MALDI-TOF MS for identifying microorganisms (Seng et al. 2009, Giebel et al. 2010). The detection and comparison of protein mass patterns has become a convenient tool for the rapid identification of bacteria, due to the highly specific mass profiles obtained by these methods. Notably, most studies of bacterial identification by MALDI-TOF MS fingerprinting are targeted towards the development of clinical diagnostics of bacterial strains associated with human infectious diseases. In contrast, only a few studies have been performed in the field of microbial food analysis by MALDI-TOF MS for the identification of foodborne pathogens and/or spoilers.

Some reports have dealt with the classification and identification of several widespread pathogens, including species causing food-borne diseases, such as *Aeromonas*, *Acrobacter*, *Campylobacter*, *Clostridium*, *Listeria*, *Salmonella*, *Staphylococcus*, *Vibrio*, *Yersinia*, *Bacillus* and species of the *Enterobacteriaceae* family (Bernardo et al. 2002, Bright et al. 2002, Barbuddhe et al. 2008, Stephan et al. 2010, Stephan et al. 2011). A more complete study, aimed at the detection of a greater number of foodborne pathogens and food spoilage bacteria was carried out by Mazzeo et al. (2006) and included genera such as *Escherichia*, *Yersinia*, *Proteus*, *Morganella*, *Salmonella*, *Staphylococcus*, *Micrococcus*, *Lactococcus*, *Pseudomonas*, *Leuconostoc* and *Listeria*. In further studies, the main pathogenic and spoilage bacterial species potentially present in seafood have been analyzed by MALDI-TOF MS (Böhme et al. 2010, Fernández-No et al. 2010, Böhme et al. 2011c, Böhme et al. 2013). These works included the most important genera involved in seafood safety and quality, such as *Acinetobacter*, *Aeromonas*, *Bacillus*, *Carnobacterium*, *Clostridium*, *Listeria*, *Pseudomonas*, *Shewanella*, *Staphylococcus*, *Stenotrophomonas*, *Vibrio* and many species belonging to the *Enterobacteriaceae* family. Studies of concrete genera were carried out on *Staphylococcus aureus* and *Streptococcus* spp. isolated from milk and dairy products (Böhme et al. 2012, Alnakip et al. 2013), as well as *Enterococcus* spp. isolated from milk and meat foodstuffs (Quintela-Baluja et al. 2013).

While these studies have been performed for the rapid identification of the main pathogenic and spoilage bacteria present in foodstuffs by MALDI-TOF MS, few studies have been conducted on bacterial differentiation in an ecological context.

The study and identification of the characteristic microbiota of a certain food product can contribute to the improvement of production, the development of organoleptic characteristics, and serve to predict and enhance the shelf life of the corresponding food product. Therefore, knowledge of the microbiota can increase the attraction of foodstuffs to consumers and ultimately its commercial value. In this regard, Ghyselinck (2011) described MALDI-TOF MS as a fast and cost-effective tool for high-throughput dereplication with the potential for automation. MALDI-TOF MS has also been applied to study and classify the characteristic microbiota in some vegetables (Hausdorf et al. 2012) and fermented foods (Nguyen et al. 2013).

Spectral profiles of some important foodborne pathogens and food spoilers are shown in Fig. 6 and spectral profiles of different species of LAB isolated from food products are shown in Fig. 7, demonstrating the high specificity of MALDI-TOF MS. Böhme et al. (2010, 2011b, 2013), Quintela-Baluja et al. (2013) and Alnakip et al. (2013) also determined characteristic biomarker peaks for several studied genera and species. Such unique or characteristic peak masses can serve to rapidly identify a bacterial genus and/or species. However, unequivocal identification cannot be achieved based on a single biomarker protein; it can only result from the consideration of a number of characteristic mass patterns, which represent the spectral fingerprint.

Furthermore, when assaying microbial mixtures, such biomarkers become more important because the presence or absence of unique peak patterns could lead to the determination of the present bacterial species. The detection of biomarker proteins by MALDI-TOF MS has been successfully applied to the identification of two bacterial species isolated from contaminated milk, water, lettuce and cotton cloth (Holland et al. 2000, Barreiro et al. 2010). Until recently, the application of MALDI-TOF MS fingerprinting for microbial mixtures had not yet been demonstrated.

Another critical challenge of MALDI-TOF MS fingerprinting is that the classification of a bacterial genus or species, as well as the determination of unique biomarker patterns, is only possible in the context of the content of the spectral reference library. However, the number of studies regarding bacterial species identification by MALDI-TOF MS in foodstuffs is continually increasing, as is its reliability. One obstacle is the relatively small number of described microbial (prokaryotic and fungal) species compared to the estimated number of existing species on earth (Wilson 2003). In the area of clinical microbiology, this ratio is less pronounced, due to the relatively high number of microbial species that have been described in detail because of their potential as pathogenic agents. Nevertheless, even in clinical microbiology there may be certain isolates than cannot be assigned to a definitive species, simply because that species has not yet been formally described. Therefore, when considering the characteristics of the global food supply, its diversity and the diversity of microbial species in

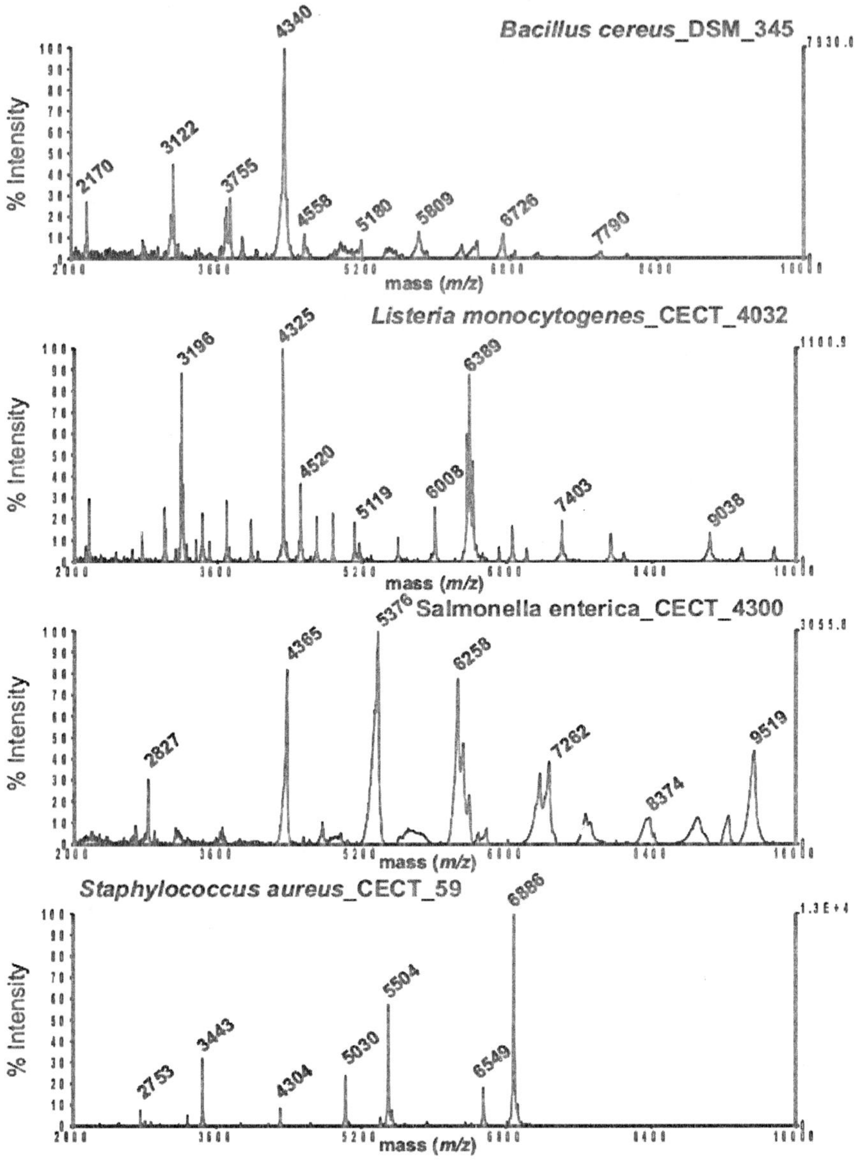

Figure 6. Spectral profiles of some important foodborne pathogens and food spoilers.

the environment, the probability that the species of any isolate would be present in a mass spectral reference database is currently relatively low.

Figure 8 shows a schematic of the entire bacterial identification process by MALDI-TOF MS in food products. Bacteria are isolated from food samples and cultivated to obtain single colonies. Afterwards, the bacterial cells are lysed by an organic solvent and a strong acid, usually acetonitrile (ACN) or

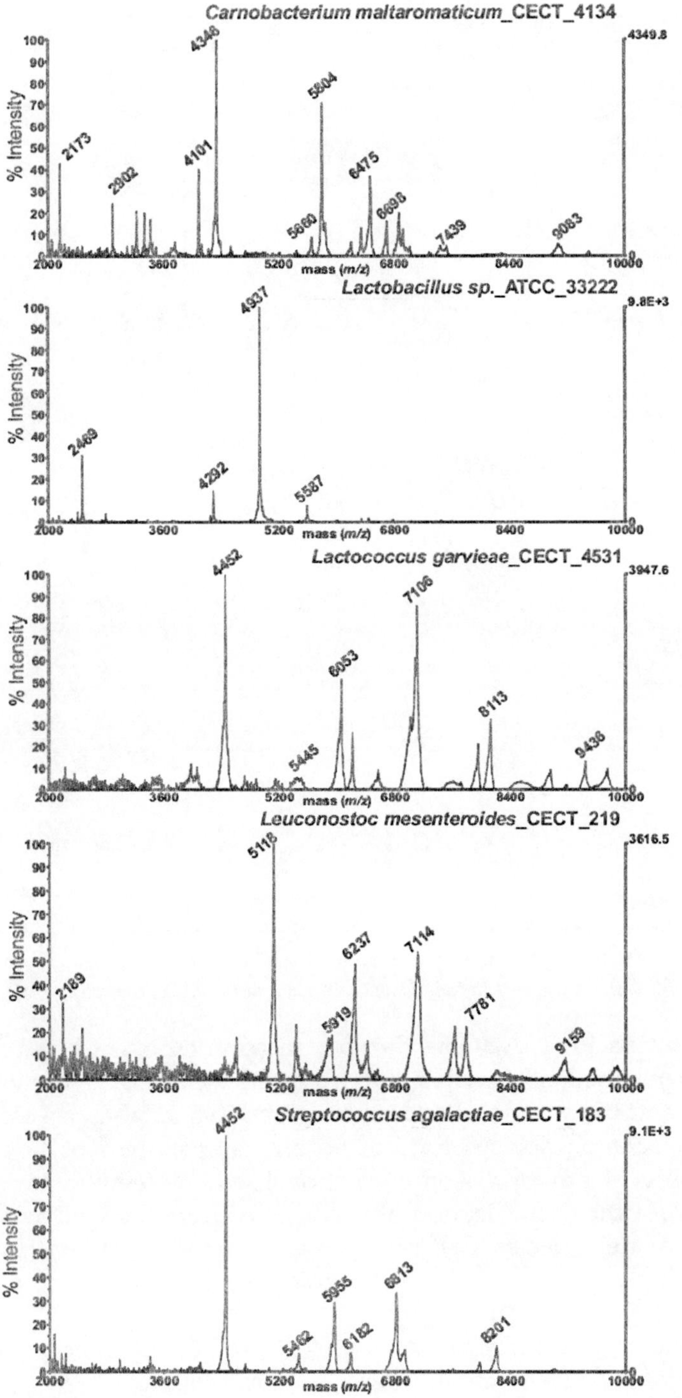

Figure 7. Spectral profiles of different species of LAB isolated from food products.

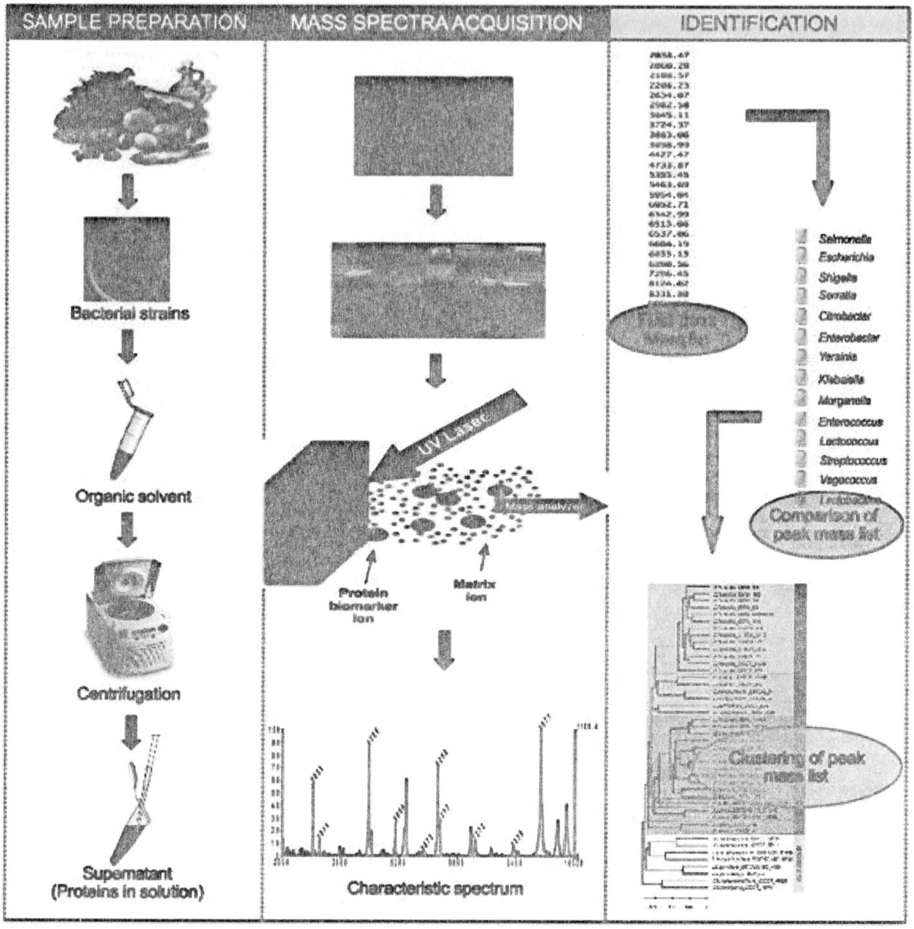

Figure 8. Process of bacterial identification by MALDI-TOF MS in food products.
Color image of this figure appears in the color plate section at the end of the book.

trifluoroacetic acid (TFA). Once the spectral profiles for each bacterial strain have been obtained, data analysis is performed to extract representative peak mass lists and the spectral data are compared to a database to differentiate among the bacterial species. Furthermore, cluster analysis of the peak mass lists will reveal phyloproteomic relationships between bacterial species and allow for the identification of unknown strains, as well as the typing of closely related species. For the sample preparation and data analysis, a number of different protocols and techniques have been described and are discussed in the following sections.

For bacterial identification by MALDI-TOF MS fingerprinting, the spectral profile of the strain of interest is compared to a spectral library of reference strains. Several private databases have been created, including spectral profiles of more than 500 bacterial strains, such as the Spectral

Archive And Microbial Identification System (Saramis TM; AnagnosTec GmbH, Potsdam, Germany) (Erhard et al. 2008) and the Microbelynx bacterial identification system (Waters Corporation, Manchester, UK) (Keys et al. 2004, Dare 2006). The MALDI Biotyper 2.0 (Bruker Daltonic) searches against an ample database of more than 1800 bacterial species and new spectral profiles are added on a daily basis. This database was demonstrated to be applicable for routine bacterial identification in the clinical sector and is a rapid, cost-effective and accurate technique that achieved correct species identification 92% of the time (Seng et al. 2009, Bizzini and Greub 2010). As previously mentioned, most studies and databases are targeted towards human pathogens causing infectious diseases. Nevertheless, these databases also include bacterial species that play an important role in food safety and quality and could represent important reference data for the identification of food pathogens and food spoilage bacterial species. However, the critical challenge of these databases is their limited availability. In this regard, it would be desirable to construct a public database for the submission of the spectral information for each species to allow for the comparison of results obtained by different researchers, favoring a more precise method for the identification of intact bacteria based on a large amount of data.

Thus, a few attempts to start a public database have been initiated. Mazzeo et al. (2006) constructed a library containing spectra of 24 food-borne bacterial species, including *Escherichia, Yersinia, Proteus, Morganella, Salmonella, Staphylococcus, Micrococcus, Lactococcus, Pseudomonas, Leuconostoc* and *Listeria*. Although, the spectral profiles and peak mass lists are freely available on the Web (http://bioinformatica.isa.cnr.it/Descr_Bact_Dbase.htm), the library only includes a few bacterial species important in food-borne diseases and/or food spoilage.

With a focus on species of interest to the food sector, a reference library containing mass spectral fingerprints of the main pathogenic and spoilage bacterial species potentially present in foodstuffs has been created (SpectraBank database), and it includes more than 250 bacterial strains from over 80 bacterial species (www.spectrabank.org, www.spectrabank.eu, www.spectrabank.es). In further studies, the library was shown to be applicable for the correct identification and classification of unknown bacterial strains isolated from commercial seafood products (Böhme et al. 2011b) and of different LAB species isolated from meat and dairy products (Alnakip et al. 2013, Quintela-Baluja et al. 2013). The constituted spectral library may easily be modified to encompass further bacterial species and strains that are of interest in the corresponding food products.

Experimental protocol of sample preparation

As with conventional methods of microbial identification, MALDI-TOF MS identification nearly always requires cultivation of the microorganism of interest. Solid agar media (Grosse-Herrenthey et al. 2008, Pennanec et al. 2010) or liquid broth media (Arnold and Reilly 1998, Wensing et al. 2010) have both been used. Cultured bacterial cells are prepared for MALDI-TOF MS analysis. Either intact cells or cell extracts are prepared for analysis. In some cases, the MALDI matrix (e.g., ferulic acid or sinapinic acid) is mixed directly with the intact cells. When preparing cell extracts, bacterial cells are lysed via physical (e.g., sonication or bead-heating) or chemical (e.g., via exposure to TFA or formic acid and an organic solvent) means to release the content of the cells into the supernatant. The supernatant is then added to the matrix and analyzed with MALDI.

A variety of mass spectrometers ranging from relatively low-resolution linear detector-only bench top units to units with much higher resolution can rapidly generate spectra from intact bacteria and cell extracts. Spectra are typically collected with a linear detector of MALDI-TOF MS units (Hettick et al. 2006, Lartigue et al. 2009, Hahn et al. 2011), but a few investigators have also used the reflector detector (available on higher resolution units) to obtain higher resolution data (Claydon et al. 1996, Bernardo et al. 2002, Vorob'eva et al. 2011). Overall though, the resolution afforded by the linear detector of most MALDI-TOF MS units appears more than adequate to obtain useful spectra of intact bacterial cells and extracts.

To allow the comparison of bacterial spectral fingerprints, the obtained spectra must be representative and reproducible. A standardized protocol to obtain representative and reproducible spectra begins with the sample preparation and includes specific instrumental parameters. It has been reported that spectral profiles are less sensitive to culture conditions, but they can show significant variability depending on the sample preparation protocol utilized (Bernardo et al. 2002, Wunschel et al. 2005).

In the first studies of bacteria by MALDI-TOF MS, protein fractions were isolated from bacterial cells but were abruptly adapted to the direct analysis of whole cells without any sample pre-treatment. This process is called intact cell mass spectrometry (ICMS). Many authors have focused on the optimization of the sample preparation protocol, with the aim of establishing a standardized protocol to obtain specific and reproducible spectral profiles in a rapid and efficient manner (Williams et al. 2003, Mazzeo et al. 2006, Liu et al. 2007). Currently, three different sample preparation protocols are commonly applied. The most rapid and efficient method is based on the direct application of the bacterial biomass obtained from culture plates to the MALDI-TOF MS sample plate by overlaying the biomass with the matrix solution (Bright et al. 2002). Although this technique is the most rapid and efficient, it has several disadvantages. It is difficult to obtain the correct amount of biomass in order apply a homogenous distribution of the sample and the matrix. Although, this technique was successfully applied for bacterial species identification in a number of studies (Keys et al.

2004, Erhard et al. 2008), spectral profiles showed more noise and less peak resolution, and no reproducible spectra were obtained (Böhme et al. 2010a). In another sample preparation method, the bacterial biomass is harvested in a solvent or directly in the matrix solution to obtain cell suspensions (Mazzeo et al. 2006, Vargha et al. 2006). Time-consuming wash steps that can lead to the loss of small soluble proteins were commonly applied in initial experiments, but they have since been eliminated from the sample preparation process (Carbonnelle et al. 2007).

In a third sample preparation method based on the latter approach, cell suspensions were centrifuged and spectra were obtained by the analysis of the supernatant. When compared to the spectra resulting from cell suspensions without a centrifugation step, the spectral profiles from the supernatant displayed better reproducibility, higher resolution and less noise. The decrease in noise was due to a lower background and the increase in resolution led to more representative and characteristic peaks for each bacterial species, thus improving the reproducibility of the spectra (Böhme et al. 2010a). It should be noted that, in general terms, a more homogeneous distribution of sample and matrix is expected with cell extracts than with cell suspensions. In addition, this sample preparation method is rapid and effortless because the extracts were obtained directly from cell cultures in only one dilution/centrifugation step. However, when working with the extracts, it is expected that small soluble protein spectral profiles will have a high number of peaks, similar to those obtained by the analysis of whole cell suspensions. Currently, most applications of MALDI-TOF MS for bacterial identification analyses employ bacterial cell extracts obtained by one dilution/centrifugation step.

As previously stated, the direct comparison of spectral profiles for bacterial differentiation requires that a strict standardized protocol is followed. It has been shown that when the same culture conditions, sample preparation method, matrix, organic solvents and MALDI-TOF MS analyzer are applied, reproducible spectral profiles can be obtained (Keys et al. 2004, Barbuddhe et al. 2008). Although different protocols can lead to high variability in the resulting spectral profiles, it should be noted that some peaks have been detected even when different protocols were applied. Such characteristic and conserved peak masses could serve as biomarker peaks for the corresponding genus and/or species.

Data analysis

The essential step for species identification is the comparison of the mass fingerprint of the unknown sample to a database containing reference mass fingerprints. Similar to other methods, the completeness and quality of such a database are crucial for the attainment of robust results. This requires that reference mass fingerprints of all species of interest and mass fingerprints of multiple strains per species are included in the database (Lartigue et al. 2009). Taxonomists have underscored the latter requirement by suggesting that an individual species should be described by studying at least 10 individual strains (Rossell-Mora and Amann 2001, Sneath 1977).

For the identification of an unknown bacterial strain by MALDI-TOF MS fingerprinting, the spectral profile is compared to a spectral library of reference strains. There are several different commercial microbial identification databases (Table 2):

- The MALDI Biotyper 2.0 (Bruker Daltonics) database uses Main Spectra as the reference spectra (i.e., consensus spectra computed from multiple spectra obtained from a single strain). The identification result (microbial species) is linked to a score computed by counting signals in sample mass spectra that match with reference mass spectra (and vice versa) and correlating signal intensities of matched signals from both mass spectra. The three scores obtained from such a procedure are multiplied and normalized to a value of 1000 and eventually log-transformed (Freiwald and Sauer 2009).
- The SARAMIS (AnagnosTec/bioMerieux) database uses SuperSpectra as the identifying reference spectra and scores are computed by weighting peaks in a consensus spectrum according to their specificity for different taxonomic levels (generally, family, genus and species). The specificity is determined empirically by comparison to a large number of individual mass spectra. Consensus spectra in turn are computed from multiple mass spectra of multiple isolates from a taxon by selecting conserved mass signals (Kallow et al. 2004). For identification, a sample's mass spectrum is matched against

Table 2. Commercial systems available for MALDI-TOF MS microorganism identification applications.

SOFTWARE AND DATABASE	Microbe Lynx System (Waters Corporation and Manchester Metropolitan University)	Maldi Biotyper (Bruker Daltonics)	SARAMIS (AnagnosTec GmbH)	MS-ID (BioMérieux)
MASS SPECTROMETER	Micro MX (Waters Corporation)	Microflex (Bruker)	Axima (Shimadzu)	Vitek MS (BioMérieux)
IDENTIFICATION	Aerobic/anaerobic bacteria	Bacteria, yeast and filamentous	Bacteria, yeast and filamentous	Bacteria, yeast and filamentous
RANGE	500–15,000 Da	2000–20,000 Da	2000–20,000 Da	2000–20,000 Da
REQUIRES SAMPLE PREPARATION	Yes (depends on the culture medium)	No	No	No
REPRODUCIBILITY	Low	High	High	High

SuperSpectra, and the sum of peak weights (unit-less numbers) is computed for matching mass signals for each SuperSpectrum and the sum of the peak weights transformed into a confidence value.

- The VITEK MS (bioMerieux) database uses an identification matrix that has been computed by the Advanced Spectrum Classifier, an algorithm based on supervised learning procedures and subjected to a data set of 25000 (binned) reference spectra. For each species, an identifier is computed in which each bin receives a weight according to its frequency within the species and its frequency among all other species in the database. In the next step, probability functions are computed for each species relating the sum of bin-weights obtained from the matching spectra, applied to an identifying matrix, and assigned a probability score. For the identification of a sample, its spectrum is matched to the identifying matrix and the sum of bin weights computed for each represented species, which in turn is transformed into a probability listed in the report.
- The MICROBELYNX (Waters Corporation) database utilizes macromolecules expressed on the surface of bacteria, which are sampled and characterized by molecular weight. The resulting mass spectrum provides a unique fingerprint for each species tested. Bacterial mass fingerprints of unknowns can be reliably matched against a database of quality-controlled reference spectra. The organisms represented in the database are currently all from the National Collection of Type Culture (NCTC) and represent a world-renowned well-characterized collection. The mass spectra are further examined for reproducibility before addition to the database. The same process can be used to collect spectra for unknown isolates, and the available software enables these isolates to be screened against a library of 2000 microbes.

Furthermore, based on the similarity scores, dendrograms can be constructed and principle component analysis carried out. The critical challenges of these databases and software programs are that both are conditioned to the corresponding instruments and are not available for other investigators without paying high fees.

Thus, in a number of studies, smaller "in house" libraries have been constructed. The difficulty in developing an "in house" database is that particular algorithms to analyze and compare obtained spectra are needed to carry out searches against the constructed reference library. Peak matching techniques eliminate the subjectivity of visual comparison. An automated peak detection algorithm has been developed to extract representative mass ions from a fingerprint and to compare spectra to fingerprints in a reference library. This algorithm carries out the identification of an unknown spectrum by calculating a degree of matching and is robust to the variability of the ion intensity (Jarman et al. 2003). Other authors have developed software (BGP-database, available on http://sourceforge.net/projects/bgp) to analyze and compare spectral profiles, allowing for rapid identification. This software determines the best match between the tested strains and the reference strains of the database, taking into account a possible error of the *m/z* value (Carbonnelle et al. 2007).

In further studies, the freely available web-based application SPECLUST (http://bioinfo.thep.lu.se/speclust.html) was used to extract representative peak masses and obtain final peak mass lists for each bacterial strain. Subsequently, required mass lists can be compared and common peak masses defined. The web interface calculates the mass difference between two peaks and accounts for a certain measurement of uncertainty (σ) and peak match score(s) (Alm et al. 2006). The resulting web program is very fast, easy to handle, and could be extended by new spectral mass lists in a simple manner. Although it was not yet possible to search an unknown spectrum directly against the library, comparison of peak mass lists could be carried out and common peaks determined with the aim of identifying a spectral profile of an unknown strain. The web-application has been successfully applied to the identification of pathogenic and spoilage bacterial strains isolated from commercial seafood products (Böhme et al. 2011b, Böhme et al. 2011c), species of biogenic amine-producing bacteria involved in food poisoning (Fernández-No et al. 2010) and different species of LAB (Alnakip et al. 2013, Quintela-Baluja et al. 2013). In addition, the program allows the rapid determination of specific biomarker peaks and includes a clustering option.

Further bioinformatics programs such as Statgraphics Plus 5.1 (Statpoint Technologies Inc., Warrenton, USA) or the WinBoot package offer a variety of functionalities. First, spectral data have to be transformed to a binary table, indicating the presence (1) or absence (0) of a peak mass. Next, various algorithms for cluster analysis can be applied, as well as Principal Component Analysis (Böhme et al. 2011c). In a different study, the BioNumerics 6.0 software (Applied-Maths, Sint-Martens-Latem, Belgium) was used for data analysis and machine learning for bacterial identification by MALDI-TOF MS (De Bruyne et al. 2011).

Taxonomic uses of MALDI-TOF MS

Taxonomy is the systematic classification of organisms based on their phenotypic, genetic and phylogenetic characteristics. Thus, various phenotypic approaches (morphology, biochemical reactions and sugar assimilation) have been used by microbiologists to classify microorganisms. However, genome analysis through sequencing of bacterial genes or the entire genome currently represents the gold standard in microbial taxonomy. MALDI-TOF MS represents an additional approach to classify microorganisms based on phenotypic and genotypic techniques. In contrast to sequencing housekeeping genes such as the 16S rRNA gene to provide taxonomical data on a single gene, MALDI-TOF MS fingerprints provide concomitant information for multiple protein components that characterize a microorganism. In addition, MALDI-TOF MS fingerprints provide data of both the presence (conservation/divergence) and the intensity (expression level) that together comprise a two-dimensional taxonomical asset, which offers better discriminative resolution for microorganism classification. Therefore, clustering of the spectral data obtained by MALDI-TOF MS represents another approach for bacterial identification and classification. Conway et al. (2001) introduced the term "Phyloproteomics" and the clustering of peak mass lists allowed for improved visualization of similarities and differences among spectral comparisons. The construction of a dendrogram based on mass spectral data is a rapid technique to analyze spectral profiles and to visualize spectral relations by grouping the obtained peak mass lists of bacterial strains. Thus, clustering has been successfully applied for the differentiation and identification of bacterial strains at the genus and species level (Wunschel et al. 2005, Vargha et al. 2006, Carbonnelle et al. 2007, Böhme et al. 2010b, Dubois et al. 2010, Böhme et al. 2011a, Alnakip et al. 2013, Quintela-Baluja et al. 2013). Additionally, clustering of mass spectral data has been applied as a typing method for the phyloproteomic study of different strains of the same species, with the aim of classifying the strains (Siegrist et al. 2007, Teramoto et al. 2007, Quintela-Baluja et al. 2013, Böhme et al. 2013). When comparing the dendrograms representing phyloproteomic relations to the phylogenetic trees, a high concordance was found by these authors. Since the peak patterns observed by MALDI-TOF MS are generally attributed to ribosomal proteins (Ryzhov and Fenselau 2001), the similarity of the MALDI-TOF MS cluster to phylogenetic trees obtained by the analysis of ribosomal genes is not surprising (Dubois et al. 2010). However, in comparison to the sequence analysis of the 16S rRNA gene that is commonly used for phylogenetic studies, the classification of bacterial strains by MALDI-TOF MS fingerprinting was more discriminating. This is important for some genera, such as *Bacillus* and *Pseudomonas* and some groups of *Enterococcus* and the *Enterobacteriaceae* family because the differentiation at the species level is difficult to obtain with 16S rRNA analysis.

Time and cost effectiveness of the MALDI-TOF MS approach in microbiology laboratories

The MALDI-TOF MS approach represents a new tool that has the potential to replace conventional identification techniques for a majority of routine isolates encountered in microbiology laboratories. The most striking differences between the MALDI-TOF MS technique and conventional identification methods are observed in the estimated time and cost required for sample identification. The cost of bacterial identification by MALDI-TOF MS was estimated to represent only 17–32% (around €1.43/sample) of the costs of conventional identification methods (around €4.6–8.23/sample) (Seng et al. 2009), which is supported by at least two other prospective studies (Bizzini and Greub 2010, Cherkaoui et al. 2010). Cherkaoui et al. (2010) have shown that the reagents required for phenotypic identification using modern automated platforms cost approximately $10 per isolate, whereas MS-required reagents do not exceed $0.50 per sample.

When compared with conventional identification methods, in most cases MALDI-TOF MS conferred a significant reduction in both technician working time (pre-analytical procedure to prepare samples) and turnaround time (automated analytical procedure to obtain results). The time required for bacterial identification from intact cells for MALDI-TOF MS was 6–8.5 min vs. 5–48 hr for conventional identification, as estimated by Seng et al. (2009). When an extraction step is required, Bizzini et al. (2010) have estimated that the extraction procedure for a single sample takes approximately 6 min, a time per sample that is further reduced with batch processing. The time reduction obtained by MALDI-TOF MS identification compared with classical identification approaches is even more apparent when several isolates are analyzed in parallel. As reported by Cherkaoui et al. (2010) the analysis of 10 isolates in parallel can be accomplished in < 15 min, with the total working time limited by the speed of the MS instrument, whereas more than 360 min would be required on an automated system with more hands-on time for each isolate. Thus, MALDI-TOF MS provides a reduction of both working and turnaround times. Nucleic acid based identification strategies suffer from several problematic limitations: (1) technical problems (inhibitory compounds, contamination, separate areas for sample preparation/amplification/analysis), (2) reagent and labor costs, (3) species identification in a single assay is often limited to a few individual species, and (4) much longer turnaround time than is required with MS. Thus, such long, costly and labor-intensive alternatives to MALDI-TOF MS are usually reserved for the identification of the small minority of isolates not identified by MS alone, or by other conventional phenotypic and/or biochemical identification approaches.

Conclusion

In order to ensure food safety and quality, the objective of this chapter is to review MALDI-TOF MS-based molecular methods that allow for the early detection and identification of the main pathogenic and spoilage bacteria.

In recent years, molecular diagnoses have become increasingly important, representing an attractive alternative to the traditional techniques of cultivation and characterization for the rapid detection and identification of microorganisms in food products that may produce foodborne illnesses. MALDI-TOF MS represents a very appealing new microbial identification technology that is efficient, rapid, cheap, and easy to use. This explains why MALDI-TOF MS can be successfully used in clinical diagnostic laboratories for microbial identification in samples from subcultures on agar plates and broth media, directly from positive blood cultures and to a lesser extent, from clinical samples such as urine. In addition to the high potential in clinical diagnostics, it is applicable to the identification of unknown bacterial strains isolated from food samples. In addition, the methodology of MALDI-TOF MS fingerprinting can be applied to different foodstuffs because bacterial strains are isolated from the corresponding matrix before analysis, and the constituent spectral library may be easily expanded by adding bacterial species and strains that are of interest in the corresponding food products. The vast amount of spectral data can be effectively examined by searching common peak masses and cluster analysis. Characteristic peak patterns can be determined and serve as biomarker proteins for the rapid identification of an unknown spectrum. The creation of a dendrogram based on phyloproteomic relations allows for the typing of closely related strains and thus extends conventional typing methods. When comparing phyloproteomic clustering with the classical taxonomy approach based on 16S rRNA gene analysis, high similarities were observed in the grouping of bacterial strains. However, genetic analysis does not always permit differentiation between species of the same genus due to the high similarity of DNA sequences; thus, spectral profiles obtained by MALDI-TOF analysis can reveal specificities for individual bacterial species. These proteomic tools and phyloproteomic clustering resulted in higher discrimination capability among different species.

The critical challenge of the MALDI-TOF MS fingerprint approach for bacterial identification in food is the limited availability of reference data. A huge public database, such as the NCBI GenBank database which facilitates the comparison of an unknown DNA sequence to millions of references strains, is not available for mass spectral data obtained by MALDI-TOF MS. Thus, the creation of an open access mass spectral database to deposit large volumes of spectral information should be a major objective in the future. An open access database of all publicly available MALDI-TOF MS bacterial spectra would represent an ample catalogue of spectra from different species and strains. Such a resource would enable accurate assignment of specific biomarkers and thus, the correct identification of unknown species or even species detection in mixed samples. The ideal database would also allow similarity searching to quickly identify an unknown strain. To create such a spectral reference library, a standardized protocol must be applied to allow comparisons of spectra from different bacterial strains and to obtain specific and reproducible spectral profiles for the subsequent identification and classification of bacterial strains. The potential exhibited by MALDI-TOF MS fingerprinting for the rapid and accurate identification of strains does not begin to illustrate the enormous amount of interesting information that could be extracted from the spectra; such information could be enhanced by the development of a database to establish relationships between the information obtained, through the use of genomic, transcriptomic, metabolomic and proteomic tools.

Recently, a new open access MALDI-TOF MS library has been created with a focus on bacterial species of interest in the food sector. It includes spectra and peak mass lists of more than 250 bacterial strains and more than 80 bacterial species (www.spectrabank.org). The SpectraBank database also describes in detail both, sample preparation and data analysis, and it also includes instructions for using the SpectraBank spectral information for microbial classification and identification.

Overall, MALDI-TOF MS will soon be available in most microbiology laboratories. Despite the significant cost of the instrument and maintenance, operation costs and consumables are much lower than those required for other conventional methods, rendering this technology a worthy quantum leap tool.

Keywords: food quality and safety, bacterial species identification, MALDI-TOF MS, phyloproteomics, phylogenetics, ecotypes

Acknowledgments

The work of Marcos Quintela Baluja is supported by the project IPT-2011-1290-010000 co-financed by the Spanish Ministry of Economy and Competitivity and the European Regional Development Fund 2007–2013 (FEDER).

References

Alm, R., P.K. Johansson, K. Hjernø, C. Emanuelsson, M. Ringner and J. Häkkinen. 2006. Detection and identification of protein isoforms using cluster analysis of MALDI-MS mass spectra. J. Proteome Res. 5: 785–792.

Alnakip, M.E., I.C. Fernández-No, M. Quintela-Baluja, K. Böhme, S. Caamaño, J. Barros-Velázquez and Pilar Calo-Mata. 2013. The differentiation of mastitis-causing *Streptococcus* spp. by MALDI-TOF MS fingerprinting. Electrophoresis (in press).

Anhalt, J.P. and C. Fenselau.. 1975. Identification of bacteria using mass spectrometry. Anal. Chem. 47: 219–225.

Arnold, R.J. and J.P. Reilly. 1998. Fingerprint matching of *E. coli* strains with matrix-assisted laser desorption/ionization time-of-flight mass spectrometry of whole cells using a modified correlation approach. Rapid Comm. Mass Spectrom. 12: 630–636.

Baar, B.L.M. 2000. Characterisation of bacteria by matrix-assisted laser desorption/ionisation and electrospray mass spectrometry. FEMS Microbiol. Rev. 24: 193–219.

Barbuddhe, S.B., T. Maier, G. Schwarz, M. Kostrzewa, H. Hof, E. Domann, T. Chakraborty and T. Hain. 2008. Rapid identification and typing of listeria species by matrix-assisted laser desorption ionization-time of flight mass spectrometry. Appl. Environ. Microbiol. 74: 5402–5407.

Barreiro, J.R., C.R. Ferreira, G.B. Sanvido, M. Kostrzewa, T. Maier, B. Wegemann, V. Böttcher, M.N. Eberlin and M.V. dos Santos. 2010. Short communication: Identification of subclinical cow mastitis pathogens in milk by matrix-assisted laser desorption/ionization time-of-flight mass spectrometry. J. Dairy Sci. 93: 5661–5667.

Bernardo, K., N. Pakulat, M. Macht, O. Krut, H. Seifert, S. Fleer, F. HüNGER and M. Krönke. 2002. Identification and discrimination of *Staphylococcus aureus* strains using matrix-assisted laser desorption/ionization-time of flight mass spectrometry. Proteomics 2: 747–753.

Bizzini, A. and G. Greub. 2010. Matrix-assisted laser desorption ionization time-of-flight mass spectrometry, a revolution in clinical microbial identification. Clin. Microbiol. Infect. 16: 1614–1619.

Blackburn, C.d.W. 2006. Food spoilage microorganisms. Woodhead Publishing Ltd. Cambridge

Böhme, K., I.C. Fernández-No, M. Pazos, J.M. Gallardo, J. Barros-Velázquez, B. Cañas and P. Calo-Mata. 2013. Identification and classification of seafood-borne pathogenic and spoilage bacteria: 16S rRNA sequencing versus MALDI-TOF MS fingerprinting. Electrophoresis 34: 877–887.

Böhme, K., S. Morandi, P. Cremonesi, I.C. Fernández No, J. Barros-Velázquez, B. Castiglioni, M. Brasca, B. Cañas and P. Calo-Mata. 2012. Characterization of *Staphylococcus aureus* strains isolated from Italian dairy products by MALDI-TOF mass fingerprinting. Electrophoresis 33: 2355–2364.

Böhme, K., I.C. Fernández-No, J. Barros-Velázquez, J.M. Gallardo, B. Cañas and P. Calo-Mata. 2011a. Rapid species identification of seafood spoilage and pathogenic Gram-positive bacteria by MALDI-TOF mass fingerprinting. Electrophoresis 32: 2951–2965.

Böhme, K., I.C. Fernández-No, J.M. Gallardo, B. Cañas and P. Calo-Mata. 2011b. Safety assessment of fresh and processed seafood products by MALDI-TOF mass fingerprinting. Food Bioprocess Technol. 4: 907–918.

Böhme, K., I.C. Fernández-No, J. Barros-Velázquez, J.M. Gallardo, P. Calo-Mata and B. Cañas. 2010b. Species differentiation of seafood spoilage and pathogenic gram-negative bacteria by MALDI-TOF mass fingerprinting. J. Proteome Res. 9: 3169–3183.

Böhme, K., I.C. Fernández-No, J. Barros-Velázquez, J.M. Gallardo, B. Cañas and P. Calo-Mata. 2010a. Comparative analysis of protein extraction methods for the identification of seafood-borne pathogenic and spoilage bacteria by MALDI-TOF mass spectrometry. Anal. Methods 2: 1941–1947.

Böhme, K., I.C. Fernández-No, J. Barros-Velázquez, J.M. Gallardo, B. Cañas and P. Calo-Mata. 2011c. Rapid species identification of seafood spoilage and pathogenic Gram-positive bacteria by MALDI-TOF mass fingerprinting. Electrophoresis 32: 2951–2965.

Bright, J.J., M.A. Claydon, M. Soufian and D.B. Gordon. 2002. Rapid typing of bacteria using matrix-assisted laser desorption ionisation time-of-flight mass spectrometry and pattern recognition software. J. Microbiol. Methods 48: 127–138.

Carbonnelle, E., J. Beretti, S. Cottyn, G. Quesne, P. Berche, X. Nassif and A. Ferroni. 2007. Rapid identification of Staphylococci isolated in clinical microbiology laboratories by matrix-assisted laser desorption ionization-time of flight mass spectrometry. J. Clin. Microbiol. 45: 2156–2161.

Cherkaoui, A., J. Hibbs, S. Emonet, M. Tangomo, M. Girard, P. Francois and J. Schrenzel. 2010. Comparison of two matrix-assisted laser desorption ionization-time of flight mass spectrometry methods with conventional phenotypic identification for routine identification of bacteria to the species level. J. Clin. Microbiol. 48: 1169–1175.

Claydon, M.A., S.N. Davey, V. Edwards-Jones and D.B. Gordon. 1996. The rapid identification of intact microorganisms using mass spectrometry. Nat. Biotechnol. 14: 1584–1586.

Conway, G.C., S.C. Smole, D.A. Sarracino, R.D. Arbeit and P.E. Leopold. 2001. Phyloproteomics: species identification of Enterobacteriaceae using matrix-assisted laser desorption/ionization time-of-flight mass spectrometry. J. Mol. Microbiol. Biotechnol. 3: 103–112.

Dare, D. 2006. Rapid bacterial characterization and identification by MALDI-TOF mass spectrometry. *In*: Advanced tecniques in diagnostic microbiology. Springer, US, pp. 117–133.

De Bruyne, K., B. Slabbinck, W. Waegeman, P. Vauterin, B. De Baets and P. Vandamme. 2011. Bacterial species identification from MALDI-TOF mass spectra through data analysis and machine learning. Syst. Appl. Microbiol. 34: 20–29.

Demirev, P.A., A.B. Feldman and J.S. Lin. 2004. Bioinformatics-based strategies for rapid microorganism identification by mass spectrometry. Johns Hopkins APL Tech. Dig. 25: 27–37.

Domin, M.A., K.J. Welham and D.S. Ashton. 1999. The effect of solvent and matrix combinations on the analysis of bacteria by matrix-assisted laser desorption/ionisation time-of-flight mass spectrometry. Rapid Commun. Mass Spectrom. 13: 222–226.

Donohue, M.J., A.W. Smallwood, S. Pfaller, M. Rodgers and J.A. Shoemaker. 2006. The development of a matrix-assisted laser desorption/ionization mass spectrometry-based method for the protein fingerprinting and identification of *Aeromonas* species using whole cells. J. Microbiol. Methods 65: 380–389.

Dubois, D., D. Leyssene, J.P. Chacornac, M. Kostrzewa, P.O. Schmit, R. Talon, R. Bonnet and J. Delmas. 2010. Identification of a variety of Staphylococcus species by matrix-assisted laser desorption ionization-time of flight mass spectrometry. J. Clin. Microbiol. 48: 941–945.

Duerden, B., K. Towner and J. Magee. 1998. Isolation, description and identification of bacteria. Topley and Wilson's Microbiology and Microbial Infections: Systematic Bacteriology. London 2: 65–84.

Elizaquível, P. and R. Aznar. 2008. A multiplex RTi-PCR reaction for simultaneous detection of *Escherichia coli* O157: H7, *Salmonella* spp. and *Staphylococcus aureus* on fresh, minimally processed vegetables. Food Microbiol. 25: 705–713.

Erhard, M., U.C. Hipler, A. Burmester, A.A. Brakhage and J. Wöstemeyer. 2008. Identification of dermatophyte species causing onychomycosis and tinea pedis by MALDI-TOF mass spectrometry. Exp. Dermatol. 17: 356–361.

Fagerquist, C.K., W.G. Miller, L.A. Harden, A.H. Bates, W.H. Vensel, G. Wang and R.E. Mandrell. 2005. Genomic and proteomic identification of a DNA-binding protein used in the "fingerprinting" of *Campylobacter* species and strains by MALDI-TOF-MS protein biomarker analysis. Anal. Chem. 77: 4897–4907.

Fao. 2003. Assuring Food Safety and Quality: Guidelines for Strengthening National Food Control Systems. FAO and WHO ISBN 92-5-1049181.

Feng, P., M. Doyle and L. Beuchat. 2007. Rapid methods for the detection of foodborne pathogens: Current and next-generation technologies 911–934.

Fernández-No, I.C., K. Böhme, J.M. Gallardo, J. Barros-Velázquez, B. Cañas and P. Calo-Mata. 2010. Differential characterization of biogenic amine-producing bacteria involved in food poisoning using MALDI-TOF mass fingerprinting. Electrophoresis 31: 1116–1127.

Freiwald, A. and S. Sauer. 2009. Phylogenetic classification and identification of bacteria by mass spectrometry. Nat. Protoc. 4: 732–742.

Ghyselinck, J., K. Van Hoorde, B. Hoste, K. Heylen and P. De Vos. 2011. Evaluation of MALDI-TOF MS as a tool for high-throughput dereplication. J. Microbiol. Methods 86: 327–336.

Giebel, R., C. Worden, S. Rust, G. Kleinheinz, M. Robbins and T. Sandrin. 2010. Microbial fingerprinting using matrix-assisted laser desorption ionization time-of-flight mass spectrometry (MALDI-TOF MS): applications and challenges. Adv. Appl. Microbiol. 71: 149–184.

Giebel, R.A., W. Fredenberg and T.R. Sandrin. 2008. Characterization of environmental isolates of *Enterococcus* spp. by matrix-assisted laser desorption/ionization time-of-flight mass spectrometry. Water Res. 42: 931–940.

Gram, L., L. Ravn, M. Rasch, J.B. Bruhn, A.B. Christensen and M. Givskov. 2002. Food spoilage—interactions between food spoilage bacteria. Int. J. Food Microbiol. 78: 79–97.

Grosse-Herrenthey, A., T. Maier, F. Gessler, R. Schaumann, H. Böhnel, M. Kostrzewa and M. Krüger. 2008. Challenging the problem of clostridial identification with matrix-assisted laser desorption and ionization–time-of-flight mass spectrometry (MALDI–TOF MS). Anaerobe 14: 242–249.

Haag, A.M., S.N. Taylor, K.H. Johnston and R.B. Cole. 1998. Rapid identification and speciation of *Haemophilus* bacteria by matrix-assisted laser desorption/ionization time-of-flight mass spectrometry. J. Mass Spectrom. 33: 750–756.

Hahn, D., B. Mirza, C. Benagli, G. Vogel and M. Tonolla. 2011. Typing of nitrogen-fixing Frankia strains by matrix-assisted laser desorption ionization-time-of-flight (MALDI-TOF) mass spectrometry. Syst. Appl. Microbiol. 34: 63–68.

Hausdorf, L., K. Mundt, M. Winzer, C. Cordes, A. Fröhling, O. Schlüter and M. Klocke. 2012. Characterization of the cultivable microbial community in a spinach-processing plant using MALDI-TOF MS. Food Microbiol. 34: 406–411

Hein, I., A. Lehner, P. Rieck, K. Klein, E. Brandl and M. Wagner. 2001. Comparison of different approaches to quantify *Staphylococcus aureus* cells by real-time quantitative PCR and application of this technique for examination of cheese. Appl. Environ. Microbiol. 67: 3122–3126.

Hettick, J.M., M.L. Kashon, J.E. Slaven, Y. Ma, J.P. Simpson, P.D. Siegel, G.N. Mazurek and D.N. Weissman. 2006. Discrimination of intact mycobacteria at the strain level: A combined MALDI-TOF MS and biostatistical analysis. Proteomics 6: 6416–6425.

Holland, R., J. Wilkes, F. Rafii, J. Sutherland, C. Persons, K. Voorhees and J. Lay. 1996. Rapid identification of intact whole bacteria based on spectral patterns using matrix-assisted laser desorption/ionization with time-of-flight mass spectrometry. Rapid Commun. Mass Spectrom. 10: 1227–1232.

Holland, R.D., F. Rafii, T.M. Heinze, J.B. Sutherland, K.J. Voorhees and J.O. Lay. 2000. Matrix-assisted laser desorption/ionization time-of-flight mass spectrometric detection of bacterial biomarker proteins isolated from contaminated water, lettuce and cotton cloth. Rapid Commun. Mass Spectrom. 14: 911–917.

Hsieh, S.Y., C.L. Tseng, Y.S. Lee, A.J. Kuo, C.F. Sun, Y.H. Lin and J.K.Chen. 2008. Highly efficient classification and identification of human pathogenic bacteria by MALDI-TOF MS. Mol Cell Proteomics. Int. J. Mass Spectrom. 7: 448–456.

Jarman, K.H., D.S. Daly, K.K. Anderson and K.L. Wahl. 2003. A new approach to automated peak detection. Chemometrics Intellig. Lab Syst. 69: 61–76.

Kallow, W., M. Erhard, R. Dieckmann and S. Sauermann. 2004. U.S. Patent Application 10/752,224.

Karas, M., D. Bachmann, U.e. Bahr and F. Hillenkamp. 1987. Matrix-assisted ultraviolet laser desorption of non-volatile compounds. Int. J. Mass Spectrom. 78: 53–68.

Keys, C.J., D.J. Dare, H. Sutton, G. Wells, M. Lunt, T. McKenna, M. McDowall and H.N. Shah. 2004. Compilation of a MALDI-TOF mass spectral database for the rapid screening and characterisation of bacteria implicated in human infectious diseases. Infect. Genet. Evol. 4: 221–242.

Klaenhammer, T., E. Pfeiler, T. Duong, M. Doyle and L. Beuchat. 2007. Genomics and proteomics of foodborne microorganisms. *In*: M.P. Doyle and L.R. Beuchat (eds.). ASM Press, Washington, pp. 935–951.

Krishnamurthy, T. and P.L. Ross. 1996. Rapid identification of bacteria by direct matrix assisted laser desorption/ionization mass spectrometric analysis of whole cells. Rapid Commun. Mass Spectrom. 10: 1992–1996.

Lartigue, M., G. Héry-Arnaud, E. Haguenoer, A. Domelier, P. Schmit, N. Van Der Mee-Marquet, P. Lanotte, L. Mereghetti, M. Kostrzewa and R. Quentin. 2009. Identification of *Streptococcus agalactiae* isolates from various phylogenetic lineages by matrix-assisted laser desorption ionization-time of flight mass spectrometry. J. Clin. Microbiol. 47: 2284–2287.

Liu, H., Z. Du, J. Wang and R. Yang. 2007. Universal sample preparation method for characterization of bacteria by matrix-assisted laser desorption ionization-time of flight mass spectrometry. Appl. Environ. Microbiol. 73: 1899–1907.

Ludwig, W. and H. Klenk. 2001. Overview: a phylogenetic backbone and taxonomic framework for procaryotic systematics, *In*: D.R. Boone, R.W. Castenholz and G.M. Garrity. Bergey's manual of systematic bacteriology springer, New York, pp. 49–65.

Lynn, E.C., M. Chung, W. Tsai and C. Han. 1999. Identification of Enterobacteriaceae bacteria by direct matrix-assisted laser desorptiom/ionization mass spectrometric analysis of whole cells. Rapid Commun. Mass Spectrom. 13: 2022–2027.

Malorny, B., E. Paccassoni, P. Fach, C. Bunge, A. Martin and R. Helmuth. 2004. Diagnostic real-time PCR for detection of *Salmonella* in food. Appl. Environ. Microbiol. 70: 7046-7052.

Mazzeo, M.F., A. Sorrentino, M. Gaita, G. Cacace, M. Di Stasio, A. Facchiano, G. Comi, A. Malorni and R.A. Siciliano. 2006. Matrix-assisted laser desorption ionization-time of flight mass spectrometry for the discrimination of food-borne microorganisms. Appl. Environ. Microbiol. 72: 1180–1189.

Mohania, D., R. Nagpal, M. Kumar, A. Bhardwaj, M. Yadav, S. Jain, F. Marotta, V. Singh, O. Parkash and H. Yadav. 2008. Molecular approaches for identification and characterization of lactic acid bacteria. J. Dig. Dis. 9: 190–198.

Nassif, X. 2009. A revolution in the identification of pathogens in clinical laboratories. Clin. Infect. Dis. 49: 552–553.

Neuhof, T., R. Dieckmann, I.S. Druzhinina, C.P. Kubicek and H. von Döhren. 2007. Intact-cell MALDI-TOF mass spectrometry analysis of peptaibol formation by the genus Trichoderma/Hypocrea: can molecular phylogeny of species predict peptaibol structures? Microbiology 153: 3417–3437.

Nguyen, T.L.D., K. Van Hoorde, M. Cnockaert, E. De Brandt, A. Maarten, B. Le Thanh and P. Vandamme. 2013. A description of the lactic acid bacteria microbiota associated with the production of traditional fermented vegetables in Vietnam. Int. J. Food Microbiol. 163: 19–27.

Pennanec, X., A. Dufour, D. Haras and K. Réhel. 2010. A quick and easy method to identify bacteria by matrix-assisted laser desorption/ionisation time-of-flight mass spectrometry. Rapid Commun. Mass Spectrom. 24: 384–392.

Quintela-Baluja, M., K. Böhme, I.C. Fernández-No, S. Morandi, M.E. Caamaño, A., S., J. Barros-Velázquez and P. Calo-Mata 2013. Characterisation of different food-isolated *Enterococcus* strains by MALDI-TOF mass fingerprinting. Electrophoresis (in press).

Rossell-Mora, R. and R. Amann. 2001. The species concept for prokaryotes. FEMS Microbiol. Rev. 25: 39–67.

Ryzhov, V. and C. Fenselau. 2001. Characterization of the protein subset desorbed by MALDI from whole bacterial cells. Anal. Chem. 73: 746–750.

Sandt, C., C. Madoulet, A. Kohler, P. Allouch, C. De Champs, M. Manfait and G. Sockalingum. 2006. FT-IR microspectroscopy for early identification of some clinically relevant pathogens. J. Appl. Microbiol. 101: 785–797.

Seng, P., M. Drancourt, F. Gouriet, B. La Scola, P.E. Fournier, J.M. Rolain and D. Raoult. 2009. Ongoing revolution in bacteriology: routine identification of bacteria by matrix-assisted laser desorption ionization time-of-flight mass spectrometry. Clin. Infect. Dis. 49: 543–551.

Severgnini, M., P. Cremonesi, C. Consolandi, G. De Bellis and B. Castiglioni. 2011. Advances in DNA microarray technology for the detection of foodborne pathogens. J. Pathog. 4: 936–953.

Siegrist, T.J., P.D. Anderson, W.H. Huen, G.T. Kleinheinz, C.M. McDermott and T.R. Sandrin. 2007. Discrimination and characterization of environmental strains of Escherichia coli by matrix-assisted laser desorption/ionization time-of-flight mass spectrometry (MALDI-TOF-MS). J. Microbiol. Methods 68: 554–562.

Sneath, P. 1977. The maintenance of large numbers of strains of microorganisms, and the implications for culture collections. FEMS Microbiol. Lett. 1: 333–334.

Stein, C., T. Kuchenmüller, S. Hendrickx, A. Prüss-Üstün, L. Wolfson, D. Engels and J. Schlundt. 2007. The global burden of disease assessments—Who is responsible? 1: e161.

Stephan, R., N. Cernela, D. Ziegler, V. Pflüger, M. Tonolla, D. Ravasi, M. Fredriksson-Ahomaa and H. Hächler. 2011. Rapid species specific identification and subtyping of *Yersinia enterocolitica* by MALDI-TOF Mass spectrometry. J. Microbiol. Methods 87: 150–153.

Stephan, R., D. Ziegler, V. Pflüger, G. Vogel and A. Lehner. 2010. Rapid genus-and species-specific identification of *Cronobacter* spp. by matrix-assisted laser desorption ionization-time of flight mass spectrometry. J. Clin. Microbiol. 48: 2846–2851.

Tauxe, R.V. 2002. Emerging foodborne pathogens. Int. J. Food Microbiol. 78: 31–41.

Teramoto, K., H. Sato, L. Sun, M. Torimura, H. Tao, H. Yoshikawa, Y. Hotta, A. Hosoda and H. Tamura. 2007. Phylogenetic classification of *Pseudomonas putida* strains by MALDI-MS using ribosomal subunit proteins as biomarkers. Anal. Chem. 79: 8712–8719.

Vargha, M., Z. Takáts, A. Konopka and C.H. Nakatsu. 2006. Optimization of MALDI-TOF MS for strain level differentiation of *Arthrobacter* isolates. J. Microbiol. Methods 66: 399–409.

Vorob'eva, L., F. Khasaeva, N. Vasilyuk and E. Trenquil. 2011. Characterization of propionic acid bacteria using matrix-assisted laser desorption/ionization (MALDI) mass spectrometry. Microbiology 80: 664–671.

Wang, Z., L. Russon, L. Li, D.C. Roser and S.R. Long. 1998. Investigation of spectral reproducibility in direct analysis of bacteria proteins by matrix-assisted laser desorption/ionization time-of-flight mass spectrometry. Rapid Commun. Mass Spectrom. 12: 456–464.

Welham, K., M. Domin, K. Johnson, L. Jones and D. Ashton. 2000. Characterization of fungal spores by laser desorption/ionization time-of-flight mass spectrometry. Rapid Commun. Mass Spectrom. 14: 307–310.

Welker, M. and E.R.B. Moore. 2011. Applications of whole-cell matrix-assisted laser-desorption/ionization time-of-flight mass spectrometry in systematic microbiology. Syst. Appl. Microbiol. 34: 2–11.

Wensing, A., S. Zimmermann and K. Geider. 2010. Identification of the corn pathogen Pantoea stewartii by mass spectrometry of whole-cell extracts and its detection with novel PCR primers. Appl. Environ. Microbiol. 76: 6248–6256.

Williams, T.L., D. Andrzejewski, J.O. Lay Jr. and S.M. Musser. 2003. Experimental factors affecting the quality and reproducibility of MALDI TOF mass spectra obtained from whole bacteria cells. J. Am. Soc. Mass Spectrom. 14: 342–351.

Wilson, E.O. 2003. The encyclopedia of life. Trends Ecol. Evol. 18: 77–80.

World Health Organization. 2007. Food safety and foodborne illness.

Wunschel, D.S., E.A. Hill, J.S. McLean, K. Jarman, Y.A. Gorby, N. Valentine and K. Wahl. 2005. Effects of varied pH, growth rate and temperature using controlled fermentation and batch culture on matrix assisted laser desorption/ionization whole cell protein fingerprints. J. Microbiol. Methods 62: 259–271.

Yamada, Y., K. Aida and T. Uemura. 1968. Distribution of Ubiquinone 10 and 9 in Acetic acid bacteria and its relation to the classification of genera. Agric. Biol. Chem. 32: 786–788.

CHAPTER 14

Monitoring Microbial Spoilage of Foods by Vibrational Spectroscopy (FT-IR & Raman)

Anthoula A. Argyri,[1,2]* *Efstathios Z. Panagou*[2] and *George–John Nychas*[2]

Introduction

Intrinsic, extrinsic, and implicit factors are involved in food quality, e.g., causing food deterioration (Nychas and Skandamis 2005). Therefore the overall effect of a combination of factors is generally much higher than the perceived effect of each individual one. The most important intrinsic parameters are water activity, acidity, redox potential, available nutrients and natural antimicrobial substances. Extrinsic parameters are related to the environment in which a food is stored, notably temperature, humidity and atmosphere composition. Physical or chemical treatments often result in changes in the characteristics of a food product determining the microbiota associated with it. Implicit factors are those physiological properties that enable particular organisms to flourish resulting from the interaction of the above mentioned factors, i.e., intrinsic, extrinsic. Thus, implicit parameters are the result of the development of a microorganism which may have a synergistic or antagonistic effect on the microbial activity of other microbial communities present in the food product (Huis in't Veld 1996).

[1] Institute of Technology of Agricultural Products, Hellenic Agricultural Organisation 'DEMETER', Sof. Venizelou 1, 14123, Lycovrissi, Attiki, Greece.
[2] Laboratory of Microbiology & Biotechnology of Foods, Dept. of Food Science & Technology, Agricultural University of Athens, Iera Odos 75, Athens 11855.
* Corresponding author: anthi.argyri@gmail.com

Understanding how different properties of a food, its environment and 'history', can influence the microbiota that develops in storage, is an important first step towards being able to make estimations concerning shelf-life, spoilage and safety. The food industry is continually creating new microbial habitats, either by developing new products and reformulating traditional ones, or by chance, as a result of the composition of raw materials or in production. Thus, the food industry needs rapid non-destructive techniques for automatic monitoring of food processes in real time on- or in-line. The advantages of using instruments based on vibrational spectroscopy such as infrared (IR) or Raman are that they are rapid, noninvasive and require little or no pre-treatment of the samples. The IR and Raman instruments can be integrated in the existing infrastructure in the food industries and consequently improve quality and safety of the product, speed the process and reduce the waste. Infrared and Raman spectroscopy are complementary techniques that both measure vibrations of the molecules but in different ways. IR works with absorption measurements, while Raman measures scattered light from a laser source, that in being scattered, is superimposed with the vibrational structure of the molecule.

Fourier Transform Infrared Spectroscopy

IR spectroscopy: principles, instrumentation, spectral interpretation

Basic principles of IR spectroscopy. Infrared spectroscopy has been an analytical technique for materials analysis in the laboratory for over seventy years, with many reported applications. The main principle of IR spectroscopy relies on the irradiation of a molecule with IR radiation and then detection of the energy frequencies absorbed which are highly specific for that molecule (Stuart 1997). As each molecule only absorbs specific frequencies of IR energy, the resultant IR absorbance spectrum can be used to identify the molecular contents of an unknown compound and thus the compound itself (Cothup et al. 1990, Banwell and McCash 1994, Stuart 1997, Schmitt and Flemming 1998, Yang and Irudayaraj 2003). Thus there is a correlation between IR band positions and chemical structures in the molecule. The vibrational modes of a molecule can be subdivided into the two main categories of stretching and bending. In stretching, the inter-atomic distance along bond axis changes. Stretching can be symmetrical or assymetrical. In bending, the angle between two bonds changes. Bending can be in-plane (rocking, scissoring or deformation vibrations) or out-of-plane (wagging and twisting vibrations) (Banwell and McCash 1994, Stuart 1997).

The infrared region (10–14000 cm^{-1}) of the electromagnetic spectrum is divided into three regions: the near-infrared (NIR) (0.75–2.5 μm or 13000–4000cm^{-1}), mid-infrared (MIR) (2.5–50 μm or 4000–400 cm^{-1}), and far-infrared (FIR) (50–1000 μm or 400-10 cm^{-1}). The MIR (400–4000 cm^{-1}) is the most commonly used region for analysis as all molecules possess characteristic absorbance frequencies and primary molecular vibrations in this range. Mid-infrared spectroscopy methods are based on studying the interaction of infrared radiation with samples. The MIR has two regions, the group region (4000–1400 cm^{-1}) and the so called fingerprint region (1400–600 cm^{-1}). Group frequencies are characteristic of molecular groups and are almost independent of the structure of the whole molecule. The amplitude of the group vibration is only significant for a small number of molecules. By contrast, the fingerprint region is derived from the interaction of the vibrational modes and it is highly characteristic of the molecule under investigation. Using information from both of these highly specific regions differences in the structure and composition of compounds can be seen, even when these are very small, enabling the identification of the specific compound (Banwell and McCash 1994, Yang and Irudayaraj 2003).

Infrared spectrometer and Fourier transform infrared spectrometer. The original infrared instruments were of the dispersive type. These instruments separated the individual frequencies of energy emitted from the infrared source. Source radiation is dispersed by the monochromator using a dispersive element such as a prism or grating, separating the components of polychromatic radiation based on their wavelength. This was accomplished by the use of a prism or grating. An infrared prism works exactly the same as a visible prism which separates visible light into its colors (frequencies). A grating is a more modern dispersive element which better separates the frequencies of infrared energy. The detector measures the amount of energy at each frequency which has passed through the sample. This results in a spectrum which is a plot of intensity vs. Frequency.

Fourier Transform Infrared (FT-IR) spectrometry was developed in order to overcome the limitations encountered with dispersive instruments. The main difficulty was the slow scanning process. A method for measuring all of the infrared frequencies simultaneously, rather than individually, was needed. The normal instrumental process is as follows: Infrared energy is emitted from a glowing black-body source. This beam passes through an aperture which controls the amount of energy presented to the sample (and, ultimately, to the detector). The beam enters the interferometer where the "spectral encoding" takes place. The resulting interferogram signal then exits the interferometer. The beam enters the sample compartment where it is transmitted through or reflected off of the surface of the sample, depending on the type of analysis being accomplished (see below 'Sampling techniques'). The beam finally passes to the detector for final measurement. The detectors used are specially designed to measure the special interferogram signal. The measured signal is digitized and sent to the computer where the Fourier transformation takes place. Fourier transformation is a mathematical technique which decodes the measured interferogram signal which cannot be interpreted directly. The final infrared spectrum is then presented to the user for interpretation and any further manipulation.

Sampling techniques. There are several sampling techniques that may be applied in FT-IR spectroscopy depending on the application and the type of the sample that is analyzed. These include the Attenuated Total Reflectance (ATR), Specular Reflectance (SR), Diffuse Reflectance Infrared Fourier Transform Spectroscopy (DRIFTS), Reflection-Absorption IR Spectroscopy (RAIRS), Transmission (TR), Photoacoustic Spectroscopy (PAS), FT-IR microspectroscopy and Multiple Internal Reflection Spectroscopy (MIR).

Mathematical treatment of the spectroscopic data. The enormous amount of molecularly specific information provided by FT-IR and Raman spectroscopy makes the data produced unmanageable to simple 'stare and compare' interpretation. Thus it is necessary to apply advanced multivariate statistical methods and intelligent machine learning methodologies for analysis. It has to be noted that it is common that the spectral data are preprocessed before application of the multivariate analysis, for noise reduction using First or Second derivatives, Wavelet Transform (WT), smoothing by WT, baseline correction by WT, data linearization, Multiplicative Signal Correction (MSC) or its extended form (EMSC), Extended Inverted Signal Correction (EISC), or Standard Normal Variate (SNV) transformation.

Consequently several unsupervised or supervised mathematical models can be applied for qualitative (classification) or quantitative purposes (regression). Qualitative methods include Principal Component Analysis, Factorial Discriminant Analysis, Discriminant Function Analysis, Cluster Analysis, Linear Discriminant Analysis, Quadratic Discriminant Analysis, Partial Least Square Discriminant Analysis, Categorical Regression, K-Nearest Neighbors Analysis, Artificial Neural Network, Support Vector Machines. Quantitative methods (calibration), include Multiple Linear Regression, Principal Component Regression, Partial Least Square Regression, Support Vector Machine Regression (with different kernel functions including linear, polynomial, radial basis and sigmoid functions), and Least- Square Support Vector Machine (LS-SVM), Artificial Neural Network, Genetic Programming, Genetic Algorithms (Jain et al. 2000, Jurs et al. 2000, Berrueta et al. 2007, Charaniya et al. 2008, Cevallos-Cevallos et al. 2009, Da-Wen 2009). The different mathematical models applied in studies relevant with the investigation of food spoilage through FT-IR spectroscopy are reported in Table 1.

Table 1. Mathematical models applied in studies relevant with investigation of food spoilage through FT-IR spectroscopy.

Modeling Method	Purpose	Food type	Reference
Principal component analysis (PCA)	Feature selection/highlighting, Unsupervised classification	Meat, cheese, weat, fruit spirits-beer	Lin et al. 2004, Ammor et al. 2009, Alexandrakis et al. 2012, Cattaneo et al. 2005, Singh et al. 2011, Lachenmeier 2007
Principal component-discriminant function analysis (PC-DFA)	Feature selection/highlighting, supervised classification	Milk	Nicolaou and Goodacre 2008, Nicolaou et al. 2011
Factorial discriminant analysis (FDA)	Supervised classification	Meat	Ammor et al. 2009
Outer product analysis (OPA)	Feature selection/highlighting	Meat	Alexandrakis et al. 2012
Artificial neural networks (ANNs)	Regression, Supervised classification	Meat	Argyri et al. 2010, 2013
Partial least-squares (PLS)	Regression	Meat, Milk	Ellis et al. 2002, 2004, Lin et al. 2004, Nicolaou and Goodacre 2008, Ammor et al. 2009, Papadopoulou et al. 2011, Nicolaou et al. 2011, Argyri et al. 2013
Genetic algorithms (GA) and Genetic programming (GP)	Feature selection/highlighting, classification, regression	Meat	Ellis et al. 2002, 2004, Argyri et al. 2013
PLS-discriminant analysis (PLS-DA)	Supervised classification	Meat	Papadopoulou et al. 2011, Alexandrakis et al. 2012
Support vector machines regression (SVR) including different kernel functions (i.e., linear, polynomial, radial basis and sigmoid)	Regression, Supervised classification	Meat	Argyri et al. 2013
kernel PLS (KPLS)	Regression	Milk	Nicolaou et al. 2011
Canonical correlation analysis (CCA)	Investigate relationships between two multidimensional variables (i.e., FTIR data and another measured variable)	Milk	Nicolaou et al. 2011

Applications for monitoring food quality

Muscle foods. FT-IR spectroscopy has the potential to be established as a technique for muscle foods authentication, as it was able to discriminate between fresh and frozen-thawed meat (Al-Jowder et al. 1997) or fish samples (Karoui et al. 2007), and between meat species (Al-Jowder et al. 1999, 1997, Downey et al. 2000, Meza-Márquez et al. 2010, Rohman et al. 2011). Yang and Irudayaraj (2001) demonstrated the dynamics of different FT-IR techniques for beef and pork quality control through a multi-layer analysis.

FT-IR measurements can serve as a rapid method to determine marine fatty acids in pork fat (Flatten et al. 2005, Olsen et al. 2008) and to discriminate carcasses of suckling lambs according to fat quality (Osorio et al. 2009). It has been applied also to study the structure and oxidation properties of extracted cooked cured meat pigment (Sun et al. 2009).

FT-IR has been used to investigate the changes of proteins in meat and meat products due to the effect of different processing factors, including aging, salting, heat treatment and oxidation (Kirschner et al. 2004, Wu et al. 2007, Wu et al. 2006, Bocker et al. 2007, Perisic et al. 2011, Sun et al. 2011). It has also been applied for monitoring the production of chitin and chitosan during the fermentation of shrimp shell with *Bacillus subtilis* (Sini et al. 2007).

Dairy foods. In the case of dairy products FT-IR has been successfully used in many applications. FT-IR spectroscopy has the potential to be established as a technique for milk authentication since it has been found able to identify and differentiate between cows, goat or sheep milk (Maudet and Taberlet 2001, Pappas et al. 2008, Nicolaou et al. 2010) and to determine the urea concentration in milk (Baumgartner et al. 2003). Moreover, except from the economic standpoint, intolerance and allergic reactions due to the cow milk component of adulterated products (Venter 2009, Chafen et al. 2010) are of importance and could be easily and quickly detected by the use of FT-IR.

FT-IR spectroscopy has been used effectively for the quantification of proteins in milk (Hewavitharana and Brakel 1997, Luginbuhl 2002, Etzion et al. 2004, Kher et al. 2007, Haque et al. 2011, Rutten et al. 2011, Wu et al. 2008) and yogurt (Khanmohammadi et al. 2009). It was also applied in milk analysis for the quantification of lactose (Cocciardi et al. 2004, Yang et al. 2007, Kittivachra et al. 2007) and of fat, protein and lactose (Lefier et al. 1996). It has been applied to determine fatty acid composition of milk (Ulberth and Henninger 1994, Afseth et al. 2010) and to discriminate butters obtained from milk creams collected in different seasons and produced from two feeding regimens (Gori et al. 2012).

It has been used also for the determination of protein and fat during cheese ripening (Chen et al. 1998) and for monitoring changes in proteins structure as a consequence of microbial proteolytic activity of *Lactococcus lactis* subsp. *lactis* and different strains of *Yarrowia lipolytica* inoculated in milk for cheese ripenning (Lucia et al. 2001). Rodriguez-Saona et al. (2006) were able to determine the moisture, protein, and fat content of cheese samples, regardless of the degree of ripening and classified the cheeses according to their manufacturer and age with FT-IR. Finally, FT-IR may be used for qualitative and semi-quantitative food contamination screening of a large number of ripened and semi-ripened cheeses by detecting cyclopiazonic acid a toxic metabolite produced by contaminating moulds (Monaci et al. 2007).

Eggs. FT-IR has been applied for studying the shell membrane in salted duck egg (Kaewmanee et al. 2011), the proteins structure in modified egg yolk powder (Ge et al. 2011) and hen egg (Herald and Smith 1992), and the protein of egg shells during storage (Perry et al. 2011).

Honey. FT-IR in combination with multivariate analysis has also been successfully used in the determination of the botanical and geographical origin of honeys (Ruoff et al. 2006, Hennessy et al. 2008).

Coffee and chocolate. Analysis of lard adulteration in chocolate and chocolate products (Che Man et al. 2005) and discrimination of coffees from different geographical origins and of different roasting degrees (Wang et al. 2011a), was feasible due to the use of FT-IR.

Fruits and vegetables. FT-IR spectroscopy was applied successfully for evaluating the changes happening during the development and maturation of olive fruit by measuring the content of olive constituents like triglycerides, water, carotenoids and phenolic compounds (López-Sánchez et al. 2010). It was found suitable for determining sugar and organic acid contents in apricot fruit slurries (Bureau et al. 2009) or simple sugars, malic acid and total phenolic compounds in apple pomace (Queji et al. 2010), as well as in rapid antioxidant capacity determination of various fruit extracts (Lam et al. 2005). It was also applied to detect adulteration in strawberry purées (Holland et al. 1998).

Drinks and beverages. FT-IR spectroscopy in combination with multivariate data analysis is introduced for the quality control and authenticity assessment of spirit drinks and beer in official food control (Lachenmeier 2007). It has been applied for the discrimination of red wine cultivars measuring the phenolic wine extracts (Edelmann et al. 2001) and of white wines of different origin measuring their polysaccharide extracts (Coimbra et al. 2002), as well as for the determination of proteins and amino acids in beers (Gorinstein et al. 1999). Cozzolino and Dambergs (2009) reported that FT-IR has been implemented for routine analysis of a large number of wine parameters such as alcohol, volatile acidity, pH, tartaric acid, lactic acid, glucose plus fructose, acetic acid, glycerol, anthocyanins, polyphenols, and polysaccharides and reported by many authors and is nowadays is commonly used by the wine industry worldwide. It has also been used in several studies to quantify ethanol and other compositional parameters in beer (Cozzolino and Dambergs 2009).

FT-IR spectroscopy was used for quantification of sugars in mango juices as a function of ripening (Duarte et al. 2002) and for juice authentication as it was able to differentiate between commercial juices (He et al. 2007). Finally, an interesting application of FT-IR has been reported as a tool for effective monitoring the red wine fermentation (Urtubia et al. 2004) and of the time course of fermentation in wild ferments (Cozzolino and Curtin 2012).

Oils. FT-IR can reliably identify adulteration of vegetable and edible oils (Ozen et al. 2003, Alexa et al. 2009), extra virgin oil with lower priced vegetable oils (Vlachos et al. 2006, Rohman and Man 2010) and hazelnut oil (Ozen and Mauer 2002). It has been also used in the past to classify different types of vegetable oils (Dahlberg et al. 1997, Ozen and Mauer 2002, Saucedo-Hernandez et al. 2011) and to discriminate extra virgin olive oils of different geographic origin (Tapp et al. 2003) and deep-frying oils of different botanic origins (Kuligowski et al. 2011). FT-IR analysis has been applied for the determination of key quality parameters of edible oils (Cooke and Billingham 1999), monitoring the oxidation process of corn oil samples undergone during heating or/and exposure to ultraviolet radiation (Vlachos et al. 2006) and quantification of moisture in edible oils (Al-Alawi et al. 2005).

Applications in discrimination and identification of food related microorganisms

Naumann and co-workers (1991) suggested this technique as a method of identification of unknown microorganisms. Using this method the infrared spectra of unknown microorganisms can be compared to an existing reference library and can be matched with the most similar spectrum of known species. The identification of molecular structures based on their vibrational spectra can be applied to microorganisms, differentiating between similar bacterial strains rapidly and accurately without the use of any reagents (Maquelin et al. 2002, Winder et al. 2004). Mariey et al. (2001) have reviewed a series of 50 studies that have applied FT-IR spectroscopy and chemometrics for the discrimination, classification and identification of food related or other microorganisms, including *Bacillus* species, lactic acid bacteria, yeasts, pseudomonads, *Enterobacteriaceae* and food borne pathogens such as *Listeria, Salmonella, Escherichia coli.*

FT-IR spectroscopy was used to analyze and identify strains of *Carnobacterium*, atypical *Lactobacillus*, *Enterococcus durans*, *Lactobacillus maltaromicus* and *Vagococcus salmoninarum* (Lai et al. 2004), a collection of facultatively heterofermentative lactobacilli isolated from plants, milk, and hard type cheeses from different geographical origins (Weinrichter et al. 2001), lactic acid bacteria belonging to the genera *Lactobacillus*, *Weissella* and *Carnobacterium* (Mouwen et al. 2011) and food-borne, predominantly fermentative yeasts (Kummerle et al. 1998). It has also been used for monitoring the population dynamics during cheese ripening through the identification of surface yeasts isolated from cheeses from different dairies (Larpin-Laborde et al. 2011), a total of 1,114 yeast and 1,201 bacterial isolates from cheese (Goerges et al. 2008) and 1307 microorganisms from the surface of different cheeses (Oberreuter et al. 2003). It was applied for the identification of geographically separated food-borne smear cheese microbial consortia of limited complexity (Wenning et al. 2006) and for monitoring the differences between the microbiota of farm and dairy tank milk (Fricker et al. 2011). The use of FT-IR was also explored as a rapid method of screening of the fermentation profiles of wine yeast (Nieuwoudt et al. 2006). Additionally, this technique has been applied to detect and identify *Alicyclobacillus* spp. that are thermophilic, acidophilic, spore-forming bacteria, some of which cause spoilage in commercially pasteurized fruit and vegetable juice products (Lin et al. 2005, Al-Qadiri et al. 2006, Lin et al. 2007, Wang et al. 2011b).

FT-IR has been used to compare and determine various food contaminating fungi (*Penicillum* ssp. and *Aspergillus niger*) based on the whole extracellular hydrolytic product profiles obtained by FT-IR (Basaran et al. 2011) and for the rapid and accurate differentiation of species in the genus *Bacillus* (*B. cereus*, *B. mycoides* and *B. thuringiensis*) that are of importance to the food and dairy industry (Beattie et al. 1998). It has also been employed effectively for the differentiation of *Streptomyces* isolates (Zhao et al. 2006) and to analyze a selection of *Acinetobacter* isolates (Winder et al. 2004).

FT-IR spectroscopy and multivariate analysis have been used to discriminate between *Enterococcus* strains (*E. faecalis*, *E. faecium*, *E. durans* and *E. hirae*) at species and strain level (Guibet et al. 2003) and *Staphylococcus aureus* from the other species of *Staphylococcus* and all the strains of *Staph. aureus* isolated from raw milk and different varieties of raw milk cheese (Lamprell et al. 2006). It has been used successfully for the differentiation of *Enterococcus* and *Streptococcus* species (Goodacre et al. 1996) and *Yersinia enterocolytica* species and subspecies (Kuhm et al. 2009). It was applied also for detection, differentiation, and quantification of different *Escherichia coli* strains (Gilbert et al. 2009), *E. coli* O157:H7 strains from ground beef (Davis et al. 2010a) or apple juice (Al-Qadiri et al. 2006), strains of *Listeria monocytogenes* and *L. innocua* (Lin et al 2004) and *L. monocytogenes* serovars and *Listeria* Species (Rebuffo-Scheer et al. 2007a,b, Davis and Mauer 2011).

Alvarez-Ordóñez et al. (2010) have studied the changes observed in *Salmonella* Typhimurium and *Salm.* Enteritidis FT-IR spectra after their growth in the presence of sublethal stress factors (growth temperature, growth medium pH, salt concentration, oxygen tension).

FT-IR was utilized to classify the packed beef samples (simulating the retail market conditions) as 'uncontaminated' and 'contaminated with *Salm.* Typhimurium (Amamcharla et al. 2010a,b), to differentiate live and heat treated '*Salm. enterica* serovars inoculated onto chicken breast (Davis et al. 2010b) and *Salm.* Enteritidis phage types (Preisner et al. 2006). In the same pattern, FT-IR was applied to detect sublethally heat-injured *Salm. enterica* serotype Typhimurium and *L. monocytogenes* strains (Lin et al. 2005), *E. coli* O157:H7 and *Salm.* Typhimurium (Burgula et al. 2006) and to detect and discriminate between *Enterobacter sakazakii* strains, *Enterobacter cloacae* strains, *E. coli* strains and *Klebsiella pneumoniae* strains (Lin et al. 2009).

Applications for monitoring food spoilage

FT-IR spectroscopy has found a variety of applications regarding spoilage evaluation of foods. Most of them have evaluated the potential of the method on muscle foods and were performed on chicken (Ellis et al. 2002, Lin et al. 2004, Alexandrakis et al. 2012), on beef (Ellis et al. 2004, Ammor et al. 2009, Argyri et al. 2010, Argyri et al. 2013) and on pork meat (Papadopoulou et al. 2011). In dairy foods, FT-IR has found applications for monitoring milk (Nicolaou and Goodacre 2008, Nicolaou et al. 2011) and cheese spoilage (Cattaneo et al. 2005). Finally in plant origin foods it has been proposed as a potential technique for monitoring spoilage on wheat (Singh et al. 2011), fruit spirits and beer (Lachenmeier 2007), and grapes (Barata et al. 2011).

Muscle foods. Spoilage of meat can be considered as an ecological phenomenon that encompasses changes in the available components (e.g., low molecular weight compounds) during proliferation of bacteria present in the microbial association of stored meat. Its high water activity and abundant nutrients make meat an excellent medium to support microbial growth, since changes that occur during spoilage take place in the aqueous phase of the meat. These substrates are catabolized by almost all bacteria in muscle food microbiota (Drosinos and Board 1995, Gill 1976, Nychas et al. 2007). The concentration of glucose, lactic acid, and certain amino acids followed by nucleotides, urea and water-soluble proteins can affect the type (e.g., saccharolytic, proteolytic), the rate of spoilage and, moreover, seems to be the principal precursor(s) of those microbial metabolite(s) that we perceive as spoilage (Nychas et al. 1998, Nychas et al. 2008).

The microbiological quality of meat depends on the physiological status of the animal at slaughter, the spread of contamination during slaughter and processing, the temperature and other conditions of storage and distribution (Nychas et al. 2008). These ecological determinants influence the establishment of a particular microbial association and determine the rate of attainment of a maximum populations known as the 'Ephemeral (specific) spoilage micro-organisms- E(S)SO (i.e., those which fill the niche by adopting ecological strategies) (Nychas and Skandamis 2005). *Pseudomonas* sp., *Brochothrix thermosphacta*, lactic acid bacteria and *Shewanella putrefaciens* are considered to be the main spoilage bacteria of low and high pH raw meat, stored in chill temperature aerobically or vacuum/map conditions (Garcia-Lopez et al. 1998, Stanbridge and Davis 1998). Cold-tolerant *Enterobacteriaceae* (e.g., *Hafnia alvei*, *Serratia liquefaciens* and *Enterobacter agglomerans*) may also occur on chilled muscle foods stored aerobically in chill storage (Borch et al. 1996, Nychas et al. 1998). Yeasts do not outgrow bacteria on muscle products unless a bacteriostatic agent is included in specific products (e.g., British fresh sausages) or the product is stored for extended periods in cold and dry environments (Nychas et al. 2007). It needs to be noted that the final composition of the microbiota eventually characterizes the type of spoilage (Nychas et al. 1998). The production of metabolic compounds that can eventually spoil muscle tissues is the outcome of substrate interactions with the developed microbial association and with the indigenous muscle enzymes as well as of nonenzymatic chemical reactions and physical changes.

Evaluation of the degree of meat spoilage and/or hygiene level is usually made either subjectively based on qualitative criteria or by time consuming, destructive microbiological analyses that do not provide the 'immediate answer required' (McMeekin et al. 2007, Nychas et al. 2008). To complement these approaches, the meat industry needs rapid analytical procedures to define and quantify spoilage indicators, in order to both determine suitable processing methods for the raw material, and to predict the remaining shelf-life of the product. In particular, Fourier transform infrared (FT-IR) spectroscopy has attracted considerable interest since it is rapid and non-destructive and has been identified as having considerable potential for applications in food and related industries, with several reports on muscle food analysis (van Kempen 2001).

Ellis et al. (2002, 2004) have been the pioneers stipulating that FT-IR spectroscopy can be used directly on the surface of meat to produce biochemical interpretable "fingerprints" (metabolic snapshots), enabling thus early detection of microbial spoilage of meat. FT-IR-ATR was exploited to measure biochemical changes within the meat substrate, enhancing and accelerating the detection of microbial spoilage of minced chicken breasts (Ellis et al. 2002) and minced beef steak rumps (Ellis et al. 2004) stored at room temperature for 24 h. Every hour, FT-IR measurements were taken directly from the meat surface using attenuated total reflectance, and the total viable counts were obtained by classical plating methods. The spectrometer was programmed to collect spectra over the wavenumber range 4000–600 cm^{-1} and the collection time for each sample spectrum was 1 min. For quantitative interpretation of the FT-IR spectra the multivariate linear regression method of partial least-squares regression PLS was used. The root mean squared errors (RMSE) in these measurements are 0.15, 0.23, and 0.27 log units for the calibration, cross validation, and independent test sets for the minced chicken (Ellis et al. 2002) and 0.43, 0.56 and 0.44 logs for the calibration, cross-validation and independent test sets for the minced beef, respectively (Ellis et al. 2004). It was demonstrated that PLS gave accurate results at $2 \cdot 10^5$ cfu/ cm^2 for microbial spoilage on beef, which is an order lower than those obtained on experiments on poultry where the limit of detection was $2 \cdot 10^6$ cfu/ cm^2. Machine learning methods in the form of genetic algorithms (GA) and genetic programming (GP) were used to elucidate the wavenumbers of interest related to the spoilage process. For GP analysis levels of 10^7 bacteria/g were set as a spoilage threshold between spoiled and fresh samples in the case of chicken experiments, whereas for beef these analyses were trained in a quantitative fashion to predict the $\log_{10}$ (TVC). Analyses of the FT-IR spectra using genetic algorithms and genetic programming in the case of minced beef selected wavenumber vibrations from within the region 1420–1400 cm^{-1}, with both evolutionary computational methods selecting the specific vibration at 1413 cm^{-1} attributable to amides, whilst other vibrations that were selected were from free amines (e.g., 1112 and 1374 cm^{-1}). In the case of minced chicken the GAs and GPs both pick the region 1,088 to 1,096 cm^{-1} as the most significant area of the FT-IR spectra for the prediction of spoilage of chicken which is attributable to amines from free amino acids.

Lin et al. (2004) evaluated the feasibility of visible and short-wavelength near-infrared (SW-NIR) diffuse reflectance spectroscopy (600–1100 nm) to quantify the microbial loads in chicken meat and to develop a rapid methodology for monitoring the onset of spoilage. Prepackaged fresh chicken breast muscle samples were stored at 21°C for 24 h and microbial analysis was conducted in terms of total aerobic plate count. Principal component analysis (PCA) and partial least squares (PLS) based prediction models were developed. PCA analysis showed clear segregation of samples held 8 h or longer compared with 0-h control and could perceiving the change of the microbial loads in chicken muscle once the viable counts increased slightly above 1 log cycle. Accurate quantification of the bacterial loads in chicken muscle could be calculated from the PLS-based prediction method (R= 0.91, Standard error of prediction=0.48 log cfu /g).

In a more recent study of this type, Ammor et al. (2009) correlated the spoilage status (microbial counts and sensory scores) of beef stored under different conditions (i.e., temperature and packaging) with the biochemical fingerprints observed with FT-IR-ATR spectroscopy. The spectra were collected in the range of 4000–400 cm^{-1} and the collection time for each sample spectrum was 2 min. Minced beef samples packaged either aerobically, under modified atmosphere (MAP: 40% CO_2/30% O_2/30%N_2) and using an active packaging (MAP with volatile compounds of oregano essential oil) were held from freshness to spoilage at 0, 5, 10, and 15°C. Frequent FT-IR-ATR measurements were collected directly from the sample surface, in parallel with the microbiological analysis (total viable counts), the sensory analysis and the pH measurements. Principal components analysis (PCA) allowed illuminating the wavenumbers potentially correlated with the spoilage process. The wavenumbers that were found to explain significantly the variance of the spectral data set and hence were considered as potential wavenumbers associated with the biochemical changes happening during the spoilage process ranged between 1714–1710 cm^{-1}, 1614–1211 cm^{-1}, and 1031–1000 cm^{-1} and were selected for further analyses. These wavenumbers corresponded mainly to the absorption of amide II, fat, amides and amines. It is noteworthy that the most significant wavenumbers explaining the variance of the data set ranged between 1402–1371 cm^{-1} and 1321–1303 cm^{-1} and corresponded to amides and amines as aforementioned. Qualitative interpretation of spectral data was carried out using factorial discriminant analysis (FDA) to corroborate sensory data and accurately determine samples freshness and packaging conditions. The classification table resulting from the FDA provided 76.3% correct classification when cross-validated (rates for each class: 69.2% of fresh samples, 80.3% of spoiled samples and 80.0% of the semi-fresh samples). Even though 23.7% of samples could not be cross-validated, no fresh sample was classified as spoiled or vice versa. The same approach was followed with packaging (air, MAP, and active packaging) constituting the dependent variable and the FDA provided 92.5% correct classification when cross-validated. The values of the slope R^2 (R is the correlation coefficient) and the root mean squared error (RMSE) indicated that the quality of TVC regression models were, respectively, 0.80 and 0.58 for the calibration and 0.73 and 0.71 for the cross-validation.

Argyri et al. (2010) used a machine learning strategy in the form of a multilayer perceptron (MLP) neural network, to correlate FT-IR spectral data with beef spoilage during aerobic storage at chill and abuse temperatures. Fresh beef fillets were packaged under aerobic conditions and left to spoil at 0, 5, 10, 15, and 20°C for up to 350 h. FT-IR-ATR spectra in the range of 4000–400 cm^{-1}, were collected directly from the surface of meat samples, whereas sensory evaluation and microbiological analysis of total viable counts were performed during storage. The FT-IR instrument and spectra collection were the same as described above for Ammor et al. (2009). A multi-layer perceptron (MLP) neural network was designed to classify beef samples to one of the three quality classes based on the biochemical profile provided by the FT-IR spectra, and in parallel to predict the microbial load (as total viable counts) on meat surface. The results obtained demonstrated that the developed neural network was able to classify with high accuracy the beef samples in the corresponding quality class using their FT-IR spectra. The network was able to classify correctly 91.7% of fresh samples, 94.1% of spoiled samples and 81.2% of the semi-fresh samples (no fresh sample was misclassified as spoiled and vice versa). The performance of the network in the prediction of microbial counts was based on graphical plots and statistical indices (bias and accuracy factors, standard error of prediction, mean relative and mean absolute percentage residuals). The plot of predicted versus observed total viable counts showed reasonably good distribution around the line of equity (y=x), with the majority of data (ca. 78%) included within the ±1 log unit area, although some over-prediction was evident in the case of fresh meat samples, especially with low observed initial counts. This is comparable with a value of ca. 0.5 log cfu/cm^2 for beef reported previously (Ellis et al. 2004). These results were also confirmed by the percent relative error index (%RE) between observed and predicted values. Based on the % relative error of prediction zone, approximately 88% of predicted microbial counts included within the ±20% RE zone. It needs to be emphasized though that the network over-estimated the bacterial population for certain fresh samples, especially at lower observed microbial counts, corresponding to low temperature (0°C) and short storage time. Based on the calculated values of the bias factor (B_f) it can be inferred that the network under-estimated total viable counts in semi-fresh and spoiled samples ($B_f < 1$), whereas for fresh samples over-estimation of microbial population was evident ($B_f > 1$). Values of B_f were close to unity indicated good agreement between predictions and observations. In addition, the values of the accuracy factor (A_f) indicated that the predicted total viable counts were 18.1%, 12.2%, and 8.4% different (either above or below) from the observed values for fresh, semi-fresh, and spoiled meat samples, respectively. It has been reported (Bäck et al. 1997) that an increase of 0.15 (15%) would be acceptable for each independent variable included in model development.

Papadopoulou et al. (2011) have investigated the feasibility of FT-IR-ATR spectroscopy to quantify biochemical changes occurring in fresh minced pork meat in the attempt to monitor spoilage. For this reason, partial least squares (PLS) models were constructed to correlate spectral data from FT-IR with minced pork meat spoilage during aerobic storage of meat samples at different storage temperatures (0, 5, 10, and 15°C). Spectral data were collected over the wavenumber range 4,000 to 650 cm^{-1} with a total integration time of 2 min, from the surface of meat samples in parallel with microbiological analysis to enumerate the population of total viable counts, *Pseudomonas* spp., *Br. thermosphacta*, lactic acid bacteria (LAB) and *Enterobacteriaceae*. Qualitative interpretation of spectral data was based on sensory evaluation, and the PLS-discriminant analysis (PLS-DA) was used to develop models allowing the discrimination of meat samples in the selected sensory classes. The obtained results demonstrated good performance in classifying meat samples in one of the three pre-defined sensory classes. The overall correct classification rate for the three sensory classes was 88.1% for the validation of the model, with a classification rate for fresh, semi-fresh and spoiled samples of 86.7, 87.5, and 88.9%, respectively. Furthermore, PLS regression models were also employed to provide quantitative estimations of microbial counts during meat storage. The performance was based on graphical plots and statistical indices (bias factor, accuracy factor, standard error of calibration, standard error of prediction, and correlation coefficient). The values of the bias factor were close to unity for all microbial groups indicating no systematic bias of the models. Moreover, the calculated values of the accuracy factor showed that the average deviation between predictions and observations was 7.5% and 7.9% for total viable counts and *Pseudomonas* spp. and 10.7% and 11.3% for LAB and *Br. thermosphacta*. Finally, correlations above 0.80 between observed and estimated counts were observed for both training and test data sets.

Alexandrakis et al. (2012), collected Near-infrared (NIR) transflectance and FT-IR-ATR spectra of intact chicken breast muscle packed under aerobic conditions and stored at 4° for 14 days and investigated their potential use in rapid non-destructive detection of spoilage. Multiplicative scatter correction transformed NIR and standard normal variate-transformed FT-IR spectra were analysed using principal component analysis (PCA), partial least-squares discriminant analysis (PLS-DA) and outer product analysis (OPA). PCA and PLS-DA regression failed to completely discriminate between days 0 and 4 samples (total viable count on days 0 = 5.23 and on day 4 = 6.75 $\log_{10}$ cfu/g) which had bacterial loads smaller than the accepted levels (8 $\log_{10}$ cfu/g) of sensory spoilage detection but classified correctly days 8 and 14 samples (TVC day 8 = 9.61 and day 14 = 10.37 $\log_{10}$ cfu/g). OPA performed on both NIR and FT-IR datasets revealed several correlations that highlight the effect of proteolysis in influencing the spectra. These highlighted areas where for the NIR spectrum area between 1,864–2,100 nm (attributed to amide), 1,396 and 1,550 nm (attributed to amide and alcohol vibrations), 1,386 and 1,570 nm (amide and amine) and 1,856–2,098 nm (amines), and for the FT-IR spectrum between 1,523–1,700 cm^{-1} (attributed to amide I and amide II), 1,571–1,695 cm^{-1} (amide I and amide II), 1,000 and 1,097 cm^{-1}

(free amino acids) and 1,000–1,147 cm^{-1} (amines). It was reported that these correlations indicate that increase in free amino acids and peptides could be the main factor in the discrimination of intact chicken breast muscle and suggested that NIR and FT-IR spectroscopy can become useful, rapid, non-destructive tools for spoilage detection.

In a later study (Argyri et al. 2013), time series spectroscopic, microbiological and sensory analysis data were obtained from minced beef samples stored under different packaging conditions, aerobic and modified atmosphere packaging (40% CO_2/30% O_2/30%N_2) stored at 5°C. The microbiological analysis included measurements on total viable counts (TVC), LAB, *Br. thermosphacta, Enterobacteriaceae, Pseudomonas* spp. and yeasts-moulds. FT-IR spectra were collected over the wavenumber range 4,000 to 650 cm^{-1} with a total integration time of 2 min. The range selected for further analysis was between 1800 and 900 cm^{-1}. These data were analyzed using machine learning and evolutionary computing methods, including partial least square regression (PLS-R), genetic programming (GP), genetic algorithm (GA), artificial neural networks (ANNs) and support vector machines regression (SVR) including different kernel functions [i.e., linear (SVR_L), polynomial (SVR_P), radial basis (RBF) (SVR_R) and sigmoid functions (SVR_S)]. Predictive models of the microbiological load and sensory assessment were calculated using these methods and their relative performance was compared (RMSE, correlation coefficient-R^2, percentage of Prediction Error-% PE). For FT-IR calibration models, regarding the predictions of the microbial counts the multivariate methods (SVM, PLS) that had similar performances, gave better predictions compared to the evolutionary ones (GA-GP, GA-ANN, GP), with better predictions obtained for TVC, LAB and *Enterobacteriaceae*. On the other hand, the GA-GP model performed better than the others in predicting the sensory scores using the FT-IR data. The % PE values of the models, indicated that for FT-IR models, PLS, SVR_L and SVR_P gave acceptable predictions (% PE > 70%) for all microbial counts, with the exception of yeasts and moulds that were underestimated and were totally outside the acceptable prediction range, regardless of the model used. The RMSE values of the latter models varied between 0.55–0.69 for PLS, 0.50–0.78 for the SVR_L and 0.41–0.68 for the SVR_P depending on the microorganism. The GA-GP, SVR_R and SVR_P models for the sensory evaluation, gave an overall performance of 87.5% correct classification, but the former gave better predictions concerning the correct classification of the fresh samples. Typical spectral data obtained from FT-IR in the range of 1800 to 900 cm^{-1} collected from minced beef stored aerobically and under MAP at 5°C are shown in Fig. 1, whereas the possible identification of each peak present in meat profile during storage is reported in Table 2. This information can be used and applied in other food systems as soon as the studied samples absorb in the same areas.

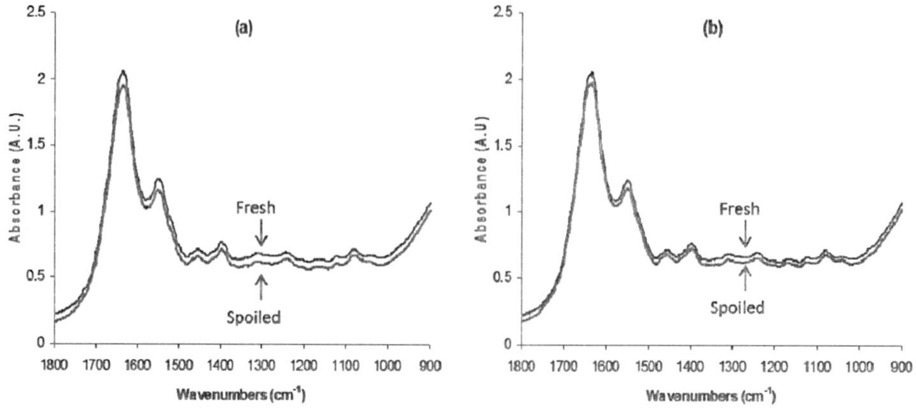

Figure 1. Raw FT-IR spectra collected from minced beef samples stored aerobically (a) and under MAP (b) at 5°C for a fresh sample (gray line) and a spoiled sample (black line) (Data from Argyri et al. 2013 modified accordingly).

Obtained results demonstrated that a FT-IR spectrum may be considered as a metabolic fingerprint and that the method in combination with the appropriate chemometric or evolutionary technique, represents a powerful, rapid, economical and non-invasive method for monitoring meat freshness regardless the storage conditions (e.g., packaging and temperature).

Dairy foods. Milk is an ideal medium for microbial growth because of its high water content and the large variety of available nutrients which can be used by microorganisms as an energy source (Gram et al. 2002). The microbiota in raw milk is considered essential for sensory richness and variety of traditional cheeses, but it can also be responsible for flavour defects of dairy products or may even constitute a health risk (Fricker et al. 2011). As in meat, the type of the spoilage varies according to the species of the microorganisms present in milk, the chemical composition of milk and the physical environment under which it is stored. In the same mode, the spoilage of the rest of the dairy products depends on the water activity, pH, salt concentration, temperature, characteristics of the lactic starter culture, types and viability of contaminating microorganisms, and characteristics and quantities of residual enzymes (Vanetti 2011). Numerous microorganisms, constitute the complex ecosystem present in milk and dairy products, and, in most situations, they are quite frequently associated with product spoilage. Even before spoilage becomes obvious, microorganisms have begun the process of breaking down milk constituents for their own metabolic needs (Vanetti 2011).

Troublesome spoilage microorganisms include aerobic psychrotrophic gram-negative bacteria (*Pseudomonas* species and of the family *Enterobacteriaceae*), yeasts, molds, heterofermentative lactobacilli (*Micrococcus, Aerococcus,* and *Lactococcus* species), and spore-forming bacteria

Table 2. FT-IR frequencies observed in spectra collected from minced beef samples and possible assignments of the vibration modes.

Frequency (cm^{-1})	Assignment	Reference
1720-1580 (1640)	water (H-O-H def vib) and amide I band (80% C=O stretch, 10% C-N stretch, 10% N-H bend)	Socrates 2001, Wu et al. 2006
1580-1520 (1550)	amide II band (40% CN str, 60% NH bend vib)	Bocker at al. 2007
1520-1480 (1520 sh)	amide II (combination of C–N stretch and N–H bend)	Ellis et al. 2004, Kaiden et al. 1987
1480-1461 (1468 sh, 1461 sh)[a]	Lipids (CH_3 asym def, CH_3 asym bend, C-H def of CH_2,CH_2 scissoring vib, C-H bend) or amines (asym CH3 def vib)	Socrates 2001, Chen et al. 1998, Bocker et al. 2007
1460-1432 (1456, 1440sh)	CH_2 bending	Bocker et al. 2007
1430-1413 (1413 sh)	amide (C–N stretch) or lipids (RCOO- sym str COO-, -CH_2-COOR def vib of CH_2)	Ellis et al. 2004, Socrates 2001
1413-1371 (1396)	amino acid side chains, lipids (CH bend or C-O stretch in carboxylates) or nitro group (NO_2 sym stretch)	Pappas et al. 2008, Pistorius 1995, Ellis et al. 2004
1370-1365 (1371sh)	free amines (C–N stretch)	Ellis et al. 2004
1365-1358 (1365 sh)	primary and secondary aliphatic amines (X-sensitive band)	Socrates 2001
1355-1344 (1344 sh)	amines -CH_2NH_2 (CH_2 twisting vib)	Socrates 2001
1344-1331 (1338)	CH_2 side-chain vibrations, free amino acids/long chain aliphatic carboxylic acids (CH stretch/CH_2 def vib)	Bocker et al. 2007, Socrates 2001
1330-1294 (1310)	amide III a-helical structures (30% C-N stretch, 30% N-H bend, 10% C=O-N bend, 20% other)	Bocker et al. 2007
1293-1275 (1284)	amide III β-turn (30% C-N stretch, 30% N-H bend,10% C=O-N bend, 20% other)	Bocker et al. 2007
1273-1263 (1263 sh)	amide III random coil (30% C-N stretch, 30% N-H bend,10% C=O-N bend, 20% other)	Socrates 2001
1262-1212 (1240)	lipids, nucleic acids (asym PO_2- stretch), amide III P=O stretch (30% C-N stretch, 30% N-H bend,10% C=O-N bend, 20% other), amines from free amino acids (C-N stretch)	Socrates 2001, Pappas et al. 2008, Pistorius 1995, Chen et al. 1998, Ellis et al. 2002

Table 2. contd....

Table 2. contd....

Frequency (cm^{-1})	Assignment	Reference
1211-1199 (1205 sh)	amide III (30% C-N stretch, 30% N-H bend,10% C=O-N bend, 20% other), amines (CN stretch)	Bocker et al. 2007, Socrates 2001
1200-1190 (1196 sh)	amines -N(CH_3)$_2$ (CH_3 rock and asym CCN stretch)	Socrates 2001
1185-1143 (1165 double)	fat related (C-O stretch), esters (C-O-C), carbohydrates (C-O stretch), –NH_2 def	Chen et al. 1998, Bocker et al. 2007, Pappas et al. 2008
1140-1110 (1124)	riboses (C-O stretch), amines (NH_2 rock/twist)	Bocker et al. 2007, Socrates 2001
1108-1065 (1078)	nucleic acids and phospholipids (PO_2 sym stretch)/C-O stretch	Bocker et al. 2007
1052-1036 (1043)	lipids, polysaccharides (C-O , C-O-P stretch)	Bocker et al. 2007
1036-1020 (1032, -^{b})	polysaccharides (C-O stretch), amines (CN stretch)	Socrates 2001, Bocker et al. 2007
1020-1008	Polyglycines	Socrates 2001
983-965	a, b, pyranose compounds (ring vib), aromatic carboxylic acids	Socrates 2001

Abbreviations: def: deformation, vib: vibration, stretch: stretching, bend: bending, sym: symmetric, asym: asymmetric, rock: rocking, twist: twisting, sh: shoulder. [a] two peaks observed at this region, with the same possible assignment. [b] second peak with no well schemed edge.

(*Bacillus* species) (Ledenbach and Marshall 2009). Although psychrotrophic bacteria are in a small part of fresh collected milk, they compose up to 80% of the population of raw refrigerated milk, and *Pseudomonas* spp. is the most important of the psychrotrophs that dominate the microbiota of raw or pasteurized milk at the time of storage. Other genera of gram-negative psychrotrophic bacteria include *Achromobater*, *Aeromonas*, *Alcaligenes*, *Chromobacterium*, *Flavobacterium*, *Serratia*, and *Enterobacter*. Thermoduric bacteria are those that survive pasteurization, and they are represented mainly by gram-positive bacteria in the genera *Bacillus* and *Clostridium* spp. and the nonsporeformers genera *Arthrobacter*, *Microbacterium*, *Streptococcus*, and *Corynebacterium* that are involved in spoilage (Vanetti 2011). Psychrotrophic bacteria can produce large amounts of extracellular hydrolytic enzymes, making these bacteria a major determinant of the remaining shelf life of a dairy product (Ledenbach and Marshall 2009, Vanetti 2011). Heterofermentative lactic acid bacteria such as lactobacilli and *Leuconostoc* (Johnson 2001) and spore-forming bacteria of the *Bacillus* and *Clostridium* genera (Meer et al. 1991, Klijn et al. 1995, Kalogridou-Vassiliadou 1992, Ternstrom et al. 1993) can develop off-flavors and gas in ripened cheeses. Yeasts and moulds are a common cause of spoilage of dairy products through the production of a variety of metabolic by-products, off-odours and flavours and visible changes in colour or texture. The growth of spoilage yeasts is favoured by the low pH in cheeses, yogurt and fermented milks (Johnson 2001, Rohm et al. 1992). Moulds can grow well on the surfaces of cheeses with low pH packaged with oxygen availability, but some moulds can grow equally well under low oxygen tension (Hocking and Faedo 1992, Ledenbach and Marshall 2009). Low water activity in some hard cheeses, sweetened condensed milk, and butter can also favor yeasts and molds spoilage (Vanetti 2011).

Nicolaou and Goodacre (2008) have applied FT-IR spectroscopy coupled with attenuated total reflectance (ATR), and high throughput (HT) FT-IR techniques on three main types of pasteurized milk—whole, semi-skimmed and skimmed—that was allowed to spoil naturally by incubation at 15°C. Multivariate statistical methods, including principal component-discriminant function analysis (PC-DFA) and partial least squares regression (PLSR), were used to investigate the relationship between metabolic fingerprints and total viable counts (TVC). According to PC-DFA results from FT-IR ATR for whole-fat milk the spectra for the first 48 h group together with TVC between 10^3–$9{\cdot}10^4$ cfu/ml (i.e., fresh samples). Spectra of samples incubated after the 48 h time point are clearly different from the earlier samples with the latter being more clear when the TVC are between 10^6–10^8 cfu/ml. In semi-skimmed milk similar results were observed, where the spectra for the first 56 h appeared in the same region, with TVC between $4{\cdot}10^2$–$2{\cdot}10^5$ cfu/ml. After this time point the spectra followed again a trend correlated to the number of bacteria. For skimmed milk the spectra for the first 56 h also appeared clustered together (TVC $4{\cdot}10^2$–$9{\cdot}10^4$ cfu/ml), after which a trend relating to the TVC from $4{\cdot}10^5$ to $2{\cdot}10^7$ cfu/ml was also observed. Similar PC-DFA results occurred for spoilage of the different milk types from FT-IR HT, with the sample points with TVC between 10^6–10^8 cfu/ml being more discrete compared to the same analysis on the ATR accessory. Preliminary modeling for all milk types was performed on the FT-IR ATR spectra and it was found that PLSR could not give accurate estimates of TVC for very low bacterial numbers. Therefore, depending on the milk type, either 48 or 56 h to 104 h were used in PLS modeling and this also corresponds to when PC-DFA could not differentiate between the early sampling points. PLS analysis on FT-IR ATR spectra from whole milk (between 56 and 104 h) gave accurate results at bacterial levels higher than $1{\cdot}10^6$ cfu/ ml. In semi-skimmed and skimmed milk, FT-IR ATR spectra from samples between 56 and 104 h, and from 48 and 104 h, respectively, were analyzed. For semi-skimmed milk the models showed a good predictive value, when the total viable counts above $2{\cdot}10^5$ cfu/ ml could be assessed. In contrast, for skimmed milk reasonable predictions were observed when the total viable counts were above $1{\cdot}10^4$ cfu/ ml. Remodeling of the whole-fat and semi-skimmed milk including the 48 h time point could not predict less than $1{\cdot}10^4$ cfu/ ml. The RMSE of the FT-IR ATR PLSR models ranged from 0.25 to 0.87 for the cross-validation and test of the model with better prediction being observed for whole, semi-skimmed and then skimmed milk, in this order. In contrast to the FT-IR ATR PLSR modeling it was possible to use the full time course of the FT-IR HT results for analysis. The PLSR gave relatively accurate results even at very low number of viable counts ($1{\cdot}10^3$ cfu/ml). For the semi-skimmed milk, PLSR again gave good predictions, but the results from skimmed milk were not quite as good as the results for the other two milk types. The RMSE of the FT-IR HT PLSR models ranged from 0.84 to 1.28 for the cross-validation and test of the model with better prediction being observed for whole, semi-skimmed and then skimmed milk, in this order. FT-IR ATR and FT-IR HT spectra for whole milk at 0 h (fresh) and 104 h (spoiled), as well as possible assignments of the peaks observed are shown in Figs. 2 and 3, respectively.

Nicolaou et al. (2011) investigated FT-IR HT in combination with multivariate analytical techniques for the detection and enumeration of *Staph. aureus* and *Lactococcus cremoris* in milk in a pure monoculture and in a combined inoculated co-culture after inoculation into ultra heated milk during the first 24 h of growth at 37°C. Plating techniques were used to obtain primary reference data for viable bacteria counts. Principal component discriminant function analysis (PC-DFA), canonical correlation analysis (CCA), partial least-squares (PLS), and kernel PLS (KPLS) multivariate statistical techniques were employed to analyze the data. The

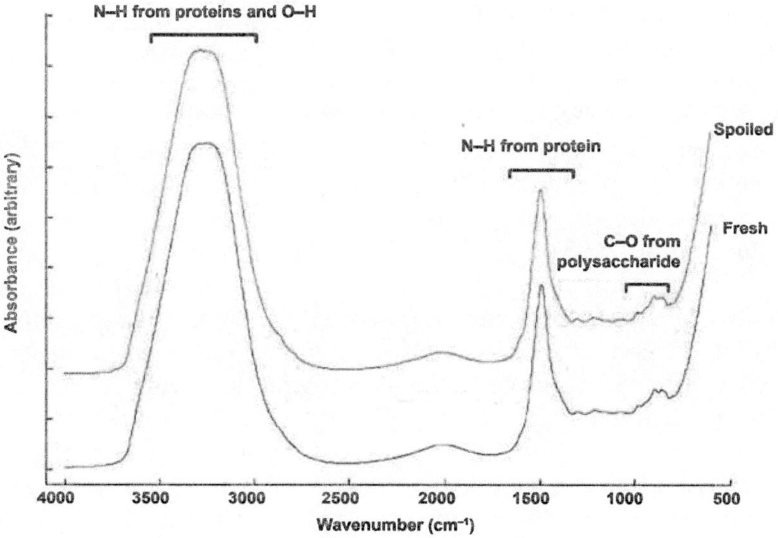

Figure 2. FT-IR ATR spectra collected from a fresh and a spoiled whole milk sample stored at 15°C (Data from Nicolaou and Goodacre, 2008 modified accordingly).

Color image of this figure appears in the color plate section at the end of the book.

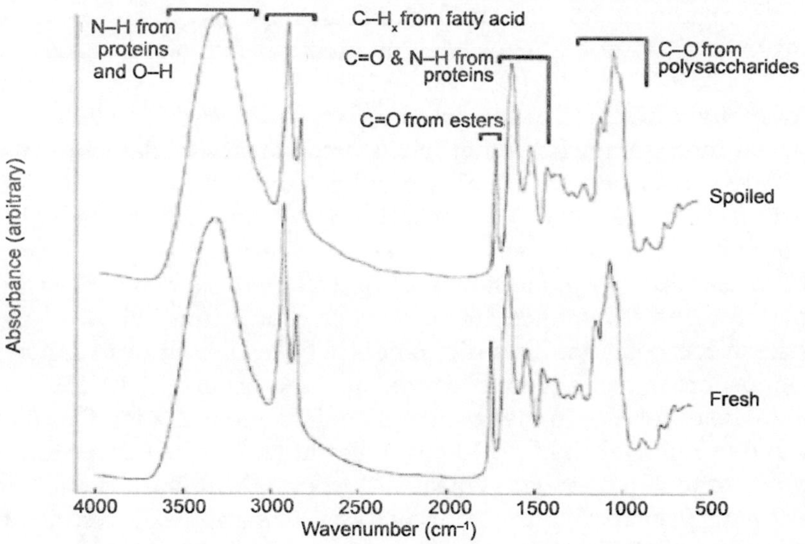

Figure 3. FT-IR HT spectra collected from a fresh and a spoiled whole milk sample stored at 15°C (Data from Nicolaou and Goodacre, 2008 modified accordingly).

Color image of this figure appears in the color plate section at the end of the book.

PC-DFA plot indicated that after 420 min of incubation there was a clear distinction between the different time points, a picture that was similar in all culture cases. The FT-IR provided very reasonable quantification results both with PLS and KPLS, the latter providing marginally better predictions, with correlation coefficients in the test set (Q^2) and training set (R^2) varying from 0.64 to 0.76 and from 0.78 to 0.88, respectively, for different bacterial sample combinations.

Cattaneo et al. (2005) applied FT-NIR and FT-IR spectroscopy to evaluate the shelf-life period in which "freshness" is maintained in Crescenza cheese. Crescenza is a fresh cheese in which "freshness" is associated with low acidity, limited proteolysis, and no bitter taste. Cheese samples were analyzed at different times for 20 days. Spectral data were collected using a FT-NIR spectrometer with an optic fibre (from 12,000 to 4,000 cm^{-1}), and a FT-IR spectrometer with an ATR cell (from 4,000 to 700 cm^{-1}), while a series of physico-chemical and chemical tests were followed to provide chemical shelf life indices, without providing other microbiological indices. Principal component analysis was able to detect the decrease of Crescenza "freshness" and to define the critical day during shelf-life. In particular, if the cheese is stored at 10°C, the original quality is maintained for about 3 days, and after 6 days an important decrease of "freshness" is detected.

Cereals. Wheat is one of the most important cereal grains grown worldwide and consumed in various forms such as bread, pasta, noodles, pastries, cakes, and cookies (Cornell and Hoveling 1998). Fungi are one of the serious causes of spoilage in stored grain including wheat. *Aspergillus* spp. is one of the most common storage fungi that spoil stored wheat. The damage caused by fungi adversely affects the quality of wheat and reduces its nutritional composition (Singh et al. 2011). In the latter study, synchrotron-based high resolution infrared imaging was used to study the compositional changes in stored wheat due to fungal damage. Spectral analysis of healthy wheat kernels and kernels damaged by *A. glaucus* group fungi showed clear differences in composition. Clear differences between healthy and damaged wheat endosperm spectra were observed at peaks around 1740, 1595, and 1250 cm^{-1}. The difference in the absorption of infrared radiation was likely caused due to reduced lipid (1740 cm^{-1}), lignin (1595 cm^{-1}) and cellulose (1250 cm^{-1}) content in damaged wheat endosperm. Multivariate statistical analysis using principal component analysis was applied to reduce the dimensionality of IR spectral data. The k-means clustering using principal component scores clearly discriminated healthy from damaged tissue.

Drinks and beverages. FT-IR spectroscopy in combination with multivariate data analysis was introduced for the quality control and authenticity assessment of spirit drinks and beer in official food control (Lachenmeier 2007). In addition, through principal component analysis (PCA) applied to spectral data, Lachenmeier (2007) attributed the differentiation of deteriorated fruit spirits to the fact that they were distilled from microbiologically spoiled mashes, and of a different group of beer samples to the fact that they were microbiologically spoiled with large bacteria counts.

Similarly, Barata et al. (2011) evaluated the effect of grape sour rot on wine fermentation and characterized the chemical composition and the sensory changes in wines produced from rotten musts. The FT-IR sour rot index showed a satisfactory correlation with the sanitary state of grape must samples and could be a reliable automated tool to evaluate sour rot grading of the incoming grapes.

Raman Spectroscopy

Raman spectroscopy: principles, instrumentation, spectral interpretation

Basic principles of Raman spectroscopy. Raman spectroscopy is a spectroscopic technique used to study vibrational, rotational, and other low-frequency modes in a system (Gardiner and Graves 1989). When a molecule is irradiated with a monochromatic light, this always results in two types of light scattering, elastic and inelastic. In elastic scattering, there occurs no change in photon frequency or without any change in its wavelength and energy and this process is known as Rayleigh scattering. Inelastic scattering is accompanied by the shift in photon frequency due to excitation or deactivation of molecular vibrations in which either the photon may lose some amount of energy or gains energy (Das and Agrawal 2011). If the molecule gains energy, scattered photons are shifted to longer wavelengths, giving rise to Stokes scattering in the Raman spectrum; otherwise, they are shifted to shorter wavelengths, giving rise to anti-Stokes scattering in the Raman spectrum. The frequency shifts of scattered light can be analyzed and presented as spectra. The Raman spectrum is produced when the molecular vibration or rotation of a molecule cause changes in its polarizability, whereas in the IR spectra when the molecular vibration or rotation causes changes of the dipole moment. It is characteristic that in Raman spectroscopy, the symmetrical vibrations present intense spectral bands and if a molecule has high symmetry, it will be frequently inactive in IR. Finally the bands observed in the spectra, represent vibrational characteristics for chemical bonds and functional groups that constitute the components in the examined samples, favoring the analysis of several kinds of samples and thus offering the basis for structural and qualitative analysis and quantitative determination (Yang and Ying 2011).

Raman spectrometer and Fourier transform Raman spectrometer. A Raman spectrometer is composed of light source, monochromator, sample holder and detector. Two major technologies are used to collect the Raman spectra, Dispersive Raman spectroscopy and Fourier transform Raman spectroscopy, with difference in their laser sources (in the visible, near infrared or ultraviolet range) and the way by which Raman scattering is detected and analyzed. Both these techniques have unique advantages and the method that best suits the sample should be preferred (Das and Agrawal 2011).

In dispersive Raman instruments, this is accomplished by focusing the Raman scattered light onto a diffraction grating, splitting the beam into its constituent wavelengths which are directed onto the detector. Dispersive Raman usually employs visible laser radiation, with typical laser wavelengths including 785, 780, 671, 633, 532, and 473 nm. The intensity of the Raman scatter is proportional to $1/\lambda^4$, so short excitation laser wavelengths deliver a much stronger Raman signal.

In place of visible excitation lasers, an FT-Raman spectrometer uses a laser in the near infrared—usually at 1064 nm. At this wavelength fluorescence is almost completely absent, however because of the $1/\lambda^4$ relationship between Raman scattering intensity and wavelength, the Raman signal is weak. An interferometer converts the Raman signal into an interferogram, permitting the detector to collect the entire Raman spectrum simultaneously. Since at low signal levels the spectral noise is predominantly detector dark noise and is independent of the intensity of the Raman signal, delivering the entire spectrum at once onto the detector greatly improves the signal-to-noise ratio. Application of the Fourier transform algorithm to the interferogram converts the results into a conventional Raman spectrum. FT-Raman has three main advantages over dispersive Raman systems: (1) reducing the laser-induced fluorescence that a number of samples exhibit; (2) facilitating the operation as with a Fourier transform infrared (FT-IR) spectrometer; and (3) showing a high spectral resolution with good wavelength accuracy (Yang and Ying 2011).

Sampling techniques. There are several sampling techniques that may be applied in Raman spectroscopy depending on the application and the type of the sample that is analyzed. These include the Resonance Raman spectroscopy, Surface Enhanced Raman Spectroscopy (SERS), Surface-Enhanced Resonance Raman Spectroscopy (SERRS), Angle Resolved Raman Spectroscopy, Hyper Raman, Surface Enhanced Hyper Raman (SEHRS), Spontaneous Raman Spectroscopy (SRS), Optical Tweezers Raman Spectroscopy (OTRS), Stimulated Raman Spectroscopy, Polarized Raman, Spatially Offset Raman Spectroscopy (SORS), Coherent anti-Stokes Raman spectroscopy (CARS), Raman optical activity (ROA), Transmission Raman, Inverse Raman spectroscopy, Tip-Enhanced Raman Spectroscopy (TERS).

Mathematical treatment of the spectroscopic data

The mathematical treatment of the Raman spectroscopic data can be performed in the same manner as with the FT-IR data (see the corresponding paragraph in FT-IR section). The different mathematical models applied in studies relevant with the investigation of food spoilage through Raman spectroscopy are reported in Table 3.

Table 3. Mathematical models applied in studies relevant with investigation of food spoilage through Raman spectroscopy.

Modeling Method	Purpose	Food type	Reference
Principal component analysis (PCA)	Unsupervised classification	Meat	Sowoidnich et al. 2012
Partial least-squares (PLS)	Regression	Meat, Milk	Nicolaou et al. 2011, Argyri et al. 2013
Genetic algorithms (GA) and Genetic programming (GP)	Feature selection/ highlighting, classification, regression	Meat	Argyri et al. 2013
Artificial neural networks (ANNs)	Regression, Supervised classification	Meat	Argyri et al. 2013
Support vector machines regression (SVR) including different kernel functions (i.e., linear, polynomial, radial basis and sigmoid)	Regression, Supervised classification	Meat	Argyri et al. 2013
Principal component-discriminant function analysis (PC-DFA)	Feature selection/ highlighting, supervised classification	Milk	Nicolaou et al. 2011
kernel PLS (KPLS)	Regression	Milk	Nicolaou et al. 2011
Canonical correlation analysis (CCA)	Investigate relationships between two multidimensional variables (i.e., FTIR data and another measured variable)	Milk	Nicolaou et al. 2011

Applications for monitoring food quality

Muscle foods. Examples of Raman spectroscopy for muscle food analysis, include studies upon the authenticity of poultry species (Ellis et al. 2005), the sensory quality of beef (Beattie et al. 2004a) and pork (Wang et al. 2012), quality screening of fish (carotenoids, collagen and fat) (Marquardt and Wold 2004, Tintchev et al. 2009) and the texture of pork muscle (Herrero et al. 2008b). It has been also applied for the determination of 'water-holding capacity' of meat after slaughter (Pedersen et al. 2003), of meat pigment as a non-invasive indirect method for determining lipid oxidation status in meat (Kathirvel et al. 2008, Sun et al. 2009), of omega-6 and omega-3 fatty acids in pork adipose tissue (Olsen et al. 2008), and to discriminate carcasses of suckling lambs according to fat quality (Osorio et al. 2009).

This analytical technique has also been used to study the changes in protein structure of various types of fish during freezing under different conditions (Careche et al. 1999, Ogawa et al. 1999, Sultanbawa and Li-Chan 2001, Badii and Howell 2002, Herrero et al. 2004, Herrero et al. 2005) and of pork due to the effect of different processing factors, including aging, salting, pressurization and heat treatment (Bocker et al. 2007, Wackerbarth et al. 2009, Xu et al. 2011).

Dairy foods. Raman spectroscopy has also found application in the dairy industry. It has been used to quantify the constituents and assess the nutritional parameters of infant formulae and milk powder (Moros et al. 2007), and has been found to be an accurate tool for screening milk powder for the inappropriate addition of melamine (Okazaki et al. 2009).

Furthermore, it has been used to characterize milk whey protein components (Liang et al. 2006), milk proteins (Alizadeh-Pasdar et al. 2004), whey protein (Ngarize et al. 2004). It was also able to quantify the conjugated linoleic acid in cows' milk fat (Meurens et al. 2005, Bernuy et al. 2008), different types of milk fatty acids (Stefanov et al. 2010, 2011) and the unsaturated fat content of clarified butter potentially for use in quality control (Beattie et al. 2004b).

Eggs. Raman spectroscopy has been applied for the analysis of egg albumen (Ngarize et al. 2004, Alizadeh-Pasdar et al. 2004) and for screening melamine contaminant in eggs (Cheng and Dong 2011).

Honey. Raman spectroscopy has been used to characterize various types of honey in different states (Oliveira et al. 2002), to determine the level of fructose and glucose in honey (Batsoulis et al. 2005) and it has been shown to be able to discriminate between honey from different floral and geographical origins, a potentially simpler and faster technique than the currently used techniques (Goodacre et al. 2002, Pierna et al. 2011).

Cereals. It was used to study the proteins in common buckwheat sample (Siu-Mei and Ching-Yung 2007), to identify specific spectral features of disubstitution and reveal the structural heterogeneity of wheat arabinoxylans (Barron et al. 2006) and to classify pure corn, cassava starch samples, and mixtures of both starches, as well as to quantify the amylose content in corn and cassava starch samples (Almeida et al. 2010). Raman spectra were used also for differentiation of rice of two geographical origins (Kim et al. 2009) and of glutinous and nonglutinous rice.

Cocoa and coffee. Raman spectroscopy was employed in analyzing the fat content of cacao (Celedon and Aguilera 2002), in the identification of a biomarker band for cacao seeds and their extracts (Edwards et al. 2005) and for fast discrimination between two coffee species (El-Abassy et al. 2011).

Fruits and vegetables. In fruit analysis, Raman spectroscopy was applied to characterize and quantify the carotenoids in tomato fruits and related products (Baranska et al. 2006, Schulz et al. 2006, Pudney et al. 2011), in fruits, vegetables, and their juices (Bhosale et al. 2004,) and in mango homogenates (Bicanic et al. 2010). It has been proposed for the discrimination of olives according to fruit quality (Muik et al. 2004), apple fruit according to surface quality (Veraverbeke et al. 2005), for sensing of orange fruit taste (Taniguchi et al. 1993) and for evaluating changes happening during the development and maturation of olive fruit by measuring the content of olive constituents like triglycerides, water, carotenoids and phenolic compounds (López-Sánchez et al. 2010).

Drinks and Beverages. Raman spectroscopy has found applications in analysis of alcoholic drinks and beverages, mainly regarding their alcohol content (Nose et al. 2004, 2005, Frausto-Reyesa et al. 2005) and in the determination of sugars is soft drinks (Silveira et al. 2009, Delfino et al. 2011).

Oils. In the oil industry, Raman spectroscopy has been shown to be a simple, rapid non-destructive tool for oil structural analysis (Weng et al. 2003). It was able to authenticate various edible oils, such as virgin olive oil (Yang and Irudayaraj 2001, Lopez-Diez et al. 2003, Baeten et al. 2005), and fats (Yang et al. 2005). Moreover, it was used to determine the free fatty acid content in olive oil and directly in olives (Muik et al. 2003), the total unsaturation of fats and oil (Barthus and Poppi 2001) and the relative oil composition and oil quality (Schulz et al. 2002).

Applications in discrimination and identification of food related microorganisms

Several studies have successfully used Raman spectroscopy for the characterization and differentiation of different microorganisms. It has been applied as a method for the identification of lactic acid bacteria (*Lactobacillus acidophilus*, *Lb. delbrueckii* ssp. *bulgaricus*, and *Streptococcus thermophilus*) from yogurt (Gaus et al. 2006), for detecting *Acinetobacter* strains (Maquelin et al. 2006), and a *Pseudomonas fluorescens* strain (Huang et al. 2007).

This technique has also been applied for the detection and identification of *E. coli* (Sengupta et al. 2006), *E. coli*, *Ps. aeruginosa*, *Acinetobacter calcoaceticus* and *Bacillus megaterium* (Zeiri et al. 2004), *Bacillus anthracis, Yersinia pestis, Burkholderia mallei, Francisella tularensis, Brucella abortus,* and ricin (Kalasinsky et al. 2007) and *E. coli, Klebsiella oxytoca, Klebsiella pneumoniae, Citrobacter freundii, Enterococcus* spp. and *Proteus mirabilis* (Jarvis et al. 2004). It could differentiate and categorize different *L. monocytogenes* strains (Oust et al. 2006). It also allowed for the discrimination between *E. coli* strains and their differentiation from other strains such as *Saccharomyces cerevisiae, Lb. casei, B. cereus, E. coli, Aspergillus niger* and *Fusarium verticilliodes* on whole apples (Yang and Irudayaraj, 2003) and of subspecies of *E. coli* and different *Pseudomonas* species inoculated in milk (Meisel et al. 2011) and for the detection of *E. coli* in milk and apple juice (Naja et al. 2010).

This technique was also used to monitor the spores of *B. cereus* and *B. megaterium*, as well as of several isogenic *B. subtilis* strains (Wang et al. 2011c) and the endospores and endospore-forming bacteria suspensions obtained from diverse *Bacillus* and *Paenibacillus* strains (De Gelder et al. 2007), as well as to analyze a closely related group of bacteria belonging to the genus *Bacillus* (Jarvis et al. 2006) and discriminate between endospore forming bacteria such as *Bacillus* and *Brevibacillus* at the species level (Lopez-Diaz and Goodacre 2004).

Applications for Monitoring Food Spoilage

In contrast to FT-IR spectroscopy, Raman spectroscopy has not received enough attention and has not been investigated for its potential on evaluating food spoilage until very recently. The reported applications regarding muscle food spoilage are from Sowoidnich et al. (2012) for pork meat and from Argyri et al. (2013) for beef meat. In the dairy foods field, Nicolaou et al. (2011) have investigated the potential of the technique through the study of detecting and enumerating inoculated bacteria in pasteurized milk.

Muscle foods. A recent study by Sowoidnich et al. (2012) on meat spoilage and Raman spectroscopy refer to Raman measurements on pork during storage at 5°C for 3 weeks. They have used a portable Raman sensor system based on a miniaturized optical bench with integrated 671 nm microsystem diode laser as excitation light source for the rapid *in situ* detection of meat spoilage. The microbial growth was measured in terms of total viable counts, setting a spoilage threshold of 10^6 cfu/cm^2 between edible and spoiled meat. The spectral changes of the Raman data set during storage were analyzed by principal component analysis (PCA). They found that the different storage times of meat were well discriminated through the mapping of the different principal components. However, this discrimination was only indirectly correlated with the microbial load, since PCA is an unsupervised method of multivariate analysis, and no further supervised method was applied to analyze the data in terms of grouping the samples according to their sensory scores or to predict the microbial load of the samples using regression models.

In a later study of Argyri et al. (2013), time series spectroscopic, microbiological and sensory analysis data were obtained from minced beef samples stored under different packaging conditions, aerobic and modified atmosphere packaging (40% CO_2/30% O_2/30% N_2) stored at 5°C. The microbiological analysis included measurements on total viable counts (TVC), lactic acid bacteria (LAB), *Br. thermosphacta*, *Enterobacteriaceae*, *Pseudomonas* spp. and yeasts-moulds. A 633 nm DeltaNu Advantage probe with a right-angled sampling attachment was used for data collection and the spectra were acquired over a Stokes Raman shift range of 200 to 3400 cm^{-1}. The collected data were analyzed using machine learning and evolutionary computing methods, including partial least square regression (PLS-R), genetic programming (GP), genetic algorithm (GA), artificial neural networks (ANNs) and support vector machines regression (SVR) including different kernel functions [i.e., linear (SVR_L), polynomial (SVR_P), radial basis (RBF) (SVR_R) and sigmoid functions (SVR_S)]. Predictive models of the microbiological load and sensory assessment were calculated using these methods and the relative performance compared. For Raman calibration models, regarding the predictions of the microbial counts the multivariate methods (SVM, PLS) that had similar performances gave better predictions compared to the evolutionary ones (GA-GP, GA-ANN, GP), with better predictions obtained for TVC, LAB and *Enterobacteriaceae*. More specifically, SVR_R and SVR_P gave acceptable predictions (% PE > 70%) for all microbial counts. The RMSE values of the latter models varied between 0.56–0.71 for the SVR_R and 0.57–0.73 for the SVR_P depending on the microorganism. On the other hand, the evolutionary methods performed better than the others in predicting the sensory scores. More specifically, the GA-ANN gave 83.33% total correct classification, with no fresh sample being misclassified as spoiled and *vice versa*. Typical spectral data obtained from Raman in the range of 3400 to 200 cm^{-1} collected from minced beef stored aerobically and under MAP at 5°C are shown in Fig. 4, whereas Table 4 presents Raman frequencies observed in spectra collected from minced beef samples during storage and possible assignments of the vibration modes. These assignments could be attributed to peaks present in other food systems spectra and are observed in the same frequencies.

Dairy foods. Nikolaou et al. (2011) have studied the accuracy of Raman spectroscopy in combination with multivariate analysis techniques, for the detection and enumeration of *Staph. aureus* in milk, as well as to study the growth interaction between *Staph. aureus* and *Lactococcus lactis* subsp.

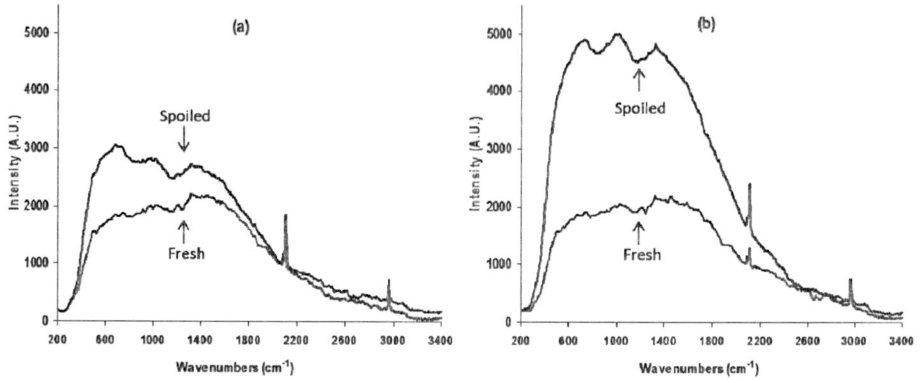

Figure 4. Raw Raman spectra collected from minced beef samples stored aerobically (a) and under MAP (b) at 5°C for a fresh sample (gray line) and a spoiled sample (black line) (Data from Argyri et al. 2013 modified accordingly).

cremoris, a common LAB. The two bacterial species were investigated both in a pure monoculture and in a combined inoculated co-culture after inoculation into ultra-heated milk during the first 24 h of growth at 37°C. Plating techniques were used to obtain primary reference data for viable bacteria counts. Raman analysis was performed using a Renishaw 2000 Raman microscope with a 785 nm diode laser. The 785 nm wavelength was used for excitation as fluorescence interference is reduced when biological materials are analyzed. Collection of Raman spectra was performed in static mode over the wavenumber shift range of 400–2,000 cm^{-1}. Principal component discriminant function analysis (PC-DFA), canonical correlation analysis (CCA), partial least-squares (PLS), and kernel PLS (KPLS) multivariate statistical techniques were employed to analyze the data. Cluster analysis using PC-DFA was subsequently performed in an attempt to gain more information regarding potential differences between the milk sustaining bacterial growth from the FT-IR and Raman spectra. The PC-DFA plot indicated good distinction between time points at 0 min as well as those after 480 min of incubation with *Lc. cremoris* monocultures. Samples collected from 720 to 1440 min followed a downward left trajectory and the mean log (TVC) values increase from 8.14 to 8.26, suggesting that while there is very little growth, the fingerprint of the milk during this time period is changing in a particular manner. Similar findings were obtained for the other monoculture (*Staph. aureus*) as well as for the co-culture. According to canonical correlation analysis with a visual inspection of the results the presence of a linear pattern in the bacterial concentration levels with the Raman data was indicated. When the Raman spectra were analyzed with PLS and KPLS the predictions of the viable bacterial load in UHT milk were not satisfactory, with large RMSEP (typically log 1.2) and RMSEC (typically

Table 4. Raman frequencies observed in spectra collected from minced beef samples and possible assignments of the vibration modes.

Frequency (cm^{-1})	Assignment	Reference
3397-3309 (num)[a]	amido acids, amino acids, carbohydrates, polypeptides, amines (NH stretch)/water, carbohydrates, lipids (OH stretch)	Socrates 2001, Yang et al. 1993
3309-3291 (3300)	amino acids/proteins/amines (NH stretch)	Socrates 2001
3284-3250 (3275, 3261)[b]	amines (N-H stretch), peracids (O-H stretch)	Socrates 2001
3250-3219 (3247, -[c])	water, lipids (OH stretch)	Socrates 2001
3228-3100 (num)	free amino acids, lipids (NH_3^+ stretch)	Socrates 2001
3100-3088	aliphatic amino acids, peptides, and proteins(C-H stretch vib)	Herrero et al. 2004
3088-3075 (3081)	polyglycines (NH_2 stretch)	Socrates 2001
3072-3041 (3066, 3050)	ketones (CH stretch)	Socrates 2001
3041-3022 (3028)	lipids (NCH_3, CH_3 stretch)	Socrates 2001
3019-3009	amines (CH_3 stretch), ketones (CH stretch), lipids (=C-H stretch)	Sarkadei and Howel 2007, Socrates 2001
3006-2991 (2994)	free amino acids (CH str)	Wong et al. 2007, Socrates 2001
2988-2947 (2966)	aliphatic amino acids,amines, proteins, lipids (CH_3 stretch)	Socrates 2001, Herrero 2008b
2947-2800 (num)	polyglycines, proteins, amines, kerones, aldehydes, lipids (CH_3, CH_2, CH stretch)	Sarkardei and Howell, 2007, Socrates 2001, Herrero et al. 2004, Greve et al. 2008
2797-2769 (2784, 2778)	aryl aldehydes (overtone CH in-plane def vib)	Socrates 2001
2769-2703 (num)	free amino acids (sym-NH_3^+ stretch), aldehydes (CH stretch, overtone CH in-plane def vib)	Socrates 2001
2700-2684 (2691)	Lipids RCOOH (O-H stretch, hydrogen bonded)	Socrates 2001
2684-2628 (2672, 2640)	free amino acids	Socrates 2001
2628-2597 (-, 2625,2602)	amido acids	Socrates 2001
2594-2547 (2591,1572,2556)	cysteinyl residues (stretch vib of the S-H group)	Herrero 2008b
2547-2263 (num)	nd	
2263-2225 (2254, -)	nitrile group (–C---N)	Thygesen et al. 2003
2244-2141 (num)	nd	

Table 4. contd....

Table 4. contd.

Frequency (cm^{-1})	Assignment	Reference
2141-1997 (num)	free amino acids (NH_3^+ stretch)	Socrates 2001
1997-1950	nd	
1950-1884 (1941,1910,-)	amido acids	Socrates 2001
1881-1866	cyclopropenones (C=O)	Socrates 2001
1866-1838	aliphatic aminoacids (C-H stretch, CH_2 groups)	Socrates 2001
1834-1803	cyclopropenones, cyclobutenediones (C=O stretch)	Socrates 2001
1791-1766	cyclobutanone derivatives	Socrates 2001
1766-1747 (1753)	ketones, free amino acids, peroxy acids, carbohydrates, nucleic acids (C=O stretch)	Socrates 2001
1747-1734	cyclopentanones, ketones, amido acids/lipids ester (C=O stretch)	Socrates 2001
1731-1708 (1708sh)	a-diketones-CO-CO	Socrates 2001
1708-1691 (1691sh)	a,b- unsat carboxylic acids (as dimer)	Socrates 2001
1691-1672 (1681)	unsuturated ketones, aldehydes, lipids (C=C stretch)	Socrates 2001, Herrero 2008a
1672-1644 (1659)	proteins with high a-helical content, amide I band (80% CO stretch, 10%CN stretch, 10%NH bend vib)	Socrates 2001, Herrero 2008a
1634-1603 (1622sh)	amino acids (ring stretch)	Herrero 2008b
1603-1588 (num)	amide II (mixtures of C–N stretch and N–H bend coordinates)	Balakrishnan et al. 2008
1494-1447 (1482, 1456)	amines (CH_3, CH_2 asym def vib), lipids (CH_2 scissoring vib, CH3 asym bend), bending vibrational modes of amino acid functional groups	Wong et al. 2007, Socrates 2001
1450-1434 (1450 sh)	aliphatic aminoacids, proteins (CH_3, CH_2, CH bend)	Herrero 2008b, Wong et al. 2007
1425-1394 (1406)	amino acids, lipids (RCOO- sym str COO- , -CH2-COOR def vib of CH_2)	Socrates 2001, Herrero 2008b
1391-1381	amide S (Ca–H bending), CH_2 (CH_2 sym bend)	Balakrishnan et al. 2008, Socrates 2001
1372-1359	amino acids (stretch and bend), aliphatic amines	Wong 2007, Herrero 2008b, Socrates 2001
1359-1347	amines	Socrates 2001

1347-1297 (1322)	amino acids, proteins, carboxylic acids (CH_3, CH_2, CH vib)	Herrero 2008b, Socrates 2001, Greve et al. 2008
1341-1313	free amino acids (CH stretch)	Socrates 2001
1313-1288	aryl aldehydes/ketones	Socrates 2001
1288-1275	amide III (α-helix structure which overlaps with the region assigned for β-turns)	Herrero 2008b
1275-1250 (1266)	amide III (random coil, β-sheets), fat (=C–H symmetric rock (cis))	Herrero 2008b, Marquardt & Wold 2007
1250-1234 (1236)	amide III (b-sheets, undefined or random coil structures)	Herrero 2008b, Wong et al. 2007
1234-1219	amide III (β-sheet structure –10% CO stretch, 30% CN stretch, 30% NH bend, 10% O=C-N bend, rest other vib), free amino acids - dicarboxylic (C-O stretch)	Socrates 2001
1219-1197 (1210)	amines, lipids (CN stretch), amino acids	Bocker et al. 2007, Socrates 2001
1194-1178	amines (CN str), carbox acids (C-O str)	Socrates 2001
1166-1138 (double)	C-C stretch, COH def, saccharide components (ring stretch)	Bocker et al. 2007, Socrates 2001
1138-1122	aliphatic amines (CN stretch)	Socrates 2001
1122-1106 (1109)	aromatic amines (NH_2 rocking/twisting), saturated primary amines, Saccharide components (ring stretch vib), PO_2- (sym PO_2- stretch)	Socrates 2001
1106-1091 (1097)	nucleic acids (sym stretch of phosphate group), lipids (PO_2-, CO stretch)	Socrates 2001
1091-1066	amines (CN stretch, CH_3 rock)	Socrates 2001
1066-1047 (1056, 1047sh)	amines (CN stretch, CH_3 rock) , lipids/proteins (CC, C-O, C-O-P stretch)	Marquardt & Wold 2004, Bocker et al. 2007, Herrero 2008b, Socrates 2001
1047-1013 (1031)	Saccharide components (ring stretch vib), C-C skeletal stretch (in aggregated b-structures)	Socrates 2001, Bocker et al. 2007
1013-1005	amino acids (ring stretch)	Herrero 2008a, Wong et al. 2007, Marquardt & Wold 2004
1003-975 (984)	aromatic carbox acids	Socrates 2001
975-953 (968)	a,b-pyranose compounds (ring vib), amino acids pyranose ring (term CH_2 def vib), lipids (CN stretch)	Socrates 2001

Table 4. contd....

Table 4. contd.

Frequency (cm^{-1})	Assignment	Reference
953-934	a-helical structure (C-C stretch)	Herrero 2008b
934-903 (915)	amino acids pyranose ring (asym ring)	Herrero 2008b, Socrates 2001
903-887 (893)	amino acids pyranose ring (anomeric C-H def), C-C stretch in a-helix	Socrates 2001, Bocker et al. 2007
887-862 (875)	CH bend, CH_2 rock, amino acids pyranose ring (equatorial C-H def other than anomeric C-H def /ring vib)	Wong et al. 2007, Bocker et al. 2007, Herrero 2008b
862-850 (856)	C–C stretch	Yang et al. 2003
850-828 (818)	sulfolipids (C-O-S stretch), lipids (P-O asym stretch)	Socrates 2001
812-784 (800)	primary aliphatic amines	Socrates 2001
784-765	amino acids pyranose ring (sym ring breathing)	Socrates 2001
765-750 (756)	amino acids side chain and C–H bend vib	Herrero et al. 2004, Wong et al. 2007
750-734 (740)	secondary amines	Socrates 2001
734-712 (734sh)	proteins amide V (N-H bend)	Socrates 2001
712-681 (696, -)	amino acid residues (C-S stretch), polyglycines (CH2 rock vib, NH def vib), aromatic carboxylic acids (CO2 in-plane def vib)	Herrero 2008b, Bocker et al. 2007, Socrates 2001
680-643br (648)	amino acid residues (C-S stretch), aliphatic aldehydes/ketones (C-C-CO in-plane def vib)	Herrero 2008b, Bocker et al. 2007, Socrates 2001
640-618 (631)	amino acid residues (C-S str), proteins- amide IV (40% O=C-N bend, rest other vib)	Herrero 2008b, Bocker et al. 2007, Socrates 2001
600-565	aldehydes/ketones (C-CO in-plane def vib), amino acids and C–H bending vibrations	Wong et al. 2007, Socrates 2001
565-518 (556)	carbox acids (CO2 out-of-plane rock def vib), free amino acids (CO2- or C-C-N def vib), aldehydes (C-CO in-plane def vib)	Socrates 2001
498-256 (num)	amines/carboxylic acids	Socrates 2001
253-221 (240)	secondary aliphatic amines (Sat.)NHCH3(CH_3 torsional vib)	Socrates 2001

Abbreviations: num: a number of peaks, def: deformation, vib: vibration, stretch: stretching, bend: bending, sym: symmetric, asym: asymmetric, rock: rocking, twist: twisting, sh: shoulder. [a] a number of peaks (more than 3) observed at this region, with the same possible assignment. [b] two peaks observed at this region, with the same possible assignment. [c] two peaks observed at this region, second peak with no well schemed edge.

log 0.75) values. Figure 5 presents Raman spectra for *Staph. aureus* and *Lc. cremoris* inoculated in UHT milk in monoculture and in co-culture after 480 min (6 h) of incubation at 37°C with possible assignment of the peaks observed.

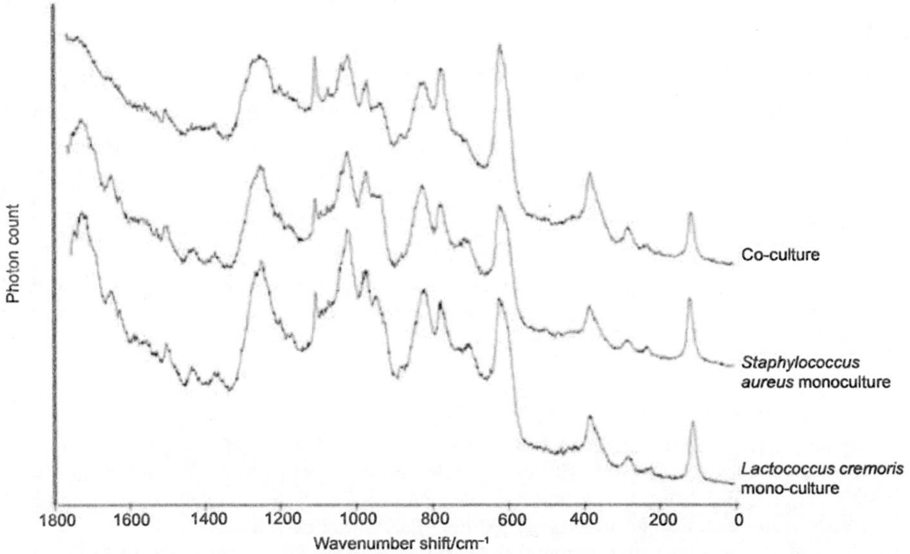

Figure 5. Raman spectra for *Staph. aureus* and *Lc. cremoris* inoculated in UHT milk in monoculture and in co-culture after 480 min (6 h) of incubation at 37°C (Data from Nicolaou et al. 2011, modified accordingly). Higher intensity of two peaks can be seen at the 1110 cm^{-1} [ν (C- C) mode] and the 610–630 cm^{-1} [ν(C-S) mode], both representing amino acids. The remaining prominent peaks in all spectra appear at the 1230–1290 cm^{-1} [amide III mode], 1050-920 cm^{-1} [ν(C-C) and ν(C-O) modes], 760–790 cm^{-1} [ν(C-S) mode] representing various amino acids, and 810–850, 380–400, and 280–300 cm^{-1} modes representing lactose.

Conclusions

According to the above mentioned studies, FTIR and Raman spectroscopies have found a wide variety of applications on food quality analysis. In the case of FTIR, several studies have been applied in the field of evaluating spoilage of foods, including animal origin foods such as meat, milk and cheese as well as plant origin foods like wheat, fruit spirits and beer and grapes. On the other hand the studies reported for evaluating food spoilage through the use of Raman spectroscopy are rather limited, including food products such as meat and milk. However the results obtained so far from both spectroscopic techniques are very encouraging. Therefore taking into advantage the current knowledge upon quality analysis of many food products with vibrational spectroscopy and having as an example the specific experiments performed so far regarding food spoilage, further studies could be performed in the latter field to expand the knowledge in other food products as well. Thus, both spectroscopic techniques could be used as a rapid, non-invasive techniques, from the food industry for automatic monitoring of food processes in real time on- or in-line, as well from the inspection authorities to provide an 'immediate' answer were needed.

Keywords: FT-IR, Raman, vibrational spectroscopy, multivariate analysis, chemometrics, microbial spoilage, food spoilage, food quality, spoilage detection

References

Afseth, N.K., H. Martens, A. Randby, L. Gidskehaug, B. Narum, K. Jørgensen, S. Lien and A. Kohler. 2010. Predicting the fatty acid composition of milk: a comparison of two Fourier transform infrared sampling techniques. Appl. Spectrosc. 64: 700–707.

Alexa, E., A. Dragomirescu, G. Pop, C. Jianu and D. Dragos. 2009. The use of FTIR spectroscopy in the identification of vegetable oils adulteration. J. Food Agr. Environ. 7: 20–24.

Alexandrakis, D., G. Downey and A.G.M. Scannell. 2012. Rapid non-destructive detection of spoilage of intact chicken breast muscle using near-infrared and Fourier transform mid-infrared spectroscopy and multivariate statistics. Food Bioprocess Technol. 5: 338–347.

Al-Alawi, A., F.R.V.D. Voort and J. Sedman. 2005. A new Fourier-transform infrared method for the determination of moisture in edible oils. Appl. Spectrosc. 59: 1295–299.

Alizadeh-Pasdar, N., E.C.Y. Li-Chan and S. Nakai. 2004. FT-Raman spectroscopy, fluorescent probe, and solvent accessibility study of egg and milk proteins. J. Agric. Food Chem. 52: 5277–5283.

Al-Jowder, O., E.K. Kemsley and R.H. Wilson. 1997. Mid-infrared spectroscopy and authenticity problems in selected meats: a feasibility study. Food Chem. 59: 195–201.

Al-Jowder, O., M. Defernez, E.K. Kemsley and R.H. Wilson. 1999. Mid-infrared spectroscopy and chemometrics for the authentication of meat products. J. Agric. Food Chem. 47: 3210–3218.

Al-Qadiri, H.M., M. Lin, A.G. Cavinato and B.A. Rasco. 2006. Fourier transform infrared spectroscopy, detection and identification of *Escherichia coli* O157:H7 and Alicyclobacillus strains in apple juice. Int. J. Food Microbiol. 111: 73–80.

Alvarez-Ordóñez, A., J. Halisch and M. Prieto. 2010. Changes in Fourier transform infrared spectra of *Salmonella enterica* serovars Typhimurium and Enteritidis after adaptation to stressful growth conditions. Int. J. Food Microbiol. 142: 97–105.

Almeida, M.R., R.S. Alves, L.B.L.R. Nascimbem, R. Stephani, R.J. Poppi and L.F.C. De Oliveira. 2010. Determination of amylose content in starch using Raman spectroscopy and multivariate calibration analysis. Anal. Bioanal. Chem. 397: 2693–2701.

Amamcharla, J.K., S. Panigrahi, C.M. Logue, M. Marchello and J.S. Sherwood. 2010a. Fourier transform infrared spectroscopy (FTIR) as a tool for discriminating *Salmonella typhimurium* contaminated beef. Sens. & Instrumen. Food Qual. 4: 1–12.

Amamcharla, J.K., S. Panigrahi, C.M. Logue, M. Marchello and J.S. Sherwood. 2010b. Application of vapour-phase Fourier transform infrared spectroscopy (FTIR) and statistical feature selection methods for identifying *Salmonella enterica* typhimurium contamination in beef. Biosyst. Eng. 107: 1–9.

Ammor, M., A. Argyri and G.-J. Nychas. 2009. Rapid monitoring of the spoilage of minced beef stored under conventionally and active packaging conditions using Fourier transform infrared spectroscopy in tandem with chemometrics. Meat Sci. 81: 507–515.

Argyri, A.A., R.M. Jarvis, D. Wedge, Y. Xu, E. Panagou, R. Goodacre and G.-J.E. Nychas. 2013. A comparison of Raman and FT-IR spectroscopy for the prediction of meat spoilage. Food Control 29: 461–470.

Argyri, A.A., E.Z. Panagou, P.A. Tarantilis, M. Polysiou and G.-J.E. Nychas. 2010. Rapid qualitative and quantitative detection of beef fillets spoilage based on Fourier transform infrared spectroscopy data and artificial neural networks. Sensor. Actuat. B-Chem. 145: 146–154.

Badii, F. and N.K. Howell. 2002. Effect of antioxidants, citrate and cryoprotectants on protein denaturation and texture of frozen cod (*Gadus morhua*). J. Agric. Food Chem. 50: 2053–61.

Bäck, T., D.B. Fogel and Z. Michalewicz. 1997. Handbook of evolutionary computation. IOP Publishing/Oxford University Press. Oxford.

Baeten, V., J.A.F. Pierna, P. Dardenne, M. Meurens, D.L. Garcia-Gonzalez and R. Aparicio-Ruiz. 2005. Detection of the presence of hazelnut oil in olive oil by FT-Raman and FT-MIR spectroscopy. J. Agric. Food Chem. 53: 6201–206.

Balakrishnan, G., C.L. Weeks, M. Ibrahim, A.V. Soldatova and T.G. Spiro. 2008. Protein dynamics from time resolved UV Raman spectroscopy. Curr. Opin. Struct. Biol. 18: 623–629.

Banwell, C.N. and E.M. Mccash. 1994. Fundamentals of Molecular Spectroscopy. McGraw-Hill. London.

Barata, A., A. Pais, M. Malfeito-Ferreira and V. Loureiro. 2011. Influence of sour rotten grapes on the chemical composition and quality of grape must and wine. Eur. Food Res. Technol. 233: 183–194.

Baranska, M., W. Schutze and H. Schulz. 2006. Determination of lycopene and β-carotene content in tomato fruits and related products: comparison of FTRaman, ATR-IR, and NIR spectroscopy. Anal. Chem. 78: 8456–461.

Barron, C., P. Robert, F. Guillon, L. Saulnie and X. Rouau. 2006. Structural heterogeneity of wheat arabinoxylans revealed by Raman spectroscopy. Carbohydr. Res. 341: 1186–1191.

Barthus, R.C. and R.J. Poppi. 2001. Determination of the total unsaturation in vegetable oils by Fourier transform Raman spectroscopy and multivariate calibration. Vibrational Spectroscopy 26: 99–105.

Basaran, P., C. Demirtaş, S. Yıkmış and A. Bülbül. 2011. Identification of food degrading fungi by using FT-IR spectroscopic analysis of hydrolytic enzyme activities. Int. J. Food Safety 13: 379–385.

Batsoulis, A.N., N.G. Siatis, A.C. Kimbaris, E.K. Alissandrakis, C.S. Pappas, P.A. Tarantilis, P.C. Harizanis and M.G. Polissiou. 2005. FT-Raman spectroscopic simultaneous determination of fructose and glucose in honey. J. Agric. Food Chem. 53: 207–210.

Baumgartner, M., M. Flock, P. Winter, W. Luf and W. Baumgartner. 2003. Evaluation of Fourier transform infrared spectrometry for the routine determination of urea in milk. Milchwissenchaft 58: 599–602.

Beattie, S.H., C. Holt, D. Hirst and A.G. Williams. 1998. Discrimination among *Bacillus cereus*, *B. mycoides* and *B. thuringiensis* and some other species of the genus *Bacillus* by Fourier transform infrared spectroscopy. FEMS Microbiol. Lett. 164: 201–206.

Beattie, R.J., S.J. Bell, L.J. Farmer, B.W. Moss and D. Patterson. 2004a. Preliminary investigation of the application of Raman spectroscopy to the prediction of the sensory quality of beef silverside. Meat Sci. 66: 903–13.

Beattie, J.R., S.E.J. Bell, C. Borgaard, A.M. Fearon and B.W. Moss. 2004b. Multivariate prediction of clarified butter composition using Raman spectroscopy. Lipids 39: 897–906.

Berrueta, L.A., R.M. Alonso-Salces and K. Héberger. 2007. Supervised pattern recognition in food analysis. J. Chromat. A. 1158: 196–214.

Bernuy, B., M. Meurens, E. Mignolet and Y. Larondelle. 2008. Performance comparison of UV and FT-raman spectroscopy in the determination of conjugated linoleic acid in cow milk fat. J. Agric. Food Chem. 56: 1159–163.

Bhosale, P., I.V. Ermakov, M.R. Ermakova, W. Gellermann and P.S. Bernstein. 2004. Resonance Raman quantification of nutritionally important carotenoids in fruits, vegetables, and their juices in comparison to high-pressure liquid chromatography analysis. J. Agric. Food Chem. 52: 3281–285.

Bicanic, D., D. Dimitrovski, S. Luterotti, C. Twisk, J.G. Buijnsters and O. Dóka. 2010. Estimating rapidly and precisely the concentration of beta carotene in mango homogenates by measuring the amplitude of optothermal signals, chromaticity indices and the intensities of Raman peaks. Food Chem. 121: 832–838.

Bocker, U., R. Ofstad, Z. Wu, H.C.Bertram, G.D. Sockalingum, M. Manfait, B. Egelandsdal and A. Kohler. 2007. Revealing covariance structures in Fourier transform infrared and Raman microspectroscopy spectra: a study on pork muscle fiber tissue subjected to different processing parameters. Appl. Spectrosc. 61: 1032–1039.

Borch, E., M.-L. Kant-Muemansb and Y. Blixt. 1996. Bacterial spoilage of meat products. Int. J. Food Microbiol. 33: 103–120.

Bureau, S., D. Ruiz, M. Reich, B. Gouble , D. Bertrand, J.-M. Audergon and C.M.G.C. Renard. 2009. Application of ATR-FTIR for a rapid and simultaneous determination of sugars and organic acids in apricot fruit. Food Chem. 115: 1133–1140.

Burgula, Y., D. Khali, S. Kim, S.S. Krishnan, M.A. Cousin, B.L. Reuhs and L.J. Mauer. 2006. Detection of *E. coli* O157:H7 and *Salmonella* Typhimurium using filtration followed by FT-IR spectroscopy. J. Food Prot. 69: 1777–1784.

Careche, M., A.M. Herrero, A. Rodriguez-Casado, M.L.D. Mazo and P. Carmona. 1999. Structural changes of hake (*Merluccius merluccius* L.) Fillets: effects of freezing and frozen storage. J. Agric. Food Chem. 47: 952–59.

Cattaneo, T.M.P., C. Giardina, N. Sinelli, M. Riva and R. Giangiacomo. 2005. Application of FT-NIR and FT-IR spectroscopy to study the shelf-life of Crescenza cheese. Int. Dairy J. 15: 693–700.

Celedon, A. and J.M. Aguilera. 2002. Applications of microprobe Raman spectroscopy in food science. Food Sci. Tech. Int. 8: 101–108.

Cevallos-Cevallos, J.M., J.I. Reyes-De-Corcuera, E. Etxeberria, M.D. Danyluk and G.E. Rodrick. 2009. Metabolomic analysis in food science: a review. Trends in Food Science and Nutrition 20: 557–566.

Chafen, J., S. Newberry, M. Riedl, D. Bravata, M. Maglione, M. Suttorp, V. Sundaram, N. Paige, A. Towfigh, B. Hulley and P. Shekelle. 2010. Diagnosing and managing common food allergies: a systematic review. Journal of the American Medical Association 303: 1848–856.

Charaniya, S., H. Wei-Shou and G. Karypis. 2008. Mining bioprocess data: opportunities and challenges. Trends Biotechnol. 26: 690–699.

Che Man, Y.B., Z.A. Syahariza, M.E.S. Mirghani, S. Jinap and J. Bakar. 2005. Analysis of potential lard adulteration in chocolate and chocolate products using Fourier transform infrared spectroscopy. Food Chem. 90: 815–819.

Chen, M., J. Irudayaraj and D.J. McMahon. 1998. Examination of full fat and reduced fat cheddar cheese during ripening by Fourier transform infrared spectroscopy. J. Dairy Sci. 81: 2791–2797.

Cheng, Y. and Y. Dong. 2011. Screening melamine contaminant in eggs with portable surface-enhanced Raman Spectroscopy based on gold nanosubstrate. Food Control 22: 685–689.

Cocciardi, R.A., A.A. Ismail, F.R. van de Voort and J. Sedman. 2004. Monitoring of lactose hydrolysis in milk by single-bounce attenuated total reflectance Fourier transform infrared spectroscopy. Milchwissenschaft 59: 403–407.

Coimbra, M.A., F.G. Alves, A.N.S. Barros and I. Delgadillo. 2002. Fourier transform infrared spectroscopy and chemometric analysis of white wine polysaccharide extracts. J. Agric. Food Chem. 50: 3405–3411.

Cooke, S. and M. Billingham. 1999. The determination of key quality parameters in the FT-IR analysis of edible oils. Spectroscopy 14: 34–7.

Cornell, H.J. and A.W. Hoveling. 1998. Wheat Chemistry and Utilization. Technomic Publishing Company Inc., Lancaster, PA.

Cothup, N.P., L.H. Daly and S.E. Wiberly. 1990. Introduction to Infrared and Raman spectroscopy. Academic Press, New York.

Cozzolino, D. and R.G Dambergs. 2009. Wine and Beer. *In:* D.-W. Sun (ed.). Infrared spectroscopy for Food Quality Analysis and Control. Academic Press, Elsevier Inc., USA, pp. 376–397.

Cozzolino, D. and C. Curtin. 2012. The use of attenuated total reflectance as tool to monitor the time course of fermentation in wild ferments. Food Control 26: 241–246.

Da-Wen, S. 2009. Infrared spectroscopy for food quality analysis and food control. Academic Press, London.

Dahlberg, D.B., S. M. Lee, S.J. Wenger and J.A. Vargo. 1997. Classification of vegetable oils by FT-IR. Appl. Spectrosc. 51: 1118–124.

Das, R.S. and Y.K. Agrawal. 2011. Raman spectroscopy: Recent advancements, techniques and applications. Vibr. Spectrosc. 57: 163–176.

Davis, R., J. Irudayaraj, B.L. Reuhs and L.J. Mauer. 2010a. Detection of E. coli O157:H7 from ground beef using Fourier Transform Infrared (FT-IR) spectroscopy and chemometrics. J. Food Sci. 75: 340–346.

Davis, R., Y. Burgula, A. Deering, J. Irudayaraj, B.L. Reuhs and L.J. Mauer. 2010b. Detection and differentiation of live and heat treated *Salmonella enterica* serovars inoculated onto chicken breast using Fourier transform infrared (FT-IR) spectroscopy. J. Appl. Microbiol. 109: 2019–2031.

Davis, R. and L.J. Mauer. 2011. Subtyping of *Listeria monocytogenes* at the haplotype level by Fourier transform infrared (FT-IR) spectroscopy and multivariate statistical analysis. Int. J. Food Microbiol. 150: 140–149.

Delfino, I., C. Camerlingo, M. Portaccio, B.D. Ventura, L. Mita, D.G. Mita and M. Lepore. 2011. Visible micro-Raman spectroscopy for determining glucose content in beverage industry. Food Chem. 127: 735–742.

De Gelder, J., P. Scheldeman, K. Leus, M. Heyndrickx, P. Vandenabeele, L. Moens and P. De Vos. 2007. Raman spectroscopic study of bacterial endospores. Anal. Bioanal. Chem. 389: 2143–2151.

Downey, G., J. Mcelhinney and T. Fearn. 2000. Species identification in selected raw homogenized meats by reflectance spectroscopy in the mid-infrared, near-infrared, and visible ranges. Appl. Spectrosc. 54: 894–899.

Drosinos, E.H. and R.G. Board. 1995. Attributes of microbial associations of meat growing as xenic batch cultures in a meat juice at 4°C. Int. J. Food. Microbiol. 26: 279–293.

Duarte, I.F., A. Barros, I. Delgadillo, C. Almeida and A.M. Gil. 2002. Application of FTIR spectroscopy for the quantification of sugars in mango juice as a function of ripening. J. Agric. Food Chem. 50: 3104–3111.

Edelmann, A., J. Diewok, K.C. Schuster and B. Lendl. 2001. Rapid Method for the discrimination of red wine cultivars based on mid-infrared spectroscopy of phenolic wine extracts. J. Agric. Food Chem. 49: 1139–1145.

Edwards, H.G.M., S.E.J. Villar, L.F.C. De Oliveira and M. Le Hyaric. 2005. Analytical Raman spectroscopic study of cacao seeds and their chemical extracts. Anal. Chim. Acta 538: 175–180.

El-Abassy, R.M., P. Donfack and A. Materny. 2011. Discrimination between Arabica and Robusta green coffee using visible micro Raman spectroscopy and chemometric analysis. Food Chem. 126: 1443–1448.

Ellis, D.I., D. Broadhurst, D.B. Kell, J.J. Rowland and R. Goodacre. 2002. Rapid and quantitative detection of the microbial spoilage of meat by Fourier transform infrared spectroscopy and machine learning. Appl. Environ. Microbiol. 68: 2822–2828.

Ellis, D.I., D.I. Broadhurst and R. Goodacre. 2004. Rapid and quantitative detection of the microbial spoilage of beef by Fourier transform infrared spectroscopy and machine learning. Anal. Chim. Acta 514: 193–201.

Ellis, D.I., D. Broadhurst, S.J. Clarke and R. Goodacre. 2005. Rapid identification of closely related muscle foods by vibrational spectroscopy and machine learning. Analyst 130: 1648–1654.

Etzion, Y., R. Linker, U. Cogan and I. Shmulevich. 2004. Determination of protein concentration in raw milk by mid-infrared Fourier transform infrared/attenuated total reflectance spectroscopy. J. Dairy Sci. 87: 2779–2788.

Garcia-Lopez, M.L., M. Prieto and A. Otero. 1998. The physiological attributes of gram-negative bacteria associated with spoilage of meat and meat products. *In:* R.G. Board and A.R. Davies (eds.). The microbiology of meat and poultry. Blackie Academic and Professional, London, pp. 1–34.

Gardiner, D.J. and P.R. Graves. (eds.). 1989. Practical Raman Spectroscopy. Springer, Berlin.

Gaus, K., P. Rosch, R. Petry, K.-D. Peschke, O. Ronneberger, H. Burkhardt, K. Baumann and J. Popp. 2006. Classification of lactic acid bacteria with UV-resonance Raman spectroscopy. Biopolymers 82: 286–90.

Ge, S.Y., M.Y. Liu, J. Zhu, F. Wang, F.Z. Ren, L.D. Zhang and H.Y. Guo. 2011. FTIR analysis of cosrelation between emulsifying properties and the secondary structure of the proteins in modified egg yolk powder. Chinese 31: 2090–3. PubMed - in process

Greve, T.M., K.B. Andersena and O.F. Nielsen. 2008. ATR-FTIR, FT-NIR and near-FT-Raman spectroscopic studies of molecular composition in human skin *in vivo* and pig ear skin *in vitro*. Spectroscopy 22: 437–457.

Gilbert, M.K., C. Frick, A. Wodowski and F. Vogt. 2009. Spectroscopic imaging for detection and discrimination of different *E. coli* strains. Appl. Spectrosc. 63: 6–13.

Gill, C.O. 1976. Substrate limitation of bacterial growth at meat surfaces. J. Appl. Bacteriol. 41: 401–410.

Goerges, S., J. Mounier, M.C. Rea, R. Gelsomino, V. Heise, R. Beduhn, T.M. Cogan, M. Vancanneyt and S. Scherer. 2008. Commercial ripening starter microorganisms inoculated into cheese milk do not successfully establish themselves in the resident microbial ripening consortia of a south german red smear cheese. Appl. Environ. Microbiol. 74: 2210–2217.

Goodacre, R., S. Vaidyanathan, W.B. Dunn, G.G. Harrigan and D.B. Kell. 2004. Metabolomics by numbers: acquiring and understanding global metabolite data. Trends in Biotechnology 22: 245–252.

Goodacre, R., E.M. Timmins, P.J. Rouney, J. J. Rowland and D.B. Kell. 1996. Rapid identification of *Streptococcus* and Enterococcus species using diffuse reflectanse-absorbance Fourier transform infrared spectroscopy and artificial neural networks. FEMS Microbiol. Lett. 140: 233–239.

Goodacre, R., B.S. Radovic and E. Anklam. 2002. Progress toward the rapid nondestructive assessment of the floral origin of european honey using despersive Raman spectroscopy. Appl. Spectr. 56: 521–527.

Gori, A., C. Cevoli, A. Fabbri, M.F. Caboni and G. Losi. 2012. A rapid method to discriminate season of production and feeding regimen of butters based on infrared spectroscopy and artificial neural networks. Journal of Food Engineering 109: 525–530.

Gorinstein, S., M. Zemser, F. Vargas-Albores, J.-L. Ochoa, O. Paredes-Lopez, C. Scheler, J. Salnikow, O. Martin-Belloso and S. Trakhtenberg. 1999. Proteins and amino acids in beers, their contents and relationships with other analytical data. Food Chem. 67: 71–78.

Guibet, F., C. Amiel, P. Cadot, C. Cordevant, M.H. Desmonts, M. Lange, A. Marecat, J. Travert, C. Denis and L. Mariey. 2003. Discrimination and classification of Enterococci by Fourier transform infrared (FT-IR) spectroscopy. Vib. Spectrosc. 33: 133–142.

Guicheteau, J., L. Argue, D. Emge, A. Hyre, M. Jacobson and S. Christesen. 2008. Bacillus spore classification with surface-enhanced raman spectroscopy and principal component analysis. Appl. Spectrosc. 62: 267–72.

Flatten, A., E.A. Bryhni, A. Kohler, B. Egelandsdal and T. Isaksson. 2005. Determination of C22:5 and C22:6 marine fatty acids in pork fat with Fourier transform mid-infrared spectroscopy. Meat Sci. 69: 433–440.

Frausto-Reyesa, C., C. Medina-Gutierrez, R. Sato-Berru and L.R. Sahagun. 2005. Qualitative study of ethanol content in tequilas by Raman spectroscopy and principal component analysis. Spectrochim. Acta A. 61: 2657–2662.

Fricker, M., B. Skånseng, K. Rudi, B. Stess and M. Ehling-Schulz. 2011. Shift from farm to dairy tank milk microbiota revealed by a polyphasic approach is independent from geographical origin. Int. J. Food Microbiol. 145: 24–30.

Haque, E., B.R. Bhandari, M.J. Gidley, H.C Deetha and A.K. Whittakerc. 2011. Ageing-induced solubility loss in milk protein concentrate powder: effect of protein conformational modifications and interactions with water. J. Sci. Food Agric. 91: 2576–2581.

He, J., L.E. Rodriguez-Saona and M.M. Giusti. 2007. Midinfrared spectroscopy for juice authentications rapid differentiation of commercial juices. J. Agric. Food Chem. 55: 4443–4452.

Hennessy, S., G. Drowney and C. O'donnell. 2008. Multivariate analysis of Attenuated Total Reflection-Fourier transfrom infrared spectroscopy data to confirm the origin of honeys. Appl. Spectrosc. 62: 1115–23.

Herald, T.J. and D.M. Smith. 1992. Heat-induced changes in the secondary structure of hen egg S-ovalbumin. J. Agric. Food Chem. 40: 1737–1740.

Herrero, A.M., P. Carmona and M. Careche. 2004. Raman spectroscopic study of structural changes in hake (*Merluccius merluccius* L.) muscle proteins during frozen storage. J. Agric. Food Chem. 52: 2147–2153.

Herrero, A.M., P. Carmona, M.L. Garcia, M.T. Solas and M. Careche. 2005. Ultrastructural changes and structure and mobility of myowater in frozenstored hake (*Merluccius merluccius* L.) muscle: relationship with functionality and texture. J. Agric. Food Chem. 53: 2558–566.

Herrero, A.M. 2008a. Raman spectroscopy a promising technique for quality assessment of meat and fish: A review. Food Chemistry 107: 1642–1651.

Herrero, A.M. 2008b. Raman spectroscopy for monitoring protein structure in muscle food systems. Critical Reviews in Food Science and Nutrition 48: 512–523.

Hewavitharana, A.K. and B. van Brakel. 1997. Fourier transform infrared spectrometric method for the rapid determination of casein in raw milk. Analyst 122: 701–704.

Hocking, S.L. and M. Faedo. 1992. Fungi causing thread mould spoilage of vacuum packaged Cheddar cheese during maturation. Int. J. Food Microbiol. 16: 123–130.

Holland, J.K., E.K. Kemsley and R.H. Wilson. 1998. Use of Fourier transform infrared spectroscopy and partial least squares regression for the detection of adulteration of strawberry purees. J. Sci. Food Agric. 76: 263–269.

Huang, W.E., M.J. Bailey, I.P. Thompson, A.S. Whiteley and A.J. Spiers. 2007. Single-cell raman spectral profiles of pseudomonas fluorescens SBW25 Reflects *in vitro* and in planta Metabolic History. Microbial Ecol. 53: 414–425.

Huis in't Veld, J.H.J. 1996. Microbial and biochemical spoilage of foods: An overview. Int. J. Food Microbiol. 33: 1–18.

Jain, A.K., R.P.W. Duin and J. Mao. 2000. Statistical pattern recognition: a review. IEEE Trans. Pattern Anal. Mach. Intell. . 22: 1–19.

Jarvis, R. M., A. Brooker and R. Goodacre. 2004. Surface-enhanced Raman spectroscopy for bacterial discrimination utilizing a scanning electron microscope with a Raman spectroscopy interface. Anal. Chem. 76: 5198–202.

Jarvis, R.M., A. Brooker and R. Goodacre. 2006. Surface-enhanced Raman scattering for the rapid discrimination of bacteria. Faraday Discussions 132: 281–92.

Johnson, M.E. 2001. Cheese products. *In:* E.H. Marth and J.L. Steele (eds.). Applied Dairy Microbiology (2nd ed.). Marcel Dekker, New York, pp. 345–384.

Jurs, P.C., G.A. Bakken and H.E. McClelland. 2000. Computational methods for the analysis of chemical sensor array data from volatile analytes. Chem. Rev. 100: 2649–2678.

Kaewmanee, T., S. Benjakul and W. Visessanguan. 2011. Effect of acetic acid and commercial protease pretreatment on salting and characteristics of salted duck egg. Food Bioprocess Technol. 342: 2423–2429.

Kaiden, K., T. Matsui, and S. Tanaka. 1987. A study of the amide III band by FT-IR spectrometry of the secondary structure of albumin, myoglobin, and γ-globulin. Appl. Spectrosc. 41: 180–184.

Kalasinsky, K.S., T. Hadfield, A.A. Shea, V.F. Kalasinsky, M.P. Nelson, J. Neiss, A.J. Drauch, G.S. Vanni and P.J. Treado. 2007. Raman chemical imaging spectroscopy reagentless detection and identification of pathogens: signature development and evaluation. Anal. Chem. 79: 2658–2673.

Kalogridou-Vassiliadou, D. 1992. Biochemical activities of Bacillus species isolated from flat sour evaporated milk. J. Dairy Sci. 75: 2681–2686.

Kathirvel, P., I.V. Ermakov, W. Gellermann, J. Mai and M.P. Richards. 2008. Resonance Raman monitoring of lipid oxidation in muscle foods. Int. J. Food Sci. Tech. 43: 2095–2099.

Karoui, R., B. Lefur, C. Grondin, E. Thomas, C. Demeulemester, J.D. Baerdemaeker and A.-S. Guillard. 2007. Mid-infrared spectroscopy as a new tool for the evaluation of fish freshness. Int. J.Food Sci. Tech. 42: 57–64.

Kim, Y., S. Lee, H. Chung, H. Choi and K. Cha. 2009. Improving Raman spectroscopic differentiation of the geographical origin of rice by simultaneous illumination over a wide sample area. J. Raman Spectro. 40: 191–196.

Kirschner, C., R. Ofstad, H.J. Skarpeid, V. Host and A. Kohler. 2004. Monitoring of denaturation processes in aged beef loin by Fourier transform infrared microspectroscopy. J. Agric. Food Chem. 52: 3920–3929.

Kittivachra, R., R. Sanguandeekul, R. Sakulbumrungsil and P. Phongphanphanee. 2007. Factors affecting lactose quantity in raw milk J. Sci. Technol. 29: 937–943.

Klijn, N., F.F.J. Nieuwendorf, J.D. Hoolwerf, C.B. van derWaals and A.H. Weerkamp. 1995. Identification of Clostridium butyricum as the causative agent of late blowing in cheese by species–species PCR amplification. Appl. Environ. Microbiol. 61: 2919–2924.

Khanmohammadi, M., A.B. Garmarudi, K. Ghasemi, S. Garrigues and M. de la Guardia. 2009. Artificial neural network for quantitative determination of total protein in yogurt by infrared spectrometry. Microchem. J. 91: 47–52.

Kher, A., P. Udabage, I. McKinnon, D. McNaughton and M.A. Augustin. 2007. FTIR investigation of spray-dried milk protein concentrate powders. Vib. Spectrosc. 44: 375–381.

Kuhm, A.E., D. Suter, R. Felleisen and J. Rau. 2009. Application of Fourier transform infrared spectroscopy (FT-IR) for the identification of *Yersinia enterocolitica* on species and subspecies level. Appl. Environ. Microbiol. 75: 5809–5813.

Kuligowski, J., D. Carrión, G. Quintás, S. Garrigues and M. de la Guardia. 2011. Sample classification for improved performance of PLS models applied to the quality control of deep-frying oils of different botanic origins analyzed using ATR-FTIR spectroscopy. Anal. Bioanal. Chem. 399: 1305–1314.

Kummerle, M., S. Scherer and H. Seiler. 1998. Rapid and reliable identification of food-borne yeasts by Fourier-transform infrared spectroscopy. Appl. Environ. Microbiol. 64: 2207–2214.

Lachenmeier, D.W. 2007. Rapid quality control of spirit drinks and beer using multivariate data analysis of Fourier transform infrared spectra. Food Chem. 101: 825–832.

Lai, S., R. Goodacre and L.N. Manchester. 2004. Whole-organism fingerprinting of the genus Carnobacterium using Fourier Transform infrared spectroscopy (FT-IR). Syst. Appl. Microbiol. 27: 186–91.

Lam, H.S., A. Proctor, L. Howard and C. Mi Jin. 2005. Rapid fruit extracts antioxidant capacity determination by Fourier transform infrared spectroscopy. J. Food Sci.70: 545–549.

Lamprell, H., G. Mazerolles, A. Kodjo, J.F. Chamba, Y. Noel and E. Beuvier. 2006. Discrimination of S. aureus strains from different species of *Staphylococcus* using Fourier transform infrared (FTIR) spectroscopy. Int. J. Food Microbiol. 108: 125–129.

Larpin-Laborde, S., M. Imran, C. Bonaïti, N. Bora, R. Gelsomino, S. Goerges, F. Irlinger, M. Goodfellow, A.C. Ward, M. Vancanneyt, J. Swings, S. Scherer, M. Guéguen and N. Desmasures. 2011. Surface microbial consortia from Livarot, a French smear-ripened cheese. Canadian Journal of Microbiology 57: 651–660.

Lin, M., M. Al-Holy, S.-S. Chang, Y. Huang, A.G. Cavinato, D.-H. Kang and B.A. Rasco. 2005. Rapid discrimination of *Alicyclobacillus* strains in apple juice by Fourier transform infrared spectroscopy. Int. J. Food Microbiol. 105: 369–376.

Lin, M., M. Al-Holy, H. Al-Qadiri, D.-H. Kang, A.G. Cavinato, Y. Huang and B.A. Rasco. 2004. Discrimination of intact and injured *Listeria monocytogenes* by Fourier Transform Infrared Spectroscopy and Principal Component Analysis. J. Agric. Food Chem. 52: 5769–5772.

Lin, M., M. Al-Holy, H. Al-Qadiri, S.-S. Chang, D.-H. Kang, B.D. Rodgers and B.A. Rasco. 2007. Phylogenetic and spectroscopic analysis of *Alicyclobacillus* isolates by 16S rDNA sequencing and mid-infrared spectroscopy. Sens. & Instrumen. Food Qual. 1: 11–17.

Lin, M., M. Al-Holy, H. Al-Qadiri, F. Kong, B.A. Rasco and D. Setiady. 2009. Detection and discrimination of *Enterobacter Sakazakii* (*Cronobacter* spp.) By mid-infrared spectroscopy and multivariate statistical analyses. J. Food Safety 29: 531–545.

Liang, M., V.Y.T. Chen, H.-L. Chen and W. Chen. 2006. A simple and direct isolation of whey components from raw milk by gel filtration chromatography and sturctural characterisation by fourier transform Raman spectroscopy. Talanta 69: 1269–1277.

Lefier D., R. Grappin and S. Pochet. 1996. Determination of fat, protein and lactose in raw milk by Fourier Transform Infrared Spectroscopy and by analysis with a conventional filter based analyser. J. Assoc. Off. Ana. Chem. Int. 79: 711–717.

Lopez-Diaz, E.C. and R. Goodacre. 2004. Characterisation of microorganisms using UV resonance Raman spectroscopy and chemometrics. Anal. Chem. 76: 585–91.

Lopez-Diez, E.C., G. Bianchi and R. Goodacre. 2003. Rapid quantitative assessment of the adulteration of virgin olive oils with hazelnut using raman spectroscopy and chemometrics. J. Agric. Food Chem. 51: 6145–150.

Lopez-Sanchez, M., M.J. Ayora-Canada and A. Molina-Diaz. 2010. Olive fruit growth and ripening as seen by vibrational spectroscopy. J. Agric. Food Chem. 58: 82–87.

Lucia, V., B. Daniela and L. Rosalba. 2001. Use of Fourier transform infrared spectroscopy to evaluate the proteolytic activity of *Yarrowia lipolytica* and its contribution to cheese ripening. Int. J. Food Microbiol. 69: 113–123.

Luginbuhl, W. 2002. Evaluation of designed calibration samples for casein calibration in Fourier transform infrared analysis of milk. Lebensm. Wiss. Technol. 35: 554–8.

Mariey, L., J.P. Signolle, C. Amiel and J. Travert. 2001. Discrimination, classification, identification of microorganisms using FTIR spectroscopy and chemometrics. Vib. Spectrosc. 26: 151–159.

Marquardt, B.J. and J.P. Wold. 2004. Raman analysis of fish: a potential method for rapid quality screening. Lebensm.-Wiss. u.-Technol. 37: 1–8.

Maudet, C. and P. Taberlet. 2001. Detection of cows' milk in goat's cheeses inferred from mithocondrial DNA polymorphism. J. Dairy Res. 68: 229–235.

Meisel, S., S. Stöckel, M. Elschner, P. Rösch and J. Popp. 2011. Assessment of two isolation techniques for bacteria in milk towards their compatibility with Raman spectroscopy. Analyst 136: 4997–5005.

Maquelin, K., C. Kirschner, L.P. Choo-Smith, N.V.D. Braak, H.P. Endtz, D. Naumann and G.J. Puppels. 2002. Identification of medically relevant microorganisms by vibrational spectroscopy. J. Microbiol. Meth. 51: 255–271.

Maquelin, K., L. Dijkhoorn, T.J.K.V.D. Reijden and G.J. Puppels. 2006. Rapid epidemiological analysis of *Acinetobacter* strains by Raman spectroscopy. J. Microbiol. Methods 64: 126–31.

McMeekin, T.A., L.A. Mellefont and T. Ross. 2007. Predictive microbiology; past present and future. *In:* S. Brul, S. van Gerwen and M. Zwietering (eds.). Modelling microorganisms in food. Woodhead, Cambridge, pp. 7–21.

Marquardt, B.J. and J.P. Wold. 2004. Raman analysis of fish: a potential method for rapid quality screening. Lebensm. Wiss. Technol. 37: 1–8.

Meer, R.R., J. Baker, F.W. Bodyfelt and M.W. Griffiths. 1991. Psychrotrophic *Bacillus* spp. In fluid milk products: a review. J. Food Prot. 54: 969–979.

Meurens, M., V. Baeten, S.H. Yan, E. Mignolet and Y. Larondelle. 2005. Determination of the conjugated linoleic acids in cow milk fat by Fourier transform Raman spectroscopy. J. Agric. Food Chem. 53: 5831– 5835.

Meza-Márquez, O.G., T. Gallardo-Velázquez and G. Osorio-Revilla. 2010. Application of mid-infrared spectroscopy with multivariate analysis and soft independent modeling of class analogies (SIMCA) for the detection of adulterants in minced beef. Meat Sci. 86511–9.

Monaci, L., R. Vatinno and G.E.D. Benedetto. 2007. Fast detection of cyclopiazonic acid in cheese using Fourier transform mid-infrared ATR spectroscopy. European Food Research Tech. 225: 585–588.

Moros, J., S. Garrigues and M.D.L. Guardia. 2007. Evaluation of nutritional parameters in infant formulas and powdered milk by raman spectroscopy. Anal. Chim. Acta 593: 30–38.

Muik, B., B. Lendl, A. Molina-Diaz and M.J. Ayora-Canada. 2003. Direct, reagent-free determination of free fatty acid content in olive oil and olives by Fourier transform Raman spectroscopy. Anal. Chim. Acta 489: 211–20.

Muik, B., B. Lendl, A. Molina-Diaz, D. Ortega-Calderon and M.J.A. Canada. 2004. Discrimination of olives according to fruit quality using Fourier transform Raman spectroscopy and pattern recognition techniques. J. Agric. Food Chem. 52: 6055–60.

Mouwen, D.J.M., A. Herman, H. Korkeala, A. Alvarez-Ordonez and M. Prieto. 2011. Applying Fourier-transform infrared spectroscopy and chemometrics to the characterization and identification of lactic acid bacteria. Vib. Spectrosc. 56: 193–201.

Naja, G., P. Bouvrette, J. Champagne, R. Brousseau and J.H.T. Luong. 2010. Activation of nanoparticles by biosorption for *E. coli* detection in milk and apple juice. Appl. Biochem. Biotechnol. 162: 460–475.

Naumann, D., D. Helm and H. Labinchinski. 1991. Microbiological characterizations by FT-IR spectroscopy. Nature 351: 81–82.

Nicolaou, N. and R. Goodacre. 2008. Rapid and quantitative detection of the microbial spoilage in milk using Fourier transform infrared spectroscopy and chemometrics. Analyst. 133: 1424–1431.

Nicolaou, N., Y. Xu and R. Goodacre. 2010. Fourier transform infrared spectroscopy and multivariate analysis for the detection and quantification of different milk species. J. Dairy Sci. 93: 5651–5660.

Nicolaou, N., Y. Xu and R. Goodacre. 2011. Fourier transform infrared and Raman spectroscopies for the rapid detection, enumeration, and growth interaction of the bacteria *Staphylococcus aureus* and *Lactococcus lactis* ssp. cremoris in milk. Anal. Chem. 83: 5681–5687.

Nieuwoudt, H.H., I.S. Pretorius, F.F. Bauer, D.G. Nel and B.A. Prior. 2006. Rapid screening of the fermentation profiles of wine yeasts by Fourier transform infrared spectroscopy. J. Microbiol. Methods 67: 248–256.

Ngarize, S., A. Adams and N.K. Howella. 2004. Studies on egg albumen and whey protein interactions by FT-Raman spectroscopy and rheology. Food Hydrocoll. 18: 49–59.

Nose, A., M. Myojin, M. Hojo, T. Ueda and T. Okuda. 2005. Proton nuclear magnetic resonance and Raman spectroscopic studies of Japanese sake, an alcoholic beverage. J. Biosci. Bioeng. 99: 493–501.

Nose, A., M. Hojo, M. Suzuki, R. Kato, K. Uehara and T. Ueda. 2004. Solute effects on the interaction between water and ethanol in aged Whiskey. J. Agr. Food Chem. 52: 5359–5365.

Nychas, G.-J.E., E.H. Drosinos and R.G Board. 1998. Chemical changes in stored meat. *In:* R.G. Board and A.R. Davies (eds.). The microbiology of meat and poultry. Blackie Academic and Professional, London, UK, pp. 288–326.

Nychas, G.-J.E. and P. Skandamis. 2005. Fresh meat spoilage and modified atmosphere packaging (MAP). *In:* J.N. Sofos (ed.). Improving the safety of fresh meat. CRC/Woodhead Publishing Limited. Cambridge, UK, pp. 461–502.

Nychas, G.-J.E., D.L. Marshall and J.N. Sofos. 2007. Meat, Poultry, and Seafood. *In:* M.P. Doyle and L.R. Beuchat (eds.). Food microbiology: fundamentals and frontiers. ASM Press, Washington D.C. pp. 105–140.

Nychas, G.-J.E., P.N. Skandamis, C.C. Tassou and K.P. Koutsoumanis. 2008. Meat spoilage during distribution. Meat Sci. 78: 77–89.

Oberreuter, H., A. Brodbeck and S.V. Stetten. 2003. Fourier-transform infrared (FT-IR) spectroscopy is a promising tool for monitoring the population dynamics of microorganisms in food stuff. Eur. Food Res. Technol. 216: 434–439.

Olsen, E.F., E.-O. Rukke, B. Egelandsdal and T. Isaksson. 2008. Determination of omega-6 and omega-3 fatty acids in pork adipose tissue with nondestructive raman and fourier transform infrared spectroscopy. Appl. Spectrosc. 62: 968–974.

Okazaki, S., M. Hiramatsu, K. Gonmori, O. Suzuki and A.T. Tu. 2009. Rapid nondestructive screening for melamine in dried milk by raman spectroscopy. Forensic Toxicology 27: 94–97.

Osorio, M.T., J.M. Zumalacarregui, R. Alaiz-Rodriguez, R. Guzman-Martinez, S.B. Engelsen and J. Mateo. 2009. Differentiation of perirenal and omental fat quality of suckling lambs according to the rearing system from Fourier transforms mid-infrared spectra using partial least squares and artificial neural networks analysis. Meat Sci. 83: 140–147.

Ogawa, M., S. Nakamura, Y. Harimoto, H. An, T. Tsuchiya and S. Nakai. 1999. Raman spectroscopic study of changes in fish actomyosin during setting. J. Agric. Food Chem. 47: 3309–18.

Oliveira, L.F.C.D., R. Colombana and H.G.M. Edwards. 2002. Fourier transform raman spectroscopy of honey. Appl. Spectr. 56: 306–11.

Oust, A., T. Moretro, K. Naterstad, G.D. Sockalingum, I. Adt, M. Manfait and A. Kohler. 2006. Fourier transform infrared and raman spectroscopy for characterisation of *Listeria monocytogenes* strains. Appl. Environ. Microbiol. 72: 228–32.

Ozen, B.F. and L.J. Mauer. 2002. Detection of hazelnut adulteration using FT-IR spectroscopy. J. Agric. Food Chem. 50: 3898–901.

Ozen, B.F., I. Weiss and L.J. Mauer. 2003. Dietary supplement oil classification and detection of adulteration using Fourier-transform infrared spectroscopy. J. Agric. Food Chem. 51: 5871–876.

Papadopoulou, O., E.Z. Panagou, C.C. Tassou and G.-J.N. Nychas. 2011. Contribution of Fourier transform infrared (FTIR) spectroscopy data on the quantitative determination of minced pork meat spoilage. Food Res. Int. 44: 3264–3271.

Pappas, C.S., P.A. Tarantilis, E. Moschopoulou, G. Moatsou, I. Kandarakis and M.G. Polissiou. 2008. Identification and differentiation of goat and sheep milk based on diffuse reflectance infrared Fourier transform spectroscopy (DRIFTS) using cluster analysis. Food Chem. 106: 1271–7.

Pedersen, D.K., S. Morel, H.J. Andersen and S.B. Englesen. 2003. Early prediction of water-holding capacity in meat by multivariate vibrational spectroscopy. Meat Sci. 65: 581–92.

Perisic, N., N.K. Afseth, R. Ofstad and A. Kohler. 2011. Monitoring protein structural changes and hydration in bovine meat tissue due to salt substitutes by Fourier transform infrared (FTIR) microspectroscopy. J. Agric. Food Chem. 59: 10052–61.

Perry, J.J., L.E. Rodriguez-Saona and A.E. Yousef. 2011. Quality of shell eggs pasteurized with heat or heat-ozone combination during extended storage. J. Food Sci. 76: 437–444.

Pierna, J.A.F., O. Abbas, P. Dardenne and V. Baeten. 2011. Discrimination of Corsican honey by FT-Raman spectroscopy and chemometrics. Biotechnol. Agron. Soc. Environ. 15: 75–84.

Pistorius, A.M.A. 1995. Biochemical applications of FTIR spectroscopy. Spectroscopy Europe 7: 8–15.

Preisner, O., R. Guiomar, J. Machado, J.C. Menezes and J.A. Lopes. 2006. Application of FT-IR spectroscopy and chemometrics for differentiation of *Salmonella Enteritidis* phage types. Appl. Environ. Microbiol. 76: 3538–3544.

Pudney, P.D.A., L. Gambelli and M.J. Gidley. 2011. Confocal Raman microspectroscopic study of the molecular status of carotenoids in tomato fruits and foods. Appl. Spectrosc. 65: 127–134.

Queji, M.D., G.Wosiacki, G.A. Cordeiro, P.G. Peralta-Zamora and N. Nagata. 2010. Determination of simple sugars, malic acid and total phenolic compounds in apple pomace by infrared spectroscopy and PLSR. Int. J. Food Sci. Tech. 45: 602–609.

Rebuffo, C.A., J. Schmitt, M. Wenning, F. von Stetten and S. Scherer. 2007b. Reliable and rapid identification of *Listeria monocytogenes* and *Listeria* Species by artificial neural network-based Fourier transform infrared spectroscopy. Appl. Environ. Microbiol. 72: 994–1000.

Rebuffo-Scheer, C.A., J. Schmitt and S. Scherer. 2007a. Differentiation of *Listeria monocytogenes* serovars by using artificial neural network analysis of Fourier-transformed infrared spectra. Appl. Environ. Microbiol. 73: 1036–1040.

Rodriguez-Saona, L.E., N. Koca, W.J. Harper and V.B. Alvarez. 2006. Rapid determination of swiss cheese composition by fourier transform infrared/attenuated Total Reflectance spectroscopy. J. Dairy Sci. 89: 1407–1412.

Rohman, A, Sismindari, Y. Erwanto and Y.B. Che Man. 2011. Analysis of pork adulteration in beef meatball using Fourier transform infrared (FTIR) spectroscopy. Meat Sci. 88: 91–95.

Rohman, A. and Y.B.C. Man. 2010. Fourier transform infrared (FTIR) spectroscopy for analysis of extra virgin olive oil adulterated with palm oil. Food Research International 43: 886–92.

Ruoff, K., W. Luginbuhl, R. Kunzli, M.T. Inglesias, S. Bogdanov, J.O. Bosset, K.V.D. Ohe, W.V.D. Ohe and R. Amado. 2006. Authentication of the botanical and geographical origin of honey by Mid-infrared spectrometry. J. Agric. Food Chem. 54: 6867–6872.

Rutten, M.J., H. Bovenhuis, J.M. Heck and J.A. van Arendonk. 2011. Predicting bovine milk protein composition based on Fourier transform infrared spectra. J. Dairy Sci. 94: 5683–90.

Sarkardei, S. and N.K. Howell. 2007. The effects of freeze-drying and storage on the FT-Raman spectra of Atlantic mackerel (*Scomber scombrus*) and horse mackerel (*Trachurus trachurus*). Food Chem. 103: 62–70

Saucedo-Hernández, Y., M.J. Lerma-García, J.M. Herrero-Martínez, G. Ramis-Ramos, E. Jorge-Rodríguez and E.F. Simó-Alfonso. 2011. Classification of pumpkin seed oils according to their species and genetic variety by attenuated total reflection fourier-transform infrared spectroscopy. J. Agric. Food Chem. 59: 4125–4129.

Schulz, H., B. Schrader, R. Quilitzsch and B. Steuer. 2002. Quantitative analysis of various citrus oils by ATR/FT-IR and NIR-FT Raman spectroscopy. Appl. Spectr. 56: 117–24.

Schulz, H., W. Schutze and M. Baranska. 2006. Fast determination of carotenoids in tomatoes and tomato products by Raman spectroscopy. Acta Horticulturae 712 II: 901–905.

Sengupta, A., M. Mujacic and E.J. Davis. 2006. Detection of bacteria by surface-enhanced Raman spectroscopy. Anal. Bioanal. Chem. 386: 1379–1386.

Singh, C.B., D.S. Jayas, F. Borondics and N.D.G. White. 2011. Synchrotron based infrared imaging study of compositional changes in stored wheat due to infection with Aspergillus glaucus. Journal of Stored Products Research 47: 372–377.

Sini, T.K., S. Santhosh and P.T. Mathew. 2007. Study on the production of chitin and chitosan from shrimp shell by using *Bacillus subtilis* fermentation. Carbohydr. Res. 16: 2423–2429.

Silveira. Jr., L. Moreira, L.M. Conceicao, V.G.B. Casalechi, H.L. Munoz, I.S. da Silva, F.F. Silva, M.A.S.R. de Souza, R.A. and M.T.T. Pacheco. 2009. Determination of sucrose concentration in lemon-type soft drinks by dispersive Raman spectroscopy. Spectroscopy 23:217–226.

Siu-Mei, C. and M. Ching-Yung. 2007. Structural characterization of globulin from common buckwheat (Fagopyrum esculentum Moench) using circular dichroism and Raman spectroscopy. Food Chem. 102: 150–160.

Schmitt, J. and H.-C. Flemming. 1998. FTIR-spectroscopy in microbial and material analysis. Biodeterior. Biodegrad. 41: 1–11.

Socrates, G. 2001. Infrared and Raman Characteristic Group Frequencies, 3rd edn. John Willey & Sons Ltd Press. West Sussex, UK.

Sowoidnich, K., H. Schmidt, H.D. Kronfeldt and F. Schwagele. 2012. A portable 671nm Raman sensor system for rapid meat spoilage identification.Vib. Spectrosc. 62: 70–76.

Stanbridge, L.H. and A.R. Davis. 1998. The microbiology of chill-stored meat. *In:* R.G. Board and A.R. Davies (eds). The microbiology of meat and poultry. Blackie Academic and Professional, London, UK, pp. 174–219.

Stefanov, I., V. Baeten, O. Abbas, E. Colman, B. Vlaeminck, B. De Baets and V. Fievez. 2010. Analysis of milk odd- and branched-chain fatty acids using Fourier Transform (FT)-Raman Spectroscopy. J. Agric. Food Chem. 58: 10804–10811.

Stefanov, I., V. Baeten, O. Abbas, E. Colman, B. Vlaeminck, B. De Baets and V. Fievez. 2011. Determining milk isolated and conjugated trans-unsaturated fatty acids using Fourier transform Raman spectroscopy. J. Agric. Food Chem. 59: 12771–12783.

Stuart, B. 1997. Biological Applications of infrared spectroscopy. John Wiley & Sons Inc. Chichester.

Sun, W.Q., G.H. Zhou, X.L. Xu and Z.Q. Peng. 2009..Studies on the structure and oxidation properties of extracted cooked cured meat pigment by four spectra. Food Chem. 115: 596–601.

Sultanbawa, Y. and E.C.Y. Li-Chan. 2001. Structural changes in natural actomyosin and surimi from ling cod (*Ophiodon elongatus*) during frozen storage in the abscence or presence of cryoprotectants. J. Agric. Food Chem. 49: 4716–25.

Sun, W., F. Zhou, M. Zhao, B. Yang and C. Cui. 2011. Physicochemical changes of myofibrillar proteins during processing of Cantonese sausage in relation to their aggregation behavior and *in vitro* digestibility. Food Chem. 129: 472–478

Taniguchi, I., Y. Yonehara, K. Masuda and Y. Hirakawa. 1993. New approach for non-destructive sensing of fruit taste. Sensor. Actuat. B-Chem. 447–450.

Tapp, H.S., Defernez, M. and E.K. Kemsley. 2003. FTIR spectroscopy and multivariate analysis can distinguish the geographic origin of extra virgin olive oils. J. Agric. Food Chem. 51: 6110–6115.

Ternstrom, A., A.-M. Lindberg and G. Molin. 1993. Classification of the spoilage flora of raw and pasteurized bovine milk, with special reference to *Pseudomonas* and *Bacillus*. J. Applied Bacteriol. 75: 25–34.

Tintchev, F., U. Kuhlmann , H. Wackerbarth, S. Toepfl, V. Heinz, D. Knorr and P. Hildebrandt. 2009. Redox processes in pressurised smoked salmon studied by resonance Raman spectroscopy. Food Chem. 112: 482–486.

Thygesen, L.G., M.M. Løkke, E. Micklander and S.B. Engelsen. 2003. Vibrational microspectroscopy of food. Raman vs. FT-IR. Trends Food Sci. Technol. 14: 50–57.

Ulberth, F. and M. Henninger 1994. Quantitation of trans fatty acids in milk fat using spectroscopic and chromatographic methods. J. Dairy Res. 61: 517–27.

Urtubia. A., J.R. Pérez-Correa, M. Meurens and E. Agosin. 2004. Monitoring large scale wine fermentations with infrared spectroscopy. Talanta 64: 778–784.

van Kempen, T. 2001. Infrared technology in animal production. Worlds Poultry Science Journal 57: 29–48.

Vanetti, M.C.D. 2011. Milk and dairy foods: Microbial Flora. *In:* L.M.L. Nollet and F. Toldra (eds.). Spoilage detection in safety analysis of foods of animal origin. Taylor and Francis Group, LLC. pp. 778–798.

Venter, C. 2009. Cow's milk protein allergy and other hypersensitivities in infants. Journal of Family Health Care 19: 128–34.

Veraverbeke, E.A., J. Lammertyn, B.M. Nicolai and J. Irudayaraj. 2005. Spectroscopic Evaluation of the Surface Quality of Apple. J. Agric. Food Chem. 53: 1046–1051.

Vlachos, N., Y. Skopelitis, M. Psaroudaki, V. Konstantinidou, A. Chatzilazarou and E. Tegou. 2006. Applications of Fourier-transfrom infrared spectroscopy to edible oils. Analysis Chimica Acta 573–4, 459–65.

Wang, N., Y. Fu and L.-T. Lim. 2011a. Feasibility study on chemometric discrimination of roasted arabica coffees by solvent extraction and Fourier Transform Infrared Spectroscopy. J. Agric. Food Chem. 59: 3220–3226.

Wang, J., T. Yue, Y. Yuan, X. Lu, J.-H. Shin and B. Rasco. 2011b. Discrimination of Alicyclobacillus strains using nitrocellulose membrane filter and attenuated total reflectance fourier transform infrared spectroscopy. J. Food Sci. 76: 137–142.

Wang, G., P. Zhang, P. Setlow and Y.-Q. Li. 2011c. Kinetics of germination of wet-heat-treated individual spores of *Bacillus* species, monitored by Raman spectroscopy and differential interference contrast microscopy. Appl. Environ. Microbiol. 77: 3368–3379.

Wang, Q., S.M. Lonergan and C. Yu. 2012. Rapid determination of pork sensory quality using Raman spectroscopy. Meat Sci. 91: 232–239.

Wackerbarth, H., U. Kuhlmann, F. Tintchev, V. Heinz and P. Hildebrandt. 2009. Structural changes of myoglobin in pressure-treated pork meat probed by resonance Raman spectroscopy. Food Chem. 115: 1194–1198.

Weinrichter, B., W. Luginnbűhl, H. Rohm and J. Jimeno. 2001. Differentiation of facultatively heterofermentative lactobacilli from plants, milk, and hard type cheeses by SDS-PAGE, RAPD, FTIR, energy source utilization and autolysis type. Lebensm. Wiss. u. Technol. 34: 556–566.

Weng, Y.-M., R.-H., Weng, C.-Y. Tzeng and W. Chen. 2003. Structural analysis of triacylglycerols and edible oils by near-infrared fourier transform raman spectroscopy. Appl. Spectrosc. 57: 413–18.

Wenning, M., V. Theilmann and S. Scherer. 2006. Rapid analysis of two food-borne microbial communities at the species level by Fourier-transform infrared microspectroscopy. Environ. Microbiol. 8: 848–857.

Winder, C.L., E. Carr, R. Goodacre and R. Seviour. 2004.The rapid identification of *Acinetobacter* species using Fourier transform infrared spectroscopy. J. Appl. Microbiol. 96: 328–339.

Wong, H.-W., D.L. Phillips and C.-Y. Ma. 2007. Raman spectroscopic study of amidated food proteins. Food Chem. 105: 784–792.

Wu, Z., H.C. Bertram, A. Kohler, U. Bocker, R. Ofstad and H.J. Andersen. 2006. Influence of aging and salting on protein secondary structures and water distribution in uncooked and cooked pork. A combined FT-IR microspectroscopy and H NMR study. J. Agric. Food Chem. 54: 8589–97.

Wu, Z., H.C. Bertram, U. Bocker, R. Ofstad and A. Kohier. 2007. Myowater dynamics and protein secondary structural changes as affected by heating rate in three pork qualities: a combined FT-IR microspectroscopic and H NMR relaxometry study. J. Agric. and Food Chem. 55: 3990–997.

Wu, D., Y. He, S. Feng and D.-W. Sun. 2008. Study on infrared spectroscopy technique for fast measurement of protein content in milk powder based on LS-SVM. J. Food Eng. 84: 124–131.

Xu, X.L., M.Y. Han, Y. Fei and G.H. Zhou. 2011. Raman spectroscopic study of heat-induced gelation of pork myofibrillar proteins and its relationship with textural characteristic. Meat Sci. 87: 159–164.

Yang, H. and J. Irudayaraj, 2001. Comparison of near-infrared, Fourier transform infrared, and Fourier transform-raman methods for detrmining olive pomace oil adulteration in extra virgin olive oil. Journal of the American Oil Chemists' Society 78: 889–95.

Yang, H. and J. Irudayaraj. 2003. Rapid detection of foodborne microorganisms on food surface using Fourier transform Raman spectroscopy. J. Mol. Struct. 646: 35–43.

Yang, H., J. Irudayraj and M.M. Paraskar. 2005. Discriminant analysis of edible oils and fats by FTIR, FT-NIR and FT-raman spectroscopy. Food Chem. 93: 25–32.

Yang, W., P. Pang, X. Gao, Q. Cai, K. Zeng and C.A. Grimes. 2007. Detection of lactose in milk samples using a wireless multi-enzyme biosensor. Sensor Lett. 5: 405–410.

Yang, D. and Y. Ying. 2011. Applications of Raman Spectroscopy in Agricultural Products and Food Analysis: A Review. Appl. Spectrosc. Rev. 46: 539–560.

Zhao, H., R.L. Parry, D.I. Ellis, G.W. Griffith and R. Goodacre. 2006. The rapid differentiation of *Streptomyces* isolates using Fourier transform infrared spectroscopy. Vibr. Spectrosc. 40: 213–218.

Zeiri, L., B.V. Bronk, Y. Shabtai, J. Eichler and S. Efrima. 2004. Surface-Enhanced Raman spectroscopy as a tool for probing specific biochemical components in bacteria. Appl. Spectrosc. 58: 33–40.

CHAPTER 15

The Potential of Hyperspectral Imaging for Monitoring Microbial Activity in Foods

Aoife A. Gowen,[1,*] *Edurne Gaston*[2] and *Vasilis Valdramidis*[3]

Introduction

Optical techniques for food quality and safety assessments

The assurance of food safety is a major public health concern and of utmost importance for the food industry. With increased expectations for products of high quality and safety, the need for accurate, fast and objective food analysis methods continues to grow. Quality and safety assurance methods used in the food industry have traditionally involved classical microbiological analysis, which is laborious and time consuming, and does not offer immediate results. Consequently, alternative analytical techniques that can be implemented on-line or can provide results in shorter time frames are currently being investigated.

[1] School of Biosystems Engineering, University College Dublin, Dublin 4, Ireland. Email: aoife.gowen@dit.ie

[2] IRIS-Innovació i Recerca Industrial i Sostenible, Avda. Carl Friedrich Gauss n° 11, 08860 Castelldefels, Barcelona, Spain.

[3] Department of Food Studies and Environmental Health, Faculty of Health Sciences, Msida MSD 2080, University of Malta, Malta.

* Corresponding author

Computer vision provides an automated, objective and non-destructive determination of the external characteristics of an object. Over the last two decades a number of computer vision techniques have been implemented in order to assess the safety of foods or food products. For example Tao et al. (1998) and Chao et al. (1999) used color imaging for laboratory inspection of chicken spleens, hearts and livers. This technique was capable of identifying (i) poultry disease conditions like leucosis, which is a result of toxins produced from cells at a localized infection or from the growth of microorganisms, and (ii) septicemia, which is caused by the presence of pathogenic microorganisms or their toxins in the bloodstream.

Vízhányó and Felföldi (2000) used a machine vision system to recognize and identify discolouration on mushrooms caused by bacterial disease and to differentiate it from natural senescence. Red-Green-Blue (RGB) values were transformed into a* and b* colour spaces. Vectorial normalization and cluster analysis of the pixels were evaluated for classification of the diseased areas of test mushrooms. The vector normalization approach was more successful than the cluster approach as this method could overcome colour intensity differences due to sample curvature.

Although, machine vision imaging could be used to detect surface features (color, size/shape, surface texture, or defects) in food inspection, it is limited in capability for detection of chemical and biological properties of materials, since these changes may not be correlated to changes in RGB colour space (Chao 2010). As a result, standard machine vision techniques may have limitations on the direct detection of specific microbial species for food safety assessments. Hyperspectral imaging has recently emerged as a non-destructive quality and safety technology that can go beyond and above machine vision and spectroscopy and their limitations in the food industry (Gowen et al. 2010). This technique will be discussed in the following sections of this chapter.

Hyperspectral imaging

Hyperspectral Imaging (HSI), also known as Chemical or Spectroscopic Imaging, is an emerging technique that integrates conventional imaging and spectroscopy to attain both spatial and spectral information from an object. It goes beyond conventional RGB imaging by obtaining images of a sample at 100's of contiguous wavebands. This technique was originally developed for remote sensing applications utilizing satellite imaging data, but recently, due to developments in sensor technology and the consequent reductions in costs, it has found application in a diverse range of disciplines, such as food, agriculture and pharmaceuticals. Hyperspectral imaging, like other spectroscopy techniques, can be carried out in reflectance, transmission, transflectance or fluorescence modes.

Hyperspectral images, known as hypercubes, are usually represented as three-dimensional blocks of data, comprising of two spatial and one wavelength dimension, as illustrated in Fig. 1. Each pixel in a hyperspectral image contains a spectrum representing the light absorbing/scattering/fluorescing properties of the spatial region represented by that pixel. The resulting spectrum acts like a fingerprint, which can be used to estimate chemical composition and or physical properties of the spatial region represented, depending on the spectral range examined. The hypercube allows for the visualization of biochemical constituents of a sample, separated into particular areas of the image, since regions of a sample with similar spectral properties tend to have similar chemical composition. By combining spectral and spatial information, HSI offers enhanced characterization of the composition and distribution of components in a product. In addition, HSI is a rapid method (typical scan time < 1 min) compared with traditional quality testing techniques such as HPLC and GC-MS which may take hours including sample preparation steps, and since it is a non-destructive and non-contact technique, samples may be further processed or tested as required. The non-destructive, rugged and flexible nature of HSI makes it an attractive tool for identification and monitoring of microbial activity in foods.

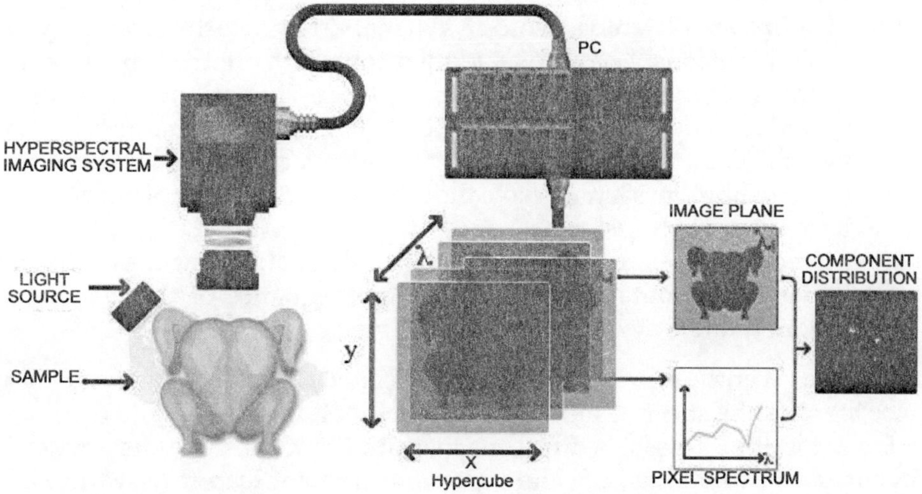

Figure 1. Schematic showing hyperspectral imaging equipment, hypercube structure, spatial axes X, Y, and wavelength axis (λ).

Color image of this figure appears in the color plate section at the end of the book.

Instrumentation

Hyperspectral images can be obtained for reflected, transmitted or emitted light coming from the UV, through the Vis-NIR and up to the short wave infrared (SWIR) regions of the electromagnetic spectrum. The camera, wavelength modulator and illumination conditions determine the wavelength range of the system. Commercially available Vis-NIR HSI systems typically range between 400 and 1000 nm, and utilize cameras with Charge Coupled Device (CCD) or Complementary Metal Oxide Semiconductor (CMOS) sensors; longer wavelength systems require more expensive IR focal-plane array detectors (e.g., Indium Gallium Arsenide (InGaAs) detectors). The sample/target is usually diffusely illuminated by a tungsten-halogen light source. At present, most commercial HSI systems are not capable of obtaining information in all three dimensions of a *hypercube* simultaneously and are limited to obtaining two dimensions at a time, then creating a three dimensional image by stacking the two-dimensional 'slices' in sequence. Three typical ways to do this are described briefly below:

- **Whiskbroom system:** a complete spectrum is acquired at a single spatial location, then the sample is repositioned and a spectrum is again acquired. This is repeated for each position at which a spectrum is desired.
- **Pushbroom system:** a two-dimensional array detector is employed to simultaneously acquire a 2-dimensional data matrix representing an image frame 'slice' with spatial *x* spectral dimensions. The second spatial dimension of the hypercube is constructed by scanning across the sample surface in a direction perpendicular to the camera imaging line. Pushbroom systems have the advantage of easy implementation in conveyor belt systems.
- **Staredown system:** each frame is a spatial *x* spatial image. A complete hypercube is obtained by collecting a sequence of these frames, acquired one wavelength band at a time.

More recently, a new hyperspectral sensor has been developed that incorporates the spectral modulation directly on the sensor chip (http://www2.imec.be). This is the first time that simultaneous three-dimensional hypercubes can be imaged, and represents a major leap forward in HSI technology. The acquisition of hyperspectral images in real time will certainly expand its adoption in industry, where speed of acquisition has previously been a limiting factor.

As the speed of data acquisition in HSI improves, the analysis and storage of this data becomes a major issue. For example, a typical image of 320 x 240 pixels in size will contain over 75,000 spectra, each with > 100 spectral data points, resulting in a file containing > 7,500,000 numbers; if each number is stored in floating point double precision (16-bytes), the resultant image will be > 100 MB in size! The following section provides a brief overview analysis of HSI data. There are several sources available in the literature that provide an in-depth description of HSI data analysis (e.g., Burger and Gowen 2011).

Data Analysis

A huge variety of techniques have been developed to analyze hyperspectral imaging data. Hyperspectral image analysis generally aims to reduce the dimension of data while retaining sufficient information to classify and or quantify important chemical or physical areas of a scene.

Pre-processing, including image calibration, is usually performed to remove non-chemical biases from the spectral and spatial information contained in a hyperspectral image and to prepare the data for further processing (Burger and Geladi 2006). Spatial operations usually carried out at the pre-processing stage include: thresholding and masking to remove redundant background information from the hypercube; image filtering (e.g., Gaussian filtering) to decrease noise and interpolation (e.g., bilinear interpolation) to decrease image size. A number of spectral pre-processing techniques also exist, including polynomial baseline correction, Savitzky-Golay derivative conversion, mean centering and unit variance normalization.

Spectral bands in hyperspectral imagery are highly correlated, thus images representing contiguous wavebands may contain redundant information. For this reason the selection of a smaller subset of spectral bands describing the variation of interest in a hypercube is a commonly applied method of dimension reduction. Another reason to carry out band selection is for the development of application-specific multispectral imaging systems that are less expensive than full range HSI systems. However, in order to quantify or classify multiple targets in a hyperspectral scene, it is usually necessary to use information from all wavelengths. Multivariate data analysis or chemometrics is commonly applied for this purpose.

Principal component analysis is the workhorse of chemometrics, and it can be used as an exploratory tool in hyperspectral imaging. It represents a computationally fast method for compressing the variance contained in the > 100 image planes of a HSI image into a smaller number (typically < 10) of principal component score images.

There is an abundance of classification and regression algorithms available for categorization and quantification of hyperspectral images; their performance tends to be goal dependent. Various unsupervised methods, including k-nearest neighbours and hierarchical clustering, can be applied in the spectral or spatial domains, or in both simultaneously, to achieve classification. Supervised classification methods, including partial least squares discriminant analysis, linear discriminant analysis and spectral angle mapping require the selection of well-defined and representative calibration and training sets for classifier optimisation. The same can be said for supervised regression methods, such as partial least squares regression.

Recent Applications of HSI for Microbial Inspection

Applications of HSI for microbial contaminated food products include studies on fish, fresh cut vegetables, fruits and meat. Some representative recent applications of HSI for microbial safety inspection are discussed hereunder. All these applications are further summarized in Table 1 providing an overview of current trends to the reader.

Fish products

In recent studies by Sone et al. (2012), HSI has been used for assessing the spoilage bacteria limiting the shelf life of salmon fillets. Hyperspectral images of fresh salmon fillets stored in air, in 60% CO_2/40% N_2 and in 90% vacuum were collected, total bacterial count and lipid oxidation were also measured. The results demonstrated spectral variations at the wavelengths 606 and 636 nm, which could be assigned to the different oxidation states of the haem proteins in the muscles. 88% correct classification of fillets according to the packaging used was achieved highlighting the potential of this technique.

Post-harvest products

Other studies focused on the detection of *Escherichia coli* contamination in fresh vegetables, e.g., spinach leaves (Siripatrawan et al. 2011). These researchers selected the microbial surrogate *E. coli* K12 as the challenge microorganism for performing studies on inoculated spinach at different initial cell concentrations (>10^5 CFU/g). Samples were analyzed using the classical colony counting and a hyperspectral imaging technique. The principle underlying this approach was based on the assumption that the metabolic activity of *E. coli* results in biochemical changes with the concurrent formation of metabolic by-products, which potentially indicate the contamination. The authors reported good results on a regression analysis (R^2 = 0.97) for predicted versus true values of the numbers of *E. coli* contaminated on the packaged spinach.

Table 1. Recent applications of hyperspectral imaging for microbial detection and classification.

Product	Wavelength (nm)	Mode	Classification	Type of bacteria	Reference
Salmon	400-1100	Interactance	PCA	Total Bacteria Count	Sone et al. 2012
Spinach	400-1000	Reflectance	PCA, ANN	*E. coli* (K12A, K12B, K12C and K12D).	Siripatrawan et al. 2011
Mushrooms	400-1000	Reflectance	PLS-DA	*Pseudomonas tolaasii*	Gaston et al. 2010
Mandarins	320-1100	Reflectance	LDA, CART	*Penicillum digitatum*	Gomez-Sanchis et al. 2008
Meat	400-1100	Reflectance	MLR	*E. coli*	Tao et al. 2012
Beef steaks	400-1100	Reflectance	MLR	Total Viable counts	Peng et al. 2011
Poultry carcass	389-753	Reflectance	Fuzzy logic	diseased or wholesome	Chao 2007
Poultry carcass	400-1000	Reflectance	Wavelength band ratio	fecal material and ingesta	Yoon 2011
Stainless steel	416-700	Fluorescence	Wavelength band ratio	Biofilms of E. coli, Pseudomonas pertucinogena, Erwinia chrysanthemi, and Listeria innocua	Jun et al. 2008
Stainless steel, HDPE, Formica	421-700	Fluorescence	PCA, single & two-band selection	Biofilms of *E.coli*, *S. enterica*	Jun et al. 2010
Model system (aluminium card)	1000-2350	Reflectance	Single bands & PLS	Series of gram+, gram	Dubois et al. 2005
Water	3100-3703	Raman scattering	Linear correlation with peak area	*Staphylococcus epidermidis* (ATCC 35984) and *Escherichia coli* (ATCC 11775)	Escoriza et al. 2006
Complex background matrix	2857-14285	Raman scattering	PCA, hierarchical cluster analysis (HCA)	*Bacillus anthracis*, *Yersinia pestis*, *Burkholderia mallei*, *Francisella tularensis*, *Brucella abortus*, and ricin.	Kalasinksy et al. 2007
Aliquot of bacteria with silver nanoparticles	5405-20000	Raman scattering	PCA	*Bacillus thuringiensis* vegetative cells ,Bacillus anthracis spores	Guicheteau (2010)
Distilled or recipe tap water	2857-20000	Raman scattering	PCA, hierarchical cluster analysis (HCA)	*Bacillus atrophaeus*, B. thuringiensis, *E. coli* ATCC 29425, ATCC 11776, and ATCC 9637, *Yersinia rhodei*, *Yersinia enterocolitica*	Tripathi et al. 2008

Gómez-Sanchis et al. (2008) proposed hyperspectral imaging system for the early detection of rot caused by *Penicillium digitatum* in mandarins. This method's success in classifying rotten fruit was above 91% and it represented a viable alternative to the operationally inefficient sorting system previously used in the citrus industry.

A team of Irish researchers demonstrated the potential of vis-NIR HSI reflectance imaging (450-950 nm) for detection of surface damage on mushrooms caused by bacterial disease (Gaston et al. 2010). Mushrooms were inoculated with *Pseudomonas tolaasii* and PLS discriminant analysis models were developed to classify the mushroom surface into one of three classes: undamaged, bruise damaged and *Brown blotch* diseased. Although bacterial and physical damage both produced similar effects on the mushroom's surface, i.e., the development of brown pigments, it was possible to differentiate between bruise and brown blotch damage with >95% classification rates reported, indicating the usefulness of this technique for microbially induced disease detection.

Meat products

Spatially-resolved hyperspectral scattering images can be obtained by shining a laser light on a sample and obtaining a hyperspectral line image at a position offset from the incident position of laser. This technique, in combination with multi-linear regression has been applied to quality evaluation of pork (Tao 2011). Using this technique in the 400–1100 nm wavelength range, it was possible to predicting *E. coli* contamination and tenderness of pork samples simultaneously. The application was validated for contamination levels of *E. coli* that varied between 10^2 up to 10^{10} cfu/ml. Spatially resolved hyperspectral scattering images were also exploited to measure biochemical changes within fresh beef due to the formation of metabolites caused by the growth and enzymatic activity of microorganisms (Peng 2011). The researchers selected feature bands and developed prediction models based on spatially resolved hyperspectral scattering profiles of beef over the visible and short-wave NIR region between 400 and 1100 nm by using a multi-linear regression (MLR) method. Experiments were performed on samples with total viable counts ranging from $10^{4.89}$ to $10^{8.89}$ cfu/g. Due to detector noise, the researchers used only the data between 560 and 980 nm in the development of TVC prediction models. The wavebands around 596 nm, 838 nm were considered to be related with oxyhaemoglobin absorption bands and the C–H third overtone, respectively.

Similar to the (indirect) safety assessment of whole chicken carcasses by the use of optical methods, multivariate analysis of hyperspectral images was also investigated for performing safety inspections of these foods. Many papers report the application of multivariate analysis of hyperspectral images of whole chicken carcasses (Chao et al. 2002, Lawrence et al. 2003, Park et al. 2003, Park 2006, Chao 2007, 2011). In particular, a Hyperspectral-Multispectral Line-Scan Imaging System was developed by Chao (2007) to assess if chicken meat was diseased or wholesome based on variations of oxymyoglobin and deoxymyoglobin, and the variations in metmyoglobin, respectively. The multispectral classification was based on the selection of 4 key wavelengths, i.e., 442, 501, 582, and 629 nm. Studies on chicken surfaces were also performed by Yoon (2011) for the detection of fecal material and ingesta (at 517, 565 and 802 nm). Fecal materials are directly linked with the presence of coliforms therefore this technique has a potential for the safety assessment of these food products. Another study by the same research group applied a multispectral imaging system using a conventional CCD camera with optical trim filters for online fecal detection (Park et al. 2004).

Other biological systems

Many applications of HSI for microbial detection or safety assessment of different biological systems (other than food) or abiotic surfaces have been reported. Although they are not directly related to food safety, their potential in this area is evident and it is further discussed hereunder.

Hyperspectral fluorescence imaging has been evaluated for the detection of microbial biofilm on stainless steel plates typically used to manufacture food processing equipment. Biofilm features were enhanced in emission bands at 444 nm and 588 nm (Jun 2008). This application could have a potential for developing a portable food safety inspection system for sanitation monitoring of food processing equipment surfaces. More recently, a hyperspectral fluorescence imaging system has been developed at the United States Department of Agriculture (USDA) (Jun et al. 2010) to detect microbial (*E. coli* and *S. enterica)* bio-films on food contact surfaces such as stainless steel, high density polyethylene (HDPE), plastic laminate (Formica), and polished granite. Hyperspectral fluorescence images were acquired from 421 to 700 nm using ultraviolet-A excitation. The use of two-wavelength band ratio images allowed for high detection rates in steel, HDPE and granite surfaces. Low bio-film growth on Formica surface led to high false positive rate.

Previous studies with NIR chemical imaging demonstrated the potential application of NIR hyperspectral imaging in differentiating bacteria based on their NIR spectra. NIR images of food-specific cards (stainless steel plates with wells containing bacteria samples) were obtained in the 1200–2350 nm region. Gram + (*L. monocytogenes, L. innocua, B. subtilis* and *B. cereus*) bacteria were successfully identified and separated from gram– (*E. coli* and *Salmonella* spp.) bacteria. Gram+ spots exhibited higher NIR intensities and both *E. coli* and gram- displayed lower intensities in the second derivative spectra (Dubois 2005).

Raman Hyperspectral Imaging has been applied for safety assessment in a variety of fields. For instance, Escoriza (2006) evaluated the suitability of this technique for enumeration of waterborne pathogens of *Staphylococcus epidermidis* (ATCC 35984) and *Escherichia coli* (ATCC 11775). This study showed that Raman hyperspectral imaging could provide quantitative information for bacterial concentration in water samples, however the signal was poor for bacterial concentrations $\leq 1 \times 10^7$ cells/membrane and sample pre-filtration was necessary prior to image analysis. Raman HSI has also been evaluated as a candidate technology for waterborne pathogen detection (Tripathi 2008). These authors managed to differentiate between Gram-positive (i.e., *Bacillus atrophaeus, B. thuringiensis*), Gram-negative organisms (i.e., *Yersinia enterocolitica, Yersinia rhodei,* and *Pseudomonas fluorescens*) by the use of a Raman Molecular Chemical Imaging System with a 532 nm laser excitation source.

Kalasinksy (2007) applied Raman HSI combining Raman spectroscopy, fluorescence spectroscopy, and digital imaging for detecting low levels of bioagent in complex background matrixes of *Bacillus anthracis, Yersinia pestis, Burkholderia mallei, Francisella tularensis, Brucella abortus,* and ricin. An important contribution to this field is the development of a comprehensive pathogen signature library by Raman Signature Development at the Armed Forces Institute of Pathology (Pennsylvania, USA) (Kalasinksy 2007). This Raman library has over 1000 end member signatures which could be used in future microbial identification studies. In other applications Raman HSI at a microscopic scale could allow for extraction of spectra emanating from single bacterial cells and spores. More specifically, Guicheteau (2010) proved that detection and identification was possible for both vegetative and spore forms of *Bacillus*.

Limitations and Future Perspectives

High set-up costs and difficulties with high-speed data acquisition and processing curently limits the wide use of hyperspectral imaging in industrial application (Yoon 2010, Gowen et al. 2007). For example, one of the main drawbacks in Raman HSI is the very long sample acquisition time needed to produce a hyperspectral image (up to a few hours). Especially, in the case of biofluorescence emission, process time may be increased as additional period is required to photobleach biological samples prior to data collection in order to reduce the sometimes overwhelming biofluorescence emission (Guicheteau 2010). Adjusting exposure time for effective high-speed imaging of moving targets is difficult for multispectral systems due to their physical shutter mechanisms and the quantum efficiencies of the charge-coupled device cameras. An exposure time setting ideal for a shorter wavelength image often causes image saturation at a higher wavelength (Park et al. 2005, Yang et al. 2005, 2006). Nevertheless, current electron-multiplying charge-coupled device (EMCCD)-based cameras, which have high quantum efficiencies and use rapid frame-transfer mechanisms and pre output signal amplification via electron multiplication, is an example of significantly improving the signal-to-readout-noise ratio (Chao et al. 2008).

Another challenge for this type of techniques is the quantification of microbes at low concentrations, which requires the use of a method that does not cause interference with detection (Escoriza 2006). This is why a lot of studies are currently focusing on assessing the spoilage of food products or identifying the microbial levels which are higher than 10^2 cfu/ml. Highly specific discriminating approach to pick out one particular organism is also another big challenge (Dubois 2005) to be considered in future developments.

Conclusions

In conclusion, hyperspectral imaging shows potential for pathogenic identification considering that in safety assessment studies the required response needed is usually binary in nature, i.e., presence/absence of a specific pathogen, rather than a profile of the flora populating the product. Towards this direction it will be imperative to create a comprehensive pathogen signature library similar to that established by Raman Signature Development at the Armed Forces Institute of Pathology (Pennsylvania, USA) (Kalasinksy 2007).

Keywords: food safety, hyperspectral imaging, reflectance, transmission, transflectance, fluorescence

References

Burger, J. and A. Gowen. 2011. Data handling in hyperspectral image analysis. Chemometr. Intell. Lab. Sys. 108: 13–22.

Burger, J. and P. Geladi.. 2006. Hyperspectral NIR image regression part I: Calibration and correction. J. Chemometr. 19: 355.

Blasco, J., N. Aleixos, J. Gomez and E. Molto. 2007. Citrus sorting by identification of the most common defects using multispectral computer vision. J. Food Eng. 83: 384–393.

Chao, K., Y. Chen, H. Early and B. Park. 1999. Color image classification systems for poultry viscera inspection. Appl. Eng. Agric. 15: 363–369.

Chao, K., P. Mehl and Y. Chen. 2002. Use of hyper- and multi-spectral imaging for detection of chicken skin tumors. Appl. Eng. Agric. 18: 113–119.

Chao, K., C.C. Yang, Y.R. Chen, M.S. Kim and D.E. Chan. 2007. Hyperspectral-multispectral line-scan imaging system for automated poultry carcass inspection applications for food Safety(1). Poultry Sci. 86: 2450–2460.

Chao, K., C.-. Yang, M.S. Kim and D.E. Chan. 2008. High throughput spectral imaging system for wholesomeness inspection of chicken. Appl. Eng. Agric. 24: 475–485.

Chao, K., C. Yang and M.S. Kim. 2011. Line-scan spectral imaging system for online poultry carcass inspection. J. Food Proc. Eng. 34: 125–143.

Chao, K., C. Yang and M.S. Kim. 2010. Spectral line-scan imaging system for high-speed non-destructive wholesomeness inspection of broilers. Trends Food Sci. Technol. 21: 129–137.

Dubois, J., E. Lewis, F. Fry and E. Calvey. 2005. Bacterial identification by near-infrared chemical imaging of food-specific cards. Food Microbiol. 22: 577–583.

Escoriza, M., J. VanBriesen, S. Stewart, J. Maier and P. Treado. 2006. Raman spectroscopy and chemical imaging for quantification of filtered waterborne bacteria. J. Microbiol. Meth. 66: 63–72.

Gaston, E., J.M. Frias, P.J. Cullen, C.P. O'Donnell and A.A. Gowen. 2010. Visible-near infrared hyperspectral imaging for the identification and discrimination of brown blotch disease on mushroom (Agaricus bisporus) caps. J. Near Infrared Spectroscop. 18: 341–353.

Gomez-Sanchis, J., E. Molto, G. Camps-Valls, L. Gomez-Chova, N. Aleixos and J. Blasco. 2008. Automatic correction of the effects of the light source on spherical objects. An application to the analysis of hyperspectral images of citrus fruits. J. Food Eng. 85: 191–200.

Guicheteau, J., S. Christesen, D. Emge and A. Tripathi. 2010. Bacterial mixture identification using Raman and surface-enhanced Raman chemical imaging. J. Raman Spectroscop. 41: 1632–1637.

Jun, W., K. Lee, P. Millner, M. Sharma, K. Chao and M.S. Kim. 2008. Detection of bacterial biofilm on stainless steel by hyperspectral fluorescence imaging. Food Processing Automation Conference, ASABE, USA 1: 1–6.

Jun, W., M.S. Kim, B. Cho, P.D. Millner, K. Chao and D.E. Chan. 2010. Microbial biofilm detection on food contact surfaces by macro-scale fluorescence imaging. J. Food Eng. 99: 314–322.

Kalasinsky, K.S., T. Hadfield, A.A. Shea, V.F. Kalasinsky, M.P. Nelson, J. Neiss, A.J. Drauch, G.S. Vanni and P.J. Treado. 2007. Raman chemical imaging spectroscopy reagentless detection and identification of pathogens: Signature development and evaluation. Anal. Chem. 79: 2658–2673.

Kleynen, O., V. Leemans and M. Destain. 2005. Development of a multi-spectral vision system for the detection of defects on apples. J. Food Eng. 69: 41–49.

Lawrence, K., W. Windham, B. Park and R. Buhr. 2003. A hyperspectral imaging system for identification of faecal and ingesta contamination on poultry carcasses. J. Near Infrared Spectroscop. 11: 269–281.

Park, B., K. Lawrence, W. Windham and R. Buhr. 2002. Hyperspectral imaging for detecting fecal and ingesta contaminants on poultry carcasses. Trans. ASAE 45: 2017–2026.

Park, B., K. Lawrence, W. Windham and D. Smith. 2006. Performance of hyperspectral imaging system for poultry surface fecal contaminant detection. J. Food Eng. 75: 340–348.

Park, B., K. Lawrence, W. Windham and D. Smith. 2005. Detection of fecal contaminants in visceral cavity of broiler carcasses using hyperspectral imaging. Appl. Eng. Agric. 21: 627–635.

Peng, Y., J. Zhang, W. Wang, Y. Li, J. Wu, H. Huang, X. Gao and W. Jiang. 2011. Potential prediction of the microbial spoilage of beef using spatially resolved hyperspectral scattering profiles. J. Food Eng. 102: 163–169.

Siripatrawan, U., Y. Makino, Y. Kawagoe and S. Oshita. 2011. Rapid detection of Escherichia coli contamination in packaged fresh spinach using hyperspectral imaging. Talanta. 85: 276–281.

Sone, I., R.L. Olsen, A.H. Sivertsen, G. Eilertsen and K. Heia. 2012. Classification of fresh Atlantic salmon (*Salmo salar* L.) fillets stored under different atmospheres by hyperspectral imaging. J. Food Eng. 109: 482–489.

Swatland, H., S. Anathanarayanan and A. Goldenberg. 1994. A review of probes and robots —implementing new technologies in meat evaluation. J. Animal Sci. 72: 1475–1486.

Tao, Y., J. Shao, K. Skeeles and Y.R. Chen. 1998. Detection of eviscerated poultry spleen enlargement by machine vision. SPIE. 3544: 138–145.

Tao, F., Y. Peng, Y. Li, K. Chao and S. Dhakal. 2012. Simultaneous determination of tenderness and Escherichia coli contamination of pork using hyperspectral scattering technique. Meat Sci. 90: 851–857.

Tao, F., Y. Peng, Y. Li, K. Chao and S. Dhakal. 2012. Simultaneous determination of tenderness and Escherichia coli contamination of pork using hyperspectral scattering technique. Meat Sci. 90: 851–857.

Tao, Y., J. Shao, K. Skeeles and Y.R. Chen. 1998. Detection of eviscerated poultry spleen enlargement by machine vision. SPIE 3544: 138–145.

Tripathi, A., R.E. Jabbour, P.J. Treado, J.H. Neiss, M.P. Nelson, J.L. Jensen and A.P. Snyder. 2008. Waterborne pathogen detection using Raman spectroscopy. Appl. Spectroscop. 62: 1–9.

Vizhanyo, T. and J. Felfoldi. 2000. Enhancing colour differences in images of diseased mushrooms. Comp. Electron. Agric. 26: 187–198.

Vote, D.J., M.B. Bowling, B.C.N. Cunha, K.E. Belk, J.D. Tatum, F. Montossi and G.C. Smith. 2009. Video image analysis as a potential grading system for Uruguayan beef carcasses. J. Animal Sci. 87: 2376–2390.

Vote, D., K. Belk, J. Tatum, J. Scanga and G. Smith. 2003. Online prediction of beef tenderness using a computer vision system equipped with a BeefCam module. J. Animal Sci. 81: 457–465.

Yang, C.-., K. Chao, M.S. Kim, D.E. Chan, H.L. Early and M. Bell. 2010. Machine vision system for on-line wholesomeness inspection of poultry carcasses. Poultry Sci. 89: 1252–1264.

Yang, C., K. Chao, Y. Chen, M. Kim and H. Early. 2006. Simple multispectral image analysis for systemically diseased chicken identification. Trans. ASABE 49: 245–257.

Yang, C., K. Chao and Y. Chen. 2005. Development of multispectral image processing algorithms for identification of wholesome, septicemic, and inflammatory process chickens. J. Food Eng. 69: 225–234.

Yoon, S.C., B. Park, K.C. Lawrence, W.R. Windham and G.W. Heitschmidt. 2011. Line-scan hyperspectral imaging system for real-time inspection of poultry carcasses with fecal material and ingesta. Comp. Electron. Agric. 79: 159–168.

Color Plate Section

Chapter 4

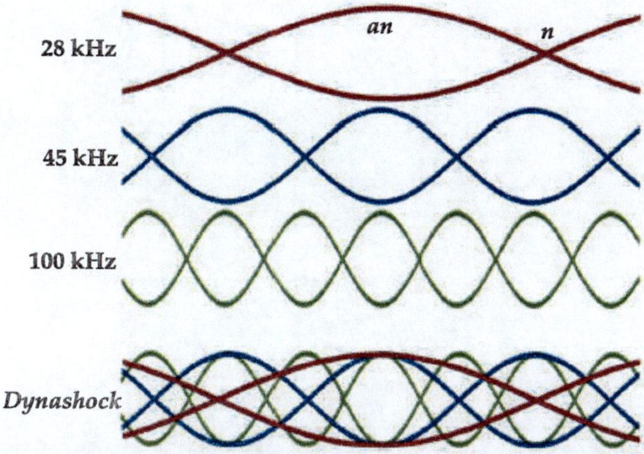

Figure 2. Difference between single frequency and Dynashock waves (Gabriel 2012).

Chapter 8

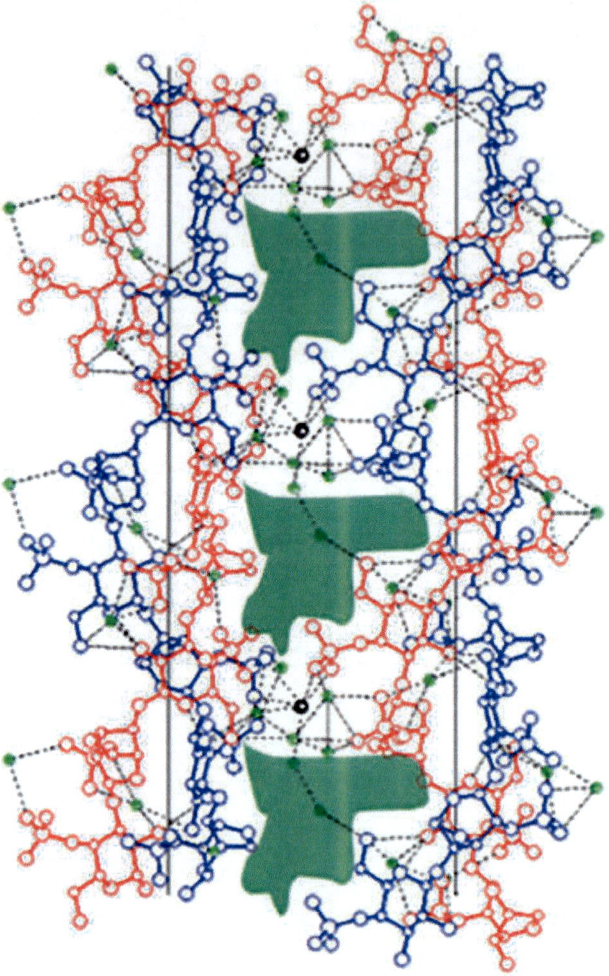

Figure 2. Association of a pair of iota-carrageenan double helices (one chain in red and other blue) in the presence of calcium ions (black circles) and water molecules (green circles). The broken lines represent hydrogen bonds. The green blocks highlight the water pockets suitable for encapsulation.

Chapter 9

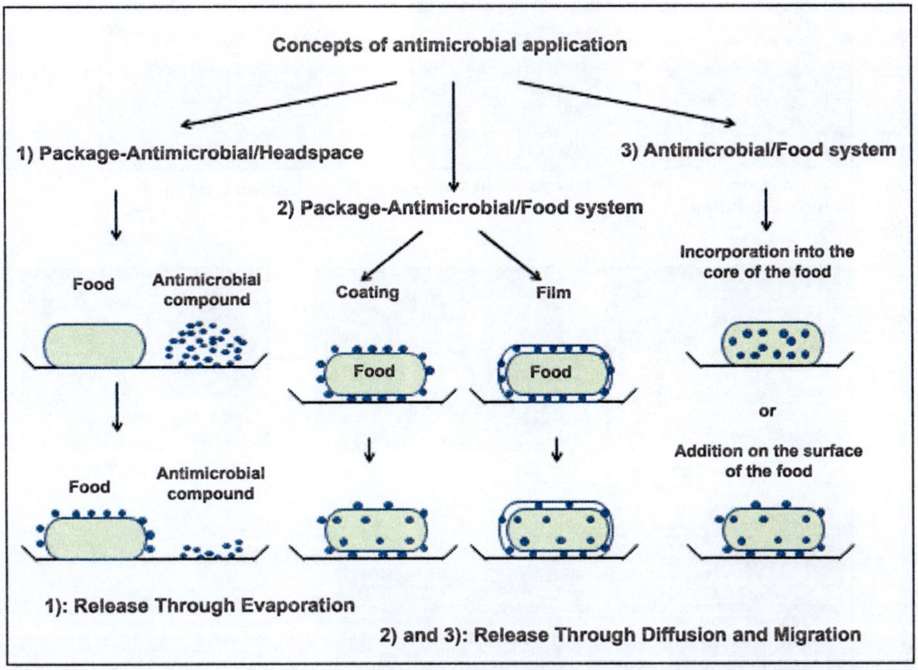

Figure 1. Presentation of the potential antimicrobial packaging concepts and their mode of action (based on Coma 2008). *Blue points correspond to antimicrobial compound.

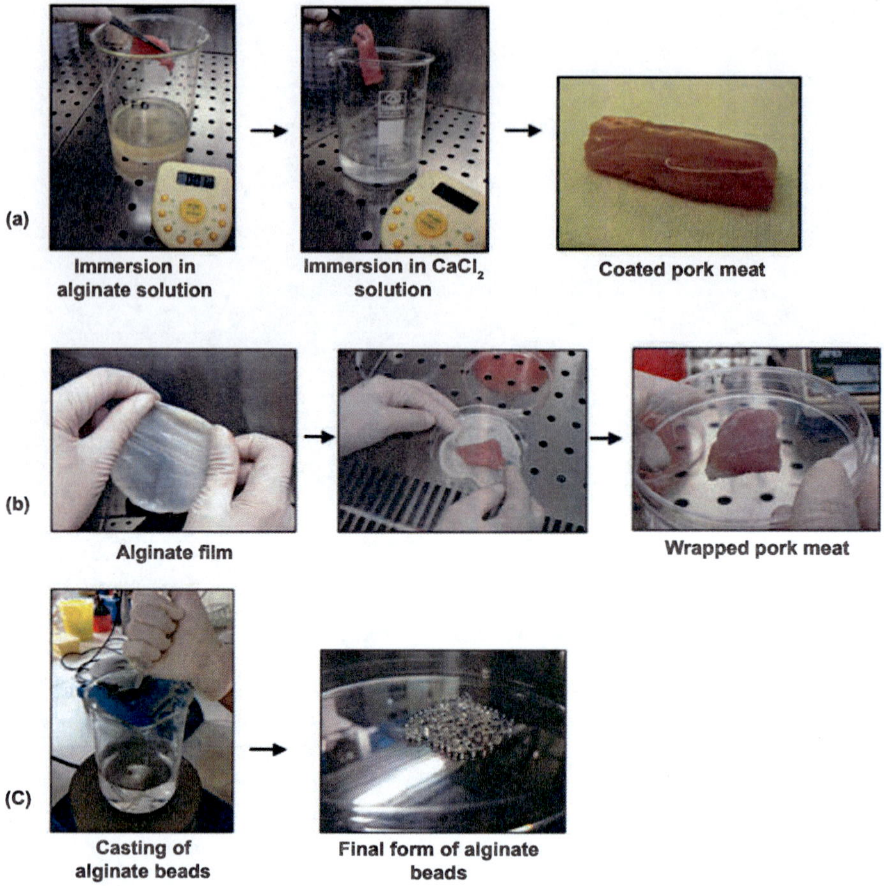

Figure 3. Lab process for casting and applying alginate based (a) coatings, (b) films, and (c) beads on pork meat.

Chapter 13

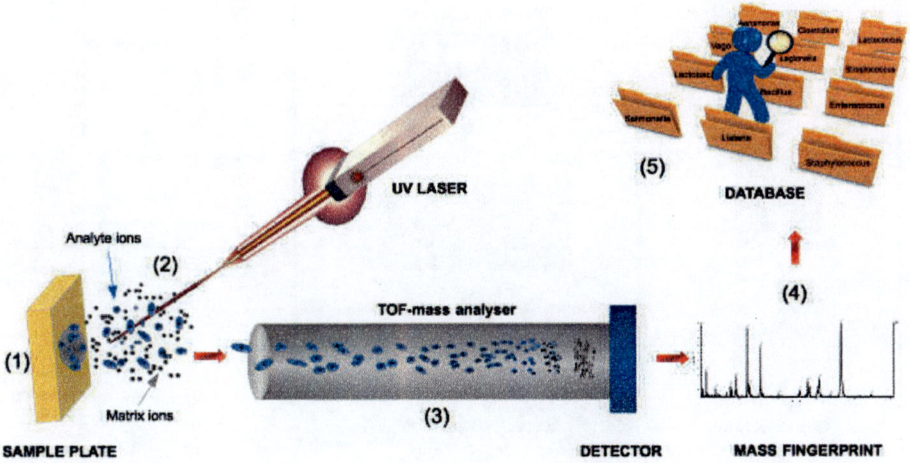

Figure 4. Technical description of MALDI-TOF MS. (1) The sample is mixed with a matrix and placed in a metal plate. (2) After crystallization of the matrix and microbial material, the metal plate is introduced into the mass spectrometer and is bombarded with brief laser pulses. (3) The desorbed and ionized molecules are accelerated through an electrostatic field and ejected through a metal flight tube subjected to vacuum until they reach a detector, with smaller ions traveling faster than larger ions. (4) Thus, bioanalytes separated according to their TOF create a mass spectrum that is composed of mass to charge ratio (m/z) peaks with varying intensities. (5) A spectrum is a microbial signature that is compared to a database for identification at the species or genus level.

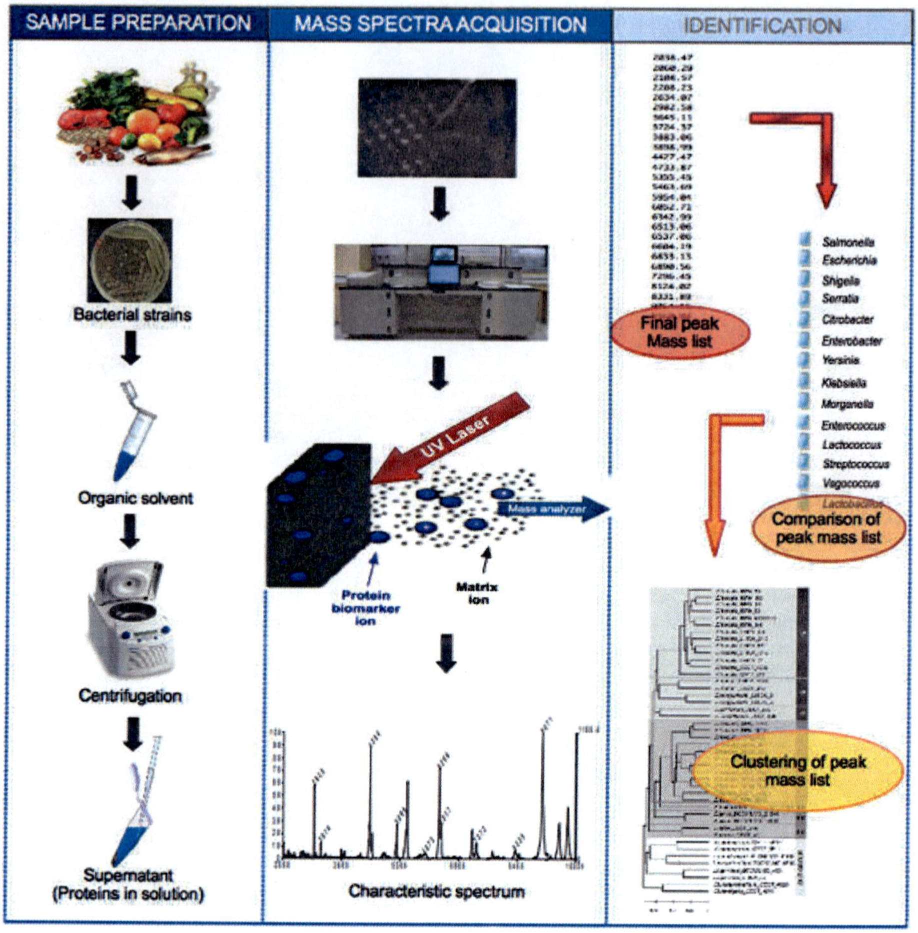

Figure 8. Process of bacterial identification by MALDI-TOF MS in food products.

Chapter 14

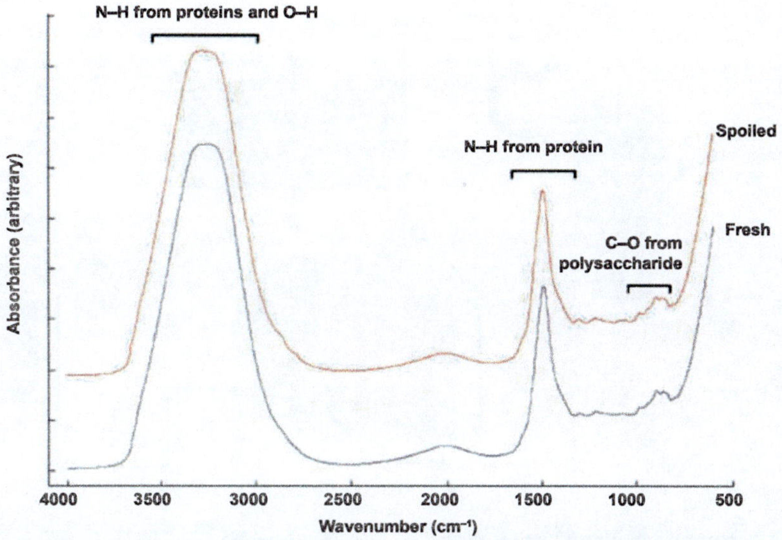

Figure 2. FT-IR ATR spectra collected from a fresh and a spoiled whole milk sample stored at 15°C (Data from Nicolaou and Goodacre, 2008 modified accordingly).

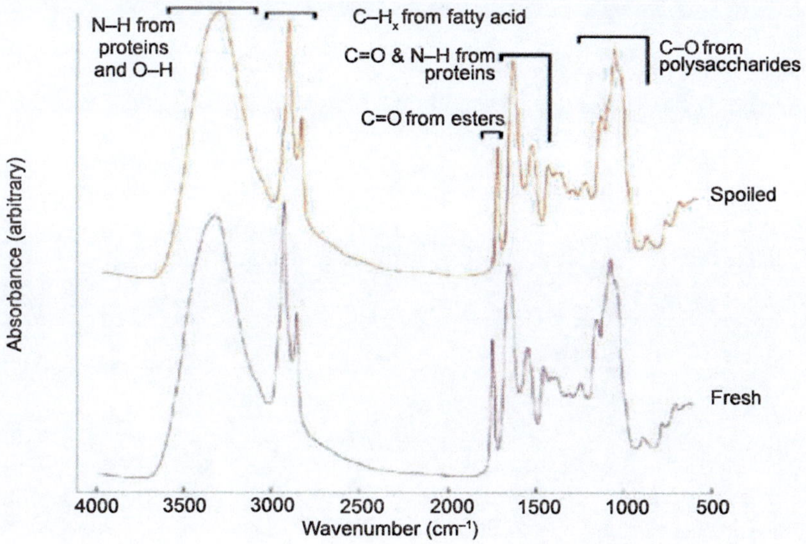

Figure 3. FT-IR HT spectra collected from a fresh and a spoiled whole milk sample stored at 15°C (Data from Nicolaou and Goodacre, 2008 modified accordingly).

Chapter 15

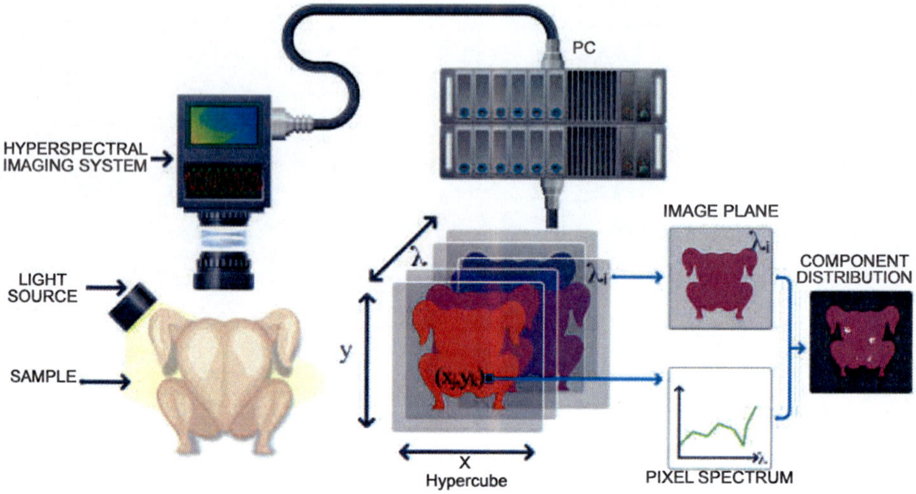

Figure 1. Schematic showing hyperspectral imaging equipment, hypercube structure, spatial axes X, Y, and wavelength axis (λ).